MICROSCOPE IMAGE PROCESSING

MICROSCOPE IMAGE PROCESSING

Second Edition

Edited by

FATIMA A. MERCHANT
Computational Biology and Medicine Laboratory, Department of Engineering Technology, Department of Electrical and Computer Engineering, Department of Biomedical Engineering, Department of Computer Science, University of Houston, Houston, TX, United States

KENNETH R. CASTLEMAN
Computational Biology and Medicine Laboratory, Department of Engineering Technology, University of Houston, Houston, TX, United States

Academic Press is an imprint of Elsevier
125 London Wall, London EC2Y 5AS, United Kingdom
525 B Street, Suite 1650, San Diego, CA 92101, United States
50 Hampshire Street, 5th Floor, Cambridge, MA 02139, United States
The Boulevard, Langford Lane, Kidlington, Oxford OX5 1GB, United Kingdom

Copyright © 2023 Elsevier Inc. All rights reserved.

No part of this publication may be reproduced or transmitted in any form or by any means, electronic or mechanical, including photocopying, recording, or any information storage and retrieval system, without permission in writing from the publisher. Details on how to seek permission, further information about the Publisher's permissions policies and our arrangements with organizations such as the Copyright Clearance Center and the Copyright Licensing Agency, can be found at our website: www.elsevier.com/permissions.

This book and the individual contributions contained in it are protected under copyright by the Publisher (other than as may be noted herein).

Notices
Knowledge and best practice in this field are constantly changing. As new research and experience broaden our understanding, changes in research methods, professional practices, or medical treatment may become necessary.

Practitioners and researchers must always rely on their own experience and knowledge in evaluating and using any information, methods, compounds, or experiments described herein. In using such information or methods they should be mindful of their own safety and the safety of others, including parties for whom they have a professional responsibility.

To the fullest extent of the law, neither the Publisher nor the authors, contributors, or editors, assume any liability for any injury and/or damage to persons or property as a matter of products liability, negligence or otherwise, or from any use or operation of any methods, products, instructions, or ideas contained in the material herein.

ISBN 978-0-12-821049-9

For information on all Academic Press publications
visit our website at https://www.elsevier.com/books-and-journals

Publisher: Mara Conner
Acquisitions Editor: Tim Pitts
Editorial Project Manager: Fernanda A. Oliveira
Production Project Manager: Prem Kumar Kaliamoorthi
Cover Designer: Christian Bilbow

Typeset by STRAIVE, India

Contents

Foreword to the First Edition

Microscope image processing dates back a half century when it was realized that some of the techniques of image capture and manipulation, first developed for television, could also be applied to images captured through the microscope. Initial approaches were dependent on the application: automatic screening for cancerous cells in Papanicolaou smears; automatic classification of crystal size in metal alloys; automation of white cell differential count; measurement of DNA content in tumor cells; analysis of chromosomes; etc. In each case, the solution lay in the development of hardware (often analog) and algorithms highly specific to the needs of the application. General-purpose digital computing was still in its infancy. Available computers were slow, extremely expensive, and highly limited in capacity. (I still remember having to squeeze a full analysis system into less than 10 kilobytes of programmable memory!) Thus there existed an unbridgeable gap between the theory of how microscope images could be processed and what was practically attainable.

One of the earliest systematic approaches to the processing of microscopic images was the CYDAC (CYtophotometric DAta Conversion) project [1], which I worked on under the leadership of Mort Mendelsohn at the University of Pennsylvania. Images were scanned and digitized directly through the microscope. Much effort went into characterizing the system in terms of geometric and photometric sources of error. The theoretical and measured system transfer functions were compared. Filtering techniques were used both to sharpen the image and to reduce noise, while still maintaining the photometric integrity of the image. A focusing algorithm was developed and implemented as an analog assist device. But progress was agonizingly slow. Analysis was done off-line, programs were transcribed to cards, and initially we had access to a computer only once a week for a couple of hours in the middle of the night!

The modern programmable digital computer has removed all the old constraints—incredible processing power, speed, memory, and storage come with any consumer computer. My 10-year-old grandson, with his digital camera and access to a laptop computer with processing programs such as iPhoto and Adobe Photoshop, can command more image-processing resources than were available in leading research laboratories less than two decades ago. The challenge lies not in processing images, but in processing them correctly and effectively. *Microscope Image Processing* provides the tools to meet this challenge.

In this volume, the editors have drawn on the expertise of leaders in processing microscope images to introduce the reader to underlying theory, relevant algorithms, guiding principles, and practical applications. It explains not only what to do, but also which pitfalls to avoid and why. Analytic results can only be as reliable as the processes used to obtain them. Spurious results can be avoided when users understand the limitations imposed by diffraction optics, empty magnification, noise, sampling errors, etc. The book not only covers the fundamentals of microscopy and image processing, but also describes the use of the techniques as applied to fluorescence microscopy, spectral imaging, three-dimensional microscopy, structured illumination, and time-lapse microscopy. Relatively advanced techniques such as wavelet and morphological image processing and automated microscopy are described in an intuitive and comprehensive manner that will appeal to readers, whether technically oriented or not. The summary list at the end of each chapter is a particularly useful feature enabling the reader to access the essentials without necessarily mastering all the details of the underlying theory.

Microscope Image Processing should become a required textbook for any course on image processing, not just microscopic. It will be an invaluable resource for all who process microscope images and who use the microscope as a quantitative tool in their research. My congratulations go to the editors and authors for the scope and depth of their contributions to this informative and timely volume.

Brian H. Mayall

Reference

[1] M.L. Mendelsohn, B.H. Mayall, J.M.S. Prewitt, R.C. Bostrum, W.G. Holcomb, Digital transformation and computer analysis of microscopic images, Adv. Opt. Elec. Microsc. 2 (1968) 77–150.

Foreword to the Second Edition

Those who would understand cellular and subcellular life tend to fall into one or the other of two camps. There are the biochemical reductionists—the "grind 'em and find 'em" brigade—who hope, by analysis of the component parts, to understand the whole. The current dominance of genomics in biological research speaks to their effectiveness. And then there are the wholistic structuralists—the "seeing is believing" folk—who seek to understand life by observing it. For them, a central problem is that cellular structures and subcellular life processes happen at a scale beyond the grasp of unaided human vision. Magnification is required to bring them within range of human perception, and for this some form of microscope is required. It is to them that this magnificent book, *Microscope Image Processing*, written and edited by Fatima Merchant and Kenneth Castleman, is addressed.

A further fundamental problem for the "seeing is believing" folk is that cellular and subcellular life processes are dynamic. Until the 1970s, investigators using the light microscope could choose either to observe living specimens at low magnification, or to examine the dead remains of biological specimens at higher magnifications, typically after chemical preservation ("fixation"), dehydration, embedding within some polymeric matrix, chemical staining to differentiate various types of structure, and finally cutting them into sections thin enough to see through.

All that has now changed. Incrementally over the past 50 years, new methodologies have been developed—testaments to the incredible imagination, ingenuity, and creativity of the individuals involved—that have pushed the boundaries of conventional light microscopy beyond the resolution limit of the light microscope as conventionally understood, have revolutionized the observation of dynamic life processes at high magnification, and have permitted the selective labeling and observation of particular structures and individual molecules within living cells. These advances have reinstated the optical microscope as an essential instrument for biological research, although often in forms that would be unrecognizable to traditional microscopists.

This multiauthor book, edited by Merchant and Castleman, who have also (with others) authored 10 of the 16 chapters, delivers much more than its title implies. While the early chapters are indeed concerned with the digital image processing and enhancement of microscope images and the quantitative analysis of their content, those in the second half of the book deal holistically with some of these new light microscopic methodologies, including recent advances in molecular localization using fluorescence microscopy, motion analysis, three-dimensional imaging, and superresolution imaging, concluding with considerations of artificial intelligence and informatics as applied to light microscopy.

Similar striking advances have been made in electron microscopy, particularly in the low-dose observation of fully hydrated cryopreserved specimens, and the three-dimensional analysis of macromolecules by computational reconstruction from two-dimensional images. While the theoretical aspects presented in this book are of general relevance, detailing the fundamental principles of imaging, resolution, and computational image processing that apply equally to the images obtained from such electron microscopy, the applications discussed in this book are restricted to the use of the light microscope for various forms of optical imaging with ever increasing clarity, particularly of biological specimens.

At each transformative stage in light microscopy image processing—from the "reality" of the original three-dimensional living specimen, via the refraction of photons to form a two-dimensional analog optical image of the specimen, the technology of the recording device employed, the sampling of the optical image required to form the pixels or voxels of the digital image, the subsequent computational processing of such a digital image, and the final conversion back to a displayed image that can be perceived by the human eye—there is potential for loss of information and corruption of the image, of which the user must be aware.

Microscope Image Processing provides detailed explanations of the principles and practicalities of the microscopy and image-processing techniques described, and the information that may be derived from resulting analyses. However, its primary purpose is to help the reader understand fully how these various transformations can be made while avoiding the potential distortions and infidelities that can be inadvertently introduced along the way, so that a realistic image of the specimen may finally be viewed and comprehended by the investigator.

This volume is the second edition of the 2008 book of the same title by Qiang Wu, Fatima Merchant, and Kenneth R. Castleman, and its creation has involved the revision of chapters from the first edition and the inclusion of new ones describing more recent advances, as explained more fully by the editors in their Preface. The volume as a whole is designed to bring the reader to the very frontiers of the field.

As the editors stress in their introductory chapter, the book is designed to be comprehensible by the nonspecialist. Of course, "comprehensible" is a relative term, since different disciplines frequently employ the same words to mean different things. To a mathematician, a function is a relation or expression involving one or more variables, while the process of division is the inverse of multiplication. To a cell biologist, however, a typical function of the cell membrane is the ingestion of specific proteins by receptor-mediated endocytosis, while cell division results in multiplication of the population! The use of specific meanings for common terms is unavoidable in any scholarly field, and for this reason the editors have included a glossary to help the reader navigate through the jargon. However, in a highly technical volume of this kind, nonmathematical readers will have to accept the presence of many mathematical equations and of sentences that might initially leave them bewildered, such as "The continuous image is a real-valued analytic function of two real variables," and "An analytic function is not only continuous but possesses all of its derivatives at every point" (both from the Summary of Chapter 1).

I can sympathize, since it was exactly in this state of partial bewilderment that I first grappled with Kenneth Castleman's seminal 1979 book *Digital Image Processing*. Castleman is a recognized authority in the field of image processing, having been involved since its inception, developing his expertise in the 1970s while working at NASA's Jet Propulsion Laboratory in Pasadena, California. In contrast, I started my research career as a "grind 'em and find 'em" protein chemist, and I remember being envious at that time of senior colleagues such as Hugh Huxley, an electron microscopist investigating muscle structure, who could actually *see* what he was studying.

My own introduction to light, fluorescence, and electron microscopy came several years later, while working with the eminent cell biologist Daniel Branton. At that time, I was privileged to be party to the first applications of video-enhanced contrast microscopy in the lab of Tom Reese, where it was being employed to visualize the movement of subresolution organelles within live nerve axons. Subsequently, I became involved in the early applications of confocal microscopy to biological specimens. These various forms of electronically enhanced light microscopy all involved digital image processing, and it was in search of a deeper understanding of the processes involved that I turned to Castleman's *Digital Image Processing*. I was not disappointed. His book remains the best on the subject, and it was only natural that I should subsequently invite him to contribute a chapter on image resolution and sampling for my 1993 multiauthor book titled *Electronic Light Microscopy*, a book that provided the first description of these and other emerging techniques.

In the subsequent 30 years, brilliant new applications of the enduring principles expounded in Castleman's chapter in *Electronic Light Microscopy* have brought forth further remarkable advances in our ability to use light microscopes and digital image processing techniques to investigate the nanoscopic world. These are described in this current volume, *Microscope Image Processing*, edited by Fatima Merchant and Kenneth Castleman, which I am sure will prove of equally enduring value to a new generation of investigators.

David Shotton

Preface to the First Edition

The digital revolution has touched most aspects of modern life, including entertainment, communication, and scientific research. Nowhere has the change been more fundamental than in the field of microscopy. Researchers who use the microscope in their investigations have been among the pioneers who applied digital processing techniques to images. Many of the important digital image processing techniques that are now in widespread usage were first implemented for applications in microscopy. At this point in time, digital image processing is an integral part of microscopy, and only rarely will one see a microscope used with only visual observation or photography.

The purpose of this book is to bring together the techniques that have proved to be widely useful in digital microscopy. This is quite a multidisciplinary field, and the basis of processing techniques spans several areas of technology. We attempt to lay the required groundwork for a basic understanding of the algorithms that are involved, in the hope that this will prepare the reader to press the development even further.

This is a book about techniques for processing microscope images. As such it has little content devoted to the theory and practice of microscopy, or even to basic digital image processing, except where needed as background. Neither does it focus on the latest techniques to be proposed. The focus is on those techniques that routinely prove useful to research investigations involving microscope images and upon which more advanced techniques are built.

A very large and talented cast of investigators has made microscope image processing what it is today. We lack the paper and ink required to do justice to the fascinating story of this development. Instead we put forward the techniques, principally devoid of their history. The contributors to this volume have shouldered their share of their creation, but many others who have pressed forward the development do not appear.

Qiang Wu
Fatima Merchant
Kenneth R. Castleman
Houston, TX, United States

Preface to the Second Edition

In the decade and a half since the first edition was published, the field of digital microscopy has progressed enormously. New developments in both hardware and algorithms have increased the usefulness of this venerable research tool and broadened its horizon of applications. Exciting optics innovations couple with brilliant new algorithms to solve problems heretofore unapproachable in medicine, industry, and research. New superresolution techniques now allow microscopists to look far deeper into much smaller structures. Structured illumination optics has also increased the reach of the microscope into the domain of the tiny. Complex neural network techniques, such as deep learning and deep belief systems, have further advanced microscope image analysis. Image informatics makes it possible to manage huge amounts of high-resolution data. And once again microscopy has led the way in digital image processing algorithm development.

This edition has been expanded to include new chapters on four topics that have become increasingly important in the last decade. Superresolution microscopy pushes the resolving power of the microscope far beyond the diffraction limit of resolution. Localization microscopy permits nanometer-scale imaging of fluorescently labeled specimens. Deep learning networks provide more accurate and robust solutions to problems in image segmentation, classification, and resolution improvement. Image informatics provides tools for organizing, analyzing, and archiving large-volume image data and experimental results.

Like the first edition, this book aims to bring together the most useful techniques in digital microscopy. We combine an explanation of basic techniques with discussions of some of the most useful new developments in this highly multidisciplinary field. As before, we devote little space to the theory and practice of microscopy or to basic digital image processing techniques that do not impact microscopy broadly. Neither do we emphasize techniques with limited application. We have integrated recent developments into the original structured framework, which we believe makes the accumulated knowledge most conveniently available to the reader.

To accommodate the inclusion of so much new material, we have streamlined the discussion of basic technology and eliminated coverage of some topics that were treated in the first edition. Still missing in all this is a history lesson that gives credit to all those who contributed to the creation of this important field of technology. Fascinating as that story is, we lack space for it, and it is covered brilliantly elsewhere. We can only describe the techniques and illustrate their use, in the hope that our readers will push the development even further.

At the same time, we understand that many of our readers are looking to apply basic image processing techniques to their microscope images. Most manufacturers of microscopy systems include dedicated image processing software. There are also many stand-alone software tools that include a majority of the basic image processing techniques described here. Importantly, there are open-source public domain software tools, such as Fiji (https://imagej.net/software/fiji/), that are specifically designed for biological image analysis. These tools not only provide a full range of image processing and analysis operations, but they are supported by a large and very knowledgeable user community. This group continues to make available tools to handle complex image manipulations while providing in-depth tutorials via web-based interactive documents and videos.

Our contributors have worked tirelessly to make this volume as useful to microscopists as possible. Beyond that, they have cooperated fully with our efforts to maintain a consistent style of presentation. We have cross-referenced the chapters, where appropriate, to make the information readily available with minimum redundancy. The aim is to make this exposition of technology more cohesive and thus more useful to our readers. Recognizing that English is a second language for many of our readers, we lean toward simple statements of fact and away from more erudite prose. Our mission is to make a wealth of multidisciplinary information readily available to a broad readership, and we take that job seriously.

Fatima Merchant
Kenneth R. Castleman
Houston, TX, United States

Acknowledgments

All of the following contributors to this volume have done important work to advance technology, in addition to their work presented herein.

Romaric Audigier
Université Paris-Saclay
Artificial Intelligence for Language and Vision
CEA-LIST
Palaiseau, France

Kenneth R. Castleman
Computational Biology and Medicine Laboratory
Department of Engineering Technology
University of Houston
Houston, Texas, United States

Alberto Diaspro
Department of Nanophysics
Istituto Italiano di Tecnologia
Department of Applied Physics
University of Genova
Genova, Italy

Oleh Dzyubachyk
Department of Radiology
Department of Cell and Chemical Biology
Leiden University Medical Center
Leiden, the Netherlands

Kevin W. Eliceiri
Departments of Medical Physics and Biomedical Engineering
Morgridge Institute for Research
University of Wisconsin at Madison
Madison, Wisconsin, United States

Christian Franke
Institute of Applied Optics and Biophysics
Abbe Center of Photonics
Jena Center for Soft Matter
Friedrich Schiller University Jena
Jena, Germany

Jiaming Guo
Department of Electrical and Computer Engineering
University of Houston
Houston, Texas, United States

Kyle I.S. Harrington
Image Data Analysis group
Max Delbrück Center for Molecular Medicine in the Helmholtz Association
Berlin, Germany

Roberto A. Lotufo
School of Electrical and Computer Engineering
University of Campinas
Campinas SP, Brazil

Rubens C. Machado
Center for Information Technology Renato Archer
Ministry of Science and Technology
Campinas, Brazil

David Mayerich
Department of Electrical and Computer Engineering
University of Houston
Houston, Texas, United States

Erik Meijering
Biomedical Image Computing Group
School of Computer Science and Engineering
University of New South Wales
Sydney, Australia

Fatima A. Merchant
Computational Biology and Medicine Laboratory
Department of Engineering Technology
Department of Electrical and Computer Engineering
Department of Biomedical Engineering
Department of Computer Science
University of Houston
Houston, Texas, United States

Jean-Christophe Olivo-Marin
BioImage Analysis Unit
Institut Pasteur
Paris, France

Ammasi Periasamy
WM Keck Center for Cellular Imaging
Departments of Biology and Biomedical Engineering
University of Virginia
Charlottesville, Virginia, United States

Andre V. Saúde
Department of Computer Science
Federal University of Lavras
Lavras, Brazil

Shishir K. Shah
Department of Computer Science
Department of Electrical and Computer Engineering
University of Houston
Houston, Texas, United States

Ihor Smal
Department of Cell Biology and Molecular Genetics
Erasmus MC—University Medical Center Rotterdam
Rotterdam, the Netherlands

Ruijiao Sun
Department of Electrical and Computer Engineering
University of Houston
Houston, Texas, United States

Yu-Ping Wang
Department of Biomedical Engineering
Tulane University
New Orleans, Louisiana, United States

Qiang Wu
Soft Imaging, LLC
Houston, Texas, United States

Ian T. Young
Department of Imaging Physics
Faculty of Applied Sciences
Delft University of Technology
Delft, the Netherlands

In addition to those who contributed chapters, the following researchers provided specific figures and examples taken from their work.

Camille Artur
Department of Electrical and Computer Engineering
University of Houston
Houston, Texas, United States

Paolo Bianchini
Department of Nanophysics
Istituto Italiano di Tecnologia
Genova, Italy

Marco Castello
Department of Nanophysics
Istituto Italiano di Tecnologia
Genova Instruments
Genova, Italy

Jose Angel Conchello
Washington University, St. Louis, Missouri, United States
Harvard University, Boston, Massachusetts, United States (Retired)

Lisa Cuneo
Department of Nanophysics
Istituto Italiano di Tecnologia
Department of Applied Physics
University of Genova
Genova, Italy

Jason L. Eriksen
Department of Pharmacological and Pharmaceutical Sciences
University of Houston
Houston, Texas, United States

Jeroen Essers
Department of Vascular Surgery
Department of Molecular Genetics
Department of Radiation Oncology
Erasmus MC, University Medical Center Rotterdam
Rotterdam, the Netherlands

Edward Evans
Morgridge Institute for Research
Madison, Wisconsin, United States

Niels Galjart
Department of Cell Biology
Erasmus MC, University Medical Center Rotterdam
Rotterdam, the Netherlands

William P. Goldman
Washington University, St. Louis, Missouri, United States
University of North Carolina at Chapel Hill, North Carolina, United States (Current)

Adriaan Houtsmuller
Department of Pathology
Erasmus MC, University Medical Center Rotterdam
Rotterdam, the Netherlands.

Deborah Hyink
Division of Nephrology
Department of Medicine
Baylor School of Medicine
Houston, Texas, United States

Jahandar Jahanipour
Department of Electrical and Computer Engineering
University of Houston
Houston, Texas, United States

Erik Manders
Confocal.Nl B.V.
Amsterdam, the Netherlands

Dragan Maric
Flow and Imaging Cytometry Core Facility
National Institute of Neurological Disorders and Stroke
National Institutes of Health
Bethesda, Maryland, United States

Irene Nepita
Department of Nanophysics
Istituto Italiano di Tecnologia
Genova, Italy

Michele Oneto
Department of Nanophysics
Istituto Italiano di Tecnologia
Genova, Italy

Wei Ouyang
School of Engineering Sciences in Chemistry, Biotechnology and Health
KTH Royal Institute of Technology
Sweden

Simonluca Piazza
Department of Nanophysics
Istituto Italiano di Tecnologia
Genova Instruments
Genova, Italy

Badrinath Roysam
Department of Electrical and Computer Engineering
University of Houston
Houston, Texas, United States

Colin J.R. Sheppard
School of Chemistry and Molecular Biosciences
University of Wollongong, Wollongong, Australia
Department of Nanophysics
Istituto Italiano di Tecnologia, Genova, Italy

Timo ten Hagen
Laboratory Experimental Oncology
Department of Pathology
Erasmus MC, University Medical Center Rotterdam
Rotterdam, the Netherlands

Gert van Cappellen
Erasmus Optical Imaging Center
Department of Pathology
Erasmus MC, University Medical Center Rotterdam
Rotterdam, the Netherlands

Tasha Womack
Department of Electrical and Computer Engineering
University of Houston
Houston, Texas, United States

Finally, several others have assisted in bringing this book to fruition. We wish to thank Kathy Pennington, Jamie Alley, Steve Clarner, Xiangyou Li, Szeming Cheng, and Vibeesh Bose.

Many of the examples in this book were developed during the course of research conducted at Advanced Digital Imaging Research, LLC. Much of that research was supported by the National Institutes of Health, under the Small Business Innovative Research program.

Fatima A. Merchant
Kenneth R. Castleman
University of Houston
Houston, TX, United States

CHAPTER ONE

Introduction

Kenneth R. Castleman and Fatima A. Merchant

1.1 The Microscope and Image Processing

Invented over 400 years ago, the optical microscope has seen steady improvement and increasing use in biomedical research and clinical medicine as well as in many other fields [1,2]. Today many variations of the basic microscope instrument are used with great success, allowing us to peer into spaces much too small to be seen with the unaided eye. More often than not, in this day and age, the images produced by a microscope are converted into digital form for storage, analysis, or processing prior to display and interpretation [3–5]. Digital image processing greatly enhances the process of extracting information about the specimen from a microscope image. For that reason, digital imaging is steadily becoming an integral part of microscopy. Digital processing can be used to extract quantitative information about the specimen from a microscope image, and it can transform an image so that a displayed version is much more informative than it would otherwise be [6,7].

1.2 The Scope of This Book

This book discusses the methods, techniques, and algorithms that have proven useful in the processing and analysis of digital microscope images. We do not attempt to describe the workings of the microscope, except to outline its limitations and the reasons for certain processes. Neither do we discuss the proper use of the instrument. These topics are beyond our scope and are well covered in other works. Instead we focus on techniques for processing microscope images.

At this time microscope imaging and image processing are of vital interest to the scientific and engineering communities. Recent developments in cellular-, molecular-, and nanometer-level imaging technologies have led to rapid discoveries and have greatly advanced knowledge in biology, medicine, chemistry, pharmacology, and other fields.

Microscopes have long been used to capture, observe, measure, and analyze images of various living organisms and structures at scales far below the limits of human visual perception. With the advent of affordable, high-performance computer and image sensor technologies, digital imaging has essentially replaced film-based photomicrography for microscope image acquisition and storage. Digital image processing has become essential

Microscope Image Processing
https://doi.org/10.1016/B978-0-12-821049-9.00014-9
Copyright © 2023 Elsevier Inc.
All rights reserved.

to the success of subsequent data analysis and interpretation of the new generation of microscope images. There are microscope imaging modalities that require digital image processing just to produce an image suitable for viewing. Digital processing of microscope images has opened up new realms of medical research and brought about the possibility of advanced clinical diagnostic procedures.

The approach used in this book is to describe image processing algorithms that have proved useful in microscope image processing and to illustrate their application with specific examples. Useful mathematical results are presented without derivation or proof, but with references to the earlier work. We have relied on a collection of chapter contributions from leading experts in the field to present detailed descriptions of state-of-the-art methods and algorithms developed to solve specific problems in microscope imaging. Each chapter provides first a summary, then an in-depth analysis of the methods, and finally specific examples to illustrate application. The insight gained from these examples of successful application should guide the reader in developing their own applications.

Although a number of monographs and edited volumes have been written on the topic of computer-assisted microscopy, most of these books focus on the basic concepts and technicalities of microscope illumination, optics, hardware design, and digital camera setups. They do not discuss in detail the practical issues that arise in microscope image processing or the development of specialized algorithms for digital microscopy.

This book is intended to complement existing works by focusing on the computational and algorithmic aspects of microscope image processing. It should serve the users of digital microscopy as a reference for the basic algorithmic techniques that routinely prove useful in microscope image processing.

The intended audience for this book includes scientists, engineers, clinicians, and graduate students working in the fields of biology, medicine, chemistry, pharmacology, and other related disciplines. It is intended for those who use microscopes and commercial or free image processing software in their work and would like to understand the methodologies and capabilities of the latest digital image processing techniques. This book is also intended for those who develop their own image processing algorithms and software for specific applications that are not covered by existing software products.

In summary, this book presents a discussion of algorithms and processing methods that complements the existing selection of books on microscopy and digital image processing.

1.3 Our Approach

A few basic considerations govern our approach to discussing microscope image processing algorithms. These are based on years of experience using and teaching digital image processing in microscope applications. They are intended to prevent many of the common misunderstandings that crop up to impair communication and confuse one seeking to understand how to use this technology productively. We have found that a detailed grasp of a few fundamental concepts does much to facilitate learning this topic, to prevent misunderstandings, and to foster successful application. We cannot claim that our approach is "standard" or "commonly used." We only claim that it makes the job easier for both the reader and the authors.

1.3.1 The Four Types of Images

To the question "Is the image analog or digital?" the answer is, "Both." In fact, at any one time we may be dealing with four separate images, each of which is a representation of the specimen that lies beneath the microscope objective lens. This is a central issue because, whether we are looking at the pages of this book, at a computer display, or through the eyepieces of a microscope, we are viewing only images and not the original object. It is only with a clear appreciation of these four images, and the relationships among them, that we can move smoothly through the design and effective use of microscope image processing algorithms. We have endeavored to use this formalism consistently throughout this book to solidify the foundation of the reader's understanding.

1.3.1.1 The Optical Image

The optical components of the microscope act to create an optical image of the specimen on the image sensor, which is most commonly a charge-coupled device (CCD) array. The optical image is actually a continuous distribution of light intensity across a two-dimensional surface. It contains some information about the specimen, but it is not a complete representation of the specimen. It is, in the common case, a two-dimensional projection of a three-dimensional object, and it is limited in resolution and subject to distortion and noise introduced by the imaging process. Though an imperfect representation, it is what we have to work with if we seek to view, analyze, interpret, and understand the specimen.

1.3.1.2 The Continuous Image

We can assume that the optical image corresponds to, and is represented by, a continuous function of two spatial variables. That is, the coordinate positions (x, y) are real numbers, and the light intensity at a given spatial position is a nonnegative real number. This mathematical representation we call the continuous image. More specifically, it is a real-valued analytic function of two real variables. This affords us considerable opportunity to use well-developed mathematical theory in the design and analysis of algorithms. We are fortunate that the imaging process allows us to assume analyticity, since analytic functions are much more well behaved than those that are merely continuous (see Section 1.3.2.1).

1.3.1.3 The Digital Image

The digital image is produced by the process of digitization. The continuous optical image is sampled, commonly on a rectangular grid, and those sample values are quantized to produce a rectangular array of integers. That is, the coordinate positions (n, m) are integers, and the light intensity at a given integer spatial position is represented by a non-negative integer. Further, random noise is introduced into the resulting data. Such treatment of the optical image is brutal in the extreme. Improperly done, the digitization process can severely damage an image or even render it useless for analytical or interpretation purposes. More formally, the digital image may not be a faithful representation of the optical image and, therefore, of the specimen. Vital information can be lost in the digitization process, and more than one project has failed for this reason alone. Properly done, image digitization yields a numerical representation of the specimen that is faithful to the original spatial distribution of light that emanated from the specimen.

What we actually process or analyze in the computer, of course, is the digital image. This array of sample values (pixels) taken from the optical image, however, is only a relative of the specimen, and a rather distant one at that. It is the responsibility of the user to ensure that the relevant information about the specimen that is conveyed by the optical image is preserved in the digital image as well. This does not mean that all such information must be preserved. This is an impractical (actually impossible) task. It means that the information required to solve the problem at hand must not be lost in either the imaging process or the process of digitization.

We have mentioned that digitization (sampling and quantization) is what generates a corresponding digital image from an existing optical image. To go the other way, from discrete to continuous, we use the process of interpolation. By interpolating a digital image, we can generate an approximation to the continuous image (analytic function) that corresponds to the original optical image. If all goes well, the continuous function that results from interpolation will be a faithful representation of the optical image.

1.3.1.4 The Displayed Image

Finally, before we can visualize our specimen again, we must display the digital image. Human eyes cannot view or interpret an image that exists only in digital form. A digital image must be converted back into optical form before it can be seen. The process of displaying an image on a screen is also an action of interpolation, this time implemented in hardware. The display spot, as it is controlled by the digital image, acts as the interpolation function that creates a continuous visible image on a screen or on paper. The display hardware must be able to interpolate the digital image in such a way as to preserve the information of interest.

1.3.2 The Result

We see that each image we work with is actually a set of four images. Each optical image corresponds to both the continuous image that describes it and the digital image that would be obtained by digitizing it (assuming some particular set of digitizing parameters). Further, each digital image corresponds to the continuous function that would be generated by interpolating it (assuming a particular interpolation method). Moreover, the digital image also corresponds to the displayed image that would appear on a particular display screen. Finally, we assume that the continuous image is a faithful representation of the specimen and that it contains all of the relevant information required to solve the problem at hand. In this book we refer to these as the optical image, the continuous image, the digital image, and the displayed image. Their relationship is shown in Fig. 1.1.

This leaves us with an option as we go through the process of designing or analyzing an image processing algorithm. We can treat the image as an array of numbers (which it is), or we can analyze the corresponding continuous image. Both of these represent the optical image, which, in turn, represents the specimen. In some cases we have a choice and can make life easy for ourselves. Since we are actually working with an array of integers, it is tempting to couch our analysis strictly in the realm of discrete mathematics. In many cases this can be a useful approach. But we cannot ignore the underlying analytic function to which that array of numbers corresponds. To be safe, an algorithm must be true to both the digital image and the continuous image. Thus we must pay close attention to both the continuous and the discrete aspects of the image.

To focus on one and ignore the other can lead a project to disaster. In the best of all worlds, we could go about our business, merrily flipping back and forth between

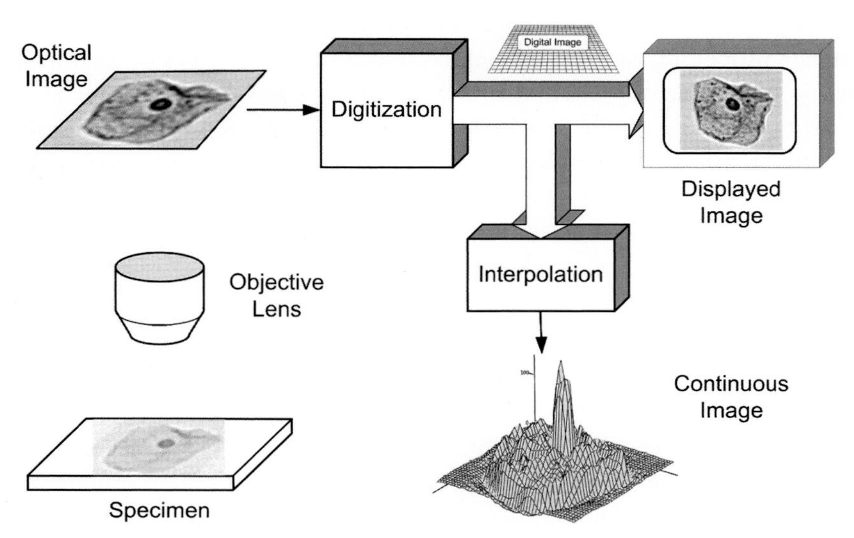

Fig. 1.1 The four images of digital microscopy. The microscope forms an optical image of the specimen. This is digitized to produce the digital image, which can be displayed and interpolated to form the continuous image. The displayed image allows the original image to be visualized.

corresponding continuous and digital images as needed. The implementations of digitization and interpolation, however, do introduce distortion, and caution must be exercised at every turn. Throughout this book we strive to point out the resulting pitfalls.

1.3.2.1 Analytic Functions

The continuous image that corresponds to a particular optical image is more than merely continuous. It is a real-valued analytic function of two real variables. An analytic function is a continuous function that is severely restricted in how "wiggly" it can be. Specifically, it possesses all of its derivatives at every point [5]. This restriction is so severe, in fact, that if you know the value of an analytic function and all of its (infinitely many) derivatives at a single point, then that function is unique, and you know it everywhere (Section 12.4.1). In other words, only one analytic function can pass through that point with those particular values for its derivatives. To be dealing with functions so tightly restricted relieves us of many of the concerns that keep pure mathematicians entertained.

As an example, assume that an analytic function of one variable passes through the origin where its first derivative is equal to 2, and all other derivatives are zero. The analytic function $y=2x$ uniquely satisfies this condition and thus is that function. Of all the functions that pass through the origin, only this one meets the stated requirements.

Thus when we work with a monochrome image, we can think of it as an analytic function of two dimensions. A multispectral image can be viewed as a collection of such functions, one for each spectral band. The restrictions implied by the analyticity property make life much easier for us than it would otherwise be. Working with such a restricted class of functions allows us considerable latitude in the mathematical analysis that surrounds image processing algorithm design. We can make the types of assumptions that are common to engineering disciplines and actually get away with them.

The continuous and digital images are even more restricted than previously stated. The continuous image is an analytic function that is bandlimited as well. The digital image is a band-limited, sampled function. The effects created by all of these sometimes conflicting restrictions are discussed in later chapters. For present purposes it suffices to say only that, by following a relatively simple set of rules, we can analyze the digital image as if it were the specimen itself. It is also true that violating any of these rules can lead to disaster.

1.3.3 The Sampling Theorem

The theoretical results that provide us with the most guidance as to what we can get away with when digitizing and interpolating images are the Nyquist sampling theorem (1928) and the Shannon sampling theorem (1949). They specify the conditions under which an analytic function can be reconstructed, without error, from its samples (Section 3.2). Although this ideal situation is never quite attainable in practice, the sampling theorems nevertheless provide us with means to keep the damage to a minimum and to understand the causes and consequences of failure, when it occurs. We cannot digitize and interpolate without the introduction of noise and distortion. We can, however, preserve sufficient fidelity to the specimen so that we can solve the problem at hand. The sampling theorem is our map through that dangerous territory. This topic is covered in detail in Chapter 3. By following a relatively simple set of rules, we can produce usable results with digital microscopy.

1.4 The Challenge

And so we are left with the following situation. The object of interest is the specimen that is placed under the microscope. The instrument forms an optical image that represents that specimen. We assume that the optical image is well represented by a continuous image (which is an analytic function), and we strive, through the choices available in microscopy, to ensure that this is the case. Further, the optical image is sampled and quantized in such a way that the information relevant to the problem at hand has been retained in the digital image. We can interpolate the digital image to produce an approximation to the continuous image or to make it visible for interpretation. We must now process the digital image, either to extract quantitative data from it or to prepare it for display and interpretation by a human observer. In subsequent chapters the model we use is that the continuous image is an analytic function that represents the specimen and that the digital image is a quantized array of discrete samples taken from the continuous image. Although we actually process only the digital image, interpolation gives us access to the continuous image whenever it is needed.

Our approach, then, is that we keep in mind that we are always dealing with two images that are representations of the optical image produced by the microscope, and this in turn represents a projection of the specimen. When analyzing an algorithm we can employ either continuous or discrete mathematics, as long as the relationship between these images is understood and preserved. In particular, any processing step performed upon the digital image must be legitimate in terms of what it does to the underlying continuous image.

1.5 Modern Microscopy

The past 400 years, and indeed the past few decades, have seen tremendous development in microscopy. Modern techniques, described later in this book, have drastically increased the utility of the microscope as a tool in research, medicine, and industry. The resolution limit found by Abbe in 1873 (Section 2.6) has been surpassed by a wide margin. Molecules can now be located with nanometer precision (see Figure 13.1). Specimens previously showing up only as blurred features can now be imaged in detail.

In addition to conventional bright-field microscopy, this book also discusses processing techniques that are applicable to confocal, fluorescence, structured illumination, and three-dimensional microscopy. The combination of precision optical equipment and advanced image processing techniques can produce extremely useful imagery.

1.6 Nomenclature

The nomenclature of recently developed techniques has not yet become standardized. Indeed, one of the challenges to understanding this complex field of endeavor is to recognize the differences, similarities, and identities among techniques that have different names. In this book we adopt a set of definitions that are common and useful, but by no means universal.

Digital microscopy consists of theory and techniques collected from several fields of endeavor. As a result, the descriptive terms used therein bear a collection of specialized definitions. Often, ordinary words are pressed into service and given specific meanings. We have included a glossary to help the reader navigate through the jargon, and we encourage its use. If a concept becomes confusing or difficult to understand, it may well be the result of one of these specialized words. As soon as that is cleared up, the pathway to understanding opens again.

1.7 Summary of Important Points

1. A microscope forms an optical image that represents the specimen.
2. The continuous image is a real-valued analytic function of two real variables that represents the optical image.
3. An analytic function is not only continuous but possesses all of its derivatives at every point.
4. The process of digitization generates a digital image from the optical image.
5. The digital image is an array of integers obtained by sampling and quantizing the optical image.
6. The process of interpolation generates an approximation of the continuous image from the digital image.
7. Image display is an interpolation process that is implemented in hardware. It makes the digital image visible.
8. The optical image, the continuous image, the digital image, and the displayed image each represent the specimen.
9. The design or analysis of an image processing algorithm must take into account both the continuous image and the digital image.

10. In practice, digitization and interpolation cannot be done without loss of information and the introduction of noise and distortion.
11. Digitization and interpolation must both be done in a way that preserves the image content that is required to solve the problem at hand.
12. Digitization and interpolation must be done in a way that does not introduce noise or distortion that would obscure the image content needed to solve the problem at hand.

References

[1] D.L. Spector, R.D. Goldman (Eds.), Basic Methods in Microscopy, Cold Spring Harbor Laboratory Press, 2005.
[2] P. Török, F.-J. Kao (Eds.), Optical Imaging and Microscopy: Techniques and Advanced Systems, vol. 87, Springer, 2007.
[3] G. Sluder, D.E. Wolf, Digital Microscopy, second ed., Academic Press, 2003.
[4] S. Inoue, K.R. Spring, Video Microscopy, second ed., Springer, 1997.
[5] D.B. Murphy, Fundamentals of Light Microscopy and Electronic Imaging, Wiley-Liss, 2001.
[6] K.R. Castleman, Digital Image Processing, Prentice-Hall, 1996.
[7] A. Diaspro (Ed.), Confocal and Two-Photon Microscopy, Wiley-Liss, 2001.

CHAPTER TWO

Fundamentals of Microscopy

Kenneth R. Castleman and Ian T. Young

2.1 The Origins of the Microscope

During the 1st century AD the Romans were experimenting with different shapes of clear glass. They discovered that, by holding a piece that was thicker in the middle than at the edges over an object, they could make that object appear larger. They also used lenses to focus the rays of the sun and start a fire. By the end of the 13th century, spectacle makers were producing lenses to be worn as eyeglasses to correct for deficiencies in vision. The word "lens" derives from the Latin word "lentil" because these magnifying chunks of glass were similar in shape to a lentil bean.

In 1590, two Dutch spectacle makers, Zacharias Janssen and his father Hans, started experimenting with lenses. They mounted several lenses in a tube, producing considerably more magnification than was possible with a single lens. This work led to the invention of both the compound microscope and the telescope [1].

In 1665, Robert Hooke, the English physicist who is sometimes called "the father of English microscopy," was the first person to see cells. He made his discovery while examining a sliver of cork. In 1674 Anton van Leeuwenhoek, while working in a dry goods store in Holland, became so interested in magnifying lenses that he learned how to make his own. By carefully grinding and polishing, he was able to make small lenses with high curvature, producing magnifications of up to 270 times. He used his simple microscope to examine blood, semen, yeast, insects, and the tiny animals swimming in a drop of water. Leeuwenhoek became quite involved in science and was the first person to describe cells and bacteria. Because he neglected his dry goods business in favor of science, and because many of his pronouncements ran counter to the beliefs of the day, he was ridiculed by the local townspeople. From the great many discoveries documented in his research papers, Anton van Leeuwenhoek (1632–1723) has come to be known as "the father of microscopy." He constructed a total of 400 microscopes during his lifetime. In 1759 John Dolland built an improved microscope using lenses made of flint glass, greatly improving resolution.

Since the time of these pioneers, the basic technology of the microscope has developed in many ways. The modern microscope is used in many different imaging modalities and has become an invaluable tool in fields as diverse as material science, forensic science, clinical medicine, and biomedical and biological research.

Microscope Image Processing
https://doi.org/10.1016/B978-0-12-821049-9.00004-6

Copyright © 2023 Elsevier Inc.
All rights reserved.

2.2 Optical Imaging

In this section we introduce the basic concept of an image-forming lens system [1–7].

2.2.1 Image Formation by a Lens

Fig. 2.1 shows an optical system consisting of a single lens. In the simplest case the lens is a thin, double-convex piece of glass with spherical surfaces. Light rays inside the glass have a lower velocity of propagation than light rays in air or vacuum. Because the distance the rays must travel varies from the thickest to the thinnest parts of the lens, the light rays are bent toward the optical axis of the lens by the process known as *refraction*.

2.2.1.1 Imaging a Point Source

A diverging spherical wave of light radiating from a point source at the origin of the focal plane is refracted by a convex lens to produce a converging spherical exit wave. The light converges to produce a small spot at the origin of the image plane. The shape of that spot is called the *point spread function* (psf). The point spread function will take on its smallest possible size if the system is in focus, that is, if

$$\frac{1}{d_f} + \frac{1}{d_i} = \frac{1}{f} \tag{2.1}$$

where f is the *focal length* of the lens. Eq. (2.1) is called the *lens equation*.

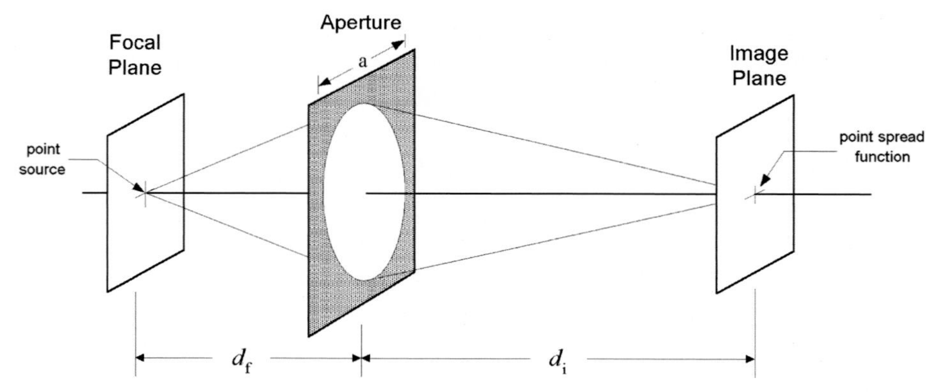

Fig. 2.1 An optical system consisting of a single lens. A point source at the origin of the focal plane emits a diverging spherical wave that is intercepted by the aperture. The lens converts this into a spherical wave that converges to a spot (i.e., the point spread function, psf) in the image plane. If d_f and d_i satisfy Eq. (2.1), the system is in focus, and the psf takes on its smallest possible dimension.

2.2.1.2 Focal Length

Focal length is an intrinsic property of any particular lens. It is the distance from the lens to the image plane when a point source located at infinity is imaged in focus. That is,

$$d_f = \infty \Rightarrow d_i = f$$

and by symmetry

$$d_i = \infty \Rightarrow d_f = f$$

The *power* of a lens, P, is given by $P = 1/f$, and if f is given in meters, then P is in diopters. By definition, the *focal plane* is that plane in object space where a point source will form an in-focus image on the image plane, given a particular d_i. Though sometimes called the "object plane" or the "specimen plane," it is more appropriately called the focal plane because it is the locus of all points that the optical system can image in focus.

2.2.1.3 Magnification

If the point source moves away from the origin to a position (x_o, y_o), then the spot image moves to a new position, (x_i, y_i), given by

$$x_i = -Mx_o \quad y_i = -My_o \tag{2.2}$$

where

$$M = \frac{d_i}{d_f} \tag{2.3}$$

is the *magnification* of the system.

Often the objective lens forms an image directly on the image sensor, and the pixel spacing scales down from sensor to specimen by a factor approximately equal to the objective magnification. If, for example, $M = 100$ and the pixel spacing of the image sensor is 6.8 μm, then at the specimen or focal plane the spacing is 6.8 μm/100 = 68 nm. In other cases, additional magnification is introduced by intermediate lenses located between the objective and the image sensor. The microscope eyepieces, which figure into conventional computations of "magnification," have no effect on pixel spacing. It is usually advantageous to measure, rather than calculate, pixel spacing in a digital microscope. For our purposes, pixel spacing at the specimen is a more useful parameter than magnification.

Eqs. (2.1) and (2.3) can be manipulated to form a set of formulas that are useful in the analysis of optical systems [8]. In particular,

$$f = \frac{d_i d_f}{d_i + d_f} = \frac{d_i}{M+1} = d_f \frac{M}{M+1} \tag{2.4}$$

$$d_i = \frac{f d_f}{d_f - f} = f(M+1) \tag{2.5}$$

and

$$d_f = \frac{f d_i}{d_i - f} = f \frac{(M+1)}{M} \tag{2.6}$$

Although it is composed of multiple lens elements, the objective lens of an optical microscope behaves as in Fig. 2.1, to a good approximation. In contemporary light microscopes, d_i is fixed by the optical tube length of the microscope. The mechanical tube length, the distance from the objective lens mounting flange to the image plane, is commonly 160 mm. The optical tube length, however, varies between 190 and 210 mm, depending upon the manufacturer. In any case, $d_i \gg d_f$ and $M \gg 1$, except when a low power objective lens ($<10\times$) is in use.

2.2.1.4 Numerical Aperture

It is customary to specify a microscope objective, not by its focal length and aperture diameter, but rather by its magnification (Eq. 2.3) and its *numerical aperture*, *NA*. Microscope manufacturers commonly engrave the magnification power and numerical aperture on their objective lenses, and the actual focal length and aperture diameter are rarely used. The *NA* is given by

$$NA = n\ \sin(\alpha) \approx n(a/2d_f) \approx n(a/2f) \tag{2.7}$$

where n is the refractive index of the medium (air, immersion oil, etc.) located between the specimen and the lens, and $\alpha = \arctan(a/2d_f)$ is the angle between the optical axis and a marginal ray from the origin of the focal plane to the edge of the aperture as illustrated in Fig. 2.1. The approximations in Eq. (2.7) assume small aperture and high magnification, respectively. These approximations begin to break down at low power and high *NA*, which normally do not occur together. One can compute and compare f and a, or the angles $\arctan(\alpha/2d_f)$ and $\text{arcsine}(NA/n)$ to quantify the degree of approximation.

2.2.1.5 Lens Shape

For a thin, double-convex lens having a diameter that is small compared to its focal length, the surfaces of the lens must be spherical in order to convert a diverging spherical entrance wave into a converging spherical exit wave by the process of *diffraction*. Furthermore, the focal length, f, of such a lens is given by the *lensmaker's equation*

$$\frac{1}{f} = (n-1)\left(\frac{1}{R_1} + \frac{1}{R_2}\right) \tag{2.8}$$

where n is the refractive index of the glass and R_1 and R_2 are the radii of the front and rear spherical surfaces of the lens [4]. For larger diameter lenses the required shape is aspherical.

2.3 Diffraction Limited Optical Systems

The previous section described the geometric optics approach to characterizing and understanding a microscope. In this section we use a more complex approach, one that emerged in the 19th century and is based upon characterizing light as a wave-based phenomenon instead of a ray-based phenomenon. It does not end there. It is also possible to construct and describe microscopes that are based upon the photon-based description of light that emerged in the 20th century and is grounded in quantum mechanics and quantum electrodynamics. It is beyond the scope of this book to examine this last class of microscopes. We refer you, instead, to the literature such as [9].

In Fig. 2.1, the lens is thicker near the axis than it is near the edges, and axial rays are delayed more than peripheral rays. In the ideal case, the variation in thickness is just right to convert the incoming expanding spherical wave into a spherical exit wave converging toward the image point. Any deviation of the exit wave from spherical form is, by definition, due to *aberration* and makes the psf larger.

For lens diameters that are not small in comparison to f, spherical lens surfaces are not adequate to produce a spherical exit wave. Such lenses do not converge peripheral rays to the same point on the z-axis as they do near axial rays. This phenomenon is called *spherical aberration*, since it results from the (inappropriate) spherical shape of the lens surfaces. High-quality optical systems employ aspheric surfaces and multiple lens elements to reduce spherical aberration. Normally the objective lens is the main optical component in a microscope that determines overall image quality.

A *diffraction-limited* optical system is one that does produce a converging spherical exit wave in response to the diverging spherical entrance wave from a point source. It is so-called because its resolution is limited only by diffraction, an effect directly related to the wave nature of light. One should understand that a diffraction-limited optical system is an idealized system and that real optical systems can only approach this ideal.

2.3.1 Linear System Analysis

It should be clear that increasing the intensity of the point source in Fig. 2.1 causes a proportional increase in the intensity of the spot image. It follows that two point sources would produce an image in which the two spots combine by addition. This means that the lens is a two-dimensional *linear system* [8]. For reasonably small off-axis distances in well-designed optical systems, the shape of the spot image undergoes essentially no change as it moves away from the origin. Thus the system can be assumed to be *shift invariant* or, in optics terminology, *isoplanatic*, as well as linear. The psf is then the *impulse response* of a shift-invariant, linear system. This implies that the imaging properties of the system can be specified by either its psf or its *transfer function* [4,8]. The *optical transfer function* is the Fourier transform of the psf. The field of linear system analysis is quite well-developed, and it provides us with very useful tools to analyze the performance of optical systems. This is developed in more detail in later chapters.

2.4 Incoherent Illumination

Incoherent illumination may be viewed as a distribution of point sources, each having a random phase that is statistically independent of the other point sources [2,5–7]. Under incoherent illumination, an optical system is linear in light intensity. The light intensity is the square of the amplitude of the electromagnetic waves associated with the light [2]. In the following, we assume narrow-band light sources. In general, light is a collection of different wavelengths where modern microscopy makes extensive use of wavelengths between 350 nm (ultraviolet) and 1100 nm (near infrared). The term *narrow-band* implies using only a small range of wavelengths, perhaps 30 nm wide around some center wavelength.

2.4.1 The Point Spread Function

The spot in the image plane produced by a point source in the focal plane is the psf. For a lens with a circular aperture of diameter a in narrow-band, incoherent light having center wavelength λ, the psf has circular symmetry (Fig. 2.2) and is given by

$$psf(r) = h(r) = \left[2 \frac{J_1\left(\pi \left[\frac{r}{r_o}\right]\right)}{\pi \left[\frac{r}{r_o}\right]} \right]^2 \tag{2.9}$$

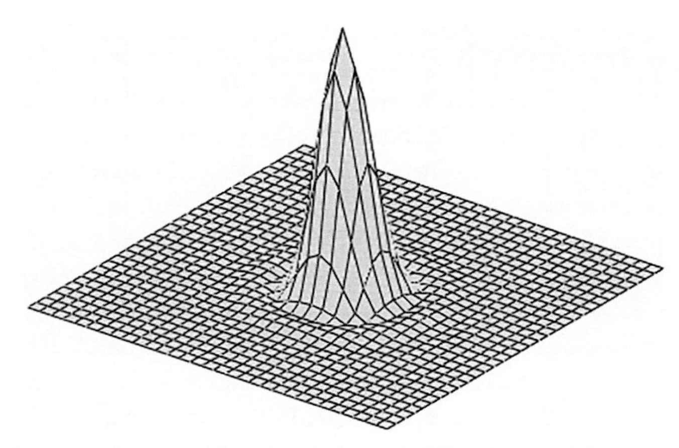

Fig. 2.2 The incoherent point spread function. A focused diffraction-limited system produces this psf in narrow-band incoherent light.

where $J_1(x)$ is the first-order Bessel function of the first kind [4]. The intensity distribution associated with this psf is called the Airy disk pattern, after Airy [10,11], and is shown in Fig. 2.2. The constant, r_o, a dimensional scale factor, is

$$r_o = \frac{\lambda d_i}{a} \tag{2.10}$$

and r is radial distance measured from the origin of the image plane, that is,

$$r = \sqrt{x_i^2 + y_i^2} \tag{2.11}$$

2.4.2 The Optical Transfer Function

Since the imaging system in Fig. 2.1 is a shift-invariant linear system, it can be specified either by its impulse response (i.e., the psf) or by the Fourier transform of its psf, which is called the *optical transfer function* (*OTF*). For a lens with a circular aperture of diameter a in narrow-band incoherent light having center wavelength λ, the *OTF* (Fig. 2.3) is given by [4].

$$OTF(q) = F\{h(r)\} = H(q) = \begin{cases} \frac{2}{\pi}\left\{\cos^{-1}\left[\frac{q}{f_c}\right] - \sin\left[\cos^{-1}\left(\frac{q}{f_c}\right)\right]\right\} & q \le f_c \\ 0 & q \ge f_c \end{cases} \tag{2.12}$$

where q is the spatial frequency variable, measured radially in two-dimensional frequency space. It is given by

$$q = \sqrt{u^2 + v^2} \tag{2.13}$$

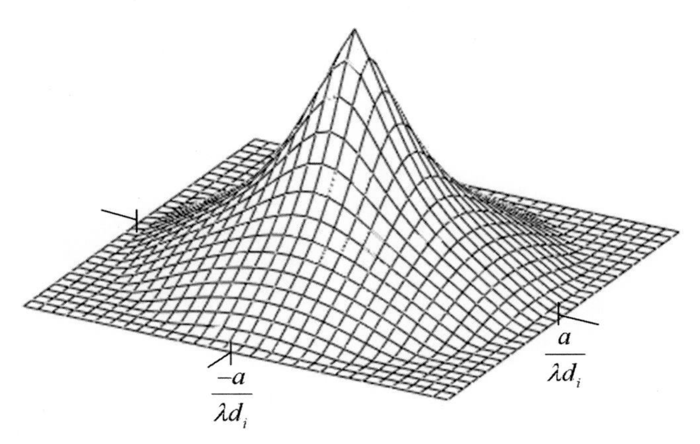

Fig. 2.3 The incoherent OTF. This is the frequency response of a focused diffraction-limited system in narrow-band incoherent light.

where u and v are spatial frequencies in the x and y directions, respectively. The parameter f_c, called the *optical cutoff frequency*, is given by

$$f_c = \frac{1}{r_o} = \frac{a}{\lambda d_i} \tag{2.14}$$

2.5 Coherent Illumination

Some microscopy applications require the use of coherent light for illumination. Lasers, for example, supply coherent illumination at high power. Coherent illumination can be thought of as a distribution of point sources whose amplitudes maintain fixed phase relationships among themselves. Diffraction works somewhat differently under coherent illumination, and the psf and *OTF* take on different forms. Under coherent illumination, an optical system is linear in complex amplitude as opposed to linear in light intensity, as in the incoherent case.

2.5.1 The Coherent Point Spread Function

For a lens with a circular aperture of diameter a in coherent light of wavelength λ, the psf has circular symmetry (Fig. 2.4) and is given by

$$h(r) = 2\frac{J_1[\pi(r/r_0)]}{\pi(r/r_0)} \tag{2.15}$$

where r_o is from Eq. (2.10) and r is from Eq. (2.11).

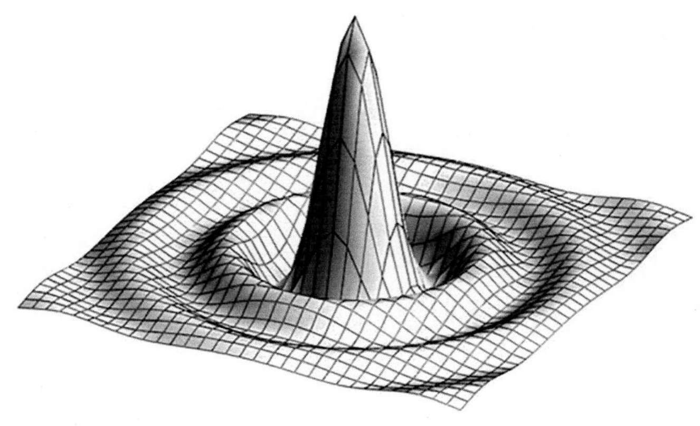

Fig. 2.4 The coherent point spread function. A focused diffraction-limited system produces this psf in coherent light.

2.5.2 The Coherent Optical Transfer Function

For a lens with a circular aperture of diameter a in coherent light of wavelength λ, the OTF (Fig. 2.5) is given by [4].

$$H(q) = \Pi\left(q\frac{\lambda d_i}{a}\right) \tag{2.16}$$

where q is from Eq. (2.13) and

$$\Pi(q) = \begin{cases} 1 & |q| < \frac{1}{2} \\ 0 & |q| > \frac{1}{2} \end{cases} \tag{2.17}$$

Notice that, under coherent illumination, the OTF is flat out to its cutoff frequency, while under incoherent illumination it monotonically decreases. Notice also that the cutoff frequency in incoherent light is twice that of the coherent case. Fig. 2.6 illustrates the relationships between the incoherent point-spread and transfer functions of diffraction-limited optical systems with circular exit pupils.

2.6 Resolution

One of the most important parameters of a microscope is its *resolution*, that is, its ability to reproduce, in the image, small structures that exist in the specimen. The optical definition of resolution is the minimum distance by which two point sources must be separated in order for them to be recognized as separate. There is no unique way to establish this distance. The psfs overlap gradually as the points get closer together, and one

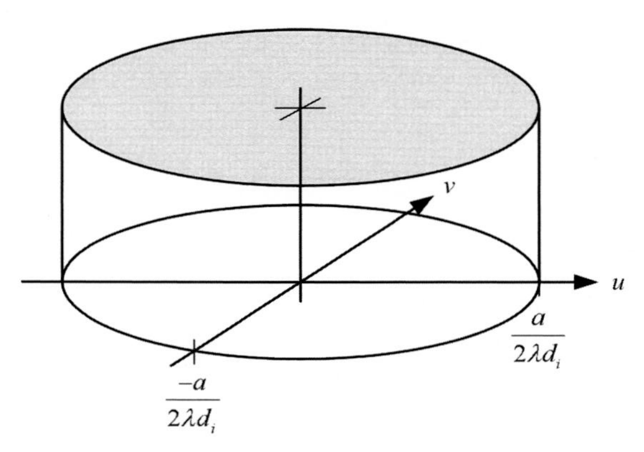

Fig. 2.5 The coherent OTF. This is the frequency response of a focused diffraction-limited system in coherent light.

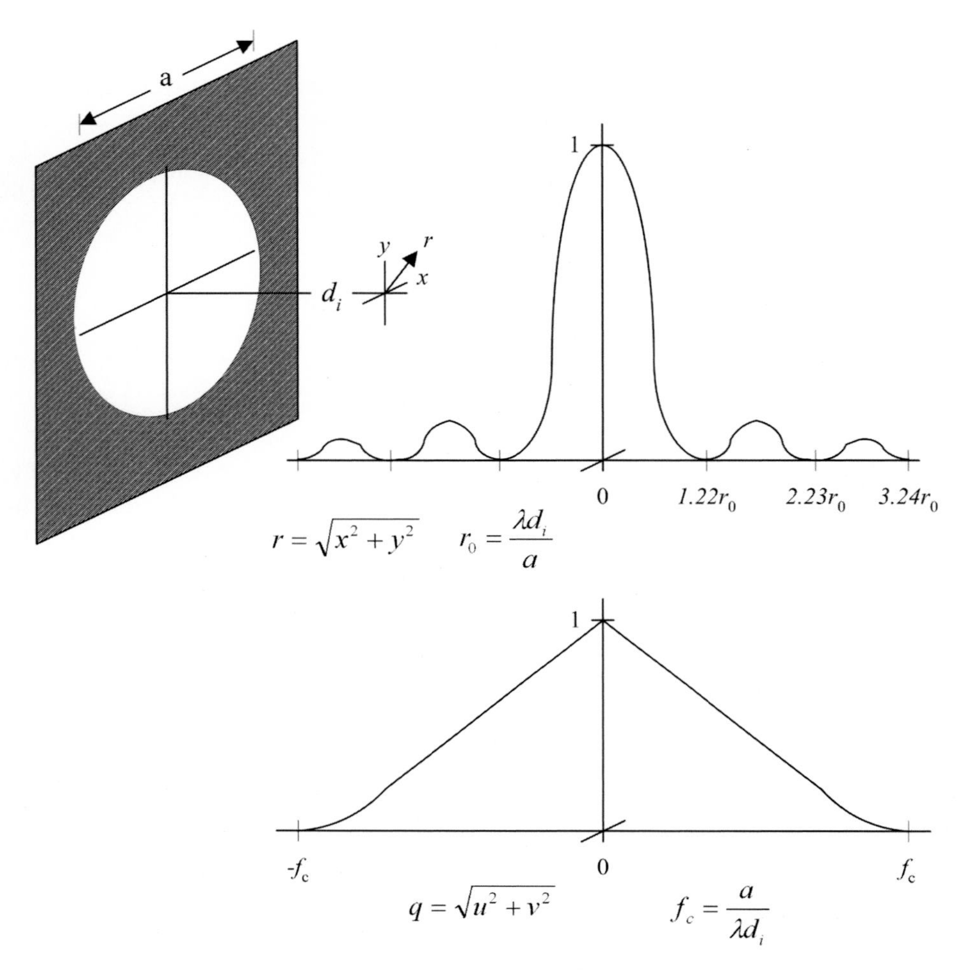

Fig. 2.6 The incoherent point spread function and transfer function. This shows how the psf and OTF depend on wavelength and aperture diameter for a diffraction-limited optical system with a circular aperture.

must specify how much contrast is required if the two objects are to be recognized as distinct. Further, the ability to distinguish between two points may be confounded by noise in the imaging system. The resolution under low-noise circumstances will almost certainly be better than high-noise situations. In this chapter, we address the former case where there are two commonly used criteria for comparing the resolving power of optical systems.

2.6.1 The Abbe Distance

To a good approximation, the half-amplitude diameter of the central peak of the image plane psf is given by the *Abbe distance* (after Ernst Abbe [12]),

$$r_{\text{Abbe}} = \frac{1}{M}\lambda\frac{d_i}{a} = \lambda\frac{d_f}{a} \approx \frac{\lambda}{2NA} = 0.5\left(\frac{\lambda}{NA}\right) \tag{2.18}$$

2.6.2 The Rayleigh Distance

For a lens with a circular aperture, the first zero of the image plane psf occurs at a radius

$$r_{\text{Airy}} = 1.22 r_o = 0.61\left(\frac{\lambda}{NA}\right) \tag{2.19}$$

which is called *the radius of the Airy disk*. According to the *Rayleigh criterion of resolution* (after Lord Rayleigh [13]), two point sources can just be resolved if they are separated, in the image, by the distance $\delta = r_{\text{Airy}}$. (See Fig. 2.7). In the terminology of optics, the Rayleigh distance defines circular *resolution cells* in the image, since two point sources can be resolved if they do not fall within the same resolution cell.

Within the framework of wave-based optics it is possible to improve upon the resolution of optical microscopes. The confocal microscope was introduced in the early 1980s after having been first developed by Marvin Minsky in 1955 [14,15]. Confocal microscopes are now in widespread use and offer a resolution improvement of about a factor of two with significantly increased image contrast as well as depth resolution producing three-dimensional images.

2.6.3 Size Calculations

In microscopy it is convenient to perform size calculations in the focal plane, rather than the image plane as previously, since that is where the objects of interest actually reside. The projection implemented by the lens involves a 180° rotation and a scaling by the

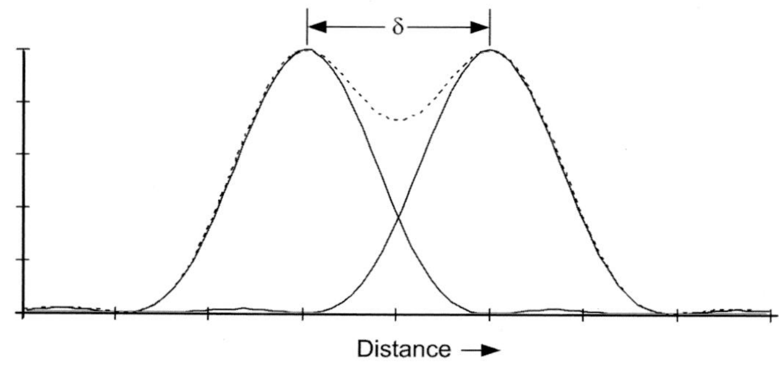

Fig. 2.7 The Rayleigh criterion of resolution. Two point sources can be just resolved if they are separated, in the image, by the Airy distance.

factor M (Eq. 2.3). The pixel spacing and resolution can then be specified in units of micrometers at the specimen. Spatial frequencies can be specified in cycles per micrometer in the focal plane.

Since $d_f \approx f$ for high-magnification lenses, the resolution parameters are more meaningful if we scale them to the focal (specimen) plane rather than working in the image plane. For a microscope objective, the incoherent optical cutoff frequency in the focal plane coordinate system is

$$f_c = \frac{Ma}{\lambda d_i} = \frac{a}{\lambda d_f} = \frac{2NA}{\lambda} \tag{2.20}$$

the Abbe distance is

$$r_{\text{Abbe}} = \frac{1}{M}\lambda\frac{d_i}{a} = \lambda\frac{d_f}{a} \approx \frac{\lambda}{2NA} = 0.5\left(\frac{\lambda}{NA}\right) \tag{2.21}$$

and the Rayleigh distance (resolution cell diameter) is

$$\delta_{\text{Rayleigh}} = 1.22 r_o = 0.61\left(\frac{\lambda}{NA}\right) \tag{2.22}$$

For $\lambda = 0.5\,\mu\text{m}$ (green light) and an NA of 1.4 (high-quality, oil-immersion lens), we have $f_c = 5.6$ cycles/μm, $r_{\text{Abbe}} = 0.179\,\mu\text{m}$, and $\delta_{\text{Rayleigh}} = 0.218\,\mu\text{m}$.

The foregoing approximations begin to break down at low power and high NA, which normally do not occur together. Again, one can compute and compare f and a, or the angles $\arctan(a/2d_f)$ and $\text{arcsine}(NA/n)$, to quantify the degree of approximation.

2.7 Aberration

Real lenses are never actually diffraction limited, but they suffer from aberrations that make the psf broader and the OTF narrower than they would otherwise be [2,6,7]. An example is spherical aberration mentioned earlier. Aberrations in an optical system can never increase the amplitude of the optical transfer function, but they can drive it negative.

2.8 Calibration

Making physical size measurements from images is very often required in the analysis of microscope specimens. This can be done with accuracy only if the pixel spacing at the focal plane is known. Making brightness measurements in the image is also useful, and it requires knowledge of the relationship between specimen brightness and gray levels in the digital image.

2.8.1 Spatial Calibration

The pixel spacing can be either calculated or measured. Calculation requires knowledge of the pixel spacing at the image sensor and the overall magnification of the optics (recall Eq. 2.3). Often it can be calculated from

$$\delta d = \frac{\Delta d}{M_o M_a} \tag{2.23}$$

where δd and Δd are the pixel spacing values at the specimen and at the image sensor, respectively, and M_o is the magnification of the objective. M_a is the magnification imposed by other optical elements in the system, such as the camera adapter. Usually this is quoted in the operator's manual for the microscope or the accessory attachment. The image sensor pixel spacing is quoted in the camera manual.

Too often, however, the numbers are not available for all the components in the system. Pixel spacing must then be measured with the aid of a calibrated target slide, sometimes called a "stage micrometer." This requires a computer program that can read out the (x, y) coordinates of a pixel in the digital image. One digitizes an image of the calibration target and locates two pixels that are a large distance apart in the image. Then

$$\delta d = \frac{D}{\sqrt{(x_2 - x_1)^2 + (y_2 - y_1)^2}} \tag{2.24}$$

where δd is the pixel spacing, D is the known distance on the calibration target, and (x_1, y_1) and (x_2, y_2) are the locations of the two pixels in the recorded image. For precision in the estimate of δd, the two points should be as far apart as possible in the microscope field-of-view. Ideally the pixel spacing would be measured in both the x- and y-directions in case these are unequal.

2.8.2 Photometric Calibration

Photometric properties that can be measured from a microscope image include transmittance, optical density, reflectance, and fluorescence intensity. Optical density calibration, as well as that for reflectance, requires a calibration target. The procedure is similar to that for spatial calibration [8]. Fluorescence intensity calibration techniques are covered in Chapter 10.

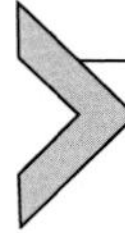

2.9 Summary of Important Points

1. Lenses and other optical imaging systems can, in most cases, be treated as two-dimensional, shift-invariant, linear systems.
2. The assumptions involved in the use of linear analysis of optical systems begin to break down as one moves far off the optical axis, particularly for wide-aperture, low-magnification systems.
3. Under coherent illumination, an optical system is linear in complex amplitude.
4. Under incoherent illumination, an optical system is linear in intensity (amplitude squared).
5. A "classical" wave-based optical system having no aberrations is called "diffraction-limited" because its resolution is limited only by the wave nature of light (diffraction effects). This is an ideal situation that real systems can only approach.
6. A diffraction-limited optical system transforms a diverging spherical entrance wave into a converging spherical exit wave.
7. The point-spread function of an optical system has a nonzero extent because of two effects: the wave nature of light (diffraction), and aberrations in the optical system.
8. The optical transfer function is the Fourier transform of the point spread function.
9. The point spread function is the inverse Fourier transform of the optical transfer function.
10. The incoherent point spread function is given by Eq. (2.9).
11. The incoherent optical transfer function is given by Eq. (2.12).
12. The coherent point spread function is given by Eq. (2.15).
13. The coherent transfer function is given by Eq. (2.16).
14. Photon-based imaging can provide spatial resolution that is 100 × better than wave-based imaging but at a cost of increased image acquisition time and money.

References

[1] W.J. Croft, Under the Microscope: A Brief History of Microscopy, World Scientific Publishing Company, 2006.
[2] M. Born, E. Wolf, Principles of Optics, sixtieth anniversary ed., Cambridge University Press, 2019.
[3] E.L. O'Neill, Introduction to Statistical Optics, second ed., Courier Dover Publications, 2015.
[4] J.W. Goodman, Introduction to Fourier Optics, fourth ed., Roberts and Company, 2017.
[5] C. Scott, Introduction to Optics and Optical Imaging, IEEE Press, 1998.
[6] E. Hecht, Optics, fifth ed., Pearson Education Limited, 2017.
[7] D.J. Goldstein, Understanding the Light Microscope, Academic Press, 1999.
[8] K.R. Castleman, Digital Image Processing, Prentice-Hall, 1996.
[9] S.J. Sahl, S.W. Hell, S. Jakobs, Fluorescence nanoscopy in cell biology, Nat. Rev. Mol. Cell Biol. 18 (2017) 685–701.
[10] G.B. Airy, On the diffraction of an annular aperture, Lond. Edinb. Dublin Philos. Mag. J. Sci. 18 (1841) 114, https://doi.org/10.1080/14786444108650232.

[11] G.B. Airy, On the Diffraction of an Object-Glass with Circular Aperture, SPIE Milestone Series, Society of Photo-optical Instrumentation Engineers, 2007.
[12] E. Abbe, Beiträge zur Theorie des Mikroskops und der mikroskopischen Wahrnehmung, Arch. Mikrosk. Anat. 9 (1873) 413–468.
[13] L. Rayleigh, On the theory of optical images, with special reference to the microscope, Philos. Mag. 42 (5) (1896) 167.
[14] M. Minsky, Memoir on inventing the confocal scanning microscope, Scanning 10 (1988) 128–138.
[15] T. Wilson, C. Sheppard, Theory and Practice of Scanning Optical Microscopy, Academic Press, 1984.

CHAPTER THREE

Image Digitization and Display

Kenneth R. Castleman

3.1 Introduction

Digitization is the process that generates a digital image, using the optical image as a guide. If this is done properly, then the digital image can be interpolated to produce (if it is also done properly) a continuous image that is a faithful representation of the optical image, at least for the content of interest. In this chapter we address the factors that must be considered in making this faithful representation happen.

Today many high-quality commercial image digitizing components are available. The design of a digital microscope system thus mainly entails selecting a compatible set of components that are within budget and adequate for the work to be done. A properly designed system, then, is well-balanced and geared to the tasks at hand. That is, no component unduly restricts image quality, and none is wastefully overdesigned.

In some cases a digitized microscope image may be analyzed quantitatively, and the resulting numerical data is all that is required for the project. In many other cases, however, a processed image must be displayed for interpretation. Indeed, more actual scientific discovery is based on viewing images than on observing data. Even in clinical applications, one usually wishes to see the specimen, if only to understand and confirm the accompanying numerical data.

Image display is the opposite of digitization. It converts a digital image back into visible form. It does so by an interpolation process that is implemented in hardware. Just as image digitization must be done with attention to the elements that affect image quality, high-quality image displays do not happen by accident. In this chapter we discuss processing steps that can help ensure that a display system presents an image in its truest or most interpretable form. Image display technologies fall outside our scope and are covered elsewhere [1–8]. Here we focus on how to select display devices and prepare digital image data for display.

3.2 Digitizing Images

The various components of the imaging system (optics, image sensor, analog-to-digital converter, etc.) act as links in a chain. Not only is this chain no stronger than its weakest link, but it is actually weaker than any of its links. In this chapter we seek to

Microscope Image Processing
https://doi.org/10.1016/B978-0-12-821049-9.00008-3
Copyright © 2023 Elsevier Inc.
All rights reserved.

establish guidelines that will lead to the design of well-balanced microscope imaging systems.

The six factors that can degrade an image in the digitizing process are (1) loss of detail (resolution), (2) noise, (3) aliasing, (4) shading, (5) photometric nonlinearity, and (6) geometric distortion. If the level of each of these is kept low enough, the digital images obtained from the microscope will be useable for their intended purpose. Different applications, however, require different levels of accuracy. Some are intrinsically more prone to noise and distortion than others. Thus the design of a system must begin with a list of the planned applications and their requirements in these six areas. In this chapter we discuss these topics individually before addressing them collectively.

This chapter addresses the various sources of degradation and how to quantify them, and the system design factors that affect overall performance. Other chapters present image processing techniques that can be used to correct these degradations. The principal goals are to preserve a suitably high level of detail and signal-to-noise ratio while avoiding aliasing, and to do so with acceptably low levels of shading, photometric nonlinearity, and geometric distortion.

3.2.1 Resolution

The term "resolution" is perhaps the most misunderstood and chronically abused word in all of digital microscopy. It is sometimes used to describe pixel spacing (e.g., 0.25 μm between adjacent pixel centers), digital image size (e.g., 1024 by 1024 pixels), test target size (e.g., 1-μm bars), and grayscale depth (e.g., 8 bits or 256 gray levels per pixel). If one does not pay careful attention to context in order to know which of these definitions is being used, considerable confusion will result.

In this book we adhere to the definition from the field of optics. *Resolution* is a property of an imaging system. Specifically, it refers to the ability of the system to reproduce the contrast of objects of different size. Notice that an object necessarily must reside upon a background. It is visible because it differs in brightness from that background, and that difference in brightness is its "contrast." Large objects are normally reproduced with their full contrast while small objects show up with reduced contrast. Below some limiting size, objects are imaged with zero contrast and are thus invisible. Resolution refers to the smallest size an object can have and still be *resolved*, that is, seen to be separate from the background and from all other objects in the image. This inherent loss of contrast with decreasing size, however, is a gradual phenomenon, so it is impossible to specify the exact size of the smallest objects that can be imaged. Instead we adopt a criterion of minimum visibility and use that to specify the size of the smallest resolvable object. This then is the "resolution" of the imaging system.

At high magnification, in a well-designed system, it is normally the optical components (principally the objective lens) that determine the overall system resolution. At low

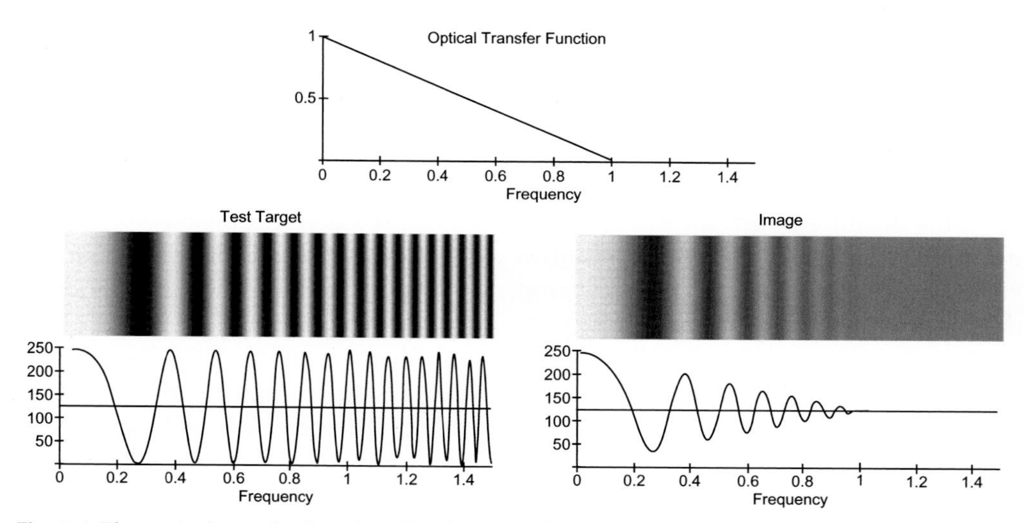

Fig. 3.1 The optical transfer function. The OTF specifies how the contrast of sinusoidal structures of different frequencies is affected by the imaging process. In this case large sinusoidal bars are passed with full contrast while smaller bars lose contrast. Where the OTF is zero, the bars disappear.

magnification, however, other components, such as interposed optical elements or the image sensor array, may set the limit of resolution.

The optical transfer function (OTF) is a plot of the reproduced contrast of an imaged object versus object size [9]. Here object size is specified as the frequency of a pattern of bars with sinusoidal profile. In Fig. 3.1 we see such a test pattern and note that the contrast of the imaged bars decreases with increasing frequency. A plot showing this decrease is the OTF.

Much to our good fortune, the OTF is the Fourier transform of the point spread function (psf), which is discussed in Chapter 2. Thus having either of these functions makes available the other, and either one is sufficient to specify the resolution of the system or of one of its components. Here we are assuming that the imaging components under consideration can be modeled as shift-invariant linear systems [9,10].

While the OTF, and equivalently, the psf, are complete specifications of a system's resolving capability, they are curves, and it is often desirable to use a single number to specify resolution. Several of these are used, such as the frequency at which the OTF drops to 10% of its zero-frequency value. More common is the Rayleigh resolution criterion [9,11]. It states that two point objects can just be resolved if they are separated by a distance equal to the radius of the first minimum of the psf, commonly called the "Airy disk."

3.2.2 Sampling

The digitization process samples the optical image to form a numerical array of sample points. These are most commonly arranged on a rectangular sampling grid with equal pixel spacing in both directions. The image intensity is averaged over a small local area at each sample point. This process can be modeled as convolution of the image with the psf of the system. Further, the image intensity is quantized at each sample point to produce an integer. This rather brutal treatment is necessary to produce image data that can be processed in a computer. If it is done properly, however, the important information in the image will remain undamaged.

The Shannon sampling theorem [9,12,13] states that a continuous function can be reconstructed, without error, from evenly spaced sample points, provided that two criteria are met. First the function must be *band limited.* That means that its Fourier spectrum is zero for all frequencies above some *cutoff frequency*, which we call f_c. This means the function can have no sinusoidal components of frequency greater than f_c. Second, the sample spacing must be no larger than $\Delta x = 1/2f_c$. This means there will be at least two sample points for each cycle of the highest frequency sinusoidal component of the function. If these two criteria are met, the function can be recovered from its samples by the process of interpolation, assuming that it is properly done.

If $\Delta x < 1/2f_c$, then we have a smaller sample spacing than what is absolutely necessary, and the function is said to be *oversampled.* The major drawbacks of oversampling are increased data file size and increased equipment cost, but reconstruction without error is still possible. If $\Delta x = 1/2f_c$ we have "critical sampling" also known as sampling at the *Nyquist rate* [1]. If $\Delta x > 1/2f_c$, then we have a larger sample spacing than that required by the sampling theorem, and the function is said to be *undersampled.* In this case interpolation cannot reconstruct the function without error, as it contains sinusoidal components of frequency greater than $1/2\Delta x$ (see Section 3.2.4).

3.2.3 Interpolation

As a further requirement for perfect reconstruction of a sampled function, the interpolating function must also be band-limited [14]. By the similarity theorem of the Fourier transform [9–11,15], a narrow function has a broad spectrum, and vice versa. This means that any suitable interpolating function (such as $\sin(x)/x$) will extend to infinity in both positive and negative x and y. Clearly we cannot implement that digitally, so we are constrained to work with truncated interpolation functions. This means that perfect reconstruction remains beyond our grasp, no matter how finely we sample. However, we can usually get close enough to produce useful results, and oversampling is a key to that.

Fig. 3.2 shows the results of interpolating a sampled cosine function with the often-used Gaussian interpolation function, shown at the upper left. Even though the sampling theorem is satisfied (i.e., $\Delta x < 1/2f_c$), in all seven cases, the inappropriate interpolation

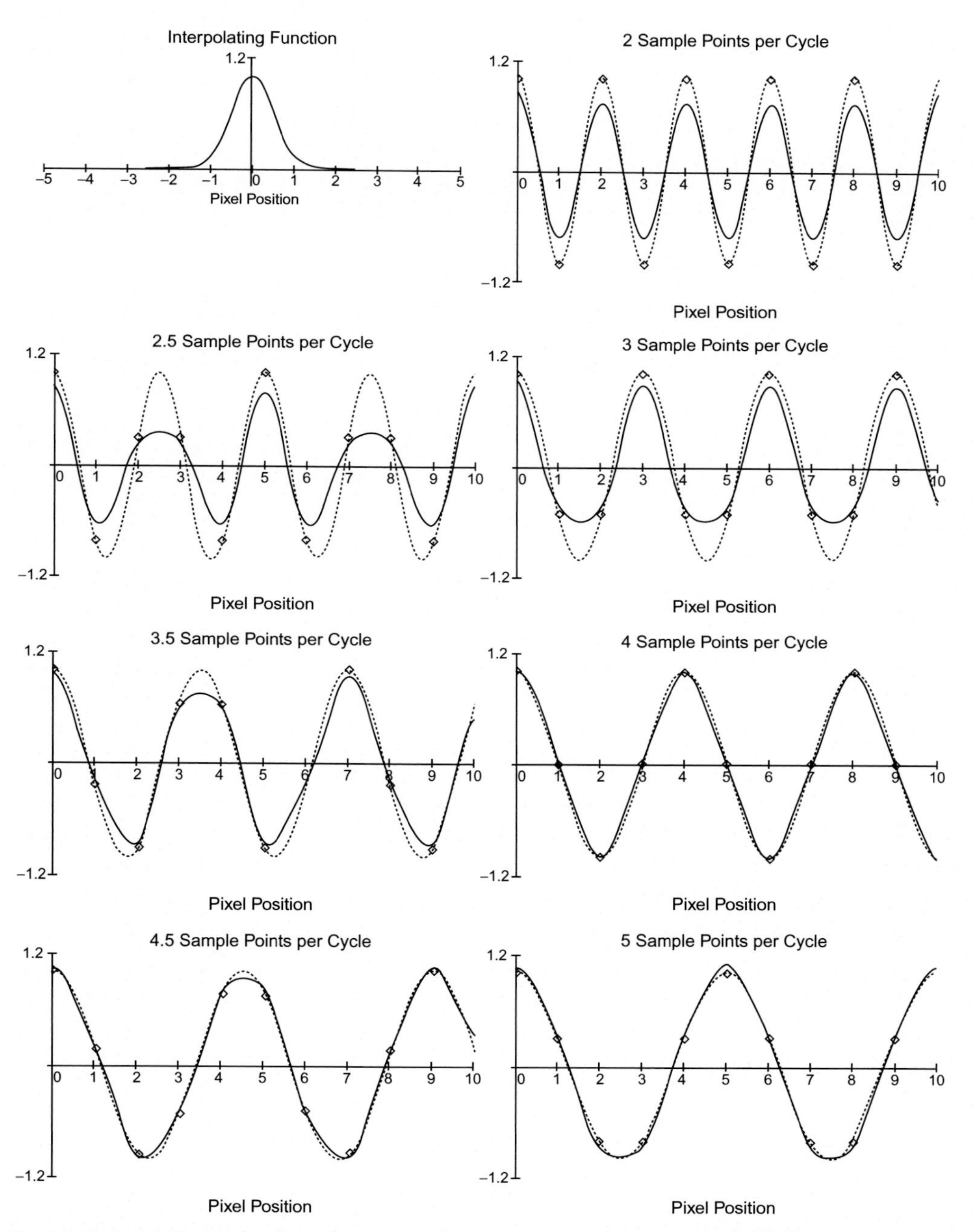

Fig. 3.2 Interpolation with a Gaussian function. Here sampled cosine functions of different frequencies are interpolated by convolution with the commonly used Gaussian function shown at the upper left. In each case the original function appears as a *dashed line*, and the interpolated function as a *solid line*. The sample points appear as *diamonds*. Notice that the inappropriate shape of the interpolation function gives rise to considerable reconstruction error, and that the amount and nature of that error varies with frequency. In general, however, the amount of interpolation error decreases as the sample spacing becomes smaller in relation to the period of the cosine. This phenomenon argues that oversampling tends to compensate for our having to use a truncated interpolation function.

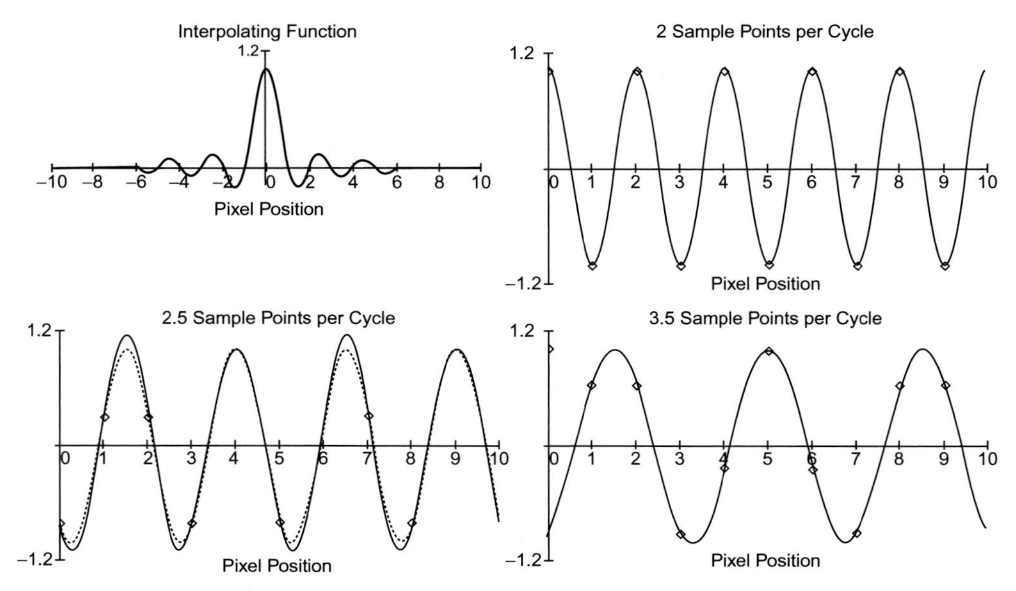

Fig. 3.3 Interpolation with a truncated sinc function. Sampled cosine functions of different frequencies are interpolated by convolution with the truncated sinc function shown in the upper left. In each case the original function is shown as a *dashed line*, and the reconstructed function as a *solid line*. The sample points appear as *diamonds*. Notice that the more appropriate shape of the interpolation function gives rise to considerably less reconstruction error than that apparent in Fig. 3.2.

function creates considerable reconstruction error. As the sampling becomes finer (more oversampling), the results of interpolation, while still imperfect, are seen to improve. The lesson is that, even though we are constrained to use inappropriate interpolation functions, judicious oversampling can compensate for much of that inadequacy.

Fig. 3.3 shows interpolation with a truncated version of the sinc function. Here the interpolation process is more complicated than with the Gaussian, but good results are obtained, all the way up to the sampling limit of $\Delta x = 1/2f_c$. This means that the continuous image is truly available to us if we implement digital interpolation using a decent approximation to the sinc function.

3.2.4 Aliasing

Aliasing is the phenomenon that occurs when an image is sampled too coarsely, that is, when the pixels are too far apart in relation to the size of the detail present in the image [9,13,14,16]. It introduces a very troublesome type of low-frequency noise. Aliasing can be a significant source of error when images contain a strong high-frequency pattern, but it can be, and should be, avoided by proper system design.

If the sample spacing, Δx, is too large (i.e., $\Delta x > 1/2f_c$), a sinusoid of frequency f_c is undersampled and cannot be reconstructed without error. When it is interpolated, even with an appropriate interpolation function, the phenomenon of aliasing introduces noise that resembles a Moiré pattern. Aliasing poses a particular problem in images that contain a high-contrast, high-frequency parallel line pattern. Most images have little contrast at the highest frequencies, so there is less of a visible effect.

Fig. 3.4 illustrates aliasing. In the upper left the cosine is oversampled and is reconstructed exactly by interpolation with the sinc function. In the other three panels the cosine is undersampled, and, while a cosine is reconstructed, it has the wrong frequency. Aliasing reduces the frequency of sinusoidal components in an image. This can be quite troublesome and should be avoided by maintaining $\Delta x < 1/2f_c$.

We mentioned in Chapter 2 that the OTF of a microscope objective lens (Eq. 2.12) goes to zero for all frequencies above the optical cutoff frequency $f_c = \lambda/2NA$ (Eq. 2.14). Thus the optics provide a built-in antialiasing filter, and we can be content simply to design for $\Delta x < 1/2f_c$, at least as far as aliasing is concerned. However, since we cannot

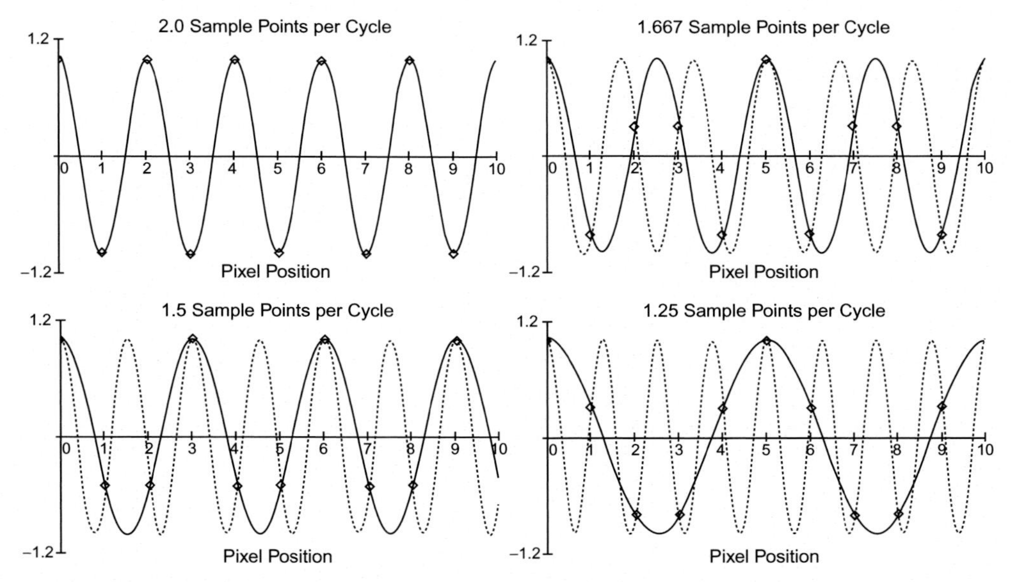

Fig. 3.4 Aliasing. Here cosine functions of different frequencies are sampled and then interpolated by convolution with the (untruncated) sinc function. In each case the original function is shown as a *dashed line*, and the reconstructed function as a *solid line*. The sample points appear as *diamonds*. Notice that violation of the sampling theorem (i.e., $\Delta x > 1/2f_c$) gives rise to a peculiar form of reconstruction error. In each case a cosine of unit magnitude is reconstructed, so amplitude and waveshape are preserved, but frequency is not. The reconstructed cosines are reduced in frequency. This phenomenon affects all sinusoidal components having frequency greater than half the sampling frequency. Notice that the reconstructed function, although of the wrong frequency, still passes through all of the sample points.

interpolate with the sinc function, as contemplated by the sampling theorem, we must compensate for this shortcoming by oversampling. Thus it is good practice to set the sample spacing below the aliasing limit by a factor of two or more.

3.2.5 Noise

The term noise is generally taken to mean any undesired additive component of an image. It can be random or periodic. The most common noise component is the random noise generated by the amplifier circuitry in the camera. Periodic noise can result from stray periodic signals (e.g., power line noise) finding their way into the camera circuits. Quantization noise results from conversion of the continuous brightness values into integers. As a rule of thumb, the overall noise level will be the root-sum-square of the various noise source amplitudes. One can control quantization noise by using enough bits per pixel. Generally quantization noise can be, and should be, kept below the other noise sources in amplitude. Eight-bit digitizers are quite common, and this puts the quantization noise level (1/256) below 0.5% of the total dynamic range. Some applications, however, require the more expensive 12-bit or even 16-bit digitizers.

3.2.6 Shading

Ideally the contrast of an object would not change as it moved around within the image. Neither would the gray level of the background area where no objects reside. In microscopy this is almost never the case. An empty field will usually show considerable variation in brightness, typically as a slowly varying pattern that becomes darker toward the periphery of the field of view. This is called "additive shading" because brightness is added to (or subtracted from) the true brightness of the object at different locations in the image. Careful adjustment of the microscope (e.g., lamp centering, condenser focusing, etc.) can minimize, but not eliminate, this effect. More subtle but equally important is "multiplicative shading." Here the contrast of the object (brightness difference from background) varies with position. The gray level is multiplied by a factor that varies with position.

Fortunately, both additive and multiplicative shading usually remain constant from one image to the next, until the microscope configuration (e.g., objective power) is changed. Thus one can usually record the shading pattern at the beginning of a digitizing session and subtract it (or divide it) from each captured image (see Chapter 10). Nevertheless, steps taken in the design phase to reduce inherent shading reap generous rewards later.

3.2.7 Photometry

Ideally the gray levels should be linearly related to some photometric property of the specimen, just as the pixel position (row and column) in the digital image is related to

(x, y) position in the specimen. Then the recorded digital image describes only the specimen and not the system that imaged it. The photometric property can be transmittance, optical density, reflectance, fluorescence intensity, etc. If the image sensor array is not linear in the desired photometric property, then a grayscale transformation (Chapter 5) may be required to bring about the desired linear relationship. Special calibration targets containing objects of known brightness are commonly used to determine the relationship. Scanning a step wedge target will produce the gray level transfer curve, and a grayscale transformation can correct the nonlinearity. Fluorescent beads of known size and brightness can be used in a similar manner.

3.2.8 Geometric Distortion

Geometric distortion is an unwanted "warping" of the image that distorts the spatial relationship among objects in the image. It can change the apparent size and shape of objects and the spacing between them. Geometric distortion can undermine the accuracy of spatial measurements such as length, area, perimeter, shape, and spacing.

Modern microscope imaging systems commonly use solid-state image sensor arrays. The pixel geometry is carefully controlled in the manufacturing process, and geometric distortion from this source is negligible. Other components in the imaging chain, however, can disturb the geometrical relationships. The imaging optics, for example, can introduce geometric distortion.

One can reduce the effects of geometric distortion by first measuring and characterizing it and then correcting it in software after the images have been digitized. Measurement requires a suitable test target of known geometry, such as a rectangular grid pattern. Correction requires a suitably defined geometric operation (see Chapter 4).

To a first-order approximation, geometric distortion is invariant from one image to the next. This means we can measure it once and for all and correct the images as a batch process. Changing the microscope configuration (e.g., the objective lens), of course, will change the distortion pattern, and separate correction is required for each different setup.

3.3 Overall System Design

No one component singlehandedly determines the quality of the images obtained from a digital microscope. It is the interaction of all of the components that establishes image quality.

3.3.1 Cumulative Resolution

Each component in the imaging chain (objective, relay lens, sensor, electronics, etc.) contributes to the overall resolution of the system. If each of these is a shift-invariant linear system, then their cumulative effect can be summarized as the system psf. The overall system psf is simply the convolution (Chapter 5) of all of the component psfs. The problem with this is that each convolution broadens the psf. Thus the system psf will be broader that any of the component psfs. Looking at the problem in the frequency domain, the overall system transfer function (TF) is the product of the transfer functions of all of the components. Since they all typically decay with increasing frequency, the system TF will be narrower than any component TF. The result of all this is that we might assemble a system with components, each of which has adequate resolution, only to find that the system itself does not. Searching for a single offending component will be futile since this chain is weaker than any of its links. Thus we must select components based on overall, not individual, performance requirements.

3.3.2 Design Rules of Thumb

Here we discuss some principles that can be used to guide the design of a system, even when this consists mainly of component selection.

3.3.2.1 Pixel Spacing

The cutoff frequency of the objective lens OTF sets the maximum sample spacing, but one is always wise to oversample generously. One should select the pixel spacing not only to avoid aliasing (Section 3.2.4) for any configuration (objective power, etc.), but also to make subsequent processing more reliable. Low-magnification configurations deserve special consideration since the objective lens may not provide an antialiasing filter (i.e., remove high-frequency detail), and aliasing becomes more likely. This is discussed further in Section 3.4.7.

3.3.2.2 Resolution

Although each component contributes, the numerical aperture (NA) of the objective lens generally establishes the resolution of the system. For high-quality lenses one can assume the diffraction-limited form of the OTF (Section 2.4.2). A measured OTF is better, however, especially for lower quality optics, which may not closely approach the diffraction limit. The objective lens OTF should pass the highest frequencies expected to be present in the specimens of interest. Stated another way, the psf (Section 2.4.1) should be less than half the size of the smallest objects of interest to be imaged.

3.3.2.3 Noise

The overall noise level is approximately the square root of the sum of squares of the amplitudes of the individual noise sources in the system. The quantization noise is the reciprocal of the number of gray levels (2^N where $N=$number of bits). Ideally it should be less than half the root-mean-square (RMS) noise level due to all other noise sources. Then this controllable source of noise will not make matters worse.

3.3.2.4 Photometry

One can measure the photometric linearity of a particular microscope configuration using a suitable calibration target (step wedge, fluorescent beads, etc.). If there is significant nonlinearity, it can be corrected with a suitable grayscale transformation (Chapter 5). This can be done as a batch job after the digitizing session is finished.

3.3.2.5 Distortion

The degree of geometric distortion in a particular microscope configuration can be measured with a calibration grid target. If it is significant, one can use a geometric transformation to correct it in each digitized image (Chapter 4). This can be done as a batch job after the digitizing session is complete, since it should be the same for each image.

3.4 Image Display

In many important microscope applications, in both clinical practice and biomedical research, the final step involves a human interpreting a displayed image. The output can be either printed on paper or displayed on a screen. The success of these cases is often tied directly to the quality of the displayed image. In this section, we discuss those characteristics which, taken together, determine the quality of a digital image display system and its suitability for particular applications. The primary characteristics of interest are the image size, the photometric and spatial resolution, the high- and low-frequency response, and the noise characteristics of the display system.

The primary job of a display system is to recreate, through interpolation, the continuous image that corresponds to the digital image that is to be displayed. Recall from Chapter 1 that any digital image is a sampled function that corresponds uniquely to a particular analytic function. The task of the display system is to reproduce that analytic function as a pattern of light on a screen, a pattern of ink on a page, or an image on film. We wish that presentation to be as accurate as possible, or at least good enough to serve the needs of the project.

Normally a display system produces an analog image in the form of a rectangular array of display pixels. The brightness of each display pixel is controlled by the gray level of the corresponding pixel in the digital image. However, the purpose of the display is to allow the human observer to understand and interpret the image content. This introduces a subjective element, and it is helpful to match the display process to the characteristics of the human eye. For example, the human eye has considerable acuity in discriminating fine detail (high-spatial-frequency information), but it is not particularly sensitive to low-frequency (slowly varying) image information [17]. Some images may be more easily understood if they are displayed indirectly, using contour lines, shading, color, or some other representation. Image display is discussed later in this chapter, and examples of displayed images appear throughout this book.

3.4.1 Volatile Displays

The most common types of volatile display are the liquid crystal display (LCD) and the thin-film transistor (TFT) flat panel monitors [18,19] The less common plasma displays are made by sandwiching a fine mesh between two sheets of glass, leaving a rectangular array of cells containing an ionizable gas [20]. By using coincident horizontal and vertical addressing techniques, the cells can be made to glow under the influence of a permanent sustaining electrical potential.

The monitor is usually driven by a display circuit in the computer that continuously transfers image data to the monitor in the proper format. Each monitor has a native image size and aspect ratio, which often does not match that of the digital image to be displayed. The circuitry in the monitor resamples the digital image vertically and/or horizontally as necessary to fill the screen. The 4:3 aspect ratio is a holdover from early television broadcast technology and has largely been replaced by modern aspect ratios such as 16:9. Note that a 4:3 image can be displayed on a 16:9 monitor by using only the central part of the screen. It can be stretched to fit the wider screen, but the image will be distorted.

Displayed grayscale formats range from 8-bit monochrome to 16-bit color (64,000 colors) and 24-bit color (16.7 million colors). The *contrast ratio* is the ratio of the intensity of the brightest white to pure black, under ambient illumination. Values of 500–5,000 and more are typical. Refresh rates vary from 30 to 120 Hz. The display can use either a progressive (line-by-line) scanning pattern, or an interlaced scan. Interlaced scanning was used in early television standards [21–29] to reduce perceived flicker. In this approach the display alternates between showing the odd-numbered lines and the even-numbered lines. Usually a high refresh rate (e.g., 60 Hz) progressive scan is preferred for best display quality. The trend is always toward physically larger displays with more pixels and faster scan rates.

3.4.2 Displayed Image Size

The image size capability of a display system has two components. First is the physical size of the display itself, which should be large enough to permit convenient examination and interpretation of the displayed images. For example, a larger screen is required when large specimens must be viewed at high resolution and for group viewing. The second characteristic is the size of the largest digital image that the display system can handle. Normally the digital image will be scaled up or down by interpolation ("resampled") to fit the full size of the display screen. The native displayed image size (e.g., 1920 by 1080 pixels) must be adequate for the number of rows and columns in the largest image to be displayed. If it is not, the digital image must be subsampled to a smaller size (with a loss of detail) or displayed in sections. The trend is toward processing larger images, and inadequate display size can reduce the usefulness of an image processing system. Fortunately, large-screen, high-resolution monitors are available at reasonable cost.

3.4.3 Aspect Ratio

Each image digitizer and each display device has a native digital image size. This is the number of pixels in the vertical and horizontal directions. The product of these two numbers specifies the total number of pixels that can be captured or displayed. Their ratio is called the "aspect ratio." Any digital image, likewise, has a size (height and width in pixels) and an aspect ratio (width in pixels divided by height in pixels).

Ideally the vertical and horizontal pixel spacings in the image digitizing camera will be equal and likewise for the display device. This situation is referred to as "square pixels." In this case digitized images can be displayed without distortion. In some devices, however, the vertical and horizontal pixel spacings are unequal, most often in the display. When an image digitized by a camera with square pixels is displayed on a monitor with unequal vertical and horizontal pixel spacing, the image becomes stretched. Round objects become oval. While this may not be of critical importance in the entertainment industry, it can be problematic in science, where quantification is more important.

Modern image display devices are often capable of operating in several different aspect ratio modes, and one can often select a mode that minimizes distortion. Software is readily available that can stretch an image horizontally or vertically by a specified amount (see Chapter 4) to match aspect ratios or compensate for unequal pixel spacing. A simple test for distortion is to generate a digital image that contains a square object and measure its height and width on the display screen or printed output.

3.4.4 Photometric Resolution

For display systems, photometric resolution refers to the accuracy with which the system can produce the correct brightness value of a pixel. Of particular interest is the number of visibly distinct shades of gray that the system can produce. This is partially dependent on the number of bits used to control the brightness of the pixels. Modern display units commonly handle 8-bit data for 256 gray levels. However, it is one thing to design a display that can accept 8-bit data, and quite another to produce a system that can reliably display 256 distinctly visible shades of gray. The effective number of gray shades is never more than the number of gray levels in the digital data, but it may well be reduced by noise in the display or saturation in black and/or white areas.

3.4.5 Grayscale Linearity

Another important display characteristic is the linearity of the gray scale. By this we mean the degree to which the displayed brightness is proportional to input gray level. Any display device has an input gray level to output brightness "transfer curve." For most uses, this curve should be reasonably linear and constant from one image to the next. On most display monitors, the transfer curve depends, in part, on the brightness and contrast settings. Often there is also an adjustable "gamma" setting that affects the shape of a nonlinear transfer curve. Thus it is possible for the user to alter the transfer curve of a monitor to suit a particular image or personal taste. In most cases, however, it is more satisfactory to allow the image processing to be done by the software and not by the display system, which should merely make the digital image visible to the operator without additional modification.

Fortunately perhaps, the human eye is not a very accurate photometer [17]. Slight nonlinearities in the transfer curve, as well as 10–20% intensity shading across the image, are seldom noticed. But if the transfer curve has a definite flattening at one end or the other ("saturation"), information may be degraded or lost in the light or dark areas.

A useful tool for determining the grayscale capability of a display system is the step target (Fig. 3.5). It is an arrangement of squares of each different gray level. If all of the boundaries between steps can be seen clearly, then the display is doing its job. One often finds that the darkest few and the lightest few steps are indistinguishable, indicating saturation at the black and/or white end of the gray scale.

A simple calibration procedure can ensure that the display properly renders the digital image. A grayscale test target, such as shown in Fig. 3.5, is displayed on the monitor or sent to the image printer. Then the various adjustments are set so that the full range of brightness is visible, with no loss of gray levels at either end. When an image processing system is in proper calibration, a print from the hardcopy recorder looks just like the

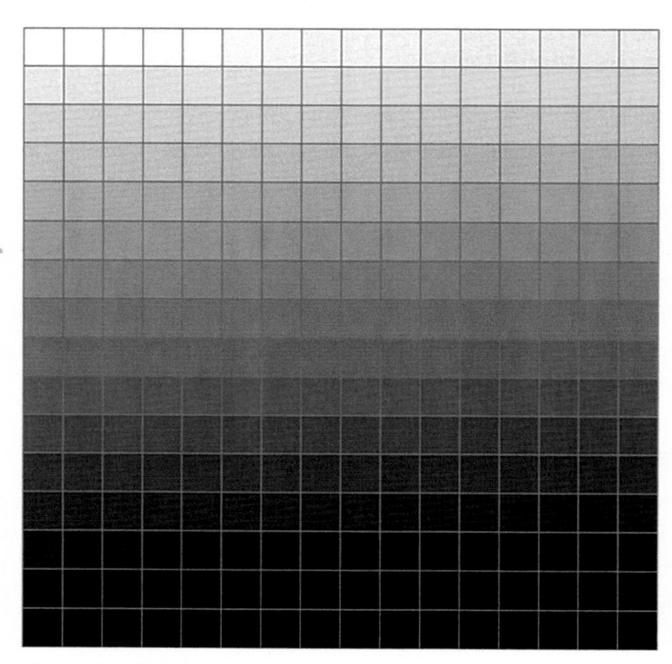

Fig. 3.5 A 256-level gray level step target.

image displayed on the screen, and this, in turn, is an accurate rendering of the digital image data.

3.4.6 Low-frequency Response

In this section, we consider the ability of a display system to reproduce large areas of constant gray level ("flat fields"). Since our goal is to minimize the visible effects of digital processing, we prefer flat fields to be displayed with uniform intensity.

A flat field can, of course, be displayed at any shade of gray between black and white. On a video monitor display, for example, a high-intensity pixel is displayed as a bright spot on an otherwise dark screen. Zero-intensity pixels leave the screen in its intrinsic dark state. In a printer or film recorder, a high-intensity pixel leaves a black spot on otherwise white paper or transparent film. Zero-intensity pixels leave the paper white or the film transparent. Thus any display system has a characteristic pixel polarity. No matter what the display polarity, zero-intensity flat fields are displayed uniformly flat. Thus flat field performance becomes an issue only at intermediate and high gray levels, and these may be either black or white, depending on the display system polarity.

Flat field performance depends primarily upon how well the pixels "fit together." Flat panel displays, such as liquid crystal displays (LCD) or thin-film transistor (TFT) units, use rectangular arrays of rectangular pixels [6,7]. Their flat field performance is affected by the size of the gaps between pixels. Cathode ray tube (CRT) devices, which are becoming increasingly rare for digital image display, use a rectangular array of circular spots [18–29]. For either type of display device, close inspection will reveal pixelization (the appearance of pixels) in bright, flat areas of the displayed image.

3.4.7 High-frequency Response

How well a display system can reproduce fine detail again depends on display spot shape and spacing. The ideal $\sin(x)/x$ spot shape is unattainable, so compromise is unavoidable. Processing steps that can improve the rendering of detail in displayed images are discussed in this section.

3.4.7.1 Sampling for Display Purposes

Displaying a digital image is actually a process of interpolation in that it reconstructs a continuous image from a set of discrete samples. We also know, from the sampling theorem, that the proper interpolation function (i.e., the display spot shape) has the form $\mathrm{sinc}(r) = \sin(r)/r$ where $r^2 = x^2 + y^2$. This is quite different from the shape of most actual display pixels, which are typically round Gaussian spots or small rectangles or squares.

Fig. 3.6 shows, in one dimension, the example of a cosine function that is sampled at a rate of 3.3 sample points per cycle. That is, the sample spacing is 30% of the period of the cosine. This sample spacing is small enough to preserve the cosine, and proper

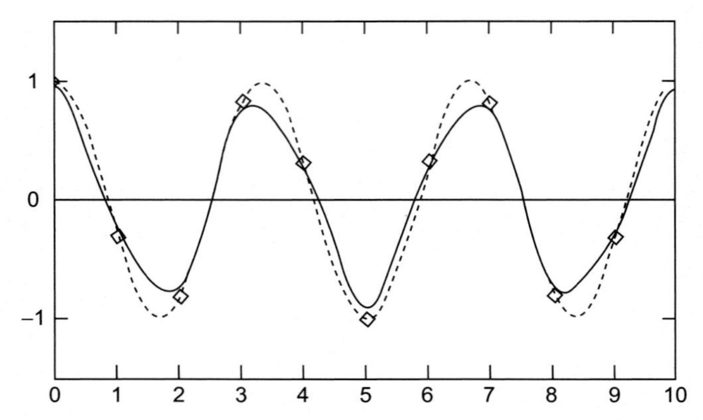

Fig. 3.6 Interpolation with a Gaussian function. The original cosine is shown as a *dashed line*, the sample points as *diamonds*, and the interpolated function as a *solid line*. In this case the sample spacing is 30% of the period of the cosine. The distortion of the reconstructed function results not from aliasing but from the inappropriate shape of the interpolation function.

interpolation (using the sinc) will reconstruct it from its samples without error (dashed line). When this sampled function is interpolated with a Gaussian display pixel, however, a distorted waveform (solid line) results. This illustrates that the display process itself can degrade an image, even one that has survived digitization and processing without damage.

The difficulties mentioned in the foregoing sections illustrate that image display using a physical display spot is a suboptimal process. While it is impractical to implement display devices with sinc-shaped display spots, there are things that can be done to improve the situation.

3.4.7.2 Oversampling

The inappropriate shape of the display spot has less effect when there are more sample points per cycle of the cosine. Thus one can improve the situation by arranging to have many pixels that are small in relation to the detail in the image. This is *oversampling*, as discussed earlier. It requires more expensive cameras and produces more image data than other system design considerations would dictate, but it solves many problems.

3.4.7.3 Resampling

Another way to improve the appearance of a displayed image is by *resampling*. This increases the size of the image by interpolation before it is displayed. For example, a 512 by 512 image might be interpolated up to 1024 by 1024 prior to being displayed. If the interpolation is properly done, the result will be more satisfactory. Note that this interpolation does not add any new information to the image, but it does help to overcome the inadequacies in the display process.

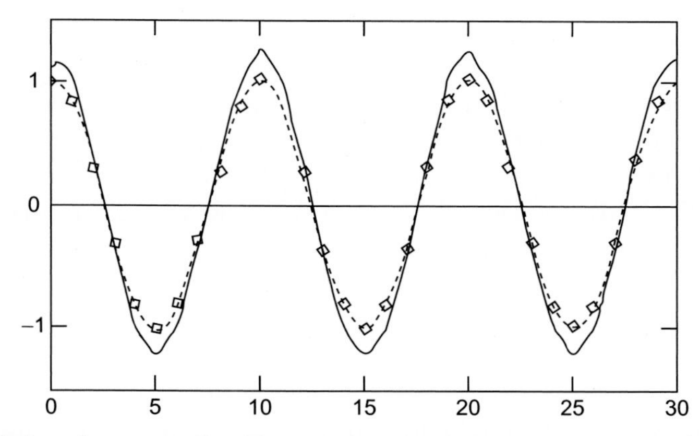

Fig. 3.7 Interpolation after resampling. The sample points in Fig. 3.6 were interpolated to place two new points between each existing pair. The sinc(*x*) function was used in that digitally implemented interpolation process. The resulting (more dense) sample points were then interpolated with a Gaussian function, as in Fig. 3.6. The result is a better reconstruction of the original cosine. Again the original cosine is a *dashed line*, the sample points are *diamonds*, and the interpolated function is a *solid line*.

Fig. 3.7 shows what happens when two extra sample points are inserted by interpolation between each pair in Fig. 3.6. The value at each new sample point is determined by placing a $\sin(\alpha x)/\alpha x$ function at each of the original sample points and summing their values at each new sample position. Here $\alpha = \pi/\tau$ where τ is the original sample spacing. This is digitally implemented interpolation using the ideal interpolation function. Fig. 3.7 shows that when the new (three times larger) sampled function is interpolated with a Gaussian display spot, the result is much more satisfactory.

Resampling a digital image by a factor of 2 or 3 increases its file size by a factor of 4 or 9, respectively, and this requires a display device that can accommodate the larger image size. It only needs to be done as the last step prior to display, however, so the burden of increased data size is not felt until that stage. When the digital image size is smaller than the native size of the screen, most modern display systems use built-in resampling to display the smaller image on the larger screen. Since this algorithm is implemented in hardware, it often lacks sophistication. Some merely display each input pixel as a small square of constant gray level or uniform color. This gives the image a blocky appearance and fails to take full advantage of the capabilities of the display device. A more satisfactory result can be obtained by first resampling the image in software up to a size that matches the native pixel resolution of the display device.

3.4.8 Noise

Random noise in the intensity channel can produce a salt-and-pepper effect that is particularly visible in flat (constant brightness) areas of the displayed image. If the noise is periodic and of reasonably high intensity, it can produce a herringbone pattern of lines superimposed on the displayed image. If the noise is periodic and synchronized with the horizontal or vertical deflection signals, it can produce a pattern of bars.

If electronic noise generated within the display system occupies more than one gray level, then the effective number of gray levels is reduced. As a rule of thumb, the root-mean-square (RMS) noise level represents a practical lower limit for grayscale resolution. For example, if the RMS noise level is 1% of the total display range from black to white, then the display can be assumed to have a photometric resolution of no more than 100 shades of gray. Even if the display system accepts 8-bit data, it still has only 100 effective gray levels because of the noise. If it were a 6-bit display system, however, it would have its full $2^6 = 64$ gray levels, in spite of the noise.

If all display noise sources, random and periodic, are kept at or below one gray level in amplitude, then the display will not reduce the effective grayscale resolution. Many systems, however, do not meet this criterion. Preprocessing is not effective at eliminating noise introduced by the display system. Only repair or replacement will improve the situation.

3.5 Summary of Important Points

1. Aliasing results when the sample spacing is greater than one-half the period of a sinusoidal component.
2. Aliasing reduces the frequency of sinusoidal components.
3. The system OTF sets an upper limit on the frequencies that can be present in an image.
4. The system OTF can act as an antialiasing filter.
5. The sampling frequency should be at least twice the OTF cutoff frequency to avoid aliasing.
6. A band-limited interpolation function is required for recovery of a sampled function without error.
7. A truncated sinc function is better for interpolation than a nonnegative pulse.
8. Oversampling tends to compensate for the use of a suboptimal interpolation function.
9. Oversampling tends to make subsequent quantitative image analysis more accurate.
10. Image display is an interpolation process done in hardware.
11. The ideal display spot for interpolation without error has the form $\sin(r)/r$.
12. Physical display spots differ significantly from the ideal shape.
13. Display quality can be improved by resampling a digital image to a larger size prior to display.
14. The horizontal and vertical pixel spacing should be equal, and the aspect ratio of the display should match that of the image, in order to avoid geometric distortion.

15. Simple image processing software can be used to prepare an image for display to minimize distortion.

16. Noise sources in a display system can reduce its effective photometric resolution.

References

[1] J.I. Pankove (Ed.), Display Devices, Topics in Applied Physics Series, vol. 40, Springer-Verlag, 1980.
[2] H. Poole, Fundamentals of Display Systems, MacMillan, 1966.
[3] H.R. Luxenberg, R.L. Kuehn (Eds.), Display Systems Engineering, McGraw-Hill, 1968.
[4] A. Cox, R. Hartmann (Eds.), Display System Optics, SPIE Press, 1987.
[5] H.M. Assenheim (Ed.), Display System Optics II, SPIE Press, 1989.
[6] W. den Boer, Active Matrix Liquid Crystal Displays: Fundamentals and Applications, Newnes, 2005.
[7] P. Yeh, Optics of Liquid Crystal Displays, Wiley-Interscience, 1999.
[8] M. Kriss (Ed.), Handbook of Digital Imaging, John Wiley & Sons, 2015.
[9] K.R. Castleman, Digital Image Processing, Prentice-Hall, 1996.
[10] R.N. Bracewell, The Fourier Transform and its Applications, McGraw-Hill, 2000.
[11] J.W. Goodman, Introduction to Fourier Optics, fourth ed., Macmillan, 2017.
[12] C.E. Shannon, A mathematical theory of communication, Bell Syst. Tech. J. XXVII (3) (1948) 379–423.
[13] A.B. Jerri, The Shannon sampling theorem—its various extensions and applications: a tutorial review, Proc. IEEE (1977) 1565–1596.
[14] A. Kohlenberg, Exact interpolation of bandlimited functions, J. Appl. Phys. (1953) 1432–1436.
[15] E.O. Brigham, The Fast Fourier Transform and its Applications, Prentice-Hall, 1988.
[16] R.R. Legault, The aliasing problems in two-dimensional sampled imagery, in: L.M. Biberman (Ed.), Perception of Displayed Information, Plenum Press, New York, 1973.
[17] B. Julesz, Foundations of Cyclopean Perception, University of Chicago Press, Chicago, 1971.
[18] K.B. Benson, Television Engineering Handbook: Featuring HDTV Systems, revised ed., McGraw-Hill, 1992.
[19] D.G. Fink (Ed.), Television Engineering Handbook, McGraw-Hill, 1984.
[20] T.J. Nelson, J.R. Wullert, Electronic Information Display Technologies, World Scientific Publications, 1997.
[21] A.F. Inglis, Video Engineering: NTSC, EDTV, & HDTV Systems, McGraw-Hill, 1992.
[22] W. Wharton, S. Metcalfe, G.C. Platts, Broadcast Transmission Engineering Practice, Focal Press, 1992.
[23] K.B. Benson, D.G. Fink, HDTV: Advanced Television for the 1990s, McGraw-Hill, 1991.
[24] J.C. Whitaker, K.B. Benson, in: Blair (Ed.), Standard Handbook of Video & Television Engineering, 2000, ISBN: 0-07-141180-1.
[25] K.B. Benson, Television Engineering Handbook, McGraw-Hill, 1992.
[26] A.C. Bovik, J.D. Gibson (Eds.), Handbook of Image and Video Processing, Academic Press, 2000.
[27] A.F. Inglis, A.C. Luther, Video engineering, IEEE Commun. Mag. (1996). IEEE Press.
[28] D.G. Fink, D. Christiansen, Electronics Engineer's Handbook, McGraw-Hill, 1989.
[29] G. Hutson, P. Shepherd, J. Brice, Colour Television Theory: System Principles, Engineering Practice & Applied Technology, McGraw-Hill, 1990.

CHAPTER FOUR

Geometric Transformations

Kenneth R. Castleman

4.1 Introduction

Geometric operations are those that distort an image spatially and change the physical relationships among the objects in an image [1–17]. This includes simple operations like translation, rotation, and scaling (magnification and shrinking), as well as more generalized actions that warp the image and move things around within it. In general, a geometric operation is simply an image-copying process, because the gray level values of pixels are not changed as they move from input image to output image. The difference is that the gray levels are copied into different pixel locations. The general definition of a geometric operation is

$$g(x, y) = f[a(x, y), b(x, y)] \tag{4.1}$$

where $f(x, y)$ is the input image, and $g(x, y)$ is the output image. The spatial transformation functions $a(x, y)$ and $b(x, y)$ specify the physical relationship between points in the input image and corresponding points in the output image. This, in turn, determines the effect the operation will have on the image. For example, if

$$g(x, y) = f[x + x_0, y + y_0] \tag{4.2}$$

then $g(x, y)$ will be a translated version of $f(x, y)$. The pixel at (x_0, y_0) moves to the origin, and everything in the image moves down and to the left by the amount $\sqrt{x_0{}^2 + y_0{}^2}$. Thus it is the spatial mapping functions, $a(x, y)$ and $b(x, y)$, that define a particular geometric operation and specify the effect it will have on the image.

The implementation of a geometric operation requires two separate algorithms. One is the algorithm that defines the spatial transformation itself, that is, $a(x, y)$ and $b(x, y)$. This specifies the "motion" as each pixel "moves" from its original position in the input image to its final position in the output image. The pixels in a digital image reside on a rectangular grid with integer coordinates, but the spatial transformation generates noninteger output pixel locations. Recall from Chapter 1 that the continuous image, an analytic function that corresponds to the digital image, can be generated by interpolation. The problem of noninteger coordinates is solved by a gray level interpolation algorithm.

Microscope Image Processing
https://doi.org/10.1016/B978-0-12-821049-9.00005-8

Copyright © 2023 Elsevier Inc.
All rights reserved.

4.2 Implementation

The output image is generated pixel by pixel, line by line. For each output pixel, $g(x, y)$, the spatial transformation functions, $a(x, y)$ and $b(x, y)$, point to a corresponding location in the input image. In general, this location falls between four adjacent pixels (Fig. 4.1). The gray level that maps into the output pixel at (x, y) is uniquely determined by interpolation among these four input pixels. Some output pixels may map to locations that fall outside the borders of the input image. In this case an arbitrary constant gray level (e.g., zero) is usually stored.

4.3 Gray Level Interpolation

There is a trade-off between simplicity of implementation and quality of results when selecting a technique for gray level interpolation.

4.3.1 Nearest Neighbor Interpolation

The simplest way to fill the output pixel is just to use the gray level of the input pixel that falls closest to the mapped position, (x, y). This technique is seldom used, however, because it creates a ragged effect in areas of the image containing detail, such as lines and edges.

4.3.2 Bilinear Interpolation

In many cases bilinear interpolation offers the best compromise between processing speed and image quality. It is a direct 2D generalization of linear interpolation in one dimension. Fig. 4.2 shows four adjacent pixels with a fractional location, (x, y), among them.

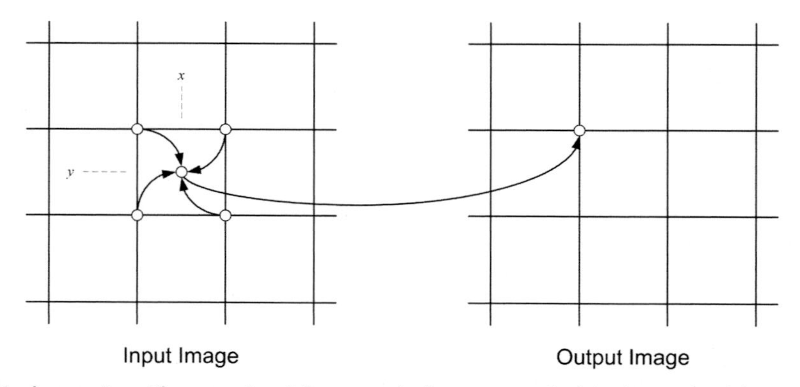

Fig. 4.1 Pixel mapping. The gray level for a particular output pixel is determined by interpolating among four adjacent input pixels. The geometric mapping specifies where in the input image the point (x, y) falls. Normally x and y take on noninteger values.

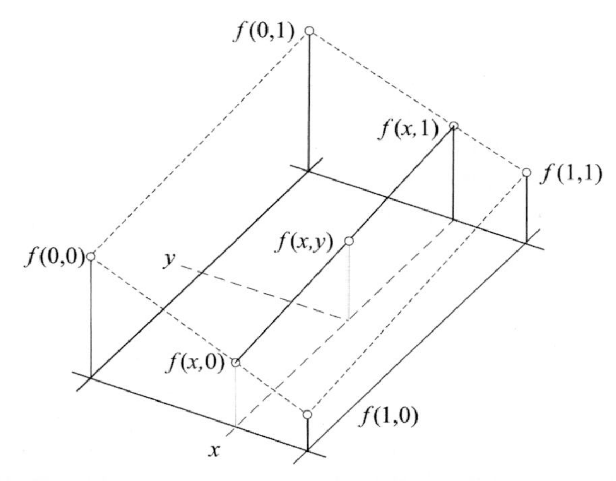

Fig. 4.2 Bilinear interpolation. We use linear interpolation first to find $f(x, 0)$ and $f(x, 1)$, and then use it again between those two points to find $f(x, y)$.

We first use linear interpolation horizontally to find the values of the continuous image at $(x, 0)$ and $(x, 1)$. We then interpolate vertically between those two points to find its value at (x, y).

Bilinear interpolation actually approximates the continuous image by fitting a hyperbolic paraboloid through the four points. The hyperbolic paraboloid surface is given by

$$f(x, y) = ax + by + cxy + d \tag{4.3}$$

where a, b, c, and d are parameters determined by the interpolation process. In particular,

$$\begin{aligned} f(x, y) = {} & [f(1,0) - f(0,0)]x + [f(0,1) - f(0,0)]y \\ & + [f(1,1) + f(0,0) - f(0,1) - f(1,0)]xy + f(0,0) \end{aligned} \tag{4.4}$$

Bilinear interpolation can be implemented with only three multiplication and six add/subtract operations per pixel and thus is only slightly more computationally expensive than nearest neighbor interpolation [1]. It guarantees that the interpolated function will be continuous at the boundaries between pixels but does not avoid slope discontinuities. In many cases, this is not a serious flaw.

4.3.3 Bicubic Interpolation

With bicubic interpolation the interpolated surface not only matches at the boundaries between pixels, but it has continuous first derivatives there as well. The formula for the interpolated surface is

$$p(x, y) = \sum_{i=0}^{3} \sum_{j=0}^{3} a_{ij} x^i y^j \tag{4.5}$$

The 16 coefficients, a_{ij}, are chosen to make the function and its derivatives continuous at the corners of the four-pixel square that contains the point (x, y). This is done by solving 16 equations in the 16 unknown coefficients at each point. The equations are derived by setting the function and its three derivatives to their known values at the four corners. Since estimating the derivatives at a pixel requires at least a 2×2 pixel neighborhood, bicubic interpolation is done over a 4×4 or larger neighborhood surrounding the point (x, y).

4.3.4 Higher-order Interpolation

In addition to slope discontinuities at pixel boundaries, bilinear interpolation has a slight smoothing effect on the image, and this becomes particularly visible if the geometric operation involves magnification. Stated differently, bilinear interpolation does not precisely reconstruct the continuous image that corresponds to the digital image. Bicubic interpolation does a better job and is becoming the standard for image processing software packages and high-end digital cameras. But even this is still imperfect. We know from the sampling theorem that the proper form for the interpolating function is $\text{sinc}(\alpha x) = \sin(\alpha x)/(\alpha x)$ (see Chapter 3). Thus an interpolation technique that better approximates that function will yield better results.

Higher-order interpolation uses a neighborhood that is larger than 2×2 to determine the gray level value at a fractional pixel position. An interpolation function is fitted through the larger neighborhood and evaluated at the fractional pixel position. This additional complexity is justified by improved performance in some applications, particularly if the geometric operation has the effect of magnifying the image. The bicubic function is given by

$$f(x, y) = \sum_{i=0}^{3} \sum_{j=0}^{3} a_{i,j} x^i y^j \tag{4.6}$$

where the coefficients $a_{i,j}$ define the surface. Since the hyperbolic paraboloid has four parameters, it can be made to fit through all four points in a 2×2 neighborhood, as in Fig. 4.2. Similarly, the bicubic function has 16 parameters and can fit to all 16 points in a 4×4 neighborhood. If the interpolating function has the same number of coefficients as the neighborhood has points, then the interpolating surface can be made to fit at every point. However, if there are more points in the neighborhood than there are coefficients, then the surface cannot fit all the points, and a curve-fitting or error-minimization procedure must be used to determine the coefficients. Fig. 4.3 shows a parabolic surface defined by six coefficients fitted through the nine points of a 3×3 neighborhood.

Alternatively, one can place a small interpolating function, such as a Gaussian or a sinc, at each pixel and then add them together at the fractional pixel location (see Resampling in Chapter 3). Higher-order interpolating functions that are widely used include cubic

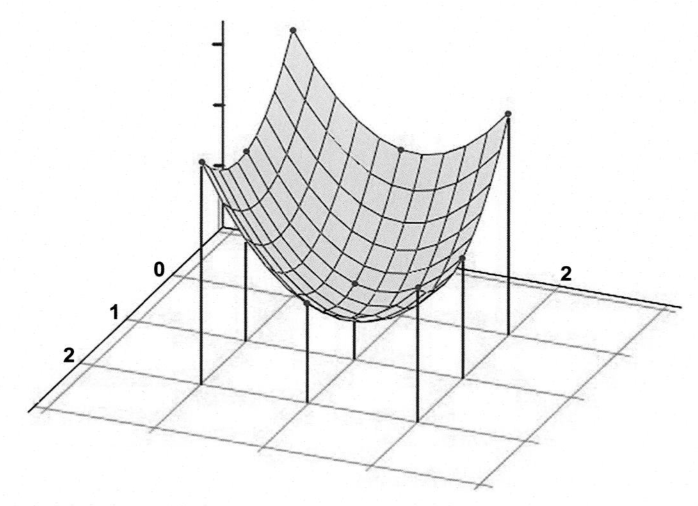

Fig. 4.3 Higher-order interpolation. Here a parabolic surface has been fitted to a neighborhood of nine pixels.

splines, Legendre centered functions, and the Lanczos kernel (truncated sinc(x) function). [18]

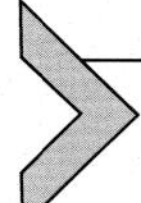

4.4 The Spatial Transformation

Eq. (4.2) specifies the translation operation. Using

$$g(x, y) = f\left[x/M_x, y/M_y\right] \tag{4.7}$$

will scale (magnify or shrink) the image by the factor M_x in the x-direction and M_y in the y-direction. Rigid rotation about the origin through an angle θ is given by

$$g(x, y) = f[x \cdot \cos(\theta) - y \cdot \sin(\theta), x \cdot \sin(\theta) + y \cdot \cos(\theta)] \tag{4.8}$$

To rotate about another point, one would first translate that point to the origin, then rotate the image about the origin, and finally translate back to its original position. Translation, rotation, and scaling can be combined into a single operation [1].

4.4.1 Control Grid Mapping

For warpings too complex to be defined by an equation, it is convenient to specify the operation using a set of *control points* [1]. This is a list of certain pixels whose positions in the input and output images are specified. The displacement values for the remaining unspecified pixels are determined by interpolation among those that have been specified. Fig. 4.4 shows how four control points that form a quadrilateral in the input image map to

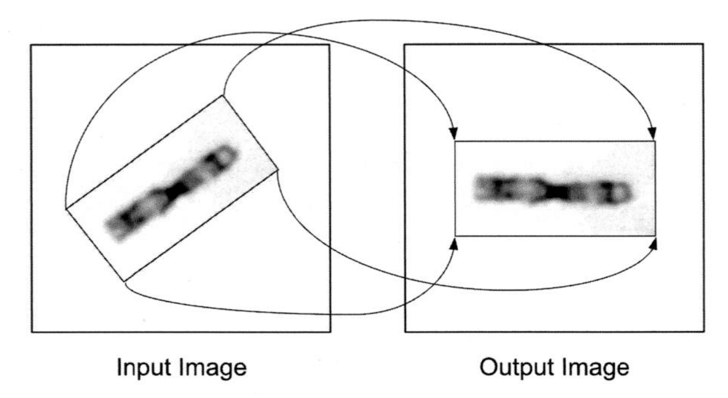

Fig. 4.4 Control point mapping. The four corners of an arbitrarily shaped quadrilateral in the input image map to the four corners of a rectangle in the output image. The mapping of each corner is specified, and the mappings of the interior points are determined by interpolation.

the vertices of a rectangle in the output image. Displacement values for points inside the rectangle can be determined with bilinear interpolation. A set of contiguous quadrilaterals that span the input image (a *control grid*) can be mapped into a set of contiguous rectangles in the output image. The specification of the transformation, then, consists of the coordinates of the rectangle vertices, along with the x, y displacement (to the corresponding control grid vertex) of each.

4.5 Applications

Geometric operations are useful in microscopy in several ways. A few of these are mentioned in this section.

4.5.1 Distortion Removal

Microscope images are sometimes affected by geometric distortion due to the optics or due to the specimen preparation. Optical distortion is usually constant from image to image and can be removed with batch processing. Physical distortion of specimens, such as that induced by slicing on a microtome, must be corrected on an image-by-image basis.

4.5.2 Image Registration

It is often necessary to align multiple images of the same specimen, as in optical sectioning (Chapter 11) and time-lapse microscopy (Chapter 14). Geometric operations are useful for these tasks [19]. Cross-correlation (see Chapter 11) is useful for determining the translation required for alignment. In some cases the images to be aligned are similar enough that translation and rotation are sufficient to effect alignment. In other cases the images to be aligned may have suffered geometric distortion (serial sections, for example). Here specific landmarks in each image must be located and brought into alignment by a geometric operation that is capable of warping images.

4.5.3 Stitching

Often it is impossible to get an entire specimen to fit within a single field-of-view. Here multiple images of the specimen can be combined into a mosaic image by the process called "stitching." Geometric operations are usually necessary to make the images match in their regions of overlap. Cross-correlation can be used in the local region of a control point to determine the displacement values for a control grid [20]. Since brightnesses may not match in areas where images overlap, it is useful to blend them together to smooth out the transition from one to the next [21]. Blending can be done using a weighted average in the overlapping areas, where the weights taper off to zero at the image borders [22]. This produces a more seamless appearance and makes the images easier to interpret. Geometric and radiometric corrections applied to the input images make automatic mosaicing possible [23].

4.6 Summary of Important Points

1. Geometric operations warp an image, changing the positions of the objects within.
2. Geometric operations include translation, rotation, and scaling, as well as more general transformations.
3. A geometric transformation can be specified by a formula or by a control grid.
4. Geometric operations are implemented by mapping output pixel positions back into the input image.
5. Since output pixels map to noninteger positions in the input image, gray level interpolation is used to estimate the underlying continuous image.
6. Bilinear interpolation is more accurate than nearest neighbor interpolation and simpler than higher-order interpolation.
7. Higher-order interpolation techniques fit an approximation of the truncated sinc(x) function through a neighborhood larger than 2×2.
8. If the interpolation function has fewer coefficients than the neighborhood has pixels, curve fitting is used to determine the coefficient values.
9. Geometric operations are useful for distortion correction, registration, and stitching images together.

References

[1] K.R. Castleman, Digital Image Processing, Prentice-Hall, 1996.
[2] G. Wolberg, Digital Image Warping, IEEE Computer Society Press, 1994.
[3] W.M. Newman, R.F. Sproul, Principles of Interactive Computer Graphics, second ed., McGraw-Hill, 1979.

[4] J.D. Foley, A. van Dam, Fundamentals of Interactive Computer Graphics, Addison-Wesley, 1982.
[5] D.H. Ballard, C.M. Brown, Computer Vision, Prentice-Hall, 1982.
[6] R. Nevatia, Machine Perception, Prentice-Hall, 1982.
[7] W.K. Pratt, Introduction to Digital Image Processing, CRC Press, 2014.
[8] R.C. Gonzales, R.E. Woods, Digital Image Processing, fourth ed., Pearson, 2017.
[9] C.A. Glasbey, K.V. Mardia, A review of image-warping methods, J. Appl. Stat. 25 (2) (1998) 155–171.
[10] R. Hartley, A. Zisserman, Multiple View Geometry in Computer Vision, second ed., Cambridge University Press, 2003.
[11] C.M. Bishop, Pattern Recognition and Machine Learning, Springer, 2006.
[12] M.M.P. Petrou, C. Petrou, Image Processing: The Fundamentals, second ed., John Wiley & Sons, 2010.
[13] R. Szeliski, Computer Vision: Algorithms and Applications, Springer, 2011.
[14] C. Solomon, T. Breckon, Fundamentals of Digital Image Processing, Wiley-Blackwell, 2011.
[15] D. Forsyth, J. Ponce, Computer Vision: A Modern Approach, second ed., Pearson, 2011.
[16] W. Förstner, B.P. Wrobel, Photogrammetric Computer Vision, Springer, 2016.
[17] P. Corke, Robotics, Vision and Control, second ed., Springer, 2017.
[18] K. Turkowski, S. Gabriel, Filters for common resampling tasks, in: A.S. Glassner (Ed.), Graphics Gems I, Academic Press, 1990. https://citeseerx.ist.psu.edu/viewdoc/download?doi=10.1.1.116.7898&rep=rep1&type=pdf.
[19] L.G. Brown, A survey of image registration techniques, ACM Comput. Surv. 24 (4) (1992) 325–376.
[20] S.K. Chow, et al., Automated microscopy system for mosaic acquisition and processing, J. Microsc. 222 (2) (2006) 76–84.
[21] V. Rankov, et al., An algorithm for image stitching and blending, Proc. SPIE 5701 (2005) 190–199.
[22] B.W. Loo, W. Meyer-Ilse, S.S. Rothman, Automatic image acquisition, calibration and montage assembly for biological X-ray microscopy, J. Microsc. 197 (2) (2000) 185–201.
[23] C. Sun, et al., Mosaicing of microscope images with global geometric and radiometric corrections, J. Microsc. 224 (2) (2006) 158–165.

CHAPTER FIVE

Image Enhancement

Yu-Ping Wang, Qiang Wu, and Kenneth R. Castleman

5.1 Introduction

Images that come from a variety of microscope technologies provide a wealth of information. The limited capacity of optical instruments combines with the noise inherent in optical imaging to make image enhancement desirable in many applications. Image enhancement is the process of altering the appearance of an image, or a subset of an image, for improved contrast or visualization of some features and to facilitate more accurate subsequent image analysis, such as classification. With image enhancement, the visibility of selected features in an image can be improved, but the inherent information content cannot be increased. The design of a good image enhancement algorithm should consider both the specific features of interest in the image and the imaging process itself. In microscopic imaging, the images are often acquired at different focal planes, different time intervals, and in different spectral channels. The design of an enhancement algorithm should therefore take full advantage of this multidimensional information.

A variety of image enhancement algorithms have previously been developed and utilized for microscopy applications. These algorithms can be classified into two categories: spatial domain and transform domain methods. The spatial domain methods include operations carried out on a whole image or on a local region selected on the basis of image statistics. Techniques that belong to this category include histogram equalization, image averaging, sharpening of important features such as edges or contours, and nonlinear filtering. The transform domain enhancement methods manipulate image information in a transform domain, such as Fourier and wavelet transform domains. In many cases, interesting image information that cannot be separated out in the spatial domain can be isolated in the transform domain. For example, one can often amplify certain Fourier coefficients and then return the image to the spatial domain to highlight interesting image content. The wavelet transform is another powerful tool that has been used for image enhancement.

In the following discussion we focus on the enhancement of two-dimensional (2D) gray level and color microscope images. Processing of multispectral and three-dimensional (3D) microscope images is discussed in Chapters 10 and 11.

Microscope Image Processing
https://doi.org/10.1016/B978-0-12-821049-9.00006-X
Copyright © 2023 Elsevier Inc.
All rights reserved.

5.2 Spatial Domain Enhancement Methods

Given a gray level image with the intensity range [0, L], a *global operation* on the image refers to an image transform, T, that maps the image, I, to a new image, g, according to the following equation:

$$g = T(I) \tag{5.1}$$

There are many examples of this type of image transform, such as contrast stretching, clipping, thresholding, grayscale reversal, and gray-level window slicing [1]. If the operation results in fractional (noninteger) values, they must be rounded to integers for the output image.

5.2.1 Contrast Stretching

Display devices commonly have a limited range of gray levels over which the image features are most visible. One can use global methods to adjust all the pixels in the image so as to ensure that the features of interest fall into the visible range of the display. This technique is also called "contrast stretching" [2]. For example, if I_1 and I_2 define the intensity range of interest, a scaling transformation can be introduced to map the image intensity I to the image g with the range of $I_{\min}$ to $I_{\max}$ as

$$g = \left[\frac{I - I_1}{I_2 - I_1} \cdot (I_{\max} - I_{\min})\right] + I_{\min} \tag{5.2}$$

This mapping is a linear stretch. A number of nonlinear monotonic pixel operations exist [2,3]. For example, the following transform maps the gray level of the image according to a nonlinear curve:

$$g = \left[\left(\frac{I - I_1}{I_2 - I_1}\right)^{\alpha} \cdot (I_{\max} - I_{\min})\right] + I_{\min} \quad 0 < \alpha < \infty \tag{5.3}$$

where α is an adjustable parameter. This image intensity scaling is commonly used for contrast stretching, clipping, display calibration, etc.

5.2.2 Clipping and Thresholding

Image clipping is a special case of contrast stretching that is useful in noise reduction when the input image, f, is known to lie in the range of 0 to L. The transform is defined in the following equation:

$$g = \begin{cases} 0 & 0 \le f < a \\ \alpha I & a \le f < b \\ L & f \ge b \end{cases} \tag{5.4}$$

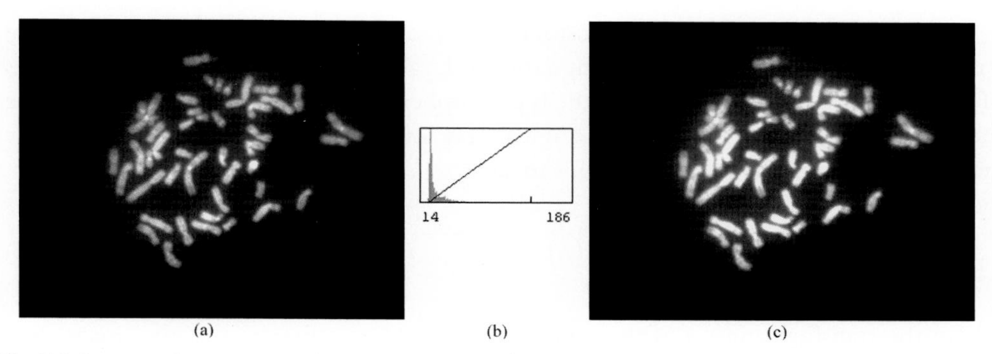

Fig. 5.1 Image enhancement using a contrast stretch. Image (c) is obtained by applying the mapping defined in (b) to image (a).

where a and b are usually obtained from the histogram of the image, and they specify the valley between the peaks of the histogram (see Fig. 5.1). When $a = b$, the transform is called thresholding, and the output is a binary image.

5.2.3 Image Subtraction and Averaging

When more than one image of a stationary object is available, averaging together N images is a simple way to improve the signal-to-noise ratio by $\sqrt{N}$. In microscopic imaging, multiple images are often obtained. For microscopic video imaging, frames from the same scene are acquired sequentially. These multiple images, if properly registered, can then be averaged to reduce noise. A geometric operation for registration may be required if the images are not already properly aligned.

Image subtraction is usually performed when two images of the same object are obtained under different conditions [3]. The image subtraction will highlight whatever has changed between the two images. Another application is background correction. In microscopy the image is often affected by a slowly varying background shading pattern. One can move the microscope stage to an empty field and acquire an image containing the background alone. When the background image is subtracted from the image containing the specimen, it removes the shading (see Chapter 10).

5.2.4 Histogram Equalization

The gray level histogram of an image is the probability of occurrence of each gray level in the image. The goal of histogram equalization is to remap the gray levels so as to obtain a uniform (flat) histogram [2]. If no prior information is available about the gray level distribution, it is often useful to distribute the intensity information uniformly over the available intensity levels. Also it is easier to compare two images taken under different conditions if their histograms match.

Mathematically, the normalized histogram $h(r_i)$ can be expressed as $h(r_i) = n_i/n$, where r_i is the ith gray level in an image having a total of L values; n_i is the number of occurrences of gray level r_i in the image, and n is the total number of pixels in the image. We can use the transformation, $T(r)$, to map the original gray levels, r_i, of the input image into new gray levels, s_i, such that, for the output image,

$$s_i = T(r_i) = \sum_{j=0}^{i} h(r_j) = \sum_{j=0}^{i} \frac{n_i}{n}, \quad i = 0, 1, \ldots, L-1 \tag{5.5}$$

where the transformation T is the cumulative distribution function of the image gray levels, which is always monotonically increasing. The resulting image will have a histogram that is "flat" in a local sense, since there are only a finite number of gray levels available (see Fig. 5.2).

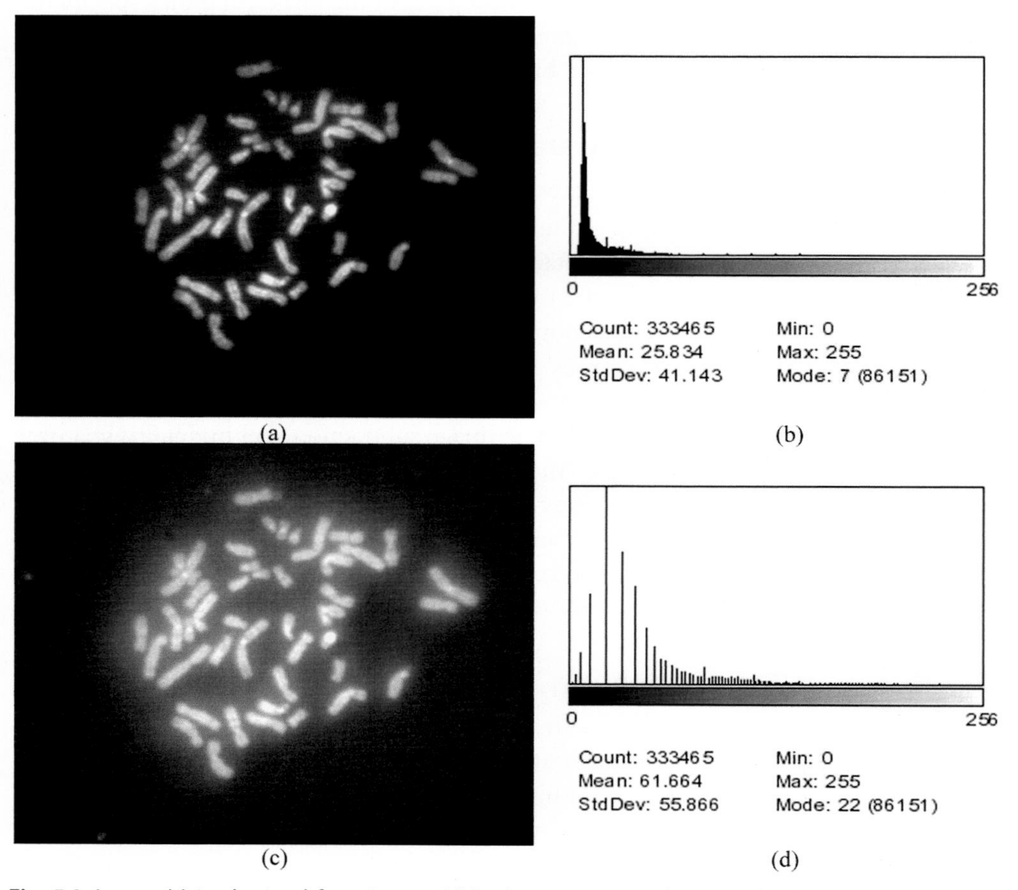

Fig. 5.2 Image (c) is obtained from image (a) by histogram equalization. The histograms of images (a) and (c) appear in (b) and (d), respectively. The histogram in (d) is flat when its sparseness in taken into account.

Local histogram equalization is a variant of the histogram equalization operation. It applies histogram equalization to small, overlapping areas of the image [4] that contain local features. This nonlinear operation can significantly increase the visibility of subtle features in the image. However, because histogram equalization is carried out in local areas, it is computationally intensive, and the complexity increases with the size of the local area used in the operation. There are also a number of other variations in image histogram transformations that account for local image properties, such as the local standard deviation [5].

5.2.5 Histogram Specification

More generally, histogram specification allows us to modify an image so that its histogram takes on a specific shape. Assume that $f(x, y)$ is the input image having histogram $h_1(r_i)$, and $g(x, y)$ is the output image with the target histogram, $h_2(z_i)$. The input image can be modified according to the equalization transformation, T, given in Eq. (5.5), to produce a flat histogram. Further, the transformation, V, will give the target image a flat distribution if

$$V(z_i) = \sum_{j=0}^{i} h_2(z_j) \quad \text{and} \quad T(r_i) = \sum_{j=0}^{i} h_1(r_j) \qquad i = 0, \ldots, L-1 \tag{5.6}$$

Then the output image, $g(x, y)$, can be computed from the input image, $f(x, y)$, using the following cascaded transformation:

$$g(x, y) = V^{-1}[T(f(x, y))] \tag{5.7}$$

Using this transformation, we can obtain an output image with the desired gray level distribution, $h_2(z_i)$.

5.2.6 Spatial Filtering

Spatial filtering involves the convolution of an image with a specific kernel operator. The gray level of each pixel is replaced with a new value that is the weighted average of neighboring pixels that fall within the window of the kernel. In the continuous form, the output image $g(x, y)$ is obtained as the convolution of the image $f(x, y)$ with the filter kernel $w(x, y)$ denoted as

$$g(x, y) = f(x, y) * w(x, y) = \iint f(u, v) w(x-u, y-v)\, du dv \tag{5.8}$$

where the integration is performed over all values of (x, y) in the image.

In the discrete form, the convolution becomes $g_{i,j} = f_{i,j} \times w_{i,j}$, and the spatial filter $w_{i,j}$ takes the form of a weight mask. Table 5.1 shows several commonly used discrete filters.

In general, an image can be enhanced by the following sharpening operation:

$$g(x, y) = f(x, y) + \lambda e(x, y) \tag{5.9}$$

Table 5.1 Examples of discrete kernel masks for spatial filtering.

Low-pass filter	High-pass filter	Laplacian filter
$w_{i,j} = \frac{1}{10}\begin{bmatrix} 1 & 1 & 1 \\ 1 & 2 & 1 \\ 1 & 1 & 1 \end{bmatrix}$	$w_{i,j} = \begin{bmatrix} -1 & -1 & -1 \\ -1 & 9 & -1 \\ -1 & -1 & -1 \end{bmatrix}$	$w_{i,j} = \begin{bmatrix} 0 & 1 & 0 \\ 1 & -4 & 1 \\ 0 & 1 & 0 \end{bmatrix}$

where $\lambda > 0$, and $e(x, y)$ is a high-pass filtered version of the image that usually corresponds to some form of the derivative of an image. The operation can be accomplished by, for example, adding gradient information to the image. A well-known gradient filter is the *Sobel* filter pair that can be used to compute an estimate of the gradient in both x and y directions. Other commonly used derivative filters include the Laplacian filter [1], which is defined as

$$e(x, y) = \nabla^2 f(x, y) = \left(\frac{\partial^2}{\partial x^2} + \frac{\partial^2}{\partial y^2}\right) f(x, y) \tag{5.10}$$

In the discrete form, the operation can be implemented as

$$\nabla^2 f_{i,j} = \left[f_{i+1,j} - 2f_{i,j} + f_{i-1,j}\right] + \left[f_{i,j+1} - 2f_{i,j} + f_{i,j-1}\right] \tag{5.11}$$

The kernel mask used in the preceding discrete Laplacian filtering is shown in Table 5.1.

To sharpen a noisy image, a Laplacian of Gaussian (LoG) filter is useful. The LoG filter first smooths the image with a Gaussian low-pass filtering, followed by the high-pass Laplacian filtering. The LoG filter is defined as

$$\nabla^2 G(x, y) = \left(\frac{\partial^2}{\partial x^2} + \frac{\partial^2}{\partial y^2}\right) G_\sigma(x, y) \tag{5.12}$$

where

$$G_\sigma(x, y) = \frac{1}{\sqrt{2\pi}\,\sigma} \exp\left(-\frac{x^2 + y^2}{2\sigma^2}\right)$$

is the Gaussian function with variance σ, which determines the size of the filter. A larger size of the filter results in more smoothing of the noise. A discrete form of the LoG filter is given in [2]. Fig. 5.3 shows the result of sharpening an image using a LoG operation. More examples of medical image enhancement using local derivative filtering can be found in references [6–8].

Image filtering operations are most commonly done globally, that is, over the entire image. However, because image properties may vary throughout the image, it is often useful to perform spatial filtering operations in local neighborhoods.

5.2.7 Directional and Steerable Filtering

Many images contain edge features in various orientations. Directional filters, such as the steerable filters [9], are used to enhance image features that lie in a particular

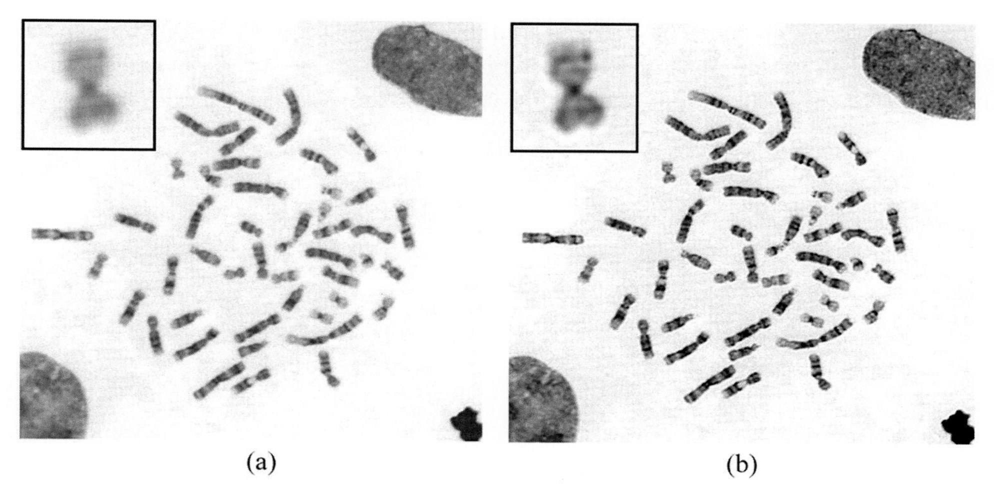

(a) (b)

Fig. 5.3 Image (b) is obtained by sharpening image (a) with a Laplacian of Gaussian operation.

direction. The filtering effect in regard to orientation can be evaluated by computing an "orientation map," which is the squared filter response as a function of filter orientation [1,6,10,11]. The concept of steerable filters [12] is based on an oriented filter that is constructed from a linear combination of a set of directionally oriented basis filters. Here the weighting factors determine the directionality of the filter. Basis filters can be derived from directional derivatives of Gaussians and used to compute local orientation maps.

The simplest example of a steerable filter is the partial derivative of a two-dimensional Gaussian. In polar coordinates, the horizontal and vertical derivatives are written as

$$G_1^{(0)}(r,\theta) = \cos(\theta)\left(-re^{-r^2/2}\right) \tag{5.13}$$

and

$$G_1^{(\pi/2)}(r,\theta) = \sin(\theta)\left(-re^{-r^2/2}\right) \tag{5.14}$$

where the subscript denotes the order of the derivative, and the superscript denotes the direction of the derivative. $G_1(r,\theta)$, at any orientation θ, can be synthesized by taking a linear combination of $G_1^{(0)}$ and $G_1^{(\pi/2)}$ as follows:

$$G_1^{(\theta)}(r,\theta) = \cos(\theta)G_1^{(\pi/2)}(r,\theta) + \sin(\theta)G_1^{(\pi/2)}(r,\theta) \tag{5.15}$$

This equation implies the steerability of these functions. The directional derivative, G_1, can be generated at any arbitrary orientation using a linear combination of the basis filters

$G_1^{(0)}$ and $G_1^{(\pi/2)}$ with coefficients $\cos(\theta)$ and $\sin(\theta)$ as the weighting functions, also known as the interpolation functions. Therefore, filtering an image with an arbitrarily oriented filter can be accomplished using a proper linear combination of the image convolved with the two basis filters.

Steerability can be extended to the higher order derivatives. The general steerability condition, for functions that are polar separable, is expressed as

$$f^{\alpha}(r,\phi) = h(\phi-\alpha)\, g(r) = \sum_{n=1}^{\hat{N}} k_n(\alpha)\, h(\phi-\alpha_n)\, g(r) \tag{5.16}$$

where $h(\phi)$ is the angular portion of the steerable filter, $g(r)$ is the radial portion, $k_n(\alpha)$ represents interpolation functions, and α_n is a fixed set of $\widehat{N}$ orientations. This equation can be satisfied by all functions with angular components that are band-limited to contain no more than $\widehat{N}/2$ harmonic terms [12]. Examples of steerable filter sets consisting of higher-order directional derivatives of a Gaussian, along with steerable approximations to their Hilbert transforms, can be found in [12]. Orientation maps can be computed as the sum of squared responses of these filters.

Steerable filters have been used to generate multiscale, self-inverting pyramid decompositions of images [12] that have the desirable properties of shift and rotation invariance. By analyzing and selectively processing the transform coefficients, image feature detection and enhancement can be achieved with designed flexibility in scale, orientation, and degree of enhancement [9].

Traditional techniques based on conventional convolution filtering and contrast stretching are limited in what they can do. By decomposing the image into several different orientated bases at multiple scales using the steerable pyramid transform, it becomes easier to selectively detect and enhance certain image features that correspond to important object structures at a particular scale, location, and orientation. Fig. 5.4 shows the result of chromosome image enhancement based on a steerable pyramid transform and selective processing of the transform coefficients [9].

5.2.8 Median Filter

The median filter is a commonly used nonlinear operator that replaces the original gray level of a pixel by the median of the gray levels of the pixels in a surrounding neighborhood. The median filter is a type of *ranking filter* [3], because it is based on the statistics derived from rank-ordering the elements of a set. This filter is often useful because it can reduce noise without blurring edges in the image [1]. The noise-reducing effect of the median filter depends on two factors: (1) the spatial extent of its neighborhood, and (2) the number of pixels involved in the median calculation. Fig. 5.5 shows an example

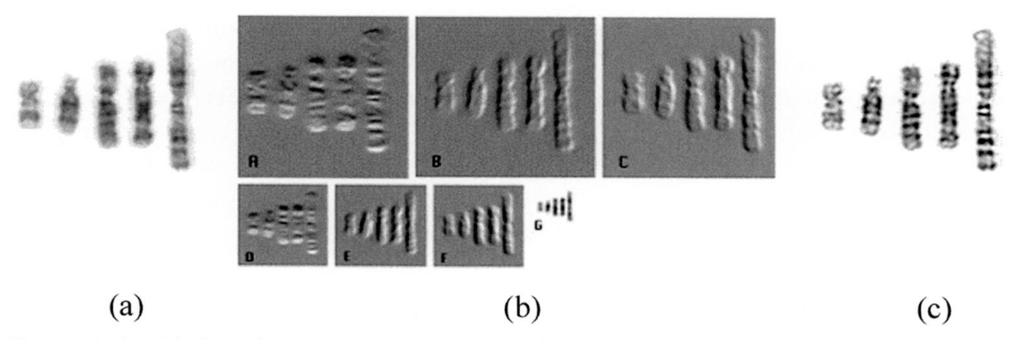

Fig. 5.4 Image (a) shows five human chromosomes in an upright orientation. Image (b) is the steerable pyramid transform of image (a). Three levels of decomposition were performed. Images A, B, and C show the three band-pass filtered images at decomposition level 2, while images D, E, and F show them at decomposition level 3. Image G shows the downsampled low-pass image at decomposition level 3. Image (c) is the result of image enhancement.

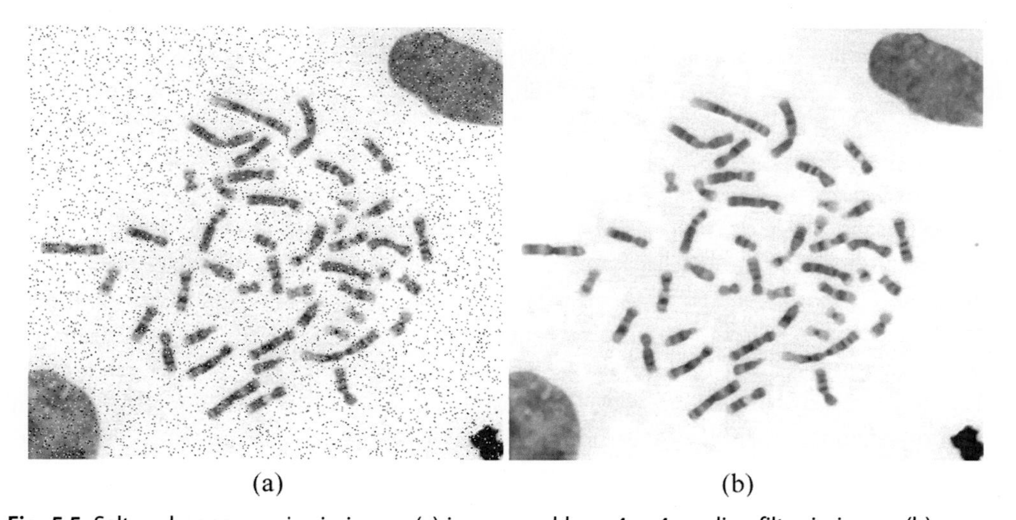

Fig. 5.5 Salt-and-pepper noise in image (a) is removed by a 4×4 median filter in image (b).

of salt-and-pepper noise removal using median filtering. This type of noise cannot be removed effectively by conventional convolution filtering.

5.2.9 Anisotropic Diffusion Filter

Another popular nonlinear filtering approach is the so-called *anisotropic diffusion*, first proposed by Perona and Malik [13] and has found successful applications in many medical imaging data enhancements [14]. The approach has the advantage of preserving edges, contours, and other significant details of the image, while smoothing out noise. This is in contrast with linear filtering such as Gaussian kernel, which smooths out noise and edges at the same time. In the anisotropic diffusion process, a family of parameterized images are generated by a combination of the original image and a filter with the size determined by the local content of the image, such as the edge. As a result, anisotropic diffusion is a *nonlinear* transformation of the original image with a variable filter to smooth noise while preserving edges.

5.3 Fourier Transform Methods

The previous section described linear filtering implemented by spatial domain convolution of the image with a chosen filter kernel. An equivalent method is to multiply the Fourier spectrum of the image by the transfer function of the desired filter. The kernel and the transfer function are related by the Fourier transform, and either one can be obtained from the other. Thus one can design a filter in the spatial or the frequency domain and then implement the filtering operation in either the same or the opposite domain. If the kernel is relatively small, as in Table 5.1, spatial domain convolution is more efficient. But frequency domain multiplication is the more practical approach if the required kernel is large.

In many cases, it is more effective to design the filter transfer function in the frequency domain because there the noise can be more easily separated from the objects of interest. In the frequency domain, large structures and smooth regions in the image correspond to low-frequency components, image features of interest typically correspond to medium-frequency components, and the high frequencies are often dominated by noise. Hence one can design filters, using knowledge of the frequency components, to sharpen the detail in an image while suppressing the noise [15,16]. A noise-reducing enhancement filter, for example, would seek to boost the amplitude of mid-frequency components and attenuate the high-frequency components. The actual filtering operation can be implemented as a convolution in the spatial domain. The convolution kernel is simply the inverse Fourier transform of the transfer function.

5.3.1 Wiener Filtering and Wiener Deconvolution

The Wiener filter is known to be optimal, in the minimum mean square error (MSE) sense, for recovering a signal that is embedded in noise [1,3,16]. The observed image, $g(x, y)$, is assumed to be resulting from the sum of the original image, $f(x, y)$, and a stationary noise source, $n(x, y)$, that is,

$$g(x, y) = f(x, y) + n(x, y) \tag{5.17}$$

where the noise is spectrally white with zero mean and variance σ^2. The transfer function of the Wiener filer is given by [2]:

$$H(u,v) = \frac{P_f(u,v)}{P_f(u,v) + \sigma^2} \tag{5.18}$$

where $P_f(u,v)$ is the power spectrum of the signal. The conventional Wiener filter has certain limitations. For instance, the minimum MSE criterion often provides more smoothing than the human eye would prefer. The Wiener filter is often outperformed by nonlinear estimators [3].

There are variants of the Wiener filter that account for the spatially variant characteristics of signals and noise [2]. One approach to making the filter spatially variant is to allow the noise parameter σ_n to vary with position in the image, and the filter changes from one pixel to the next. Another variant is the noise-adaptive Wiener filter [17], which models the signal as a locally stationary process. Image recovery using noise-adaptive Wiener filtering is given by

$$\tilde{f}(x,y) = m_f(x,y) + \frac{\sigma_f^2(x,y)}{\sigma_f^2(x,y) + \sigma_n^2(x,y)}\big(g(x,y) - m_f(x,y)\big) \tag{5.19}$$

where m_f is the local mean of the signal f, and σ_f^2 is the local signal variance.

Another limitation of the Wiener filter is that it only accounts for the second-order statistics of an input image. By incorporating nonlinearity into the processing, this limitation can be overcome. A modified adaptive filter can be constructed as a linear combination of the stationary Wiener filter H and an identity operation [18].

$$H_\alpha = H + (1-\alpha)(1-H) \tag{5.20}$$

The modified adaptive filter equals the Wiener filter when $\alpha = 1$, whereas when $\alpha = 0$ it becomes the identify (null) transformation. Based on a study of the human vision system, an anisotropic component was introduced to improve this filter [19]:

$$H_{\alpha,\gamma} = H + (1-\alpha)\big(\gamma + (1-\gamma)\cos^2(\varphi - \theta)\big)(1-H) \tag{5.21}$$

where the parameter γ controls the degree of anisotropy, φ is the angular direction of the filter, and θ defines the orientation of the local image structure. In this way, the more dominant the local orientation is, the smaller the γ value and the more anisotropic the filter. The local direction and level of anisotropy can be estimated using three oriented Hilbert transform pairs. The weighting function $\cos^2(\varphi - \theta)$ was imposed by its ideal interpolation properties. The directed anisotropy filter can also be implemented as a steerable filter [12].

Fig. 5.6 shows an example of image deblurring using conventional Weiner deconvolution.

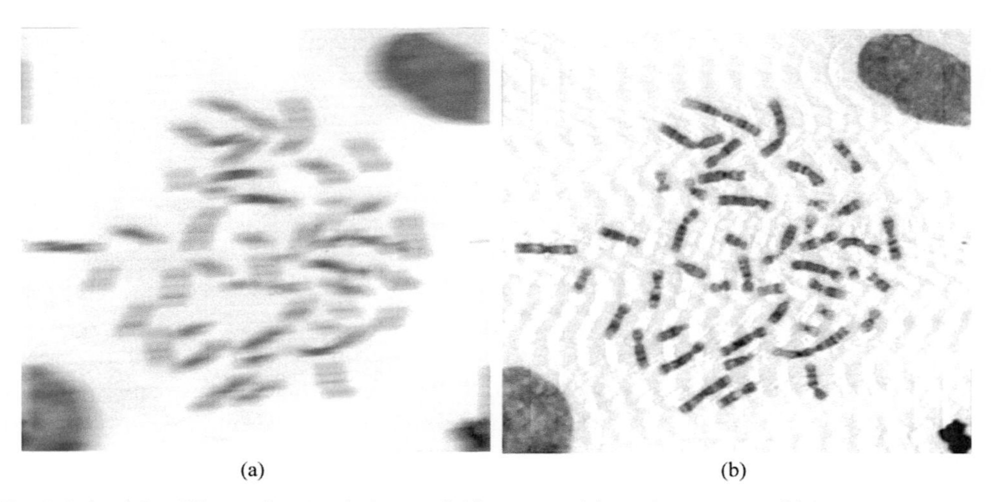

Fig. 5.6 Applying Wiener deconvolution to deblur image (a) produces image (b).

5.3.2 Deconvolution Using a Least Squares Approach

The observed image, g, can be expressed in a matrix form as

$$\mathbf{g} = \mathbf{Hf} + \mathbf{n} \tag{5.22}$$

where $\mathbf{g}$, $\mathbf{f}$, and $\mathbf{n}$ are $N^2 \times 1$ column vectors, $\mathbf{H}$ is an $N^2 \times N^2$ matrix that stands for blurring, $\mathbf{f}$ is the original image, and $\mathbf{n}$ is the noise. When the blurring is shift-invariant, the matrix $\mathbf{H}$ becomes a block-circulant matrix. If $\mathbf{n} = 0$, we can find the approximate solution by minimizing the mean square error

$$e(\hat{\mathbf{f}}) = \|\mathbf{g} - \mathbf{H}\hat{\mathbf{f}}\|^2 = (\mathbf{g} - \mathbf{H}\hat{\mathbf{f}})^t(\mathbf{g} - \mathbf{H}\hat{\mathbf{f}}) \tag{5.23}$$

by setting the derivative of $e(\hat{\mathbf{f}})$ with respect to $\hat{\mathbf{f}}$ equal to zero

$$\frac{\partial e(\hat{\mathbf{f}})}{\partial \hat{\mathbf{f}}} = -2\mathbf{H}^t(\mathbf{g} - \mathbf{H}\hat{\mathbf{f}}) = 0 \tag{5.24}$$

The solution for $\hat{\mathbf{f}}$ becomes

$$\hat{\mathbf{f}} = (\mathbf{H}^t\mathbf{H})^{-1}\mathbf{H}^t\mathbf{g} = \mathbf{H}^{-1}\mathbf{g} \tag{5.25}$$

If $\mathbf{n}$ is nonzero, the problem can be formulated as the solution of constrained optimization

$$e(\hat{\mathbf{f}}) = \|\mathbf{Q}\,\hat{\mathbf{f}}\|^2 + \lambda\left(\|\mathbf{g} - \mathbf{H}\hat{\mathbf{f}}\|^2 - \|\mathbf{n}\|^2\right) \tag{5.26}$$

where the first term is a regularization term, such that the solution is smooth, and the matrix $\mathbf{Q}$ is usually taken to be the first or second difference operation on $\hat{\mathbf{f}}$. λ is a constant called a Lagrange multiplier. Similarly, we can set the derivative of $e(\hat{\mathbf{f}})$ with respect to $\hat{\mathbf{f}}$ equal to zero

$$\frac{\partial e(\hat{\mathbf{f}})}{\partial \hat{\mathbf{f}}} = 2\,\mathbf{Q}^t\mathbf{Q}\,\hat{\mathbf{f}} - 2\,\lambda\mathbf{H}^t(\mathbf{g} - \mathbf{H}\,\hat{\mathbf{f}}) = 0 \tag{5.27}$$

and find the solution for $\hat{\mathbf{f}}$ as

$$\hat{\mathbf{f}} = (\mathbf{H}^t\,\mathbf{H} + 1/\lambda\,\mathbf{Q}^t\mathbf{Q})^{-1}\mathbf{H}^t\mathbf{g} \tag{5.28}$$

It turns out that this is the general form of the solution for the deconvolution problem.

5.3.3 Low-Pass Filtering

As mentioned earlier, an alternative to spatial domain filtering is to implement the filtering operation in the Fourier domain. To accomplish this, a 2D filter transfer function $H(u, v)$ is multiplied with the Fourier transform $G(u, v)$ of the image

$$\hat{F}(u, v) = H(u, v)\; G(u, v) \tag{5.29}$$

where $\hat{F}(u, v)$ is the Fourier transform of the filtered image $f(x, y)$ that we wish to recover. Here $f(x, y)$ can be obtained by taking the inverse Fourier transform.

Since noise often contaminates the high frequencies of an image's spectrum, a low-pass filter can suppress the noise in an image. An ideal low-pass filter is designed by assigning a frequency cut-off value

$$H(u, v) = \begin{cases} 1 & \text{if } D(u, v) \le D_0 \\ 0 & \text{otherwise} \end{cases} \tag{5.30}$$

where $D(u, v)$ is the distance of a point from the origin in the Fourier domain. However, since the rectangular passband in the ideal low-pass filter causes ringing artifacts in the spatial domain, usually filters with smoother roll-off characteristics are used instead. For example, the following Butterworth low-pass filter of nth order [20] is often used for this purpose:

$$H(u, v) = \frac{1}{1 + [D(u, v)/D_0]^{2n}} \tag{5.31}$$

When the order, n, increases, the roll-off characteristics of the band-pass filter become sharper. Hence a first-order Butterworth filter will produce fewer ringing artifacts in the filtered image than a higher-order filter.

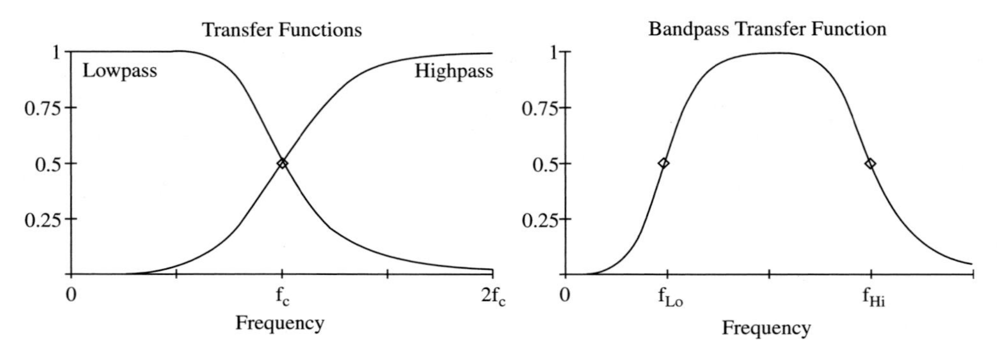

Fig. 5.7 Butterworth Filter Transfer Functions. The cut-off frequency and roll-off rate of each are adjustable. A low-pass and a high-pass filter can be combined to form a band-pass filter. These filters are useful because they are flat in the passband, they have adjustable cut-off frequencies, and the sharpness of the roll-off is controlled by the order parameter.

5.3.4 High-pass and Band-pass Filtering

While a low-pass filter can suppress noise and smooth an image, a high-pass filter can accentuate edge information and sharpen the detail in the image. An ideal high-pass filter with cut-off frequency D_0 is given by

$$H(u,v) = \begin{cases} 1 & \text{if } D(u,v) \geq D_0 \\ 0 & \text{otherwise} \end{cases} \tag{5.32}$$

As with the ideal low-pass filter discussed earlier, the sharp cut-off characteristics of a rectangular window function in the frequency domain can cause ringing artifacts in the filtered image. Therefore, one can use a filter with smoother roll-off characteristics, such as the Butterworth high-pass filter, given by

$$H(u,v) = \frac{1}{1 + [D_0/D(u,v)]^{2n}} \tag{5.33}$$

This is the Butterworth high-pass filter of nth order. Note Eq. (5.33) has the same form as Eq. (5.31), except that the terms D_0 and $D(u,v)$ in the denominator are interchanged.

The low-pass and high-pass filters can be concatenated to form a Butterworth band-pass filter. The transfer functions of these Butterworth filters appear in Fig. 5.7. Making the order of the filters higher makes the band edges sharper. Butterworth filters offer a convenient way to design flexible filters for image enhancement and noise suppression.

5.4 Wavelet Transform Methods

Human visual perception is known to function at multiple scales. Wavelet transforms were developed for the analysis of multiscale image structures [21]. Unlike traditional transform domain methods, such as the Fourier transform, wavelet-based methods not only dissect signals into their component frequencies, but also enable the analysis of the component frequencies across different scales. As a result, these methods are often more suitable for applications such as image data compression, noise reduction, and edge detection.

5.4.1 Wavelet Thresholding

The application of wavelet-based methods to image enhancement has been studied extensively. A widely used technique, known as wavelet thresholding, performs enhancement through the manipulation of wavelet transform coefficients so that object signals are boosted while noise is suppressed. Wavelet transform coefficients are modified using a nonlinear mapping. Hard-thresholding and soft-thresholding functions [12] are representative of such nonlinear mapping functions. For example, the soft-thresholding function is given by

$$\theta(x) = \begin{cases} x - T & if \quad x > T \\ x + T & if \quad x < -T \\ 0 & if \quad |x| \leq T \end{cases} \tag{5.34}$$

Small coefficients (below the threshold T or above $-T$) normally correspond to noise and are reduced to a value near zero. Usually, the thresholding operation of Eq. (5.34) is performed in the orthogonal or biorthogonal wavelet transform domain. A translation invariant wavelet transform [22] may be more appropriate in some cases. Enhancement schemes based on nonorthogonal wavelet transforms are also used [10,23,24]. Nonlinear mapping functions can be used in these schemes to accomplish multiscale image sharpening.

5.4.2 Differential wavelet transform and multiscale pointwise product

Edge sharpening is an essential part of image enhancement, and edges can be detected and characterized by differential operators. A particular family of differential wavelets has been used for this purpose [11,24]. In this case the differential wavelet transform approximation and detail coefficients of an image f are defined as $S_{2^j} f$ and $W_{2^j} f$, and the wavelet transform is computed using the following equations:

$$\begin{cases} S_{2^j} f &= S_{2^{j-1}} f \times h_{\uparrow 2^{j-1}} \\ W_{2^j} f &= S_{2^{j-1}} f \times g_{\uparrow 2^{j-1}} \end{cases}, \quad 1 \leq j \leq J \tag{5.35}$$

where h and g are the low-pass and high-pass filters, and $\uparrow 2^{j-1}$ is the upsampling operation by putting $2^{j-1} - 1$ zeros between each pair of adjacent samples in the filter [11]. This differential wavelet transform facilitates a desirable image representation for the extraction of edges at multiple scales. Since edge patterns are correlated spatially across multiple scales, one can take advantage of this property during the identification of

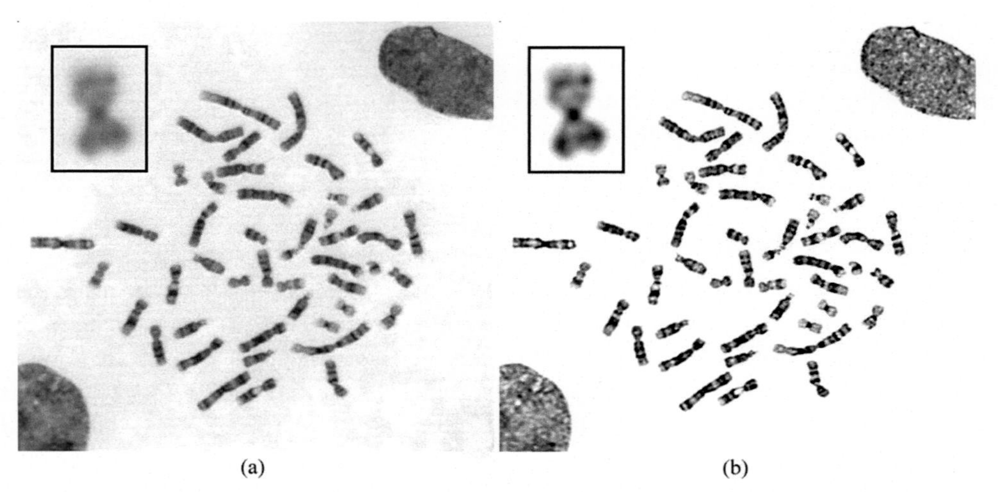

Fig. 5.8 The left image, showing chromosomal banding patterns, is enhanced by the differential wavelet transform, applying the MPP criterion to produce the image on the right.

the edges and subsequent enhancement. A *multiscale point-wise product* (MPP) can be employed to measure the cross-scale correlation of the differential wavelet transform coefficients [24]. The MPP is defined as

$$P_K(n) = \prod_{j=1}^{K} W_{2^j} f(n) \tag{5.36}$$

where $\{W_{2^j} f\}$ are the detail coefficients defined in Eq. (5.35). Because the maxima of $W_{2^j} f(n)$ represent edges in the signal, and $f(n)$ tends to propagate across scales, whereas the maxima of $W_{2^j} f(n)$ caused by noise do not, $P_K(n)$ reinforces the responses from edges in the wavelet domain rather than from noise. Experimental observation of edge patterns shows that the MPP has a built-in ability to suppress isolated and narrow impulses while preserving edge responses across different scales [24].

Based on the preceding consideration, the following nonlinear mapping function $\theta(x)$ can modify the wavelet coefficient x subject to the MPP criterion [24]:

$$\theta(x) = \begin{cases} \lambda x & \text{if } |P_K(n)| \geq \mu \\ 0 & \text{otherwise} \end{cases} \tag{5.37}$$

where λ is an adjustable constant associated with the degree of enhancement desired. The threshold parameter, μ, can be empirically determined. A larger value of μ results in a higher denoising effect, and vice versa. The choice of μ also depends on the noise level in the image [24]. Fig. 5.8 shows an example of chromosomal banding pattern enhancement using this approach.

5.5 Color Image Enhancement

Color image processing in microscopy applications usually deals with the tricolor images acquired with modern color imaging devices. This topic is discussed in more detail in Chapter 10, which addresses the subject of multispectral image processing. Among many color coordinate systems, the *RGB* and *HSI* formats are commonly used. The RGB format is the most straightforward as it directly deals with red, green, and blue images that are captured by color cameras and are closely associated with human vision. The HSI (hue, saturation, intensity) format [25] is a system popularly used among artists. Hue and saturation can be best described by the use of a color circle [8]. The hue of a color refers to the spectral wavelength that it most closely matches. The saturation is the radius of the point from the origin of the color circle and represents the purity of the color. The RGB and HSI formats can be easily converted from one to the other [8]. One can also convert a color image to a monochrome image by averaging the RGB components together, which discards all of the chrominance information.

When processing the components of a color image, one must exercise caution to avoid improperly changing the color balance. Essentially all of the image enhancement techniques discussed previously can be applied to the intensity component of an image in HSI format since this component encodes contrast, detail, and edge information. The color information is encoded in the hue and saturation components of the image. Enhancements of these color components should be approached with caution, as they are likely to upset the color balance and give objects in the image an unnatural appearance.

5.5.1 Pseudo-Color Transformations

A pseudo-color image transformation involves a mapping from a single-channel (monochrome) image to a three-channel (color) image. It is primarily used as a display technique to aid human visualization and interpretation of grayscale images, since humans can discern the combinations of hue, saturation, and intensity better than shades of gray alone. The technique of intensity slicing and color coding is a simple example of pseudo-color image processing. If an image is interpreted as a 3D terrain model, this method can be viewed as one of painting each elevation with a different color. Pseudo-color techniques are also useful for projecting multispectral image data down to three channels for display purposes.

5.5.2 Color Image Smoothing

The difference between color and gray-level image smoothing is that, for color processing, the smoothing is performed in each of the three RGB channels using conventional grayscale neighborhood processing [19]. This is shown in Eq. (5.38), where S_{xy}

denotes the neighborhood of a pixel at (x, y). Equivalently, if the HSI color format is used, one should apply the smoothing operation only to the intensity component of the image.

$$\tilde{f}_c(x,y) = \begin{bmatrix} \frac{1}{N} \sum_{(x,y)\in S_{xy}} f_R(x,y) \\ \frac{1}{N} \sum_{(x,y)\in S_{xy}} f_G(x,y) \\ \frac{1}{N} \sum_{(x,y)\in S_{xy}} f_B(x,y) \end{bmatrix} \tag{5.38}$$

5.5.3 Color Image Sharpening

Similar to its gray-level counterpart, color image sharpening is accomplished by extracting and accentuating edge information of an image. The Laplacian operator provides an example. For a three-component color vector $f_c(x,y) = (f_R(x,y), f_G(x,y), f_B(x,y))$, the Laplacian of a vector is defined as a vector whose components are equal to the Laplacian of each of the individual scalar components of the input vector. Specifically, the Laplacian of the vector $f_c(x,y)$ is given by

$$\nabla^2 f_c(x,y) = \begin{bmatrix} \nabla^2 f_R(x,y) \\ \nabla^2 f_G(x,y) \\ \nabla^2 f_B(x,y) \end{bmatrix} \tag{5.39}$$

which means that one can compute the Laplacian of an RGB color image by simply computing the Laplacian of each component image separately [19]. Likewise, applying the Laplacian operator only to the intensity component of an HSI image accomplishes the same objective. Fig. 5.9 shows an example of applying a Sobel edge enhancement operator to the RGB channels of a color image for sharpening.

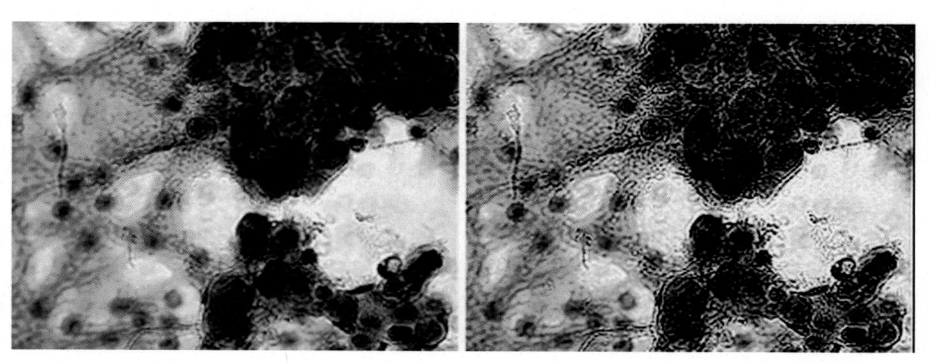

Fig. 5.9 The right image is the result of applying image sharpening to the color image on the left.

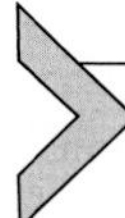

5.6 Summary of Important Points

1. Image enhancement is the process of enhancing the appearance of an image, or a subset of the image, to improve contrast or visualization of image features of interest or to facilitate more accurate subsequent image analysis.
2. Image enhancement can be achieved using computational methods either in the spatial domain or in the transform domain.
3. Spatial domain methods accomplish image enhancement using either global operations on the whole image, or local operations acting on a neighborhood of each pixel.
4. The operations used to increase contrast in the image include contrast stretching, clipping and thresholding, image subtraction and averaging, and histogram equalization and specification.
5. The operations used to sharpen image features and reduce noise include spatial band-pass filtering, directional and steerable filtering, and median filtering.
6. If image noise is a stationary random process, variants of the Wiener filter can be used to reduce the noise.
7. Nonlinear filters, such as the median filter, can reduce noise without blurring edges.
8. Transform domain methods accomplish image enhancement based on computations performed in a transform domain, such as the Fourier or wavelet transform domain. Often salient image features can be more easily isolated and extracted in the transform domain than in the spatial domain.
9. Commonly used Fourier domain image enhancement methods include Wiener filtering, least-squares deconvolution, and band-pass filtering. The Wiener filter is optimal for noise removal in the sense of minimum mean square error.
10. Wavelet domain image enhancement methods leverage the advantages of multiscale image representation and nonlinear filtering. Since image edges tend to correlate spatially across multiple scales, whereas noise does not, one can exploit this property and use nonlinear filtering to accentuate edge structures while suppressing noise in an image.
11. Essentially all of the techniques developed for the enhancement of monochrome images can be applied to enhance color images by performing the operations on their multiple intensity components or on their luminance component, where image detail and edge information are encoded.

References

[1] A.K. Jain, Digital Image Processing, Prentice-Hall, 1989.
[2] R. Fisher, et al., Hypermedia Image Processing Reference, 2003. http://homepages.inf.ed.ac.uk/rbf/HIPR2/.

[3] D.A. Agard, et al., Fluorescence microscopy in three dimensions, Methods Cell Biol. 30 (1989) 353–377.
[4] R.B. Paranjape, Fundamental enhancement techniques, in: I.N. Bankman (Ed.), Handbook of Medical Imaging, Academic Press, 2000.
[5] A. Laine, W. Huda, Enhancement by multiscale nonlinear operators, in: I.N. Bankman (Ed.), Handbook of Medical Imaging, Academic Press, 2000.
[6] A. Polesel, G. Ramponi, V.J. Mathews, Image enhancement via adaptive unsharp masking, IEEE Trans. Image Process. 9 (3) (2000) 505–510.
[7] Y. Xu, et al., Wavelet domain filters: a spatial selective noise filtration technique, IEEE Trans. Image Process. 3 (11) (1994) 747–757.
[8] K.R. Castleman, Digital Image Processing, Prentice Hall, 1996.
[9] Q. Wu, M.A. Schulze, K.R. Castleman, Steerable pyramid filters for selective image enhancement applications, in: Proceedings of ISCAS, 1998.
[10] B.M. Sadler, A. Swami, Analysis of multiscale products for step detection and estimation, IEEE Trans. Inf. Theory 45 (3) (1999) 1043–1051.
[11] Y. Wang, Image representations using multiscale differential operators, IEEE Trans. Image Process. 8 (12) (1999) 1757–1771.
[12] W.T. Freeman, E.H. Adelson, The design and use of steerable filters, IEEE Trans. Pattern Anal. Mach. Intell. 13 (9) (1991) 891–906.
[13] P. Perona, J. Malik, Scale-space and edge detection using anisotropic diffusion, IEEE Trans. Pattern Anal. Mach. Intell. 12 (7) (1990) 629–639.
[14] G. Gerig, O. Kubler, R. Kikinis, F.A. Jolesz, Nonlinear anisotropic filtering of MRI data, IEEE Trans. Med. Imaging 11 (2) (1992) 221–232.
[15] A. Beghdadi, A.L. Negrate, Contrast enhancement technique based on local detection of edges, Comput. Vis. Graph. Image Process. 46 (1989) 162–174.
[16] W. Carrington, Image restoration in 3D microscopy with limited data, Proc. IEEE 1205 (1990) 72–83.
[17] A.P. Dhawan, Medical Image Analysis, John Wiley & Sons, 2003.
[18] S. Mallat, S. Zhong, Characterization of signals from multiscale edges, IEEE Trans. Pattern Anal. Mach. Intell. 14 (7) (1992) 710–732.
[19] R.C. Gonzalez, R.E. Woods, Digital Image Processing, fifth ed., Prentice Hall, 2019.
[20] S. Butterworth, On the theory of filter amplifiers, Exp. Wirel. Wirel. Eng. (1930) 536–541.
[21] H. Knutsson, R. Wilson, G.H. Granlund, Anisotropic non-stationary image estimation and its applications part I, restoration of noisy images, IEEE Trans. Commun. 31 (3) (1983) 388–397.
[22] J.S. Lee, Digital image enhancement and noise filtering by local statistics, IEEE Trans. Pattern Anal. Mach. Intell. 2 (1980) 165–168.
[23] R.R. Coifman, D.L. Donoho, Translation-invariant de-noising, in: A. Antoniadis, G. Oppenheim (Eds.), Wavelets and Statistics, Springer-Verlag, 1995.
[24] Y. Wang, et al., Chromosome image enhancement using multiscale differential operators, IEEE Trans. Med. Imaging 22 (5) (2003) 685–693.
[25] A.H. Munsell, A pigment color system and notation, Am. J. Psychol. 23 (1912) 236–244.

CHAPTER SIX

Morphological Image Processing

Roberto A. Lotufo, Romaric Audigier, André V. Saúde, and Rubens C. Machado

6.1 Introduction

This chapter presents the main concepts of morphological processing (MP) for the microscope image analyst. Morphological processing has application in such diverse areas of image processing as filtering, segmentation, and pattern recognition, applied both to binary and grayscale images. It represents an alternative approach to the classical methods discussed in Chapter 5 (Image Enhancement), Chapter 7 (Image Segmentation), Chapter 8 (Object Measurement), and Chapter 9 (Object Classification). One of the advantages of MP is its being well suited for discrete image processing, as its operators can be implemented in digital computers with complete fidelity to their mathematical definitions. Another advantage of MP is its inherent building-block structure where complex operators can be created by the combination of a few primitive operators. Further, each of these primitive operators has an intuitive physical analogy that greatly aids understanding the effects it can produce in an image.

This chapter introduces the most-used concepts as they are applied to real situations in microscopy imaging, with explanations that appeal to the intuition wherever possible, while remaining true to the underlying mathematical theory. The motivated reader can find many texts where the details and formalisms are treated in-depth [1–5]. This chapter gives the main mathematical equations and several algorithms, without stressing efficient implementations. Source code of implementations of these equations are available on the internet [6]. This chapter illustrates how to solve image analysis problems using a combination of the primitive MP operators.

Morphological image processing is based on probing an image with a *structuring element* and either filtering or quantifying the image according to the manner in which the structuring element fits (or does not fit) within each object in the image. A binary image is made up of foreground and background pixels, and connected sets of foreground pixels make up the *objects* in the image. A set of pixels is said to be *connected* if any two pixels in the set can be linked by a sequence of neighbor pixels that are also in the set. In Fig. 6.1 we see a binary image and a circular structuring element (probe) that is placed in two different positions. In one location it fits within the object, but in the other it does not. By marking those locations at which the structuring element does fit within the object, we derive structural information about that image. This information depends on both the size

Microscope Image Processing
https://doi.org/10.1016/B978-0-12-821049-9.00012-5
Copyright © 2023 Elsevier Inc.
All rights reserved.

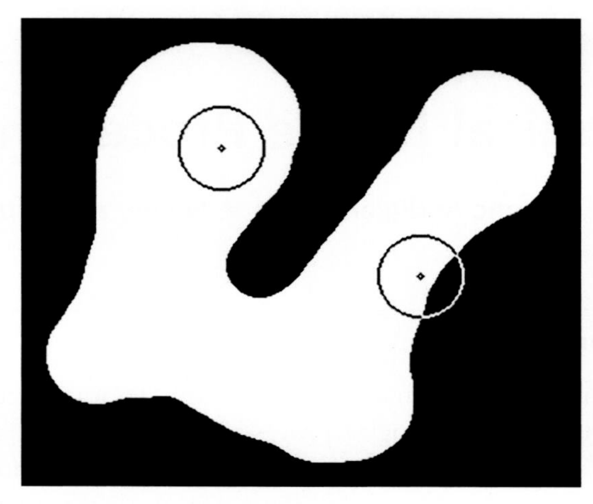

Fig. 6.1 Probing an image with a structuring element.

and shape of the structuring element. Although this concept is rather simple, it is the basis of the majority of the operations presented in this chapter (erosion, dilation, opening, closing, morphological reconstruction, etc.), as applied to both binary and grayscale images. Common measurements that can be derived from this concept are the largest disk that fits inside the object and the area of the object.

In the general case, morphological image processing operates by passing a "structuring element" over the image in an activity similar to convolution (Fig. 6.2). Like the convolution kernel, the structuring element can be of any size, and it can contain any complement of 1 and 0s, or grayscale values. One of the pixels of the structuring element is specified as its origin. The structuring element is sequentially positioned at all possible locations in the image, where it is compared with the corresponding neighborhood of

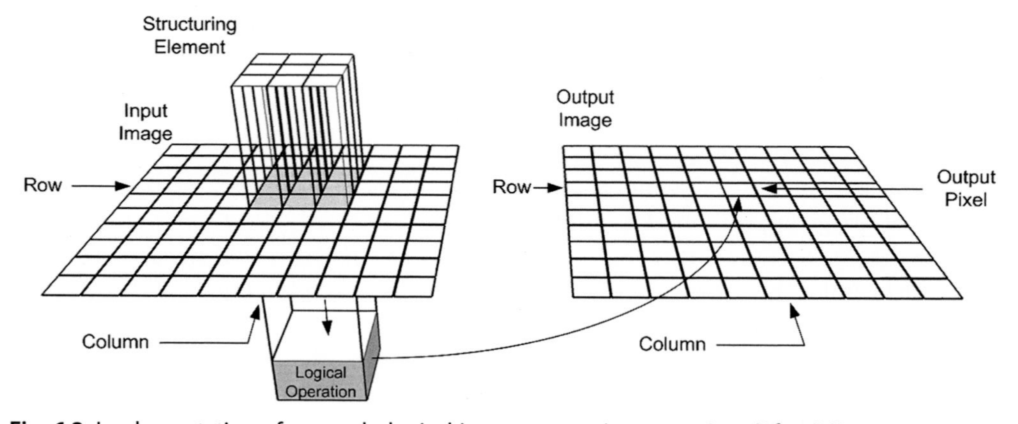

Fig. 6.2 Implementation of a morphological image processing operation. *(After [2])*.

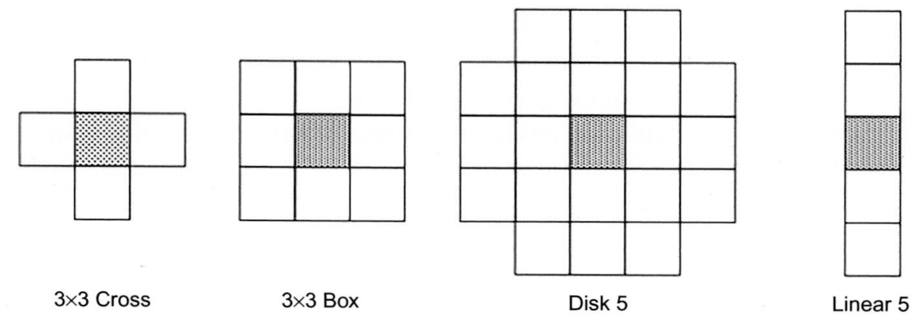

Fig. 6.3 Four different structuring elements that may be used in morphological image processing operations.

pixels, as defined by the size and shape of the structuring element. At each position, a specified logical (for binary images) or nonlinear (for grayscale images) operation is performed between the structuring element and the underlying image. The result of that operation is stored in the output image at the pixel position corresponding to the *origin* of the structuring element. The effect that the operation creates depends upon the size, shape, and content of the structuring element and upon the nature of the logical or nonlinear operation.

A common practice in MP is to use structuring elements of odd dimensions with the origin defined as the center pixel. In this chapter, only structuring elements that are symmetric about the origin are used. When the structuring element is not symmetric, care must be taken, as only some properties are valid for a reflected structuring element. This chapter uses four structuring element types (Fig. 6.3) in the illustrations: (1) the *elementary cross*, a 3×3 structuring element with the central pixel and its direct four neighbors; (2) the *elementary box*, which has the central pixel and its eight neighbors; (3) the *disk* of a given radius; and (4) the *linear structuring element* of a given length and orientation.

6.2 Binary Morphology

A *binary image* consists of foreground and background pixels, having only 2 gray levels, 1 and 0, respectively (generation of binary images is discussed in Chapter 7, Image Segmentation). A connected set of foreground pixels forms an object in the binary image. Some of these objects will correspond to physical structures on the microscope slide, while others may be due to artifacts or noise. Morphological processing was first applied to binary images. In 1968, Matheron and Serra founded the Centre de Morphologie Mathématique at the École des Mines de Paris in Fontainebleau, France. There they developed much of the early theory of "mathematical morphology."

6.2.1 Binary Erosion and Dilation

The basic fitting operation of morphology is the *erosion* of an image by a structuring element. Erosion is done by scanning the image with the structuring element. When the structuring element fits completely inside the object, the probe position is marked (set to 1). The erosion output consists of all scanning locations where the structuring element fits inside the object. The eroded object is typically a shrunken version of the original object, and the shrinking effect is controlled by the structuring element size and shape. The erosion of set A by set B is defined by

$$A \ominus B = \{x : B_x \subset A\} \tag{6.1}$$

where $\subset$ denotes the subset relation and $B_x = \{b + x : b \in B\}$ is the *translation* of set B by a point x.

In morphology, for every operator that changes the foreground in a particular way, there is a *dual operator* that changes the background in the same way. The *complement* of an image is formed by reversing the foreground and background pixels. The dual of the erosion operator is the *dilation* operator. Dilation involves fitting a probe into the complement of the image. Thus it represents a filtering on the outside of the object, whereas erosion represents a filtering on the inside of the object, as depicted in Fig. 6.4. Formally, the dilation of set A by B is defined by

$$A \oplus B = \left(A^c \ominus \breve{B}\right)^c \tag{6.2}$$

where A^c denotes the complement of A and $\breve{B} = \{-b : b \in B\}$ is the *reflection* of B, that is, a 180° rotation of B about the origin.

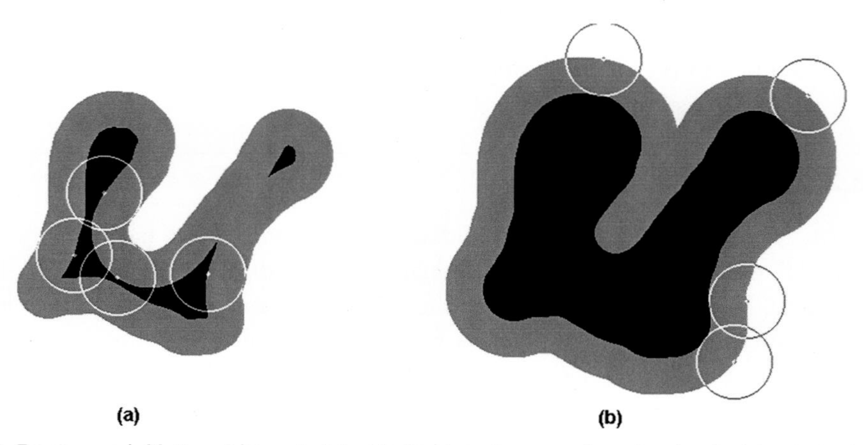

Fig. 6.4 Erosion and dilation: (a) input object in *black* and *gray* and erosion in *black* (region where the probe can fit); (b) input object in *black* and dilation in *black* and *gray*.

The foreground is usually labeled with white color, while the background is labeled black, but the inverse convention is also used.

An alternative way to compute dilation is by "stamping" the structuring element on the location of every foreground pixel in the image. For instance, Fig. 6.16d was obtained by stamping small arrows (the structuring element) on the centroids of the detected spots in Fig. 6.16c.

Formally, the dilation can be defined by

$$A \oplus B = \bigcup_{a \in A} B_a \tag{6.3}$$

Dilation has the effect of expanding the object, filling in small holes and intrusions into the object (Fig. 6.4b), while erosion has a shrinking effect, enlarging holes and eliminating small extrusions (Fig. 6.4a).

Since dilation by a disk expands an object, and erosion by a disk shrinks an object, they can be combined to find object boundaries in binary images. The three possibilities are: (1) the *external boundary* (dilation minus the image), (2) the *internal boundary* (the image minus the erosion), and (3) the *morphological gradient* (dilation minus erosion), which is the boundary that straddles the actual boundary. The morphological gradient is often used as a practical way of displaying the boundary of segmented objects, as in Fig. 6.25f.

6.2.2 Binary Opening and Closing

Besides the primary operations of erosion and dilation, two more operations play key roles in morphological image processing; they are *opening* and, its dual, *closing*. The opening of an image A by a structuring element B, denoted by $A \circ B$, is the union of all the structuring elements that fit inside the image, as depicted in Fig. 6.5a.

$$A \circ B = \cup \{B_x : B_x \subset A\} \tag{6.4}$$

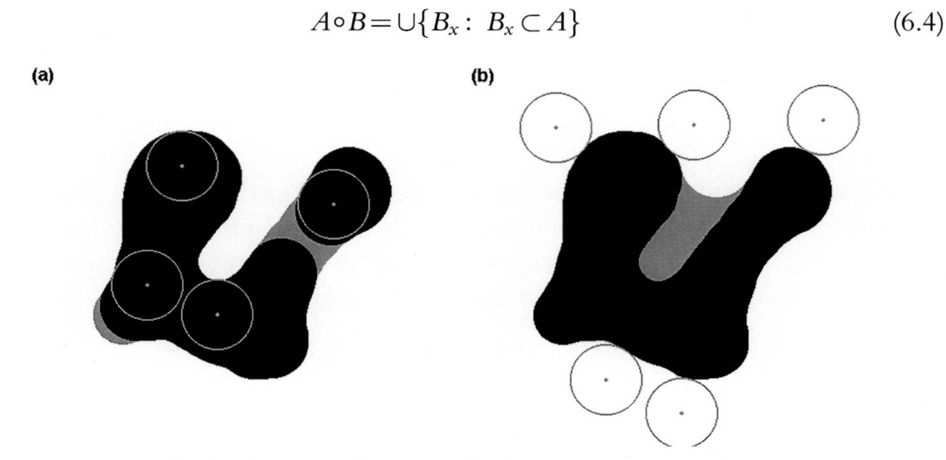

Fig. 6.5 Opening and closing: (a) input image in *black* and *gray* and opening in *black* (region where the probe fits); (b) input image in *black* and closing in *black* and *gray*.

Fig. 6.7c through Fig. 6.7f show how the result of opening depends on the structuring element when using different linear structuring elements to open the same image.

$$A \circ B = (A \Theta B) \oplus B \tag{6.5}$$

The dual of opening is closing, defined by

$$A \bullet B = \left(A^c \circ \breve{B}\right)^c$$
$$A \bullet B = (A \oplus B) \ominus B \tag{6.6}$$

Fig. 6.5b shows an example of closing. Note that, while the position of the origin relative to the structuring element has a role in both erosion and dilation, it plays no role in opening or closing.

Opening and closing have two important properties [7]. First, once an image has been opened (or closed), successive openings (or closings) using the same structuring element produce no further effect. Second, an object in an opened image is contained in the original object which, in turn, is contained in the closed object, as illustrated in Fig. 6.5. As a consequence of this property, we can consider the subtraction of the opening from the input image, called the *opening top-hat* operation, and the subtraction of the image from its closing, called the *closing top-hat* operation, respectively, defined by

$$A \bullet B = \left(A^c \circ \breve{B}\right)^c$$
$$A \bullet B = (A \oplus B) \ominus B \tag{6.7}$$

The opening top-hat and closing top-hat results correspond to the gray parts in Fig. 6.5a and b, respectively.

Opening can clean the boundary of an object by eliminating small extrusions, but it does this in a much finer manner than erosion. The net effect is that the opened object is a much better replica of the original than the eroded object (compare Fig. 6.5a with Fig. 6.4a). Analogous remarks apply to closing and the filling of small intrusions (compare Fig. 6.5b with Fig. 6.4b).

Binary images can have both additive noise (extraneous foreground pixels in the background) and subtractive noise (extraneous background pixels in the foreground). One strategy for correcting this is to open the image to eliminate additive noise and then close to fill subtractive noise. The close-open filter is given by

$$ASF_{co}^{n}(S) = (((((S \bullet B) \circ B) \bullet 2B) \circ 2B) \ldots \bullet nB) \circ nB \tag{6.8}$$

and the open-close filter by

$$ASF_{oc}^{n}(S) = (((((S \circ B) \bullet B) \circ 2B) \bullet 2B) \ldots \circ nB) \bullet nB \tag{6.9}$$

The open-close strategy works when noise components are small but fails when attempts to remove large noise components destroy too much of the original object. In this case, a better strategy is to employ an *alternating sequential filter* (ASF). Here open-close (or close-open) filters are performed iteratively, beginning with a very small structuring element and then proceeding with ever-larger structuring elements.

6.2.3 Binary Morphological Reconstruction From Markers

One of the most important operations in morphological image processing is *reconstruction from markers*. The basic idea is first to mark certain image components and then reconstruct the portion of the image that contains the marked components.

6.2.3.1 Connectivity

As mentioned earlier, a region (set of pixels) is said to be *connected* if any two pixels in the set can be linked by a sequence of neighbor pixels that are also in the set (see also Chapter 7). If the region is *4-connected*, the linking involves only vertically and horizontally adjacent pixels. In an *8-connected* region, the linking involves diagonally adjacent pixels as well.

Every binary image can be expressed as a union of connected regions. A region is *maximally connected* if it is not a proper subset of a larger connected region in the image. The maximally connected regions are called the *connected components* of the image. A connected object has only one component. The union of all connected components C_k recovers the input image A and the intersection of any two connected components is empty.

$$A = \bigcup_{k=1}^{n} C_k \tag{6.10}$$

To find all the connected components of an image, one can iteratively (1) choose any foreground pixel of the image to use as a starting point, (2) use it to reconstruct its connected component, (3) remove that component from the image, and (4) iteratively repeat the same extraction until no more foreground pixels are found in the image. This operation is called *labeling*, and it decomposes an image into its connected components (objects). These can be numbered sequentially as they are found. The result of the labeling operation can be conveniently stored as a grayscale image ("object membership map") in which the gray level of each pixel is its object number. An example of labeling can be seen in Fig. 6.8c.

6.2.3.2 Markers

For an object, the *morphological reconstruction* of an image A from a *marker* M (a subset of A) is denoted by $A \Delta M$ and defined as the union of all connected components of image A that intersect marker M. This filter is also called a *component filter*.

$$A \Delta M = \cup \{ C_k : C_k \cap M \neq \varnothing \} \tag{6.11}$$

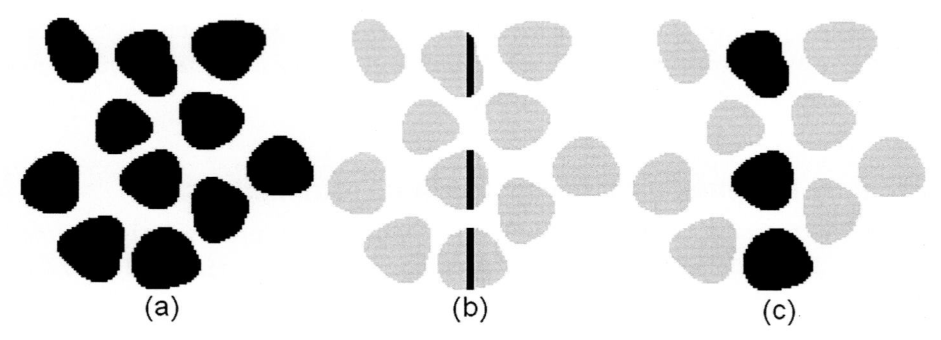

Fig. 6.6 Reconstruction from markers: (a) input image; (b) marker image; (c) reconstructed image.

The reconstruction operation requires the input image, the marker, and a selection of the type of connectivity. The marker specifies which component of the input image is to be extracted. The choice of connectivity usually depends on the structuring element. For example, 4-connectivity is useful for the elementary cross, and 8-connectivity for the elementary box.

An example of reconstruction from markers, using 8-connectivity, is shown in Fig. 6.6. Fig. 6.6a is the input image, which is a collection of grains. Fig. 6.6b is the marker image, that is, a central vertical line intersecting the grains. Fig. 6.6c shows the reconstruction from the markers, which extracts the three central components from the original image.

There are typically three ways to select the marker placement for the component filter: (a) a priori selection, (b) selection from the opening, and (c) selection by means of a more complex operation. An example of reconstruction from an a priori marker can be seen in Fig. 6.26e where the background component, in black, was selected by placing a marker at the top left of the image in Fig. 6.26d. In this case, we work on the complement of the image. This dual reconstruction has the overall effect of filling in bounded regions.

6.2.3.3 A Priori Selection Using the Image Border for Marker Placement

The *edge-off* operation, particularly useful for removing objects that touch the image border, combines reconstruction and top-hat concepts. The objects touching the border are selected by reconstructing the image using its boundary as an a priori marker. The objects not connected to the image frame are then selected by subtracting the input image from the reconstructed image. The result is an image containing only those objects that do not touch the border.

Fig. 6.24 illustrates a variant of the edge-off operation applied to grain boundaries. Here we want to keep only the boundaries of those grains that do not touch the image border in Fig. 6.24f. The boundaries connected to the border cannot simply be removed because that would also remove the boundaries of the neighboring grains that do not directly touch the border. So, the strategy is first to fill in all the bounded grains that do not directly touch the border by reconstruction of the border from the complement of image in Fig. 6.24f (see Fig. 6.24g). Then we remove the thin grain boundaries that do touch the border by applying an opening. The result in Fig. 6.24h is obtained by intersecting the remaining inner grains and the boundaries in Fig. 6.24f.

6.2.3.4 Reconstruction From Opening

With marker selection by opening, the marker is found by opening the input image with a particular structuring element. The result of the reconstruction detects all the connected components into which that structuring element fits. By using the mechanism of reconstruction from the opening to detect objects with particular geometric features, one can design more complex techniques to find the markers from combined operators. At the last step, the reconstruction reveals those objects that exhibit those geometric features.

The biomedical application illustrated in Fig. 6.7 detects overlapping human chromosomes by using the intersection of four reconstructions from openings. Because chromosome length is highly variable, the structuring element length is a critical parameter for

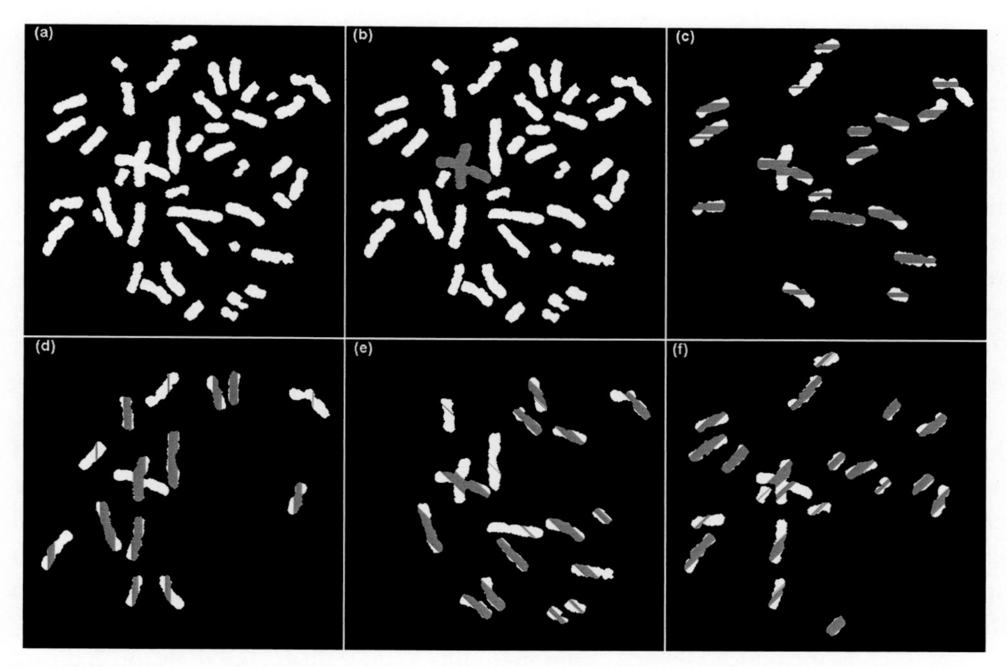

Fig. 6.7 Detecting overlapping chromosomes: (a) input image; (b) intersection *(in gray)* of four reconstructions from openings; (c) opening *(in gray)* by *horizontal line* and its reconstruction; (d) opening *(in gray)* by *vertical line* and its reconstruction; (e) opening *(in gray)* by 45° line and its reconstruction; (f) opening *(in gray)* by −45° line and its reconstruction.

good filter performance. A length of 30 pixels was used in this example. To identify overlapping chromosomes, as shown in Fig. 6.7b, only the objects (connected components) into which all four of the linear structuring elements can fit are chosen. This is achieved by performing four reconstructions from opening operations and intersecting them. The four linear structuring elements used are vertical, horizontal, and ±45° (Fig. 6.7c–f).

The top-hat concept can be applied to reconstruction by opening, producing the *reconstruction from opening top-hat* operation. This is the image minus its reconstruction. In this case the operator reveals the objects that do not exhibit a specified fitting criterion. For instance, to detect thin objects, one can use a disk of diameter larger than the thickest of those objects. Then only objects too thin to contain the disk remain.

6.2.4 Reconstruction Using Area Opening and Closing

Another common criterion for selection of connected components is area. This is achieved by an *area opening* that removes all connected components C_i with area less than a specified value α:

$$A \circ (\alpha)_E = \cup \{ C_i, \mathit{area}(C_i) \geq \alpha \} \tag{6.12}$$

Fig. 6.25e shows how the area opening can behave as a filter to remove small artifacts from the image in Fig. 6.25d and select only the large objects (epithelial cells) having area greater than 1000 pixels.

The next example targets cytogenetic images of human metaphase cells. This is a classical application of area opening. The task is to preprocess the image by segmenting out the chromosomes from the nuclei, stain debris, and the background. Fig. 6.8 shows the input image (a), the thresholded (binary) image (b), the labeling (c) of the identified connected components, and the result (d), with the components classified by area. The components with area less than 100 pixels are background noise, those with area greater than 10,000 pixels are nuclei (shown in dark gray), and the rest are the chromosomes, shown in light gray.

These operations are not restricted to openings. Analogous dual operations can be developed to form sup-*reconstruction from closing, sup-reconstruction from closing top-hat* and *area closing*, respectively. The names of these operations are derived from the word *supremum* (least upper bound). Fig. 6.25d shows how area closing can be used to fill small aggregate holes with area less than 200 pixels (c) that are specific to a texture, to form significant clusters of interest. Similarly, small holes in Fig. 6.32g, having area less than 100 pixels, are closed and merged to recover the objects of interest (Fig. 6.32h).

This section introduced several component filters, their duals, and their top-hat versions. Filters of these types are only able to select out connected components in the image when the selection is based on an area or shape-fitting criterion. The following section addresses operations that are less restricted.

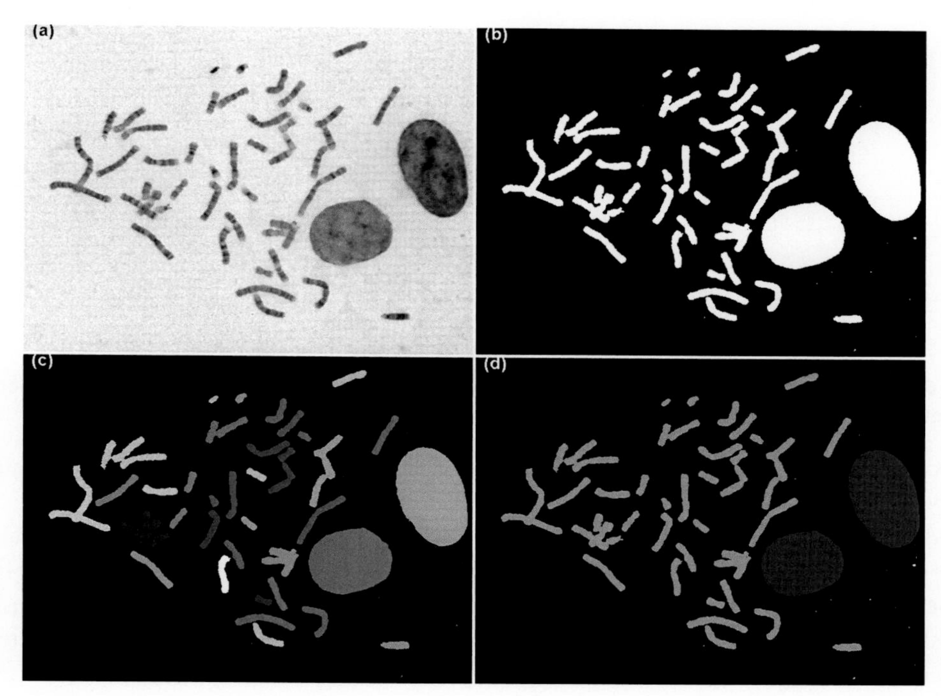

Fig. 6.8 Preprocessing chromosome spreads using area opening: (a) input image; (b) thresholded image; (c) labeled (grayscale) image; (d) objects classified by area: residues in *white* (area <100), chromosomes in *light gray* (area between 100 and 10,000), and nuclei in *dark gray* (area $>10,000$).

6.2.5 Skeletonization

A commonly encountered image processing problem is finding a thinned replica of a binary image to use either in a recognition algorithm or for data compression. An often-used thinning procedure is skeletonization, a technique based on the concept of maximal disks. In microscopy images, skeletons have been successfully used for feature extraction for classification purposes. Given a point interior to a Euclidean binary image, there exists a largest disk having the point at its center and lying completely within the image. Regarding the largest disk at a particular point, there are two possibilities: either there exists another disk lying within the image and properly containing the given disk, or there does not. Any disk satisfying the second condition is called a *maximal disk*. The centers of all maximal disks comprise the *medial axis* of the image.

As an illustration, consider the isosceles triangle in Fig. 6.9, whose skeleton is depicted in part (a) of the figure. Part (b) shows a maximal disk $D(x)$ centered at point$x$, so that x lies on the skeleton. In part (c), $D(w)$ is the largest disk centered at$w$; however, it is not

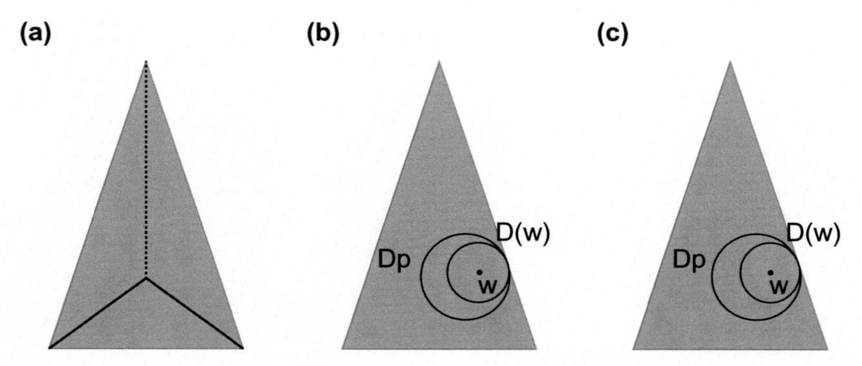

Fig. 6.9 Triangle medial axis: (a) medial axis; (b) maximal disk $D(x)$ for skeleton point x; (c) w is not a skeleton point as $D(w)$ is contained in disk D_p.

maximal since it is properly contained in D_P, which itself lies within the triangle. Thus, $w$does not lie in the skeleton.

As might be expected, adaptation of the skeleton to the digital setting requires some care since Euclidean disks are defined in a continuous space. More efficient algorithms for generating the Euclidean medial axis exist, but implementations based on discrete disks are still in use. We begin with some "disk-like" digital primitive, and the actual skeleton obtained will depend on the choice of that primitive. As an example, let B denote the 3×3 square structuring element, and let nB be defined by the iterated dilation of Eq. (6.13).

$$nB = B \oplus B \oplus \ldots \oplus B \tag{6.13}$$

where n is a positive integer. This equation implies that $2B = B \oplus B$, $3B = B \oplus B \oplus B$, and so on. The notion of maximal disk is put into the digital setting by considering "disks" chosen from among $0B$, B, $2B$, $3B\ldots$, where $0B$ is simply the origin.

The discrete skeleton can be characterized morphologically. For $n = 0, 1, \ldots$, we define the *skeletal subset Skel(S; n)* to be the set of all pixels x in S such that x is the center of a maximal disk nB. Then it is evident from the definition of the skeleton that the skeleton is the union of all skeletal subsets:

$$Skel(S) = \bigcup_{n=0}^{\infty} Skel(S; n) \tag{6.14}$$

It can be shown that the skeletal subsets are given by

$$Skel(S; n) = (S \ominus nB) - [(S \ominus nB) \circ B] \tag{6.15}$$

Together Eqs. (6.14) and (6.15) yield *Lantuejoul's formula* for the skeleton:

$$Skel(S) = \bigcup_{n=0}^{\infty} (S \ominus nB) - [(S \ominus nB) \circ B] \tag{6.16}$$

The drawbacks of such discrete skeletons are illustrated further in a comparison with Euclidean skeletons. Lantuejoul's formula is not applicable to Euclidean disks, but there are efficient algorithms for Euclidean skeletons available, and we can define such skeletons. We denote by $d(x, y)$ the Euclidean distance between points x and y. The usual definition of an n-dimensional discrete Euclidean disk of radius R and center point x is the set $DE(x, R)$ given by

$$DE(x, R) = \{y \in \mathbb{Z}^n, d(x, y) < R\} \tag{6.17}$$

The *discrete Euclidean medial axis* is thus defined by the same terms as its continuous counterpart, but with discrete Euclidean disks. Similarly, to the continuous medial axis, each maximal disk that composes the discrete medial axis can be reconstructed from its center and its radius, and the original object can be reconstructed by the union of all such disks. However, topology preservation is no longer guaranteed. For instance, the discrete medial axis of a single connected set may be composed by many disconnected subsets. This problem is solved by the combination of the discrete medial axis with a topology-preserving thinning process.

A basic topology-preserving thinning process is based on the notion of *simple point*. A simple point is a point that can be removed from the object without changing its topology. The sequential removing of simple points from the object, with the constraint of not removing points of the object's medial axis, constitutes a thinning process that preserves the object topology and guarantees the object reconstruction. To preserve some characteristics of the object's geometry, the thinning process can be guided by a *priority function* based on geometric information.

Following, we present the algorithm of a thinning process guided by the Euclidean distance transform. In a binary image, the *distance transform* is defined for any point in a given set as the distance from this point to the complement of the set. In a guided thinning, when the thinned object has no simple point, the process stops, and the resulting subset is called the *Euclidean skeleton*. For skeleton analysis, one may need to prune spurious branches that result from boundary irregularities and noise. There are many thinning and pruning algorithms described in the literature [8–10]. The following function computes the Euclidean skeleton:

```
Function sk = EuclSkel(f,T)
   f: input image
   T: threshold value for pruning the medial axis
 1. Medial axis extraction and pruning
      M <- discrete medial axis of f, with all disks radii
      pM <- x in M, x is center of a disk with R > T
 2. Initialize guided thinning
      DT <- distance transform of f
      for all x in DT
```

```
            if DT(x) = 1 then inHFQ(x,DT(x))
3. Propagation
        while HFQ is not empty:
            p <- outHFQ
            for each q neighbor of p:
                inHFQ(q,DT(q))
            if p is simple and M(p) = 0 (p not in the medial axis)
            then
                f(p) <- 0 (p is deleted)
```

In this function, the hierarchical FIFO queue (HFQ) includes the following operations: inHFQ(p,v), insert pixel p with priority v; outHFQ, remove the pixel with the lowest priority with the FIFO policy for pixels at the same priority.

The detection of overlapping chromosomes, performed earlier by other means, can now be done by detecting crossing points in the chromosomes' skeletons. In the following, we illustrate the skeletons of overlapping chromosomes. In order to detect crossing points, a further mask matching operation is required. Fig. 6.10 compares the Euclidean skeleton with the discrete skeleton nB. Fig. 6.10a shows the Euclidean skeleton, obtained by a topology-preserving thinning process guided by the Euclidean distance, with the constraint of not removing the points of the exact Euclidean medial axis. The skeleton in Fig. 6.10b was produced by a topology-preserving thinning process guided by the 8-neighbor distance, with the constraint of not removing the points of the exact medial axis

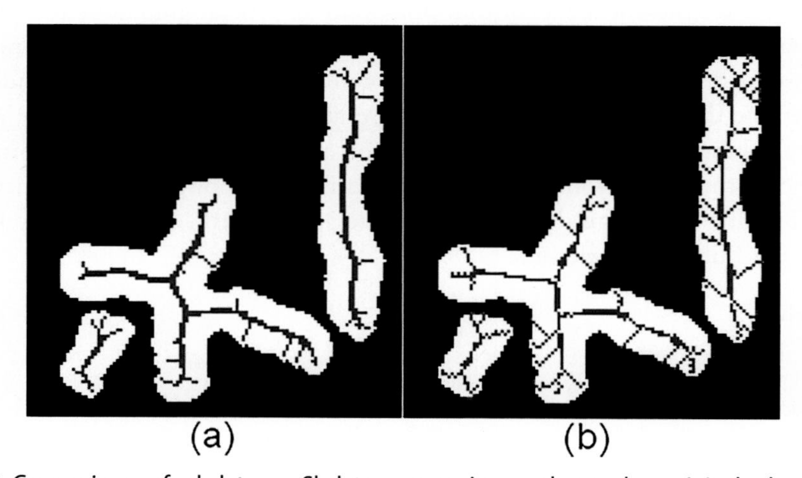

Fig. 6.10 Comparison of skeletons. Skeletons superimposed on the original chromosomes: (a) Euclidean skeleton with topology preservation, containing the exact Euclidean medial axis; (b) Discrete skeleton with topology preservation, containing the exact medial axis by nB disks.

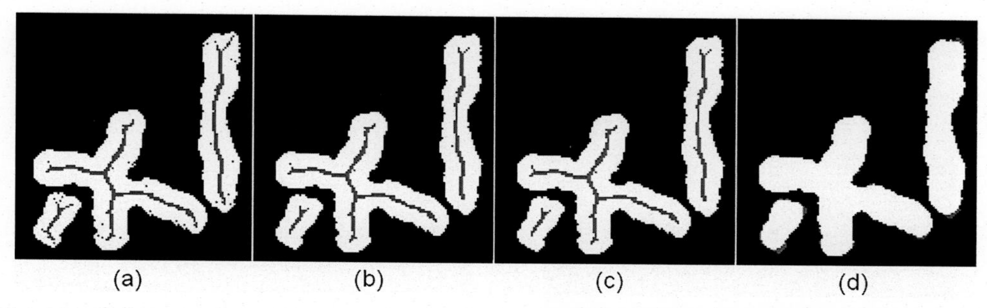

Fig. 6.11 Skeletonization and pruning. The skeletons are superimposed on the original chromosomes. (a) The exact Euclidean medial axis (center of maximal discrete Euclidean disks); (b) pruned medial axis, disks with (squared) radius less than 36 have been removed; (c) Euclidean skeleton with topology preservation, containing the pruned medial axis shown in (b); (d) points on the boundaries that could not be reconstructed by the reverse process of the pruned skeleton.

with nB disks. Although the Euclidean skeleton has many branches, the skeleton in Fig. 6.10b has more, illustrating that the discrete skeleton nB is more sensitive to noisy boundaries.

The effect of skeleton pruning is shown in Fig. 6.11, where different skeletons based on Euclidean disks are superimposed on the chromosomes. The exact Euclidean medial axis is shown in Fig. 6.11a. Some isolated points result from the noisy chromosome boundaries, and these may be discarded without affecting the quality of the resulting skeleton. In Fig. 6.11b the points that represent centers of maximal disks with squared radii less than 36 are removed from the medial axis. This filtering of small radius disks is one of the simplest skeleton pruning techniques. To obtain the Euclidean skeleton of Fig. 6.11c, we perform a topology-preserving thinning process guided by the Euclidean distance with the constraint of not removing the points of the pruned medial axis. Finally, the reverse process (reconstruction of the object) is performed on the pruned skeletons, and Fig. 6.11d shows the details of the boundaries that could not be reconstructed.

6.3 Grayscale Operations

Morphological processing can be performed on grayscale images as well as binary images. In so doing, it is useful to think of a grayscale image as a surface. Fig. 6.12 shows a grayscale image made up of three Gaussian-shaped peaks of different heights and widths. The image is depicted in four different graphical representations: (a) the pixel values displayed as gray levels: low values are dark and high values are bright gray tones; (b) the pixel values again mapped to the gray scale but in a reverse order; (c) the same image but as a top view of a shaded surface; and (d) a mesh plot of the pixel values.

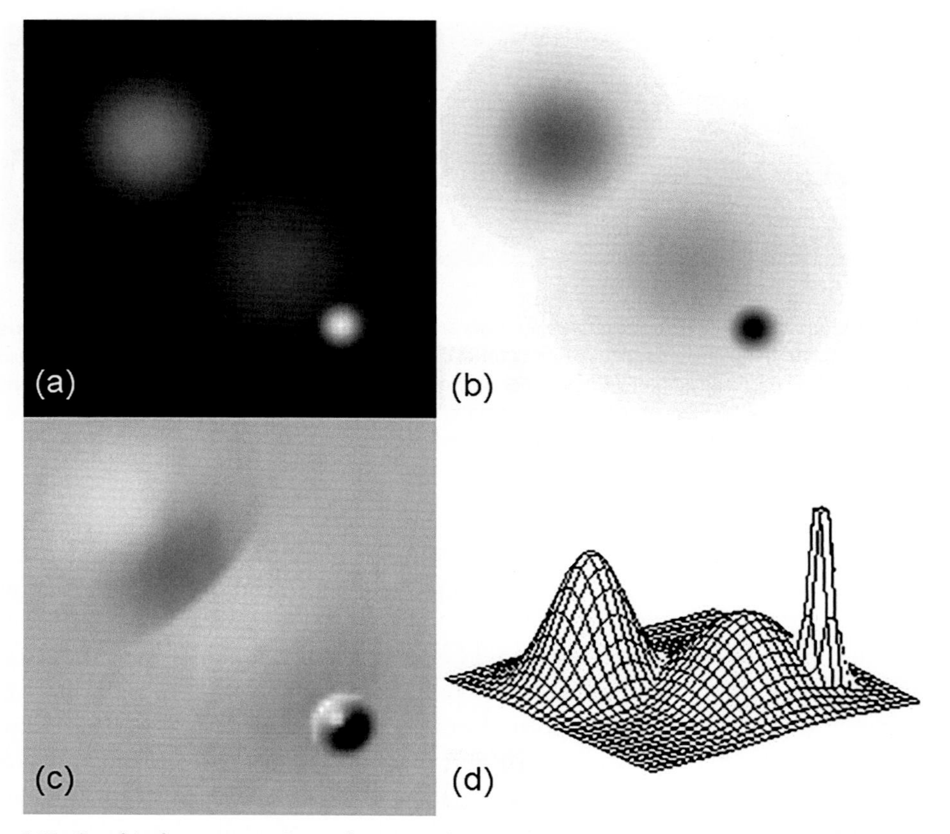

Fig. 6.12 Graphical representations of a grayscale image: (a) grayscale mapping: zero is dark and 255 is bright; (b) reverse grayscale mapping: zero is bright and 255 is dark; (c) top-view shaded surface; (d) surface mesh plot.

6.3.1 Threshold Sets and Level Sets

The *threshold sets* of a grayscale image contain all binary images obtained by thresholding the image at each of the (e.g., 256) gray levels. *Threshold decomposition* of a grayscale image is the process that creates the threshold sets. A *level component* in a grayscale image is a connected set of pixels in a threshold set of the image at a particular gray level. The *level set* for that image is the set of boundaries of all level components present in that threshold set. A grayscale image, then, can be thought of as a *cardboard landscape model*. This is a stack of thin, flat pieces of cardboard, each cut into the shape of a binary object (level component) from the threshold sets. The pieces are then stacked up to form the topography. Within that topography an object is a set of overlapping level components. Each cardboard piece, then, corresponds to a level component of one threshold set for that grayscale image. The image can be characterized uniquely by its threshold sets.

Recovering an image from its threshold sets is called *stack reconstruction*. Note that, for stack reconstruction to be possible, the threshold sets must satisfy the *stack property*. This means that each level component at a gray level must contain the level component above it and must be contained in the level component below.

The threshold decomposition of a gray-scale image, f, is the collection of all the threshold sets, $X_t(f)$, obtained at each gray level t

$$X_t(f) = \{z : f(z) \geq t\} \tag{6.18}$$

The image can be characterized uniquely by its threshold decomposition collection. The image can be recovered from its threshold sets by stack reconstruction.

$$f(x) = \max\{t : x \in X_t(f)\} \tag{6.19}$$

In all the grayscale morphological operations presented herein, we use *flat structuring elements*, that is, structuring elements that have no grayscale variation, and are the same as those used in the case of binary images. We use the term "flat structuring elements" to avoid confusing them with their grayscale analogs. This restriction greatly simplifies the use of grayscale operators as an extension of the binary case. Care must be taken, however, when using a grayscale structuring element, because the erosion (dilation) is not a moving-minimum (moving-maximum) filter and the threshold decomposition property does not hold for the primitive operators, nor does it hold for grayscale morphological reconstruction. As mentioned earlier, only symmetric structuring elements are discussed in this chapter.

6.3.2 Grayscale Erosion and Dilation

Grayscale erosion (*dilation*) of an image, f, by a flat structuring element D is equivalent to a moving-minimum (moving-maximum) filter over the window defined by the structuring element. Thus, erosion $f \ominus D$ and dilation $f \oplus D$ in this case are simply special cases of order-statistic filters:

$$\begin{aligned} (f \ominus D)(x) &= \min\{f(z) : z \in D_x\} \\ (f \oplus D)(x) &= \max\{f(z) : z \in D_x\} \end{aligned} \tag{6.20}$$

An example of erosion by a disk on a grayscale image is shown in Fig. 6.13. The two images on the left, input and eroded, are represented in grayscale, and the two on the right are the same images represented as top-view surfaces. Note how well the term "erosion" fits this example. The eroded surface appears to have been created by a pantograph engraving machine equipped with a flat disk milling cutter. The pantograph follows the original surface while shaping the eroded surface with the flat disk milling cutter.

The intuitive interpretation for grayscale erosion is the following: slide the structuring element along beneath the surface, and at each point record the highest altitude to which

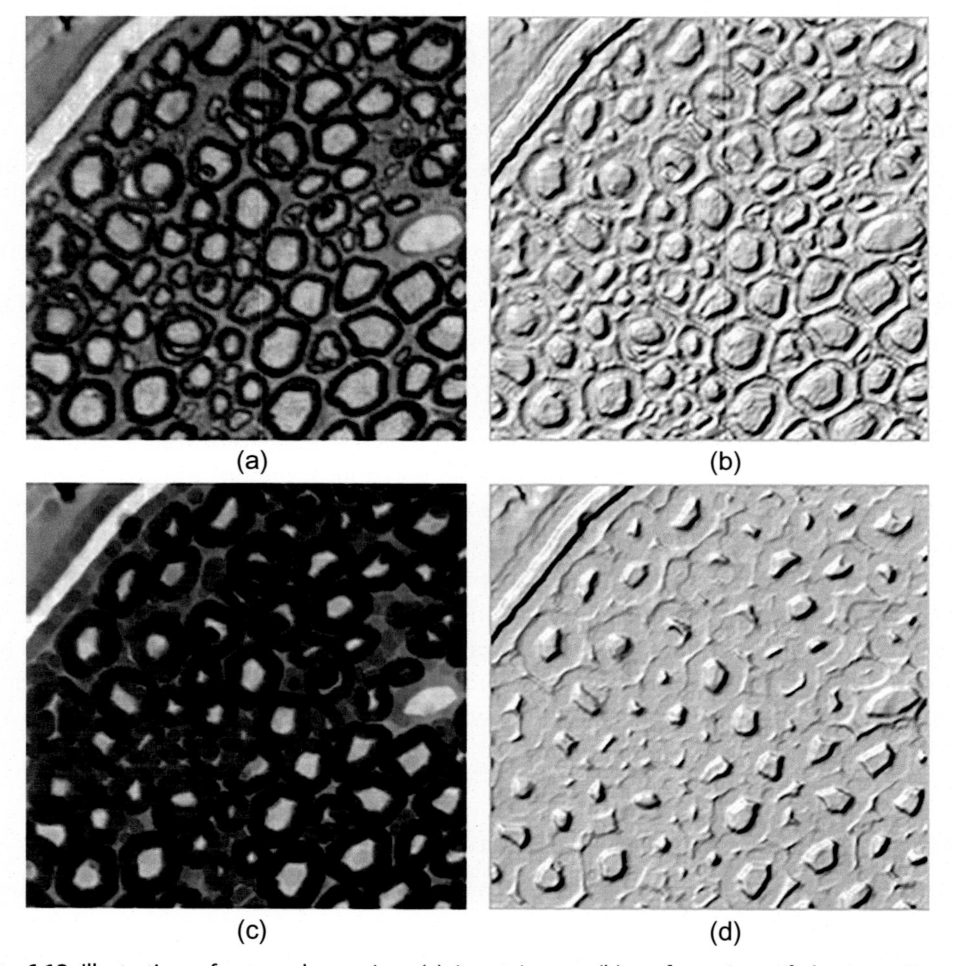

Fig. 6.13 Illustration of grayscale erosion: (a) input image; (b) surface view of the input image; (c) erosion by a disk; (d) surface view of the eroded image.

the structuring element can be translated while still fitting beneath the surface. Alternatively, one can simply compute the erosion (dilation) of a grayscale image by computing the threshold decomposition of the image, applying binary erosion (dilation) to the threshold sets, and following up with stack reconstruction.

Fig. 6.14 illustrates grayscale erosion by means of threshold decomposition. At the right of the grayscale images (original (a) and eroded (e)), there are three threshold sets, at gray levels 80, 120, and 180, respectively. Note that the binary images shown in (f), (g), and (h) are eroded versions of the binary images shown in (b), (c), and (d).

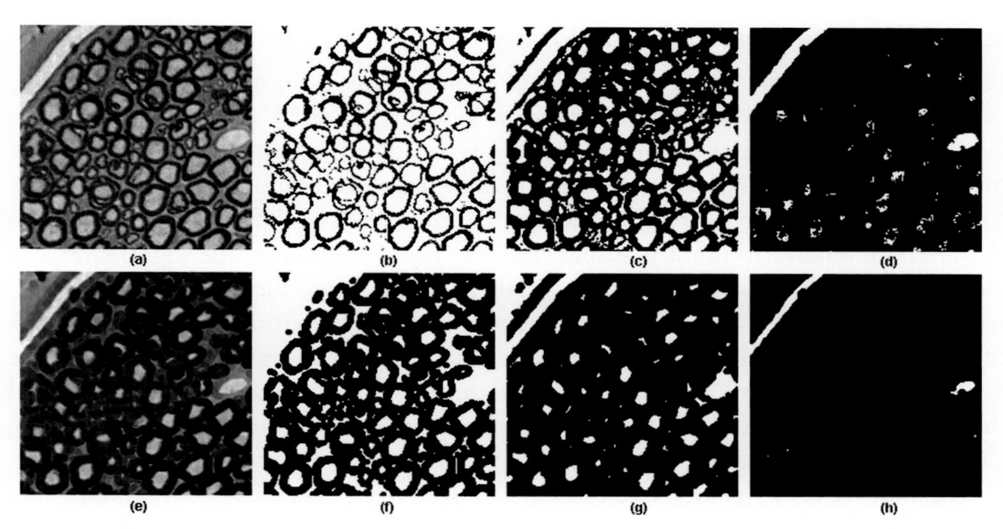

Fig. 6.14 Illustration of grayscale erosion by means of threshold decomposition: (a) the input image; (e) the eroded grayscale image; (b), (c), and (d) show the input image thresholded at 80, 120, and 180 respectively; (f), (g), and (h) show the eroded image thresholded at 80, 120, and 180, respectively.

The filters that can be implemented by threshold decomposition are called *stack filters*. A stack filter can be built from any binary filter as long as it does not prevent stack reconstruction. Dilation and erosion by a flat structuring element are, in this sense, stack filters. So too is the median filter. A practical characteristic of a stack filter is that it stores all results of filtering the input thresholded images. So, when dealing with stack filters, instead of thresholding the image and then applying the filter, one might prefer first to apply the filter, keeping the image in grayscale, and threshold the result later.

6.3.2.1 Morphological Gradient

The morphological gradient (dilation minus erosion), previously described for binary pictures, is directly extensible to grayscale morphology if grayscale erosions and dilations are used. At each point the morphological gradient yields the difference between the maximum and minimum values, over the neighborhood, at the point determined by the flat structuring element.

Grayscale morphological gradient is often used as one step of a more complex process such as segmentation. The segmentation of urology specimens in Fig. 6.32a and the chromosomes in Fig. 6.32b also uses the gradient computation (Figs. 6.32b and 6.29c, respectively).

6.3.3 Grayscale Opening and Closing

As an extension of the binary case, *grayscale opening* (*closing*) can be achieved simply by threshold decomposition, followed by binary opening (closing), and stack reconstruction. The intuitive interpretation for opening is the following: slide the structuring element beneath the surface and, at each point, record the maximum altitude to which the structuring element can be translated while still fitting beneath the surface. The position of the origin relative to the structuring element is irrelevant. Note the slight difference between the opening and the erosion. While in the opening, the maximum altitude is recorded for all points of the structuring element, and in the erosion, only the location of the structuring element is recorded. Fig. 6.28b shows the grayscale opening of the input image (a) by a disk of radius 20. Note the way the white dots are removed, and the image is filtered from the bottom in accordance with the shape of the structuring element.

An intuitive interpretation of closing can be seen from the duality relation. Opening filters the image from below the surface, whereas closing filters it from above. By duality, closing is an opening of the negated image. Hence, to apply closing, simply flip the image upside down (invert), filter by the opening, and then flip it back. The filtering effect of closing is illustrated in Fig. 6.32c, which shows the closing of image (b) by a 7×7 box structuring element.

6.3.3.1 The Top-Hat Concept

Grayscale opening and closing have the same properties as their binary equivalents [7], and the top-hat concept is also valid. *Grayscale opening top-hat* is the subtraction of the opened image from the input image, and *grayscale closing top-hat* is the subtraction of the image from its closing. By choosing an appropriately sized structuring element, one can use the top-hat transform to mark narrow peaks while not marking wider peaks in the image. In some applications it is impossible to separate desirable from undesirable bright spots simply by using an appropriately sized structuring element, but it is possible to separate them by an appropriately chosen threshold.

The following application illustrates the use of grayscale closing and grayscale closing top-hat. The input is a grayscale image of a microelectronic circuit. In this image irregularities in the vertical metal stripes are to be detected. This procedure uses the grayscale closing top-hat, followed by filtering (by size threshold) of the residues. The top part of Fig. 6.15 shows the grayscale images, while the lower part shows their surface views. The input image is shown in (a), and its closing, by a vertical line of length 25 pixels, appears in (b). Then the top-hat result is the subtraction of the original from the closing (c), revealing dark defects in the metal stripes. It shows the discrepancies of the image where the structuring element cannot fit the surface from above. In this case, it highlights vertical depressions longer than 25 pixels. Thresholding the top-hat image, followed by the

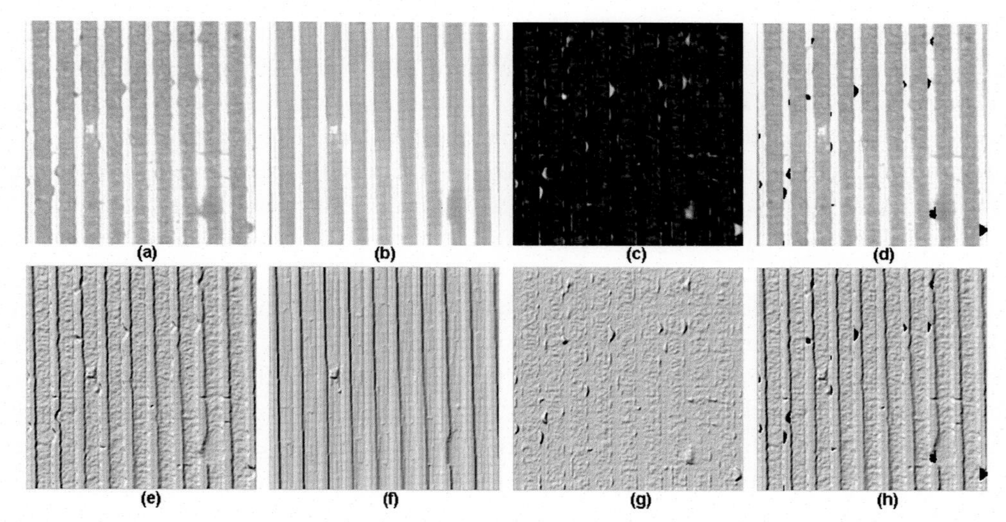

Fig. 6.15 Illustration of grayscale closing and closing top-hat: (a) input image; (b) closing by a vertical structuring element; (c) closing top-hat; (d) thresholded top-hat (*black* areas) overlaid on original; (e)–(h) surface view of the preceding images.

elimination of small objects by an area opening of 5 pixels, results in the detected regions with irregularities that are pointed by arrows overlaid on the original image (d).

Open top-hat is very useful as a preprocessing step to correct uneven illumination encountered in fluorescence microscopy (see Chapter 10) before applying a threshold, thereby implementing an adaptive thresholding technique (see Chapter 7). The following application illustrates this. For a real-world biological application, we consider fluorescent in situ hybridization (FISH) imaging, which is discussed in Chapter 10. A DNA probe labeled with a fluorophore hybridizes to a matching sequence of DNA in the cell, and the dye fluoresces at a particular wavelength when excited by illumination in a microscope.

Fig. 6.16 shows the open top-hat transform applied to a FISH image: (a) the FISH image; (b) open top-hat of the FISH image by a disk of radius 4; (c) binary image resulting

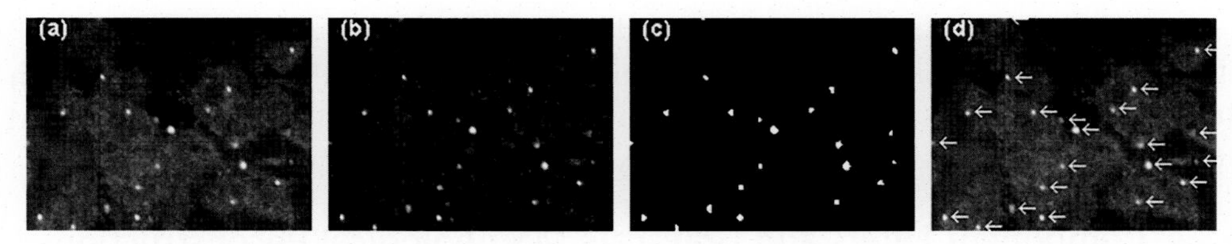

Fig. 6.16 Grayscale open top-hat example: (a) input image; (b) opening top-hat by a disk of radius 4; (c) thresholded area open (by 2 pixels); (d) dilation of centroids by arrow, for illustration, overlaid on original.

from thresholding the top-hat image at a gray level of 50; and (d) final result with an arrow indicating the position of each detected spot. Due to noise, the top-hat methodology typically yields a number of very small extraneous spots in the thresholded top-hat image. These can be eliminated by filtering the image with a grayscale area opening operation of two pixels. The arrows were overlaid automatically by a dilation of the centroids of the detected spots with an arrow-shaped structuring element having its origin translated slightly from the arrow tip so as not to disturb the visualization of the original spots in the image.

An illustration of the detection of thin structures using the closing top-hat can be seen in Figs. 6.24c and d. In (c) we see crystals surrounded by a dark contour and a bright oriented shade. The application of a closing top-hat by a disk with diameter larger than the thickness of the dark contours (see Fig. 6.24c) removes the bright shade, leaving only the bright contours of the crystals. To detect both peaks and valleys, one can apply the open top-hat transform, threshold to find peak markers, apply the close top-hat transform, threshold to find valley markers, and then form the union of the two marker images.

6.3.3.2 Grayscale Image Filtering

Grayscale opening can be employed to filter positive noise spikes from an image, and closing can remove negative noise spikes. Typically, one encounters both, and, as long as the noise spikes are sufficiently well separated, they can be suppressed by an opening and closing or by a closing and opening operation. However, selection of an appropriate structuring element size is crucial. If the spacing between noise spikes is uneven, and they are not sufficiently separated, one can employ an *alternating sequential filter* (ASF). This is a sequence of alternating opening and closing operations with increasingly larger structuring elements.

Fig. 6.17 shows an example of grayscale image filtering using ASF. A single stage close-open filter (shown in (b)) is the result of the closing followed by an opening using a 3×3 diamond-shaped structuring element. For the second stage, another close-open operation is concatenated using a 5×5 diamond structuring element. In (c) a three stage ASF was applied with the last stage being processed by a 7×7 diamond structuring element.

For correction of uneven illumination, the background can be estimated by an ASF filter. An example of this technique appears in Fig. 6.24. Part (a) is the input image, which shows a strong uneven illumination component. In part (b) the background is estimated by a 10-stage close-open ASF using a family of different sized octagonal disks.

6.3.4 Component Filters and Grayscale Morphological Reconstruction

The concept of a connected component filter that was introduced in the binary morphology section can be extended to grayscale morphology. Such a filter can be constructed

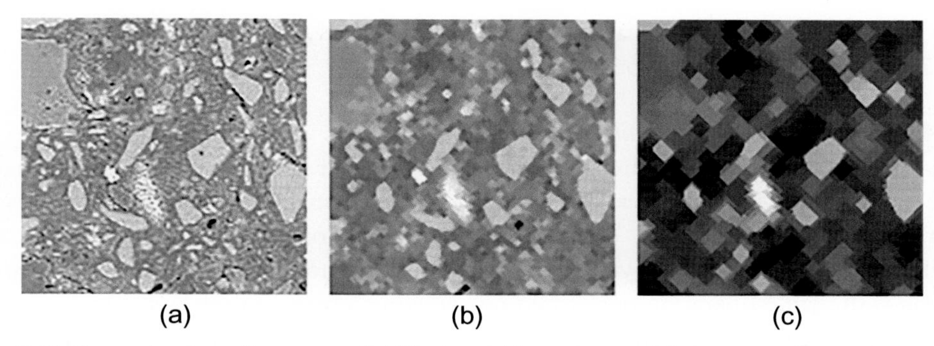

Fig. 6.17 Grayscale alternating sequential filtering: (a) input image; (b) close-open by an elementary cross; (c) ASF close-open with three stages.

from (1) reconstruction from markers operations, (2) reconstruction from opening operations, and (3) area opening operations. These grayscale operators can be constructed from their corresponding binary counterparts by using threshold decomposition, described earlier (Section 3.1). A grayscale component filter is an operator that removes only a few level components (cardboard pieces) in such a way that the stack reconstruction property is not violated, that is, a level component is removed only if all the level components above it are also removed. One important property of a component filter is that it never introduces a false edge, so it is one of a family of edge-preserving smoothing filters. Recall that the definition of component filters requires the specification of a connectivity convention, which can be either 4- or 8-neighbor connectivity.

6.3.4.1 The Reconstruction Process

Morphological reconstruction is one of the most often used tools to build component filters [11–13]. As with binary reconstruction, grayscale reconstruction proceeds from markers. The morphological reconstruction of an image from a marker can be obtained by (1) threshold decomposition of the image and the marker, followed by (2) binary reconstruction from markers done at each gray level, and (3) stack reconstruction of the results. This can be interpreted intuitively by using the cardboard landscape model of the image. Imagine that the cardboard pieces are stacked, but not glued. The markers are seen as needles that pierce the model from bottom to top. If one shakes the model while holding the needles, those cardboard pieces not pierced by the needles will be lost. The remaining cardboard pieces constitute a new model, possibly with fewer objects, and that corresponds to the final result of grayscale morphological reconstruction. Note that the marker can be a grayscale image, with the pixel gray level corresponding to the height that the needles reach as they pierce the model. This process is also called inf-*reconstruction*. By duality, the sup-*reconstruction* works in the same manner on the complemented image.

The names of these operations are derived from the words *supremum* (least upper bound) and *infimum* (greatest lower bound).

6.3.4.2 Grayscale Area Opening and Closing

Grayscale area opening is another type of component filter [14]. It is defined analogously to the binary case. The size-*a* area opening of a grayscale image can be modeled as a stack filter in which, at each level, only level components containing at least *a* pixels are retained. This operation removes all cardboard pieces whose area is less than *a*. An area closing is the same operation performed on the complement image.

An important class of component filters is composed of those generated from alternating reconstructions from openings and closings. These are called *alternating sequential component filters* (ASCFs). Fig. 6.18 shows two examples of grayscale ASCF using, as input, the image shown in Fig. 6.17. A three-stage close-open filter (shown in (b)) is the result of the sup-reconstruction from closing followed by a reconstruction from opening using a 3×3, 5×5, and 7×7 diamond structuring element in the first, second, and last stage, respectively. In (c) an ASCF area close-open operation is applied using 30 pixels as the area parameter. One can compare the fidelity of the edges between the results of this component filter and the results of the ASF filters shown in Fig. 6.17.

As with the binary case, in the morphological reconstruction from markers, there are mainly three ways to design the marker image: (a) with an a priori selection, (b) using selection from opening (grayscale reconstruction from opening) [8], and (c) determination from more complex processing (see the following).

6.3.4.3 Edge-Off Operators

The grayscale edge-off operator can be easily derived from the binary case, and it is useful in many situations. As with the binary case, the edge-off operator is the top-hat of the reconstruction from a marker placed at the image border (i.e., an example of the case

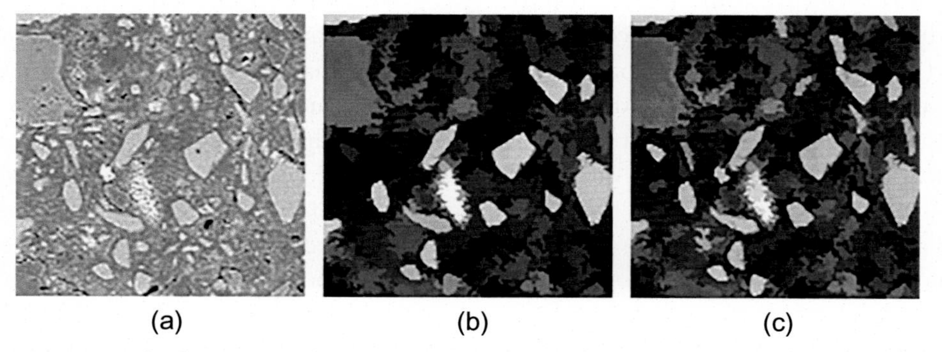

Fig. 6.18 Grayscale alternating sequential component filtering: (a) input image; (b) reconstructive close-open by a 3×3 diamond structuring element of stage 3; (c) area close-open with area parameter of 30 pixels.

where the marker is placed a priori). In the cardboard landscape model (threshold decomposition), all of the cardboard pieces that touch the image border are removed, leaving only those cardboard pieces that form domes inside the image.

The following application illustrates the use of area close-open ASCF as a preprocessing filter followed by the edge-off operator to enhance pollen grains. It is known a priori that the pollen grains have areas ranging from 5,000 to 150,000 pixels at the resolution of these images. Fig. 6.19a shows the input image containing two grains. Fig. 6.19b is the area close-open ASCF. It was used with area parameter 1,500 pixels, first filling holes of less than this size and then removing level components with area less than 1,500 pixels. Note that there are deep dark areas around the pollen grains, and these make it useful to apply the edge-off operator, shown in Fig. 6.19c. The process of enhancing the pollen grain images was done entirely in the grayscale domain. Finally, a segmentation was obtained by thresholding Fig. 6.19c at a gray level of 5 (Fig. 6.19d).

6.3.4.4 h-*Maxima and* h-*Minima Operations*

The reconstruction of an image, f, after subtracting h from itself, is called the *h-maxima* operation.

$$HMAX_{h,D}(f) = f\,\Delta_D(f-h) \tag{6.21}$$

This is a component filter that removes any object with height less than or equal to h, and it decreases the height of the other objects by h. The intuitive interpretation of the h-maxima operation based on the threshold decomposition concept is that the height attribute associated with a particular level component (cardboard piece) is one plus the maximum number of levels that exist above it in that object. The h-maxima filter

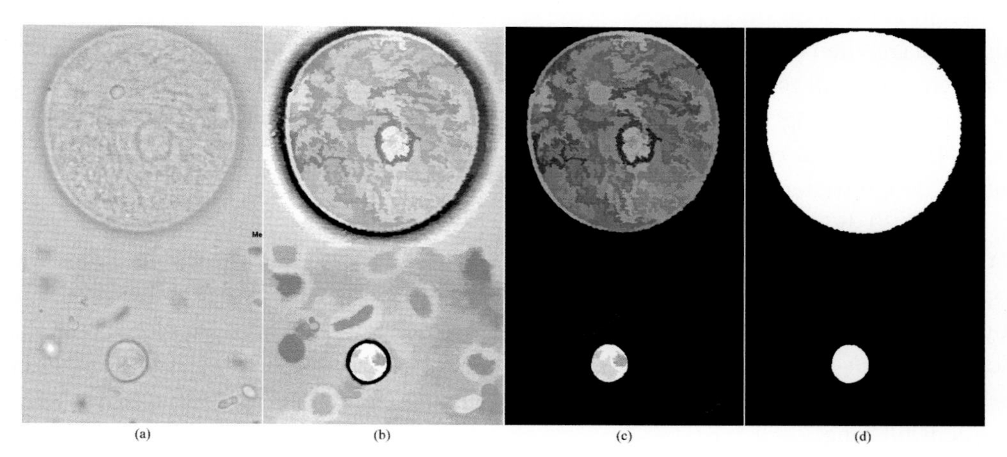

Fig. 6.19 Segmenting pollen grains with grayscale component filters: (a) input image; (b) area close-open ASCF; (c) edge-off operation; (d) thresholding.

removes all the cardboard pieces with height attribute less than or equal to h. The dual operator of h-maxima is called *h-minima*, $HMIN_{h,\,D}(f)$. It fills in any basin with depth less than h and decreases the depth of the other basins by h.

6.3.4.5 Regional Maxima

Considering the threshold decomposition of a grayscale image, *regional maxima* are level components with height attribute equal to 1, i.e., there are no other level components above them. These are the cardboard pieces at the top of their respective peaks. For instance, Fig. 6.31c shows the regional maxima of the image in Fig. 6.31b. All regional maxima can be found by subtracting the h-maxima with $h=1$ from its original image, f,

$$RMAX_D(f) = f - HMAX_{1,D}(f) \tag{6.22}$$

By duality, a *regional minimum* is a flat connected region that is at the bottom of a basin. Regional maxima and minima are generically called *extrema* of the image.

$$RMIN_D(f) = HMIN_{1,D}(f) - f \tag{6.23}$$

6.3.4.6 Marker Extraction

The watershed transform, described in Section 6.4, is an important morphological segmentation process. One of the crucial steps in watershed transform segmentation is marker extraction. A marker must be placed inside every object that needs to be extracted. The regional maxima (minima) can be used as markers for watershed segmentation. Marker finding, using the regional maxima, is most effective when done on filtered images. One advantage of this method is its independence of any grayscale thresholding values.

Typically, an image presents a large number of false regional maxima because of noise inherent in the acquisition process. If the regional maximum operator is applied to a gradient image, then the situation is even worse. Filtering the image with a component filter can remove small regional maxima, which are most likely due to noise. This is usually accomplished using (1) opening, (2) reconstruction from opening, (3) area opening, and (4) h-maxima, or combinations thereof. The choice of which filter to use is a part of the design strategy.

Fig. 6.20 shows the regional maxima of an input image following four different filters: (a) input image; (b) regional maxima without filtering; (c) regional maxima following opening by a disk of radius 3; (d) regional maxima following reconstruction from opening by the same disk; (e) regional maxima following area opening; and (f) regional maxima following an h-maxima operation. Note how the oversegmentation (excessive number of separate markers) produced by the direct regional maxima in (b) is reduced by the subsequent filtering.

In the next section these markers are used to segment an image using the watershed transform (see Fig. 6.23). Analogously, in Figs. 6.25c and 6.24e, the h-minima operator is

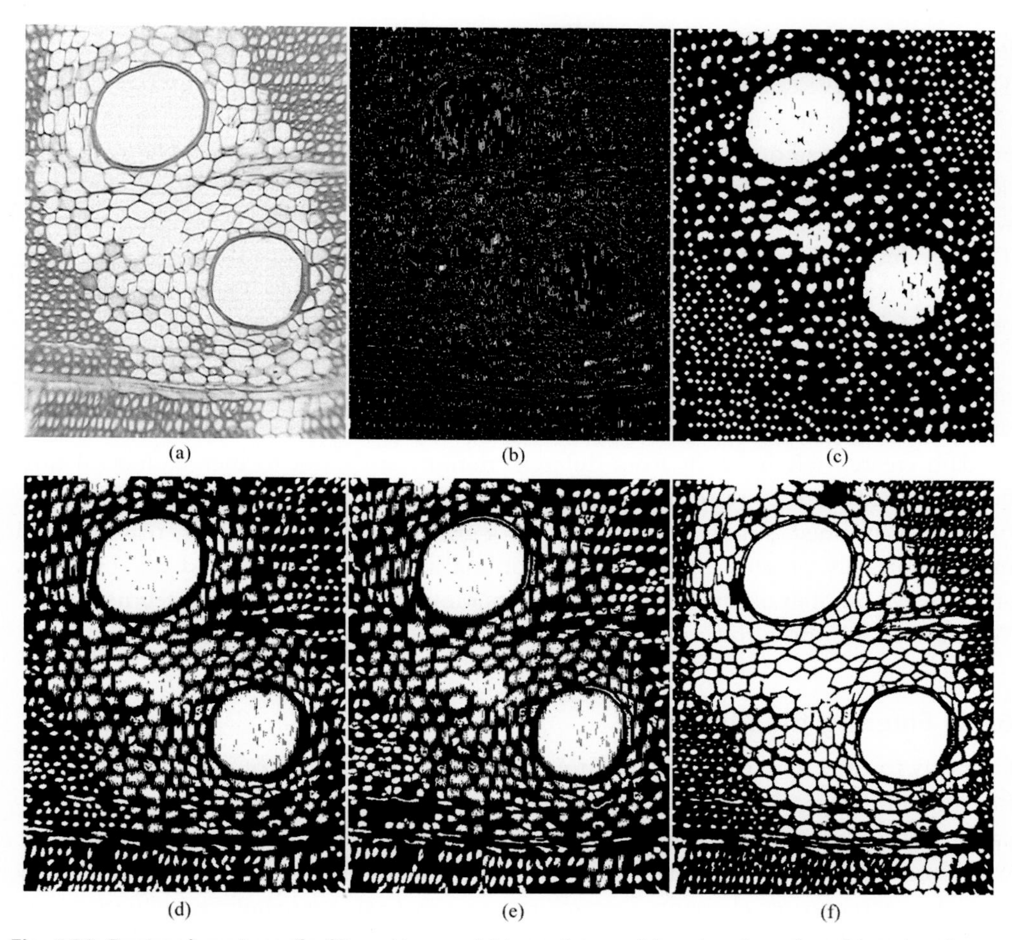

Fig. 6.20 Regional maxima of a filtered image: (a) input image; (b) regional maxima; (c) regional maxima after opening by a disk of radius 3; (d) regional maxima after reconstruction from opening by the same disk; (e) regional maxima after an area open of 100 pixels; (f) regional maxima after *h*-maxima filtering with $h = 20$.

used to filter minima in the images of Figs. 6.25b and 6.24d, respectively, prior to the watershed transform.

6.4 Watershed Segmentation

The watershed transform is a key building block for the morphological segmentation of images [15]. In particular, the watershed transform is applied to the morphological gradient of the image to be segmented. A number of different watershed segmentation algorithms have been implemented [16,17].

6.4.1 The Classical Watershed Transform

The most intuitive description of the *watershed transform* is based on a flooding simulation. Consider the input grayscale image as a topographic surface (recall Fig. 6.12). A hole is punched at each regional minimum in the image. The topography is slowly flooded from below by allowing water to rise through these holes at a uniform rate. When the rising water from two distinct minima is about to merge, a dam is built to prevent the merger. The flooding will eventually reach a stage where only the tops of the dams are visible above the water's surface, and these establish the *watershed lines*. The final segmented regions arising from the various regional minima are called *catchment basins*.

Fig. 6.21 illustrates this flooding process on a one-dimensional signal with four regional minima generating four catchment basins. The figure shows some steps of the process: (a) input image, (b) holes punched at minima and initial flooding, (c) dam created when waters from different minima are about to merge, and (d) final flooding yielding three watershed lines and four catchment basins. For image segmentation, the watershed is most often applied to a gradient image. However, real digitized images normally present many regional minima in their gradients, and this produces an excessive number of catchment basins, a situation called watershed oversegmentation. We now address how this can be prevented.

6.4.2 Filtering the Minima

One way to cope with oversegmentation is to reduce the number of regional minima by filtering the image, thereby creating fewer catchment basins. Fig. 6.22 shows the application of the classical watershed transform. The input image (a) is a small synthetic image with three different size Gaussian peaks. Part (b) shows its morphological gradient using a 3×3 box-structuring element. Part (c) shows a typical watershed result with oversegmentation due to spurious regional minima, each one giving rise to a separate catchment basin. By filtering the gradient image with the h-minima operator, with $h=9$, the watershed gives the desired result, shown in Fig. 6.22d. This filtering is effective but very subtle as the spurious regional minima that are eliminated are quite small.

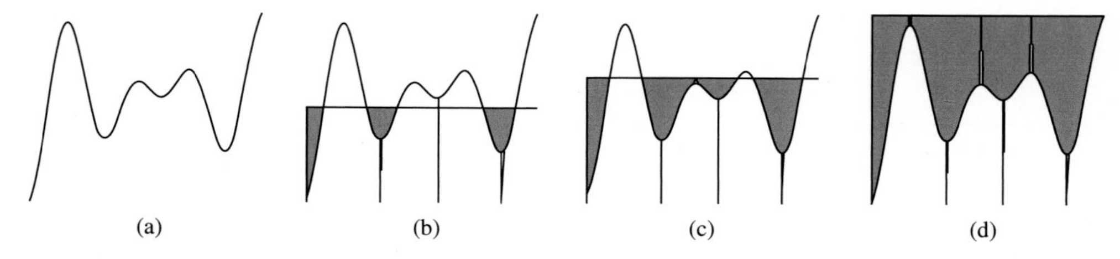

Fig. 6.21 Flooding simulation of the watershed transform: (a) input signal; (b) punched holes at minima and initial flooding; (c) a dam is created when waters from different minima are about to merge; (d) final flooding, three watershed lines and four catchment basins.

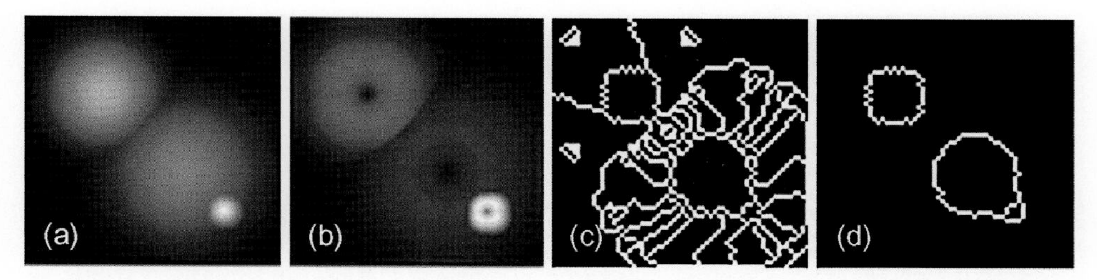

Fig. 6.22 Classical watershed segmentation with regional minimum filtering: (a) small synthetic input image (64×64); (b) morphological gradient; (c) watershed on the morphological gradient; (d) watershed on the h-minima ($h = 9$) filtered morphological gradient.

Fig. 6.23 illustrates reducing watershed oversegmentation by reducing the number of minima in the input image with several different filters. In Fig. 6.23 (a) is the input image; (b) is the watershed of (a) without filtering; (c) is when it is filtered with a closing operation with a disk of radius 3; (d) is when it is filtered with sup-reconstruction from closing with the same disk; (e) is when it is filtered with area closing; and (f) is when it is filtered with the h-minimum operator. This example is equivalent to the regional maxima simplification shown in Fig. 6.20. If we compute the regional minima in the filtered images, we get the same results as that figure. Note that to filter regional minima, we use filters that operate on the valleys, such as closings and h-minima operators. Applying filters that operate on peaks does not reduce the number of minima or the number of catchment basins found by the watershed algorithm.

The next application we consider is an electron micrograph of silver-halide T-grain crystals embedded in emulsion. Automated crystal analysis involves segmentation of the grains for size measurement. This segmentation problem has several things that make the segmentation challenging. The image has a strong illumination gradient, the interior gray level values for the crystal grains are the same as the background, the image has strong white "shadow" noise, it has a wide range of grain sizes, and there are overlapping and touching grains. Despite all these problems, watershed segmentation can produce very good results. The two key points here are the background correction and the enhancement of the dark contours using a close top-hat operation with a disk diameter larger than the thickness of those contours. The crystals are surrounded by dark contour lines that are used for the watershed-based segmentation. The preprocessing stage is intended to enhance only those lines.

The input image is shown in Fig. 6.24a. The illumination gradient is estimated with a 10-stage alternating sequential filtering using an octagon structuring element in (b). The input image is then divided by this background estimate, normalized by its minimum value in (c). This division ensures that the dark contours have the same depth, so in the dark regions of the image, the depths of the dark contours are increased by this

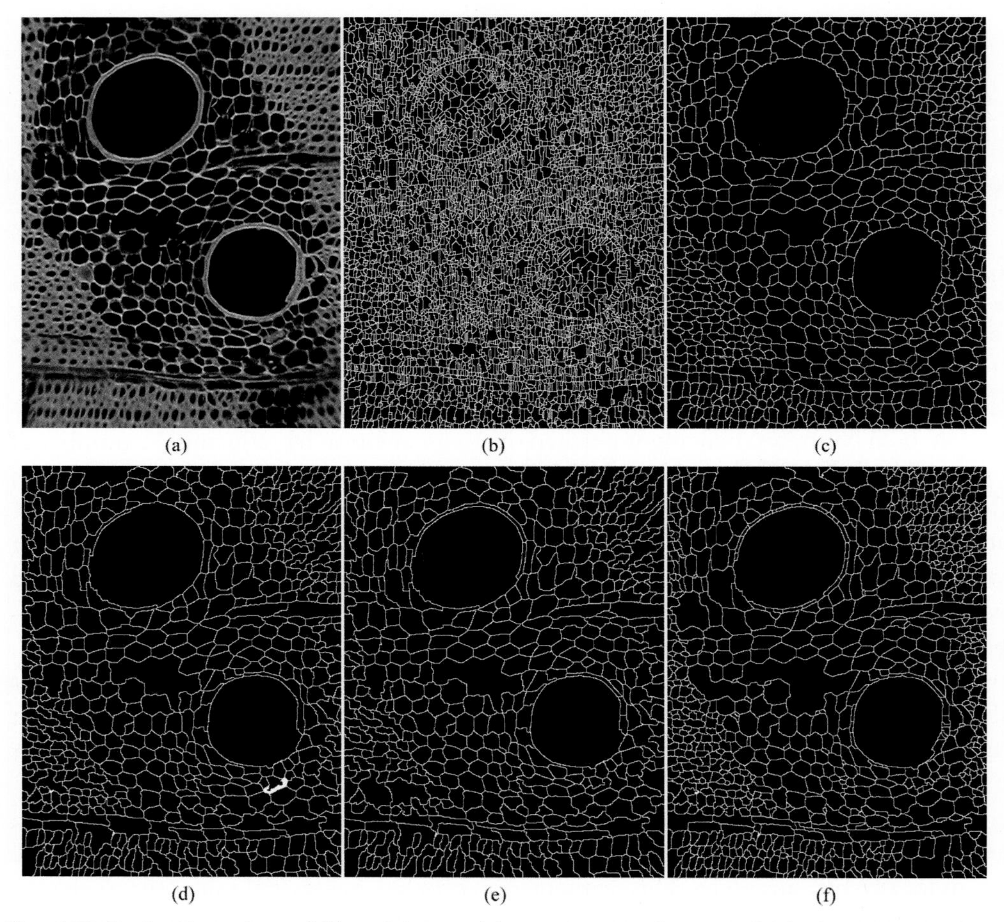

Fig. 6.23 Regional maxima of filtered image: (a) input image; (b) watershed of the input image; (c) watershed of the input image after closing with a disk of radius 3; (d) watershed of the input image after sup-reconstruction from closing by the same disk; (e) watershed of the input image after area open of 100 pixels; (f) watershed of the input image after *h*-minima filtering with $h = 20$.

procedure. This is necessary to yield uniform segmentation when applying the *h*-minima filtering later. A classical closing top-hat filter detects the dark contours in (d). The size of the structuring element must be larger than the thickness of the dark contours. Note that the white areas of the image do not affect the detection of the dark contours.

The watershed transform detects the dividing lines between the dark regions. In this case, it is not necessary to compute the gradient since the contour lines have already been enhanced. We apply an area closing followed by an *h*-minima operation in (e).

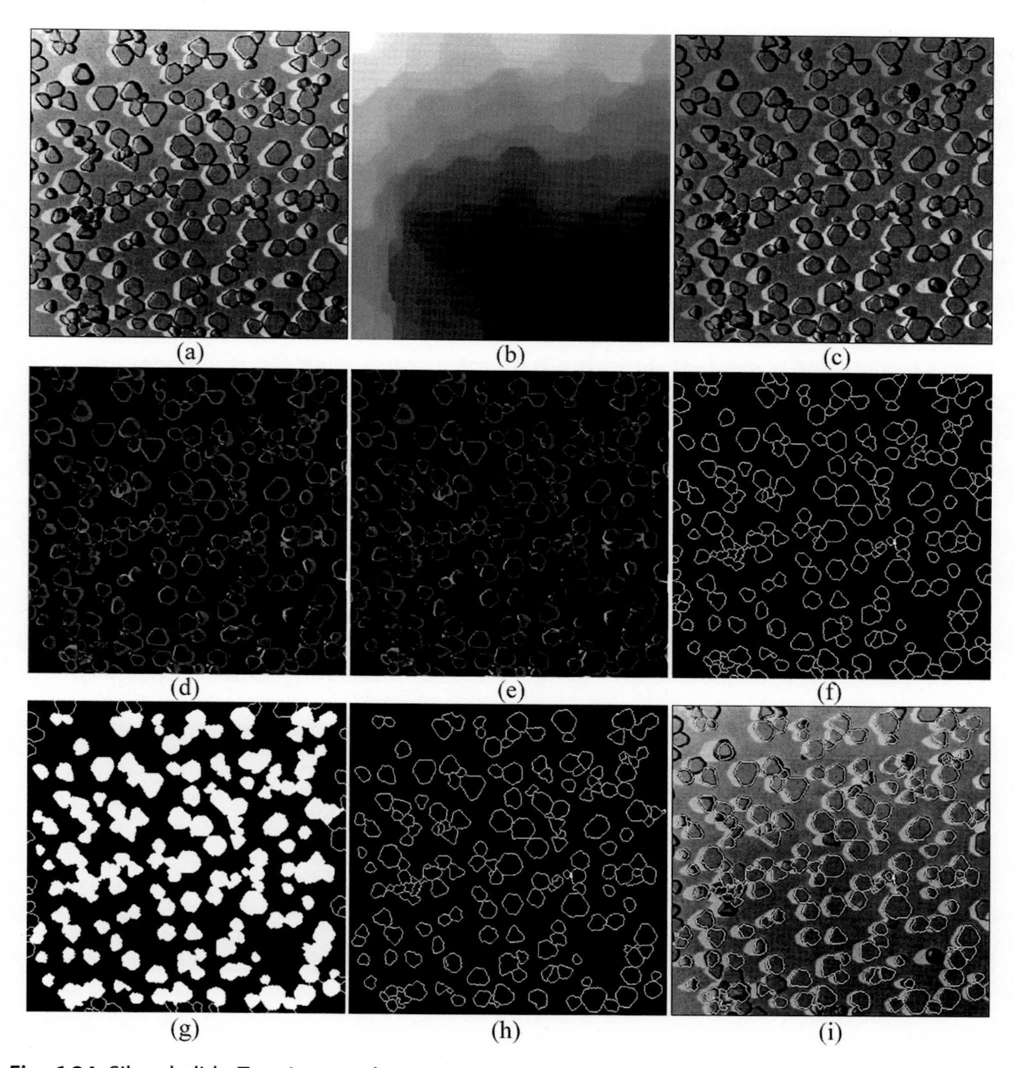

Fig. 6.24 Silver-halide T-grain crystal segmentation: (a) input image; (b) alternating sequential filter applied; (c) input image divided by (b); (d) contour enhancement; (e) *h*-minima filter; (f) watershed; (g) removal of grains touching the border; (h) watershed lines not touching the border; (i) final result.

The choices of parameters for the filters used in this example are critical and have been found by trial and error. Application of the watershed on the simplified contour image gives the watershed lines shown in (f). Grains connected to the image border are removed in (g). After that, the intersection of the watershed lines and the grains not connected to the image border gives the final boundaries of grains not touching the border in (h). For display purposes, the contours are overlaid on the original image.

6.4.3 Texture Detection

Oversegmentation, usually seen as a drawback of the watershed transform, can be useful to separate homogeneous from textured regions of an image. Indeed, the watershed transform will create large catchment basins in rather homogeneous regions and very small catchment basins in textured regions.

The following example, illustrated in Fig. 6.25, segments epithelial cells in a monochrome image. The cells are nearly transparent, so the approach is to use the watershed to detect small catchment basins due to the texture present in the interior of these cells. The catchment basins are area-closed and then opened to remove noise. Although the oversegmentation of the watershed algorithm is useful, a component filtering operation is still required to detect only those catchment basins inside the cells. An h-minima operation on the gradient is used. In Fig. 6.25, (a) is the original image containing the almost transparent target cells; (b) is the morphological gradient; (c) shows the watershed lines of the h-minima filtered gradient; (d) shows the merging of small catchment basins by area closing; (e) is the result of the area opening to detect large objects; and in (f) the contours (gradient) of segmented objects are overlaid on the input image.

The example that follows shows an image analysis technique to detect anhydrous phase and aggregate in a polished concrete section, imaged by the scanning electron microscope (SEM) in Fig. 6.26a as homogeneous medium-gray grains. The steps in this

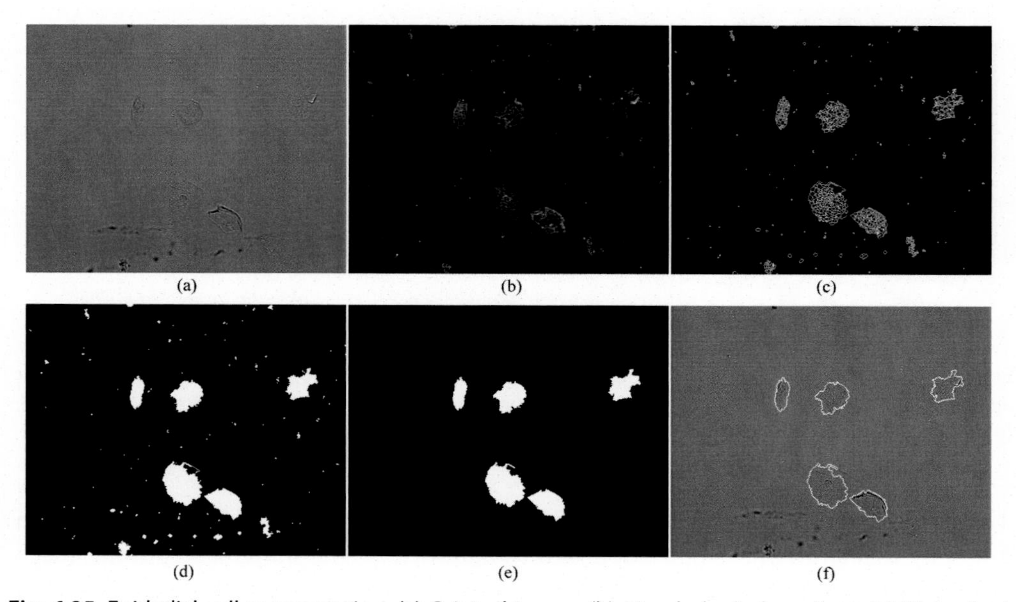

Fig. 6.25 Epithelial cell segmentation: (a) Original image; (b) Morphological gradient; (c) Watershed lines (with oversegmentation) on h-minima ($h=6$) filtered gradient; (d) Area closing to merge small catchment basins; (e) Area opening to select only the large objects; (f) Outline (in *white*) of the segmented objects.

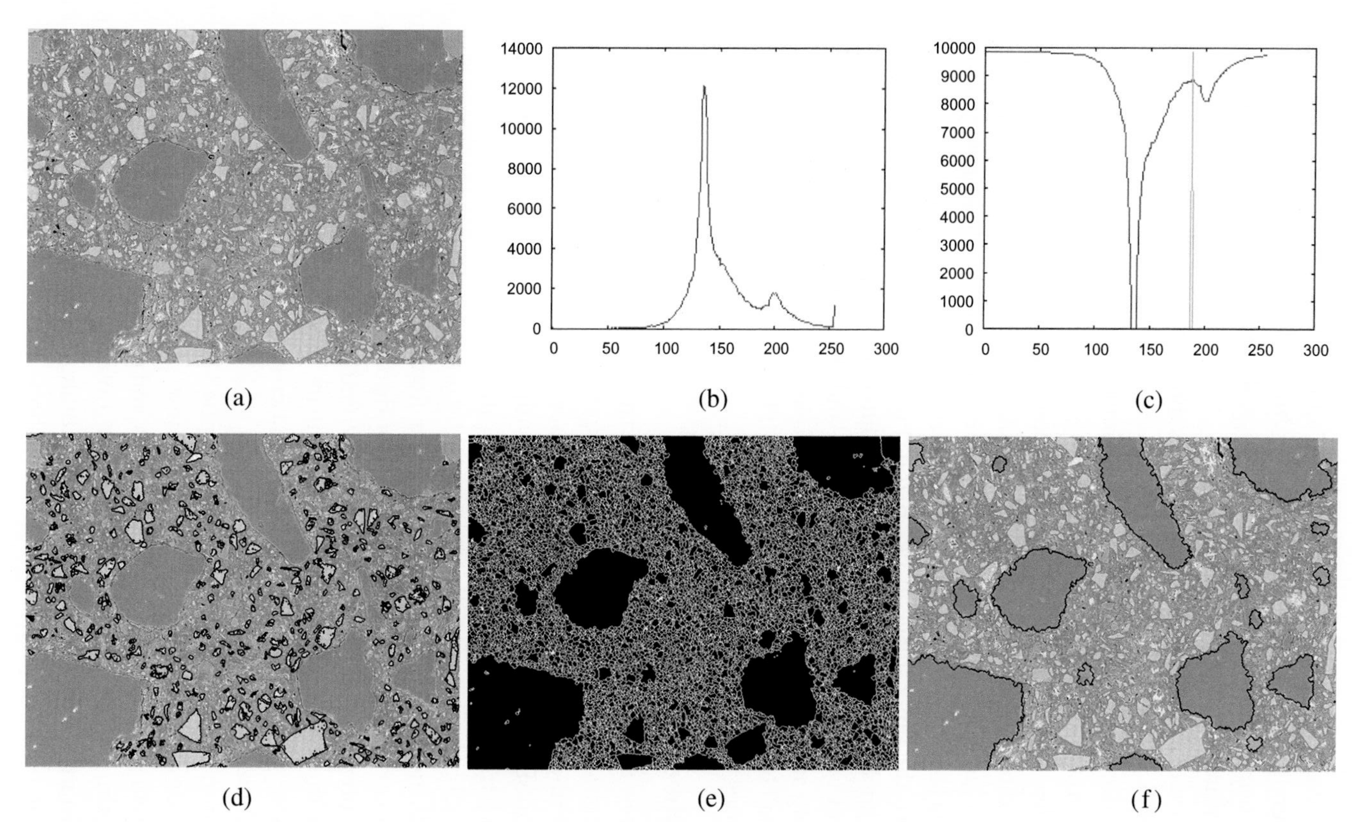

Fig. 6.26 SEM image of aggregate and anhydrous phase extraction from a concrete section: (a) input image; (b) gray level histogram; (c) automatic threshold from histogram using watershed; (d) contour (in *black*) of the anhydrous regions from automatic thresholding; (e) watershed lines from filtered regional minima of the gradient; (f) contour (in *black*) of the aggregate regions obtained from area open-close of the watershed.

analysis are: (1) anhydrous detection by automatic threshold analysis, (2) homogeneous grain detection using the watershed technique, and (3) identification of aggregates as homogeneous grains that are not from the anhydrous phase. The automatic threshold analysis is done using one-dimensional morphological processing of the gray level histogram with the watershed algorithm.

The histogram in Fig. 6.26b has a small peak in the white region due to the anhydrous phase. The threshold value is automatically determined with the 1D watershed technique. This is done by computing the one-dimensional watershed on the filtered, negated histogram. The filter is based on a closing of 5 points followed by an h-minima operation with $h=10$ (see Fig. 6.26c). The threshold parameter is taken as the position of the watershed point, and it corresponds to the minimum between the two peaks.

Anhydrous regions are detected by applying the automatic threshold and by removing objects with area less than 20 pixels. Their contour, computed from the gradient, is overlaid (in black) on the input image in Fig. 6.26d. Fig. 6.26e shows the watershed applied on the filtered gradient of the input image. The filter is the h-minima with $h=10$. The larger catchment basin regions are the aggregate and the anhydrous. These regions are filtered out using an area opening of 300 pixels followed by an area closing of 50 pixels to eliminate small holes. The aggregate, contoured in black in (f), is obtained by removing the anhydrous phase already computed.

6.4.4 Watershed From Markers

The *watershed from markers* technique is a very effective way to reduce oversegmentation if one can place markers within the objects to be segmented. The watershed from markers can also be described as a flooding simulation process. In this case, holes are punched at the marker locations. Each marker is associated with a color. The topography is flooded from below by letting colored water rise through the hole associated with its color, this being done for all holes at a uniform rate across the image. If the water reaches a catchment basin with no marker in it, then the water floods that catchment basin without restriction. However, where the rising waters of distinct colors are about to merge, a dam is built to prevent the merger. The final colored regions are the catchment basins associated with the various markers. To differentiate these marker-derived catchment basins from those obtained with the classical watershed transform, we call the latter "primitive catchment basins."

Fig. 6.27 illustrates the flooding of the watershed from markers in one dimension. There are markers placed into the two rightmost primitive catchment basins. Part (a) shows the two holes punched at the markers and some initial flooding. When the water rises, a primitive catchment basin without a marker is flooded without creating a dam, as shown in part (b). In part (c), a dam is built to prevent the merging of waters coming from the two markers. Finally, part (d) shows the final flooding with only one watershed line separating the two marked regions.

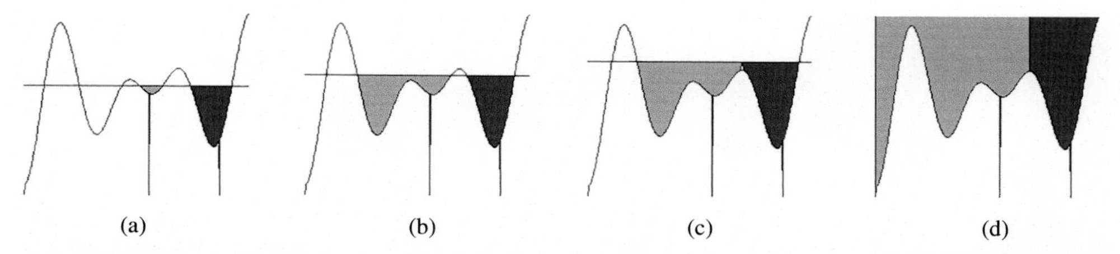

Fig. 6.27 Flooding simulation of the watershed from markers: (a) punched holes at markers and initial flooding; (b) flooding a primitive catchment basin without a marker; (c) a dam is created when waters coming from different markers are about to merge; (d) final flooding, only one watershed line.

The classical watershed transform can be constructed using the watershed from markers technique, and vice versa. If we place the markers at all regional minima of the input image, then the watershed from markers technique gives the classical watershed transform result. To obtain the watershed from markers result from the standard watershed transform, we must apply the classical watershed transform to the sup-reconstruction of the image from the markers.

6.4.5 Segmentation of Overlapped Convex Cells

Application of the watershed transform often involves a certain spatial decomposition. Given a set of isolated points (grains) in a binary image, its *Voronoi diagram* is composed of lines that partition the plane into regions, each consisting of the points that are closest to one particular grain. More generally, the grains can consist of connected components of arbitrary sets, instead of isolated points. In this case, the Voronoi regions are called "influence zones," and the Voronoi diagram is called a *skeleton by influence zones* (SKIZ).

In a binary image, the *distance function* or *distance transform* is defined, for any point inside an object, as the distance from that point to the nearest point outside the object. The watershed transform is a useful method to compute the Voronoi diagram and the SKIZ. The idea is to compute the classical watershed transform of the distance transform of the background of the objects. The catchment basins are the influence zones, and the watershed lines compose the SKIZ.

This concept can address one of the earliest uses of the watershed transform, the problem of binary-image segmentation of images with touching and overlapping objects. For instance, in the segmentation of Fig. 6.28c, there appear to be seven cells overlapping to form a single connected component. Our goal is to segment that component in a manner consistent with the integrity of each cell. A key to this problem, and with many segmentation problems, is to find markers for each of the objects.

In Fig. 6.28, after the input image (a) is filtered using an opening with a disk (b), the image is thresholded in (c). Using the watershed transform to segment the overlapped cells works in the following way. The distance transform is computed on the binary

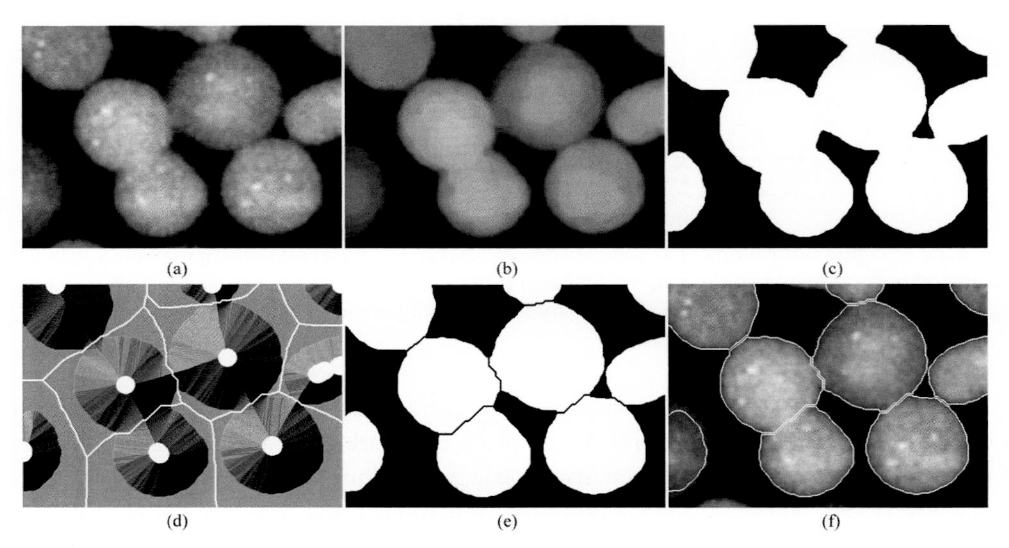

Fig. 6.28 Segmentation of overlapping convex cells: (a) input image; (b) preprocessing by opening with a disk of radius 20; (c) threshold of the preprocessed image; (d) distance transform of complement image (in *gray*) overlaid by marker *(white)* and watershed lines *(white)*; (e) watershed lines used to cut apart the binary overlapped cells; (f) cell boundaries overlaid on the input image.

image, and one marker is required for each cell. In the case of rounded cells, these markers can be extracted from the regional maxima of the distance function. Depending on the difficulty of this extraction, it may require filtering the distance function. This can be done using an opening according to the methodology of marker extraction using regional maxima, which is discussed in the previous section. In this case the distance function was filtered by an opening with a disk of radius 15. The lines from the watershed transform on the negated distance function from the markers are used to cut the input binary image (e). The shaded distance function, displayed in part (d) of Fig. 6.28, is overlaid with the marker (in white) and the watershed lines (in white). In part (f), the input image is overlaid by the morphological gradient of the binary image.

Overlapping cells may appear in complex images such as cytogenetic specimens prepared with FISH techniques. Fig. 6.29 illustrates such an application, where the task is to find and segment chromosomes, interphase cell nuclei, and DNA-probe dots in the images: (a) is the original color image; (b) is the blue channel; (c) is the morphological gradient; (d) is the watershed transform calculated on (c), after an h-minima filtering with height 7; in (e) the objects are filled out using reconstruction; in (f) cut nuclei cells have been filtered from reconstruction from open by a circle; (g) is the red channel image; in (h) spots were detected by thresholding the h-maxima; (i) shows the segmented components. Note the classical binary-image segmentation problem in (f), where the touching round nuclei have been separated.

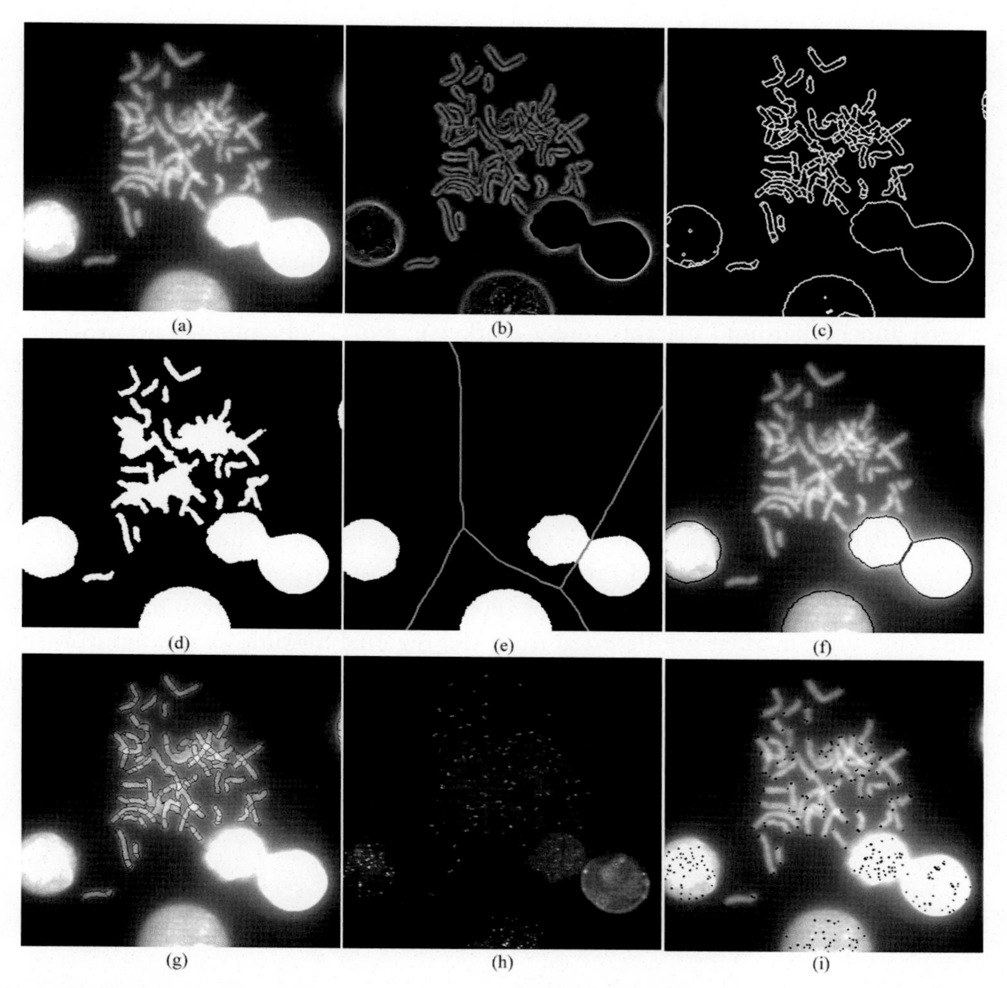

Fig. 6.29 Fluorescence in situ hybridization image: (a) blue channel image; (b) morphological gradient; (c) watershed calculated on (b), after an h-minima filtering with height 7; (d) objects filled out using reconstruction; (e) cut nuclei cells filtered with reconstruction from open by a circle; (f) outline (in *black*) of the segmented nuclei overlaid on blue channel; (g) outline (in *black*) of the detected chromosomes overlaid on blue channel; (h) red channel image; (i) spots (in *black*) detected by thresholding h-maxima and overlaid on the blue channel image.

We now discuss the other operations that were performed in this application. The sup-reconstruction of the watershed result, which selects the whole components from (e), is tricky: the marker of the sup-reconstruction is a white image with a single black point. To detect the red dots from the original color image the red channel is selected, and a simple h-maxima filter, with a high-dynamics parameter (e.g., 20), selects the red dots. The *dynamic* of a regional maximum is the height we must climb down from that maximum in order to reach another maximum of higher elevation [18]. Similarly, the dynamic of a minimum is the minimum height we must climb from that regional minimum in order to reach a lower regional minimum.

6.4.6 Inner and Outer Markers

A typical watershed-based segmentation problem is to segment cell-like objects in a grayscale image. The general approach commonly used to solve these problems is threefold: (1) preprocessing using a smoothing filter, (2) extraction of object markers (inner markers) and background markers (outer markers), and (3) obtaining watershed lines of the morphological gradient from the markers. Usually the most critical part is the extraction of object markers, since an object not marked properly will be missed in the final segmentation.

Fig. 6.30 shows a typical application of the watershed from markers technique using inner and outer markers. The input image for this example is the same one used in Fig. 6.22. Parts (a), (b), and (c) are the same images presented in the illustration of the classical watershed in Fig. 6.22. In this case, however, oversegmentation is avoided by using the inner and outer markers concept applied to the watershed from markers technique. The inner markers are detected from the regional maxima of the input image

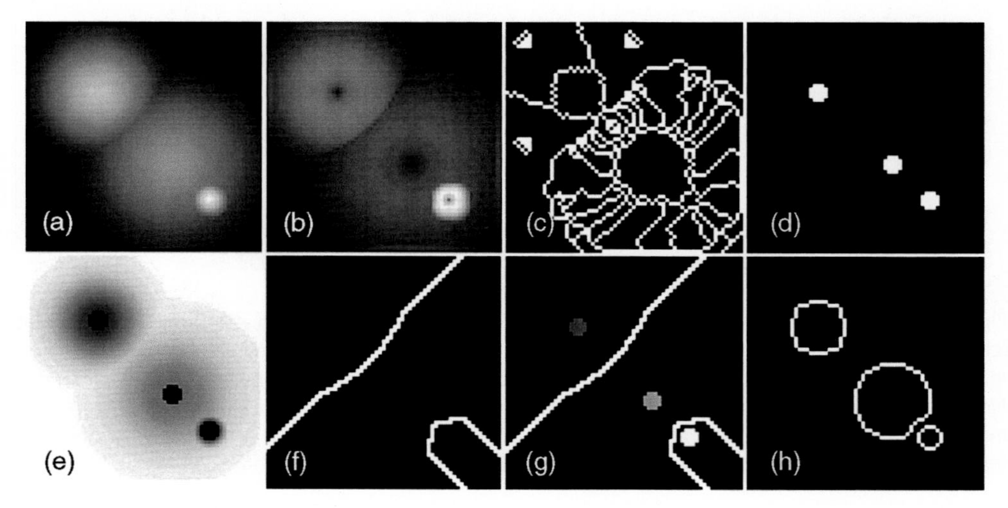

Fig. 6.30 Watershed from markers: (a) small synthetic input image (64×64); (b) morphological gradient; (c) oversegmentation using watershed on the morphological gradient; (d) detection of inner markers; (e) inner markers overlaid on negated input image; (f) outer marker as the watershed from markers; (g) inner and outer markers (labeled); (h) watershed from markers applied to the morphological gradient.

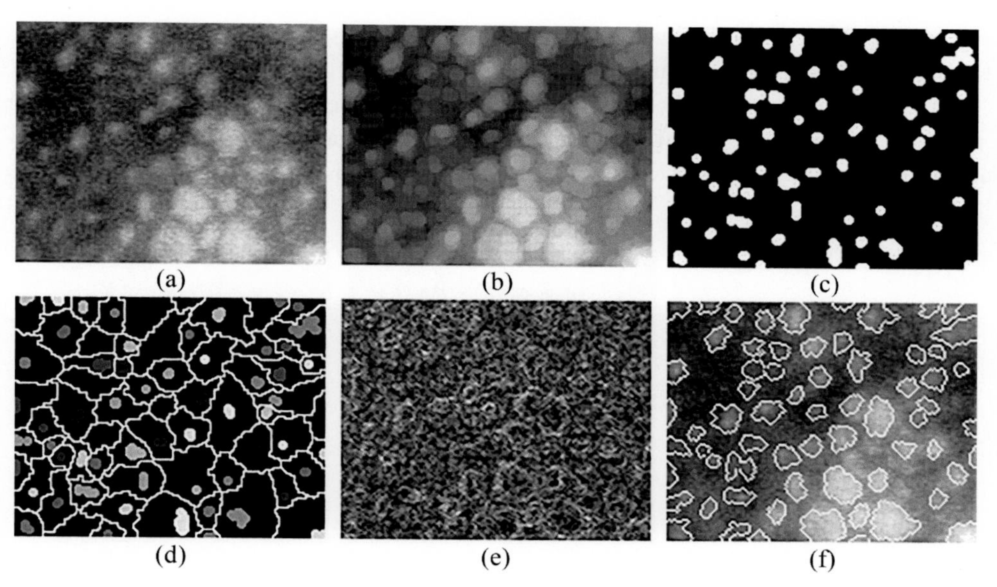

Fig. 6.31 Segmentation of cornea cells from a noisy image: (a) input image; (b) filtered by an opening operation; (c) regional maxima of the opening (inner markers); (d) inner and outer markers (watershed lines of the negated input image from the inner markers); (e) morphological gradient of the original image; (f) final watershed lines overlaid on the input image.

opened with a small disk (see part (d)). For the outer markers, we first negate the input image (e) and then compute the watershed transform on it. In the negated input image, the peaks become basins, and taking the watershed transform of the inner markers yields the skeletons of the influence zones of the basins. These compose the background (outer) marker (f). Care is required in combining both markers as they can touch each other at some points. We first label the inner markers with integers and then label the outer marker with 1 greater than the maximum inner-marker label. In (g) the different markers are painted with different colors. The final segmentation is in (h).

To illustrate watershed segmentation using inner and outer markers we consider the poor-quality microscopic image of a cornea tissue shown in Fig. 6.31a. The cell markers are extracted as the regional maxima of the opening with a disk operation performed on the input image. The criterion used with regional maxima is mainly topological. We can model each cell as a small hill, and we want to mark the top of each hill that has a base larger than the disk used in the opening. Parts (b) and (c) of the figure show the opened image and its regional maxima, respectively. The inner and outer markers (d) are detected by the same procedure as in Fig. 6.30. The regional maxima constitute the inner markers, and the outer markers are obtained by a watershed transform on the negated input image. After labeling the markers, the morphological gradient is computed in (e). Although it is a very noisy gradient, the final watershed lines, which are overlaid on the input image in (f), provide a satisfactory segmentation.

The watershed transform is most often applied to a gradient image, a top-hat image, or a distance function image, but in other cases the input image itself is suitable for application of the watershed transform. Fig. 6.32 illustrates this on a brightfield image of a

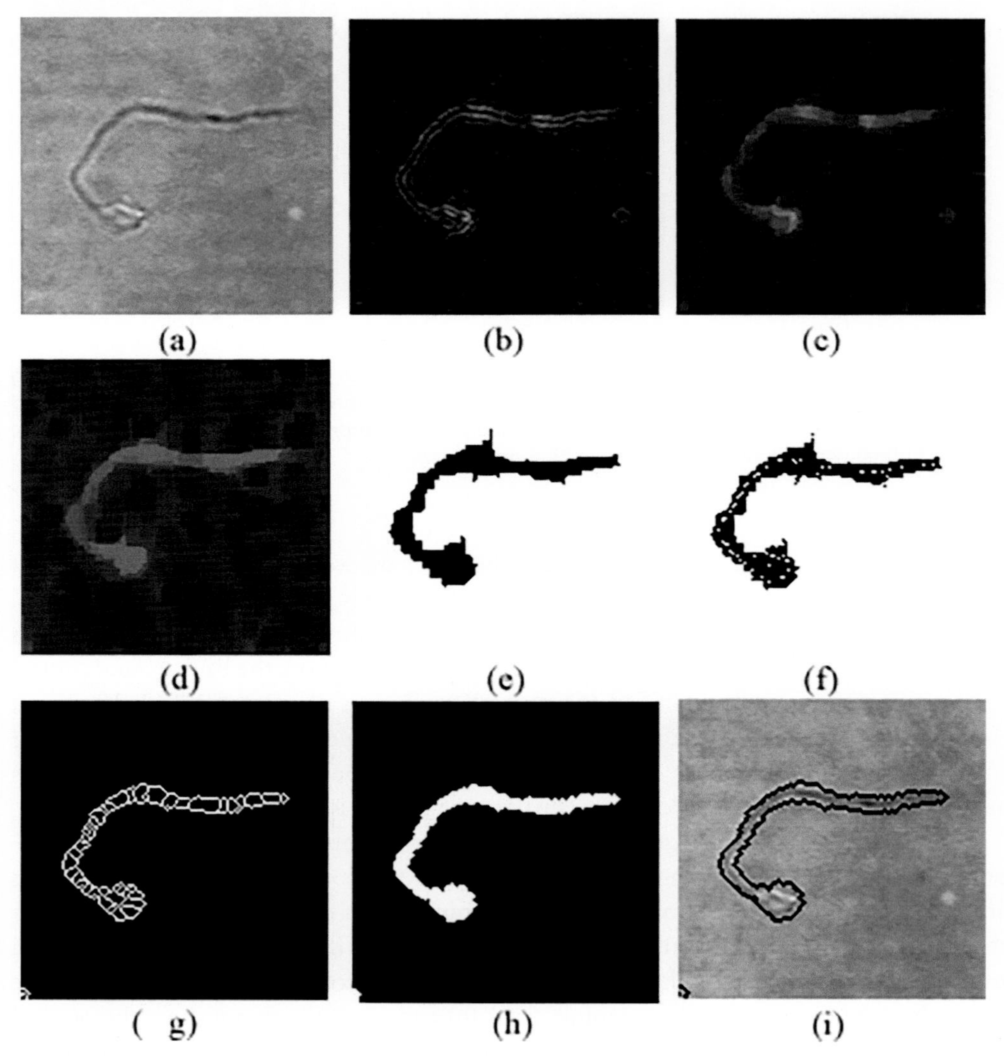

Fig. 6.32 Segmentation of a urology specimen: (a) input image; (b) gradient; (c) closing with a box of size 7×7; (d) area opening by 120 pixels; (e) thresholding (presegmentation); (f) markers given by the regional minima of the input image, masked by the presegmentation; (g) watershed from markers of the input image; (h) area closing; (i) contour (gradient) overlaid on the input image.

urology specimen where the task is to find the boundaries of the low-contrast objects in the image. The approach is to segment the low-contrast structures in (a) by a watershed from markers operation applied directly to the input image, as the input image already has a gradient-like structure. For this to work, it is necessary to detect the markers only on the urology specimens. So, the first step is to find a mask image roughly larger than the specimens. We achieve this by thresholding the filtered gradient image. Part (b) shows the gradient of the input image, and part (c) is the closing of the gradient by a box of size 7×7 followed by an area opening of 120 pixels in (d). These parameters were chosen based on the thickness of the objects. Part (e) is the thresholding of the filtered image which, after a union operation with the regional maxima of the input image, forms the markers for the watershed, as seen in (f). Area closing the catchment basins can be seen in (g), and the contours computed from the gradient of part (g) are overlaid on the input image in (h).

6.5 Summary of Important Points

1. Morphology processing has to do with the fitting, or not fitting, of a structuring element inside the objects in an image.
2. Morphology processing can be directly and efficiently applied to discrete images.
3. Morphology processing can be applied to both binary and grayscale images.
4. Morphology algorithm design is based on a building-block concept, where complex operators are built up from sequences of simple ones.
5. Erosion and dilation are primitive operators of morphology processing, and they are duals.
6. The erosion of an image consists of all locations where the structuring element fits inside an object.
7. Opening and closing are created from the composition of erosion and dilation. They are dual operations.
8. An image contains its opening while it is contained within its closing.
9. The opening of an image by a structuring element is the union of all the structuring elements that fit inside the objects. It is a way of marking objects that have certain specified morphological properties.
10. An alternating sequential filter (ASF) combines openings and closings with increasing structuring element sizes to filter out iteratively both additive and subtractive noise components.
11. The top-hat concept consists of subtracting the input image from the output image of a morphological filter, or the inverse.
12. The labeling process decomposes an image into its connected components.

13. The morphological reconstruction of an image from a marker is the union of all connected components of the image that intersect that marker.
14. The area opening operation removes all connected components with area less than a specified value.
15. Skeletonization is a classic tool for image processing that shrinks an object's boundary to a thin line. It has been widely used in microscopy, primarily for segmentation and shape analysis.
16. Grayscale erosion (or dilation) by a flat structuring element is equivalent to a moving-minimum (moving-maximum) filter over the window defined by the structuring element.
17. The morphological gradient is defined as the subtraction of the erosion from the dilation.
18. When a flat structuring element is used, grayscale erosion, dilation, opening, closing, ASF, morphological reconstruction, alternating sequential component filters (ASCF), area opening, and closing are stack filters. They can be implemented by operating on the threshold sets with their binary equivalent operator, followed by stack reconstruction.
19. The h-maxima operator is a component filter that removes any peaks with height less than or equal to h and decreases the height of the remaining peaks by h. The h-minima operator is its dual.
20. A regional maximum (or minimum) is a flat connected region that is on top of a peak (or at the bottom of a basin).
21. Intuitively, the watershed transform is a flood simulation where the watershed lines separate waters flooded in from different regional minima of the image.
22. Normally the watershed is computed on a gradient image to detect contour lines around homogeneous regions.
23. In noisy images there are often very many minima, resulting in a watershed oversegmentation effect.
24. Watershed oversegmentation is effectively eliminated by filtering the image with closing, area closing, or h-minima operators.
25. The watershed oversegmentation phenomenon can be used to separate homogeneous from textured regions.
26. Another way to eliminate watershed oversegmentation is by using the watershed from markers technique, where water can flood only from markers instead of from all regional minima.
27. The watershed-based segmentation from markers technique typically requires inner markers for the objects, and outer markers for the background.
28. The classical watershed transform can be constructed using the watershed from markers technique, and vice versa.

References

[1] E. Dougherty, R. Lotufo, Hands-on Morphological Image Processing, SPIE, 2003.
[2] P. Soille, Morphological Image Analysis: Principles and Applications, second ed., Springer-Verlag, 2003.
[3] A. Ramesh, Morphological Image Enhancement (Paperback—Illustrated), Lambert Academic Publishing, 2014.
[4] F.Y. Shih, Image Processing and Mathematical Morphology: Fundamentals and Applications, CRC Press, 2011.
[5] S.B. Shaik, Implementation of Morphological Image Processing Operations (Paperback), Lambert Academic Publishing, 2015.
[6] SDC Morphology Toolbox for MATLAB or Python, URL: https://sdc-morphology-toolbox-for-matlab.software.informer.com/.
[7] R.M. Haralick, S.R. Sternberg, X. Zhuang, Image analysis using mathematical morphology, IEEE Trans. Pattern Anal. Mach. Intell. 9 (1987) 532–550.
[8] E.J. Breen, R. Jones, Attribute openings, thinnings and granulometries, Comput. Vis. Image Underst. 64 (3) (1996) 377–389.
[9] A. Rosenfeld, A characterization of parallel thinning algorithms, Inf. Control. 29 (1975) 286–291.
[10] T. Pavlidis, A thinning algorithm for discrete binary images, Comput. Graphics Image Process. 13 (2) (1980) 142–157.
[11] P. Salembier, A. Oliveras, L. Garrido, Anti-extensive connected operators for image and sequence processing, IEEE Trans. Image Process. 7 (1998) 555–570.
[12] P. Salembier, J. Serra, Flat zones filtering, connected operators and filter by reconstruction, IEEE Trans. Image Process. 3 (8) (1995) 1153–1160.
[13] L. Vincent, Morphological grayscale reconstruction in image analysis: efficient algorithms and applications, IEEE Trans. Image Process. 2 (1993) 176–201.
[14] L. Vincent, Grayscale area opening and closings, their efficient implementation and applications, in: J. Serra, P. Salembier (Eds.), Mathematical Morphology and its Applications to Signal Processing, UPC Publications, 1993, pp. 22–27.
[15] S. Beucher, F. Meyer, The morphological approach to segmentation: The watershed transformation, in: E.R. Dougherty (Ed.), Mathematical Morphology in Image Processing, Marcel Dekker, 1993.
[16] R. Lotufo, A. Falcão, The ordered queue and the optimality of the watershed approaches, in: J. Goutsias, L. Vincent, D. Bloomberg (Eds.), Mathematical Morphology and its Application to Image and Signal Processing, Kluwer Academic Publishers, 2000.
[17] L. Vincent, P. Soille, Watersheds in digital spaces: an efficient algorithm based on immersion simulations, IEEE Trans. Pattern Anal. Mach. Intell. 13 (1991) 583–598.
[18] M. Grimaud, New measurement of contrast: dynamics, in: P. Gader, E. Dougherty, J. Serra (Eds.), III Image Algebra and Morphological Image Processing, SPIE-1769, 1992.

Image Segmentation

Qiang Wu and Kenneth R. Castleman

7.1 Introduction

Image segmentation is a task of fundamental importance in digital image analysis. It is the process that partitions a digital image into disjoint (nonoverlapping) regions, each of which typically corresponds to one object in the field of view. Once isolated, these objects can be measured and classified, as discussed in Chapters 8 and 9, respectively. Unlike human vision, where image segmentation takes place without effort, digital processing requires that we laboriously isolate the objects by breaking up the image into regions, one for each object [1]. Errors in the segmentation process almost certainly lead to inaccuracies in subsequent analysis. Furthermore, the exact location of object boundaries is subject to interpretation, and different segmentation algorithms often produce different results.

Traditionally, image segmentation is approached from one of two different but complementary perspectives. One seeks to identify the regions in the image that correspond to individual objects. The other seeks to find the boundaries that separate the objects in the image [2,3]. A region is a connected set of (adjacent) pixels. In the region-based approach, we consider each pixel in the image and assign it to a particular region or object. In the boundary-based approach we either attempt to directly locate the boundaries that exist between the regions, or we seek to identify edge pixels and then link them together to establish the boundaries. Segmentations resulting from these two approaches may not be exactly the same, but both approaches are useful to understanding and solving image segmentation problems, and their combined use can lead to improved performance [2–4].

With recent AI research in deep learning networks, a new generation of image segmentation methods have been developed over the past few years. Unlike the traditional approaches that explicitly analyze visual features pertaining to regions and boundaries, such as gray level homogeneity and edges, deep learning models (covered in Chapter 15) process an input image directly with concatenated neural networks to generate a segmentation map where pixels are classified semantically. This paradigm employs complex multilayer neural network structures with a large number of parameters that must be computed and fine-tuned through training using annotated image datasets. The architecture of neural networks is loosely inspired by the connectivity pattern of

Microscope Image Processing
https://doi.org/10.1016/B978-0-12-821049-9.00003-4
Copyright © 2023 Elsevier Inc.
All rights reserved.

neurons in animal brains [5]. Readers interested in neural networks and deep learning are referred to [6,7] for additional information. Applications of deep learning models for microscope image segmentation [8,9] can be found in Chapter 15.

In this chapter we limit our scope to traditional image segmentation approaches and describe a number of techniques for locating and isolating objects in an image. We focus the discussion on the segmentation of 2D gray level images, but most of the techniques can be extended to multispectral images (see Chapter 10) and 3D images (see Chapter 11). Segmentation is also treated in Chapter 6, as it is an important application of morphological image processing. Real-world applications in digital microscopy often pose very challenging segmentation problems. Variations and combinations of the basic techniques presented here often must be tailored to the specific application in order to produce acceptable results.

7.1.1 Pixel Connectivity

Before introducing various methods for image segmentation, it is important to understand the concept of connectivity of pixels in a digital image (see also Chapters 5, 6, and 9). A set of connected pixels is a set in which all the pixels are adjacent to or touching another member of the set [10]. Between any two pixels in a connected set there exists a connected path wholly within the set. A *connected path* is one that always moves between neighboring pixels. Thus, in a connected set, one can trace a connected path between any two pixels without ever leaving the set.

There are two rules of *connectivity*. If only laterally adjacent pixels (up, down, right, left) are considered to be connected, we have "4-connectivity," and the objects are "4-connected." Thus a pixel has only four neighbors to which it can be connected. If diagonally adjacent (45° neighbor) pixels are also considered to be connected, then we have "8-connectivity," the objects are "8-connected," and each pixel has eight neighbors to which it can be connected. Either connectivity rule can be adopted as long as it is used consistently. Any region that is 4-connected is also 8-connected, but the converse is not necessarily true. Overall, 8-connectivity is more commonly used, and it produces results that are closer to one's intuition.

7.2 Region-Based Segmentation

Region segmentation methods partition an image by grouping similar pixels together into identified regions. Image content within a region should be uniform and homogeneous with respect to certain attributes, such as intensity, rate of change in intensity, color, texture, etc. Regions are important in interpreting an image because they typically correspond to objects or parts of objects in a scene. In this section we discuss several widely used techniques that fall into this category.

7.2.1 Thresholding

Thresholding is an essential region-based image segmentation technique that is particularly useful for scenes containing solid objects resting upon a contrasting background. It is computationally simple and never fails to define disjoint regions with closed, connected boundaries. The operation is used to distinguish between the objects of interest (also known as the foreground) and the background upon which they lay. The output at each pixel location is the label of either "object" or "background," which can be represented as a Boolean variable. In general, a gray level thresholding operation can be described as

$$G(x, y) = \begin{cases} F, & \text{if } I(x, y) \geq T \\ B, & \text{if } I(x, y) < T \end{cases} \tag{7.1}$$

where $I(x, y)$ is the original image; T is the threshold; $G(x, y)$ is the thresholded image; F corresponds to the foreground labeled with either a designated gray level value or the original gray level, $I(x, y)$; and B corresponds to the gray level chosen to label the background. Thus all pixels at or above the threshold are assigned to the foreground and all pixels below the threshold are assigned to the background. The boundary is then that set of interior points each of which has at least one neighbor outside the object. It should be noted that this formulation assumes that we are interested in high gray level objects on a low gray level background. For the converse, one can simply invert the image and the discussion here is still applicable.

Thresholding works well if the objects of interest have uniform interior gray level and rest upon a background of unequal but uniform gray level. If the objects differ from the background by some property other than gray level (color, texture, etc.), one can first use an operation that converts that property to gray level. Then gray level thresholding can segment the processed image. Thresholding can also be generalized to multivariate classification operations in which the threshold becomes a multidimensional discriminant function classifying pixels based on several image properties. Detailed discussion of multivariate image thresholding and clustering can be found in [11–13].

7.2.1.1 Global Thresholding

In the simplest implementation of thresholding, the threshold value is held constant throughout the image. If the background gray level is reasonably constant over the image, and if the objects all have approximately equal contrast above the background, then the gray level histogram is bimodal, and a fixed global threshold usually works well, provided that the threshold, T, is properly selected. In most cases the threshold is determined from the gray level histogram of the image to be segmented. In general, the choice of the threshold, T, has considerable effect on the boundary position and overall size of segmented objects. This, in turn, affects the data obtained from subsequent object measurement. For this reason, the value of the threshold must be determined carefully.

7.2.1.2 Adaptive Thresholding

Due to uneven illumination and other factors, the background gray level and the contrast between the objects and the background often vary within an image. In such cases, global thresholding is unlikely to produce satisfactory results, since a threshold that works well in one area of the image might work poorly in other areas. To cope with this variation, one can use an adaptive or variable threshold that is a slowly varying function of position in the image [14].

One approach to adaptive thresholding is to partition an $N \times N$ image into nonoverlapping blocks of $n \times n$ pixels each ($n < N$), analyze gray level histograms of each block, and then form a thresholding surface for the entire image by interpolating the resulting threshold values determined from the blocks. The blocks should be large enough that there are a sufficient number of foreground and background pixels in each block to allow reliable estimation of the histogram and setting of an accurate threshold [15].

Adaptive thresholding can also be implemented as a two-pass operation [15,16]. Before the first pass, a threshold is computed based on the histogram of each block by choosing, for example, the value located midway between the background and object peaks. Blocks containing unimodal histograms can be ignored. In the first pass, the object boundaries are defined using a gray level threshold that is constant within each block but differs for the various blocks. The objects so defined are not extracted from the image, but the interior mean gray level of each object is computed. On the second pass, each object is given its own threshold, which lies midway between its interior mean gray level and the background gray level of its principal block.

Fig. 7.1 shows an example of applying thresholding for segmentation of human chromosomes in a microscope image. In this example, the background gray level varies due to nonuniform illumination, and contrast varies from one chromosome to the next. In

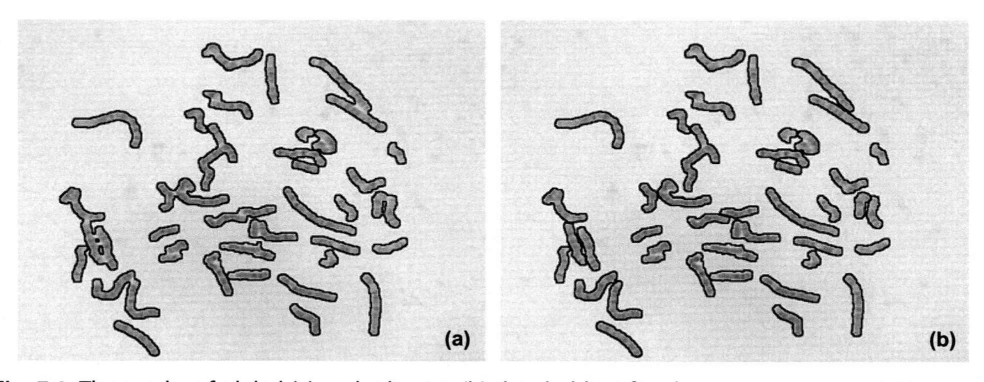

Fig. 7.1 The results of global (a) and adaptive (b) thresholding for chromosome segmentation.

Fig. 7.1a, a global threshold has been used on the image to isolate the chromosomes. Each isolated chromosome is displayed with a boundary. In Fig. 7.1b an adaptive threshold has been used instead. This results in fewer segmentation errors, that is, cases where multiple chromosomes are stuck together or individual chromosomes are broken up. The accuracy of the area measurement for chromosomes is improved by adaptive thresholding as well [15,17].

7.2.1.3 Threshold Selection

The selection of threshold value is crucial to the success of a thresholding operation. Unless the object in the image has very steep sides, any variation in threshold value can significantly affect the boundary position and thus the overall size of the extracted object. This means that subsequent object measurements, particularly the area measurement, are quite sensitive to the threshold value. While there is no universal methodology for threshold selection that works on all kinds of images, many techniques have been developed to facilitate the determination of threshold values under different circumstances [18,19].

An image containing an object on a contrasting background normally has a bimodal gray level histogram (Fig. 7.2). The two peaks correspond to the relatively large numbers of pixels that belong to the object and to the background. The dip between the peaks corresponds to the relatively few pixels located around the edge of the object. When a threshold value, T, is chosen, the area of the object is given by

$$area = \int_0^T H(I)dI \tag{7.2}$$

where $H(I)$ is the gray level histogram of the image, and the object has a lower gray level than the background. Notice in Fig. 7.2 that increasing the threshold from T to $T+\Delta T$ causes only a slight change in area if the threshold is placed at the dip in the histogram; hence the threshold chosen at or near the dip minimizes the sensitivity of the object area measurement to small variations in the threshold value.

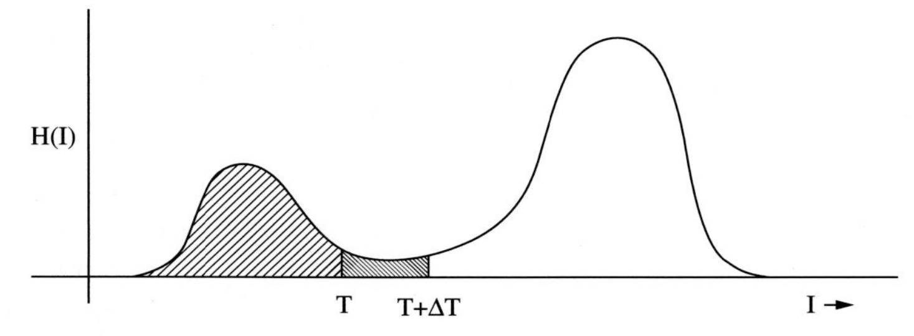

Fig. 7.2 A bimodal histogram. The shaded areas show the effect that threshold variation has on the area of the object.

Histogram Smoothing

If an image containing the object is small and noisy, the histogram itself will be noisy. Unless the dip is unusually sharp, the noise can make its location obscure and unreliable. This can be overcome to some extent by smoothing the histogram using either a convolution filter or a curve-fitting procedure.

A simple yet effective smoothing operation is the convolution of the input histogram with a moving-average filter, also known as a "box filter":

$$H_{Output}(i) = \frac{1}{W} \sum_{j=-(W-1)/2}^{(W-1)/2} H_{Input}(i-j) \tag{7.3}$$

where W is an odd number, typically chosen to be 3 or 5. This operation is designed to reduce small fluctuations without shifting the peak positions. If the two peaks are unequal in size, smoothing may tend to shift the position of the dip in the histogram, making it difficult to locate uniquely. The peaks, however, are easy to locate and relatively stable under reasonable amounts of smoothing. Therefore, placing the threshold at some designated position relative to the two peaks can be more reliable than trying to place it at the dip. In this section we introduce several methods that are based on different threshold selection criteria.

The ISODATA Algorithm

The *ISODATA* algorithm is an iterative threshold selection technique [20]. Initially, the histogram is divided into two parts by a starting threshold $T^{(0)}$ placed midway between the maximum and minimum gray level. Next we compute the sample mean $\mu_F^{(0)}$ of the gray level values associated with the foreground pixels and the sample mean $\mu_B^{(0)}$ of the gray level values associated with the background pixels, respectively. A new threshold value $T^{(1)}$ is then obtained as the average of these two sample means. This process is repeated using the new threshold, until the threshold value no longer changes, that is

$$T^{(k)} = \frac{\mu_F^{(k-1)} + \mu_B^{(k-1)}}{2}, \quad \text{until } T^{(k)} = T^{(k-1)},\ k > 0. \tag{7.4}$$

The Background Symmetry Algorithm

The *background symmetry* algorithm works under the assumption that there is a dominant background peak in the histogram, and it is symmetrical about its maximum [21]. Preprocessing the histogram with a smoothing operation may help improve the performance of this algorithm. The position of the peak maximum $I_{\max}$ is determined by searching the entire histogram. Then the algorithm searches on one side of the background peak that is farther away from the foreground to locate a certain percentile point I_p. Using the symmetry assumption, the threshold value is then selected on the other side of the background peak at an equal displacement from the background peak, that is,

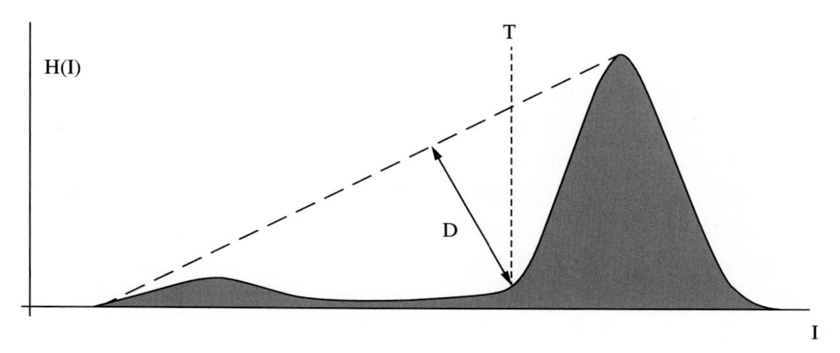

Fig. 7.3 An illustration of the triangle algorithm.

$$T = I_{max} - \left(I_p - I_{max}\right). \tag{7.5}$$

This algorithm can be adapted to the case where the object peak, rather than the background peak, is symmetrical and dominant.

The Triangle Algorithm

The *triangle* algorithm is particularly effective when the object pixels produce only a weak peak in the histogram [22]. It is illustrated in Fig. 7.3, where low gray level objects reside on a high gray level background. One first finds the maximum peak of the histogram. A line is then constructed to connect the maximum peak point $[I_{max}, H(I_{max})]$ to the lowest point $[I_{lowest}, H(I_{lowest})]$ in the histogram. This is followed by the computation and comparison of the distances between that line and all the histogram points $H(I)$, with I ranging from I_{lowest} to I_{max}. The threshold value, T, is taken as the value of I where this distance is greatest.

Gradient-Based Algorithms

A variant of the preceding methods is the construction of a histogram of only those pixels having relatively high gradient magnitude [23] (see Section 7.3.5.1). This eliminates a large number of interior and exterior pixels from consideration and may make the dip in the histogram easier to locate [24]. One can also divide the histogram by the average gradient of pixels at each gray level to further enhance the dip [23], or simply average the gray level of high-gradient pixels to determine a threshold [24].

7.2.1.4 Thresholding Circular Spots

In many important cases it is necessary to find objects that are roughly circular in shape. Suppose an image $I(x, y)$ contains a single spot. By definition, this image contains a point (x_0, y_0) of maximum gray level. Using polar coordinates centered upon (x_0, y_0), the image can be represented as $I(r, \theta)$ and we have

$$I(r_1, \theta) \geq I(r_2, \theta), \; r_2 > r_1 \tag{7.6}$$

for all values of θ. If equality is not allowed in Eq. (7.6), $I(x, y)$ is a "monotone spot." An important special case occurs if all contours of a monotone spot are circles centered on (x_0, y_0). Such a special case is referred to as a *concentric circular spot* (CCS). To a good approximation, this model can be used to represent the noise-free images of certain types of cells in a microscope. For a CCS, the function $I(r, \theta)$ is independent of θ, and it serves as the 1D spot profile function. This function is useful for threshold selection. For example, one can locate the inflection point and select the gray level threshold to place the boundary at the point of maximum slope. Other unique points on the profile, such as the maximum magnitude of the second derivative [24], can also be used. If we threshold a monotone spot at a gray level T, we define an object with a certain area and perimeter. As we vary T throughout the range of gray levels, we generate the threshold area function, $A(T)$, and the perimeter function, $P(T)$. Both of these functions are unique for any spot. They are both continuous for monotone spots, and either is sufficient to specify a CCS completely. If two spots have identical perimeter functions or identical histograms, they are known as p-equivalent or h-equivalent, respectively. It turns out that h-equivalent spots have identical threshold area functions [24].

Analytical expressions relating the profile function to the threshold area function and the perimeter function of a CCS can be derived to guide the selection of threshold. The radius of the circular object obtained by thresholding a CCS at gray level T is

$$r(T) = \left[\frac{1}{\pi} A(T)\right]^{1/2} = \left[\frac{1}{\pi} \int_0^T H(I) dI\right]^{1/2} \tag{7.7}$$

For a monotone spot, the histogram $H(I)$ is nonzero between its minimum and maximum gray levels. This means that the area function $A(T)$ is monotonically increasing, and so is $r(T)$. Thus the inverse function of $r(T)$ exists, and it is the spot profile. We can therefore compute the area-derived profile of a CCS by integrating the histogram to obtain the area function, followed by taking the square root and then the inverse function. Similarly, the profile may also be obtained from the perimeter function through the relationship

$$r(T) = \frac{1}{2\pi} P(T) \tag{7.8}$$

7.2.1.5 Thresholding Noncircular and Noisy Spots

For an image containing a noise-free CCS, we can easily obtain the profile simply by taking the gray levels along the scan line that contains the peak. Even for near-circular spots and noisy spots, that analysis can still be useful. For example, one can use the histogram of a near-circular spot to obtain the profile of the h-equivalent CCS and select the threshold gray level that maximizes the slope at the boundary. On the other hand, it is useful to measure the perimeter function and determine the profile of the p-equivalent

CCS. Either technique could be used to select thresholds suitable for the image under consideration. In some microscope images the noise level is so high that differentiating a single scan line cannot reliably identify the inflection point on the profile. Nonetheless, since the area-derived and perimeter-derived profiles are computed using most or all of the edge pixels in the object, the noise is reduced by averaging in the process.

Further noise reduction can be achieved by smoothing the histogram or perimeter function before profile computation, or by smoothing the profile function itself. The area-derived profile is easier to compute, and it has greater noise reduction effect. Random noise in the image usually makes the spot boundary jagged. While this may have little effect on the area function, it tends to make the perimeter function erroneously large. Even though the error can be reduced by boundary smoothing built into the perimeter measurement routine, the area-derived profile clearly has the advantage of computational simplicity.

In the study described in [24], nine methods of threshold selection were compared, including two based on the area-derived profile (maximum magnitudes of the first and second derivative) for measuring the diameter of fluorescent microspheres. Generally, the method based on maximum magnitude of the second derivative was found to be the most accurate of the nine for spheres of different sizes and intensities. It also performed well for cells in culture [24,25]. Other methods tended to underestimate object size.

Noncircular Spots

For highly noncircular spots, the h-equivalent and p-equivalent CCS profiles may no longer be acceptable for placing the gray level threshold. For objects of arbitrary shape, we can examine the average gradient around the boundary as a function of the threshold gray level that defines the boundary [15]. Suppose a noncircular monotone spot is thresholded at gray levels of I and $I + \Delta I$, as shown in Fig. 7.4. At some point a on

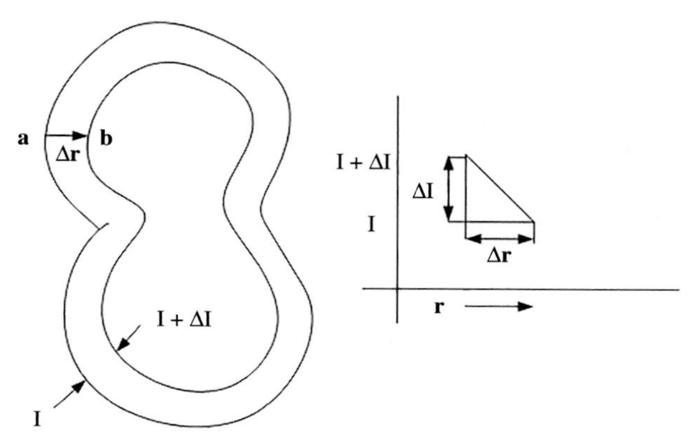

Fig. 7.4 Threshold selection for a noncircular object.

the outer boundary, Δr is the perpendicular distance to the inner boundary. Since Δr is perpendicular to a contour line, it lies in the direction of the gradient vector at point a. The magnitude of the gradient vector at point a on the outer boundary is given by

$$|\nabla \boldsymbol{I}| = \lim_{\Delta I \to 0} \frac{\Delta I}{\Delta r} \tag{7.9}$$

To obtain the average gradient around the boundary, we can simply average $|\nabla \boldsymbol{I}|$ around the outer boundary. If Δr is small compared to the perimeter, the area between the two boundaries is approximately

$$\Delta A = P(I)\overline{\Delta r} \tag{7.10}$$

where $\overline{\Delta r}$ is the average perpendicular distance from the outer to the inner boundary, and $P(I)$ is the perimeter function. To obtain the average gradient around the boundary, we substitute $\overline{\Delta r}$ for Δr in Eq. (7.9) and get

$$|\nabla \boldsymbol{I}| = \lim_{\Delta I \to 0} \frac{\Delta I}{\Delta A} \cdot P(I) = \frac{P(I)}{H(I)} \tag{7.11}$$

Hence the average boundary gradient turns out to be the ratio of the perimeter function to the histogram. This function is not difficult to compute, and it readily identifies the threshold gray level that maximizes the slope at the boundary. For noisy images, it may be beneficial to smooth the perimeter function and the histogram before this computation.

Objects of General Shape

Objects of arbitrary shape that are nonmonotonic and relatively flat on top (without a unique peak) usually have sides that slope uniformly down toward the background. The point spread function of microscope optical systems prohibits sides of infinite slope in real images. On the sides of the objects, the contour lines are closed and generally convex curves, but they may have local concavities. We can assume that each threshold gray level defines a single closed contour for each object. Under these conditions we need to consider only the range of gray levels corresponding to the sloping sides of the object, and the ways to establish the maximum slope threshold can be summarized as follows:

(1) Select T at a local minimum in the histogram. This is the easiest technique, and it minimizes the sensitivity of the area measurement to small variations in T.

(2) Select T corresponding to the inflection point in the h-equivalent CCS profile function. This is a simple computation, and it involves considerable averaging for noise reduction.

(3) Select T to maximize the average boundary gradient. This involves computing the perimeter function but requires no approximation regarding equivalent spot images.

(4) Select T corresponding to the inflection point in the p-equivalent CCS profile function.

For large-scale studies, one may use one of these methods to characterize the objects under study. Then a shortcut method can be implemented for efficiency. If a profile analysis shows, for example, that the optimal threshold gray level for isolated cells in microscope images occurs midway between the peak and the background gray level, then this simplified technique can be employed for routine use.

7.2.2 Morphological Processing

After thresholding, a given image is segmented into a binary image of object (foreground) and background. If this initial segmentation is not satisfactory, a set of morphological operations or the procedures based on these operations and their variants can be utilized to improve the segmentation results. The techniques of morphological processing provide versatile and powerful tools for image segmentation. The design of particular algorithms involves using one's knowledge of what effect each of the primitive operations has on an image and combining them appropriately to obtain the desired result. For a more thorough discussion of morphological operations, see Chapter 6.

Many of the binary morphological operations can be implemented as 3×3 neighborhood operations. In a binary image, any pixel, together with its eight neighbors (assuming 8-connectivity), represents 9 bits of information. Thus there are only 512 possible configurations for a 3×3 neighborhood in a binary image. Convolution of a binary image with the 3×3 kernel

$$\begin{bmatrix} 1 & 2 & 4 \\ 8 & 16 & 32 \\ 64 & 128 & 256 \end{bmatrix}$$

generates a 9-bit (512 gray level) image in which the gray level uniquely specifies the configuration of the 3×3 binary neighborhood centered on that pixel. Neighborhood operations thus can be implemented with a 512-entry look-up table with one-bit output. Whether the operation is implemented in software or in specially designed hardware, it is often much more efficient to use a look-up table for fast "pipeline processing" [26–28] than other ways of implementation.

In the general case, morphological image processing operates by sliding a structuring element over the image, manipulating one neighborhood of pixels at a time, similar to convolution (Fig. 7.5). Like the convolution kernel, the structuring element can be of any size, and it can contain any complement of 1's and 0's. At each position, a specified logical operation is performed between the structuring element and the underlying binary image. The binary result of that logical operation is stored in the output image

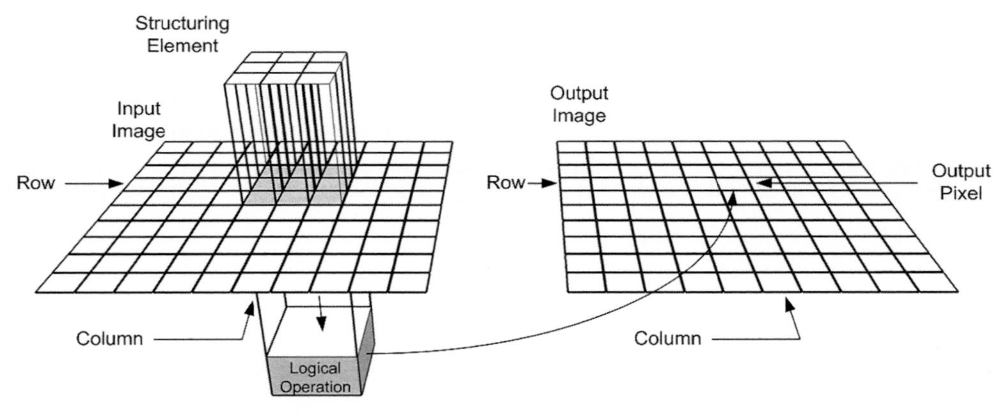

Fig. 7.5 Implementation of a morphological image processing operation.

at that pixel position. The effect created depends upon the size and content of the structuring element and upon the nature of the logical operation.

By definition, a boundary point is a pixel located inside the object but that has at least one neighbor that is outside the object. Binary erosion eliminates all the boundary points from an object, leaving it smaller in area by one pixel all around the perimeter. If the object is circular, its diameter decreases by two pixels with each erosion. If it narrows to less than three pixels thick at any point, it will become disconnected (broken into two objects) at that point. Objects no more than two pixels thick in any direction are eliminated. Binary erosion is useful for removing from a thresholded image objects that are too small to be of interest.

Conversely, binary dilation is the process of incorporating into the object all the background points that touch the object, leaving it larger in area by that amount. If the object is circular, its diameter increases by two pixels with each dilation. If two objects are separated by less than three pixels at any point, they will become connected (merged into one object) at that point.

7.2.2.1 Hole Filling

Dilation-based propagation (also known as reconstruction) can be used, for example, to fill interior holes of segmented objects in a thresholded image. Fig. 7.6 shows an example of such a procedure. Starting from the binary segmented image of the object shown in Fig. 7.6a, one inverts this image to create a mask. Then the border of the image is used as the marker of a propagation (reconstruction) inward toward the mask. This generates the image shown in Fig. 7.6b. Inverting this propagated image produces the desired result that contains the object with all interior holes filled (Fig. 7.6c).

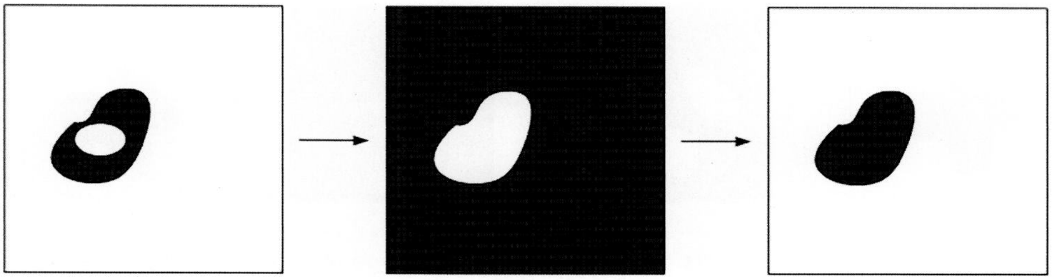

(a) (b) (c)

Fig. 7.6 Filling interior holes of segmented objects using binary morphological operations.

7.2.2.2 Border Object Removal

Another useful procedure is the removal of border-touching objects. In quantitative microscopy the objects that are connected to the image border are partially obscured and usually unsuitable for subsequent analysis. In such cases one can use the procedure illustrated in Fig. 7.7 to eliminate border-touching objects. Here the binary thresholded image (in Fig. 7.7a) is used as the mask, and the border of the image is used as the marker. A propagation from the border inward toward the mask generates the image shown in Fig. 7.7b. Then computing the logical exclusive OR (XOR) operation of the propagated image and the mask image produces the image shown in Fig. 7.7c.

7.2.2.3 Separation of Touching Objects

Binary morphological processing can also be used to separate slightly touching objects that result from the segmentation process. As illustrated in Fig. 7.8, the procedure works as follows. Starting from a binary segmented image (Fig. 7.8a), compute a few erosions that are enough to separate the touching objects (Fig. 7.8b). Invert the resulting image and compute the skeleton (Fig. 7.8c). This is known as the "exoskeleton" because it is the skeleton of the background outside the objects [21]. It is then followed by computing the

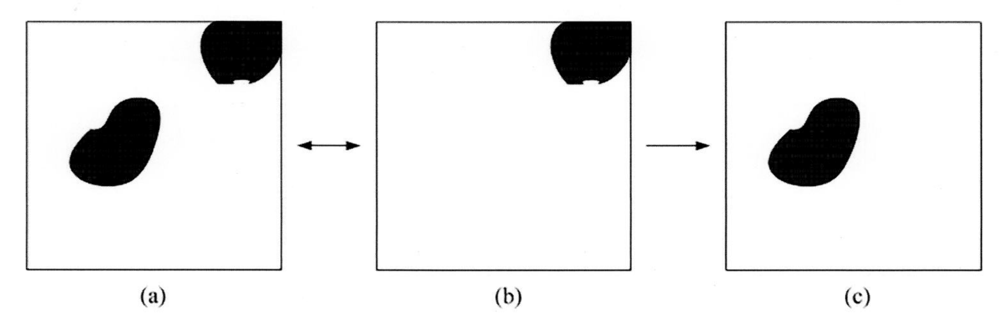

Fig. 7.7 Removing border objects using binary morphological operations.

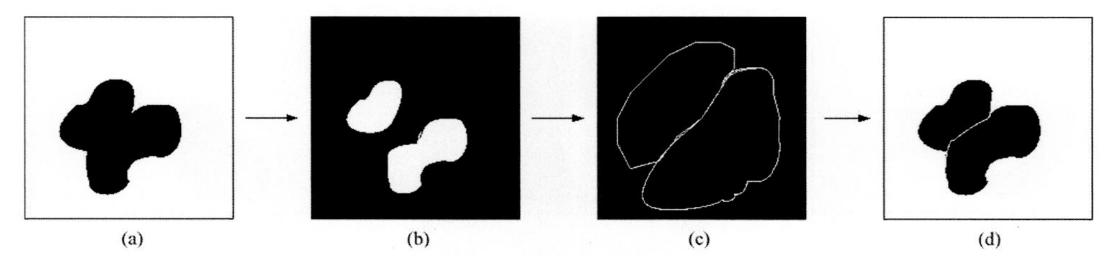

Fig. 7.8 Separating touching objects. (a) A binary segmented image. (b) After a few erosions and inversion. (c) The exoskeleton. (d) Separated objects resulting from applying XOR between the images in (a) and (c).

logical AND operation of the original binary image and the inverted skeleton image. The final result is shown in Fig. 7.8d where the touching objects are separated.

7.2.2.4 The Watershed Algorithm

Perhaps the best-known morphological processing technique for image segmentation is the watershed algorithm. This topic is discussed extensively in Chapter 6. Fig. 7.9 shows a 1D illustration of how this approach works. For this example, we assume the objects are of low gray level, on a high gray level background. Fig. 7.9 shows the gray levels along one scan line that cuts through two objects lying close together. The image is initially thresholded at a low gray level, one that segments the image into the proper number of objects. Then the threshold is raised gradually, one gray level at a time. The object boundaries will expand as the threshold increases. When they touch, however, they are not allowed to merge. Thus these points of first contact become the final boundaries between adjacent objects. The process is terminated before the threshold reaches the gray level of the background.

Rather than simply thresholding the image at the optimum gray level, the watershed approach begins with a threshold that is low enough to properly isolate the individual

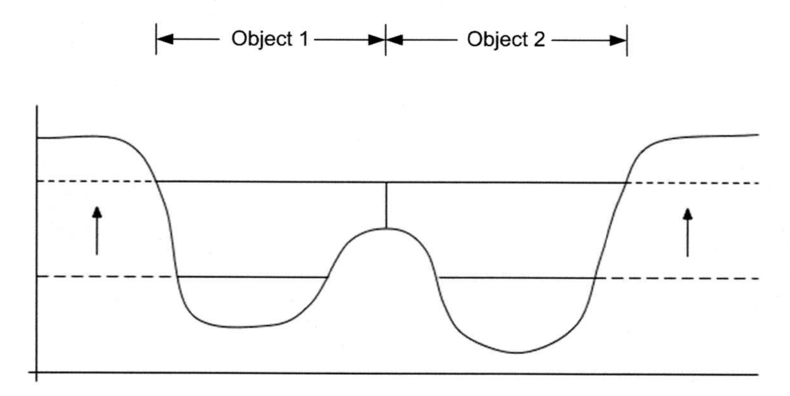

Fig. 7.9 An illustration of the watershed algorithm.

objects. Then the threshold is gradually raised to the optimum level but merging of objects is not allowed. This can solve the problem posed by objects that are either touching or too close together for global thresholding to work. The final segmentation will be correct if and only if the segmentation at the initial threshold correctly isolates the individual objects. Both the initial and final threshold gray levels must be well chosen. If the initial threshold is too low, objects will be oversegmented and low contrast objects will be missed at first, and then merged with nearby objects as the threshold increases. If the initial threshold is too high, objects will be merged from the start. The final threshold value determines how well the final boundaries fit the objects. The threshold selection methods discussed in this chapter can be useful in setting these two values.

7.2.3 Region Growing

The fundamental limitation of histogram-based region segmentation methods, such as thresholding, is that the histograms only describe the distribution of gray levels without providing any spatial information. Region growing [4,29–31] is an approach that exploits spatial context by grouping adjacent pixels or small regions together into larger regions. Homogeneity is the main criterion for merging the regions together. With this approach, one begins by dividing an image into many small regions. These initial regions can be small neighborhoods or even individual pixels known as the "seeds." In each region we compute suitably defined functions of image property parameters that reflect on its membership in an object. The parameters that distinguish different objects may include average gray level, texture, color, etc. Thus the first step assigns to each region a set of parameters whose values reflect the object to which it belongs. Next, all boundaries between adjacent regions are examined. A measure of boundary strength is computed utilizing the differences of the parameters of the adjacent regions. A given boundary is strong if the parameters differ significantly on either side of that boundary, and it is weak if they do not. Strong boundaries are allowed to stand, while weak boundaries are dissolved, and the adjacent regions are merged. This process is iterated by alternately recomputing the object membership parameters for the enlarged regions and once again dissolving weak boundaries, until it reaches a point where no boundaries are weak enough to be dissolved.

Region growing methods often produce good segmentation results that correspond well to the visually apparent edges of objects in the image. Observing this procedure gives the impression of regions in the interior of an object growing and merging until their boundaries reach the edge of the object. Although region growing algorithms are computationally more expensive than the simpler techniques, the methods are able to utilize several image parameters directly and simultaneously in determining the final boundary location. Fig. 7.10 shows an example of region growing.

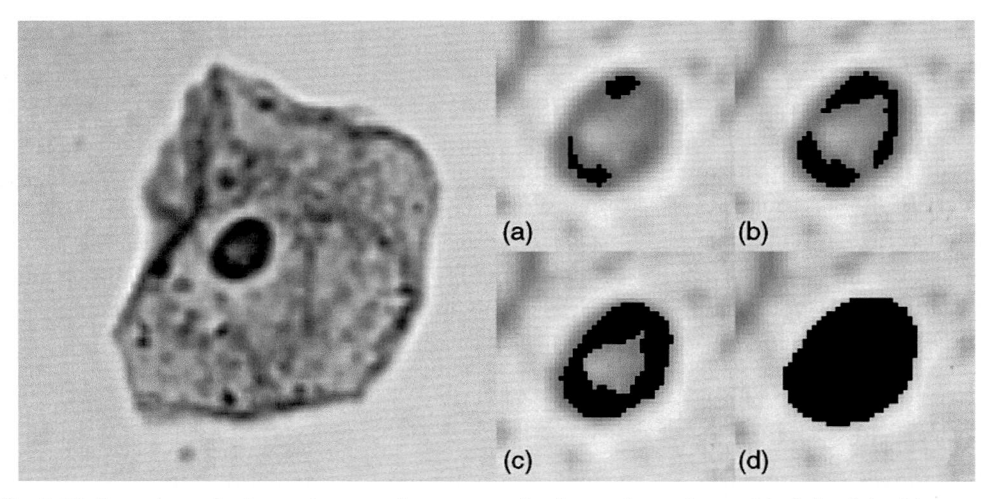

Fig. 7.10 Four stages in the region growing process for the nucleus of an epithelial cell. In this example, low gradient was the sole region membership parameter. Image (d) shows the final segmentation.

7.2.4 Region Splitting

Opposite to the "bottom-up" approach of region growing, region splitting is a "top-down" operation. The basic idea of region splitting is to break the image into disjoint regions within which the pixels have similar properties. In a sense, the morphological procedures discussed earlier, such as the exoskeleton and watershed algorithms, can be viewed as region splitting methods since they work to separate touching objects. However, morphological techniques are generally more suitable for segmenting regularly shaped objects and for those cases where there are distinct bottleneck connections between touching objects.

Region splitting usually starts with the whole image as a single initial region. It first examines the region to decide if all pixels contained in it satisfy certain homogeneity criteria of image properties. Region splitting methods generally use homogeneity criteria similar to those that region growing methods use, and they differ only in the direction of application. In region splitting, if the criterion is met, then the region is considered homogeneous and hence left unmodified in the image. Otherwise, the region is split into subregions, and each of the subregions, in turn, is considered for further splitting. This recursive process continues until no further splitting occurs.

The most commonly used region splitting algorithms employ a pyramid image representation known as the "quadtree." Regions are square-shaped and correspond to the nodes of the quadtree. After region splitting the resulting segmentation may contain neighboring regions that have identical or similar image properties. Hence a merging process can be used after each split to compare adjacent regions and merge them if

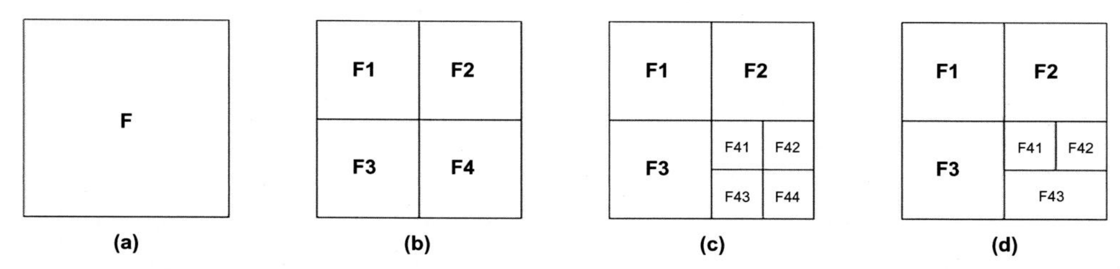

Fig. 7.11 Illustration of region splitting and merging: (a) whole image; (b) first split; (c) second split; (d) merge.

necessary. Such combined operations lead to the methods known as "region splitting and merging algorithms" that exploit the advantages of both approaches [32]. Fig. 7.11 illustrates the basic idea of these methods. We use **F** to denote the whole image (Fig. 7.11a) and suppose that not all the pixels in **F** meet the chosen criterion of homogeneity. Thus the region is split as in Fig. 7.11b. We then assume that all pixels within regions **F1**, **F2**, and **F3** are homogeneous, respectively, but those in **F4** are not. Hence **F4** is split next, as in Fig. 7.11c. Now, let's assume that all pixels within each resulting region are homogeneous with respect to that region, and after comparing all the regions, regions **F43** and **F44** are found to be similar or identical. These regions are therefore merged together, as in Fig. 7.11d.

7.3 Boundary-Based Segmentation

The region-based methods discussed in the previous section aim to segment an image by partitioning the image into sets of interior and exterior pixels according to the similarity of certain image properties. Boundary-based techniques, on the other hand, seek to extract object boundaries directly, based on identifying the edge pixels located at the boundaries in the image. In this section we discuss several methods that fall into this category.

7.3.1 Boundaries and Edges

In the simplest cases, for scenes containing isolated solid objects on a contrasting background, image segmentation can be readily done by thresholding. To obtain the boundaries of these objects, one can perform a dilation and an erosion of the binary segmented image separately. Then subtracting the eroded image from the dilated one will result in the object boundaries. In practice, however, input images are less ideal, and the localization of object boundaries requires more sophisticated gray level computation.

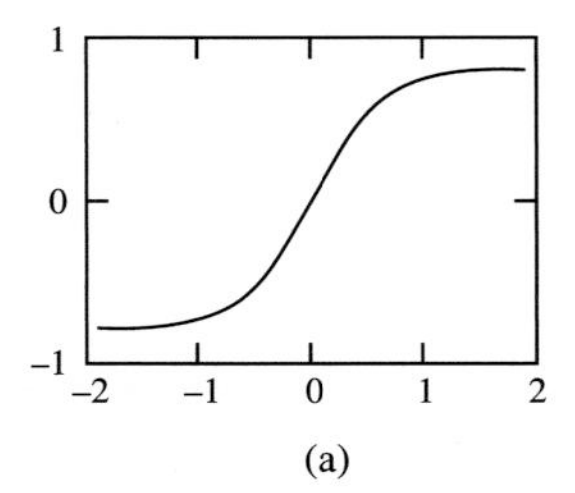

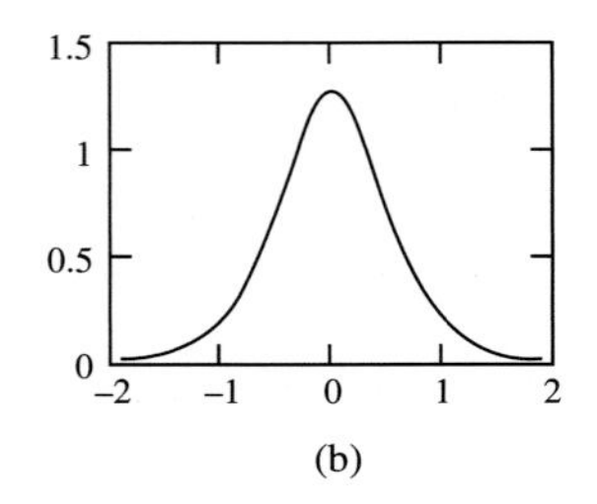

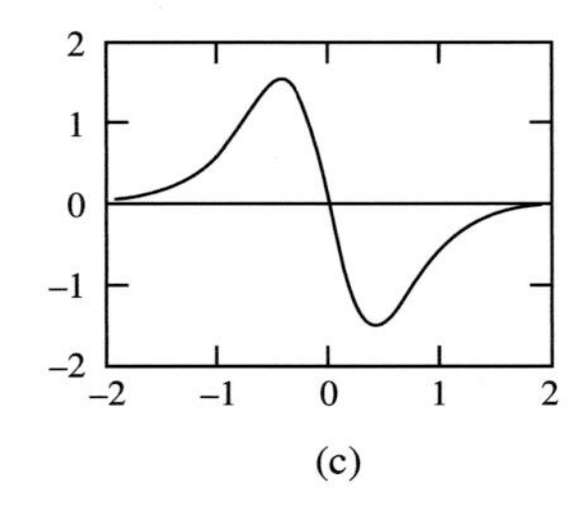

Fig. 7.12 An edge and its first and second derivatives.

In general, edges correspond to those points in an image where the gray level changes sharply. Such sharp changes or discontinuities usually occur at object boundaries. Pixels exhibiting the characteristics of an edge can be detected and used to establish the boundaries of the objects. One can locate these pixels by computing the derivatives of the image. This is illustrated for the one-dimensional case in Fig. 7.12. Theoretically, we can detect edges by applying either a high-pass filter in the Fourier frequency domain, or by convolving the image with an appropriate derivative operator in the spatial domain. In practice, however, edge detection is usually performed in the spatial domain because it is computationally less expensive and often yields better results. There are many derivative operators designed for 2D edge detection, most of which can be categorized as gradient-based or Laplacian-based methods. The gradient-based methods detect the edges by looking for the maximum in the first derivative of the image. The Laplacian-based methods search for zero-crossings in the second derivative of the image to find edges.

7.3.2 Boundary Tracking Based on Maximum Gradient Magnitude

As object or region boundaries are associated with high gradient magnitudes, one can track the boundaries based on the information in a gradient magnitude image. Suppose a gradient magnitude image is computed from a noise-free input image that contains a single object on a contrasting background. We can start the boundary tracking on this gradient image from the highest gray level pixel as the first boundary point since it is certainly on the boundary. If several points have the maximum gray level, then we choose arbitrarily. Next, we search the 3×3 neighborhood centered on the first boundary point and take the neighbor with the maximum gray level as the second boundary point. If two neighbors have the same maximum gray level, we choose arbitrarily. From this point on, we begin the iterative process of finding the next boundary point, given the current and last boundary points. Working in the 3×3 neighborhood centered on the current boundary point, we examine the neighbor diametrically opposite the last boundary point and the neighbors on either side of it (Fig. 7.13a). The next boundary point is one of those three that has the highest gray level. If all three or two adjacent boundary points share the

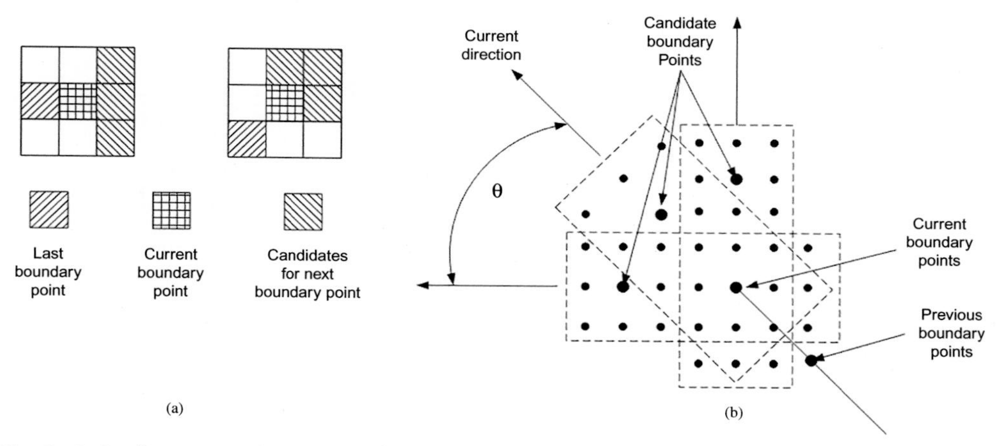

Fig. 7.13 An illustration of the boundary tracking process. (a) boundary tracking with a 3×3 window. (b) boundary tracking bug.

highest gray level, then we choose the middle one. If the two nonadjacent points share the highest gray level, we choose arbitrarily.

With the assumption of a noise-free image of a monotone spot, this algorithm will trace out the maximum gradient boundary nicely. However, if noise is present, the tracking is likely to move away from the boundary. Noise effects can be reduced by smoothing the gradient image before tracking or by implementing a "tracking bug," which works as follows. First a rectangular averaging window, usually having uniform weights, is defined to embody the bug (Fig. 7.13b). The last two or last few boundary points define the current boundary direction. The rear portion of the bug is centered on the current boundary point, with its axis oriented along the current direction. The bug is subsequently oriented at an angle θ to either side, and the average gradient under the bug is computed for each position. The next boundary point is then taken as one of the pixels under the front portion of the bug when it is in the highest average gradient position.

Essentially, the tracking bug is a spatially larger implementation of the boundary tracking procedure described earlier. The larger size of the bug implements smoothing of the gradient image and makes it less susceptible to noise. It also limits how sharply the boundary can change directions. The size and shape of the bug may be adjusted for best performance. The "inertia" of the bug can be increased by reducing the side-looking angle θ.

In practice, boundary tracking on gradient magnitude images is useful only in low-noise cases. The tracking algorithms do not guarantee closed boundaries and can even get lost in cases where the noise level is high.

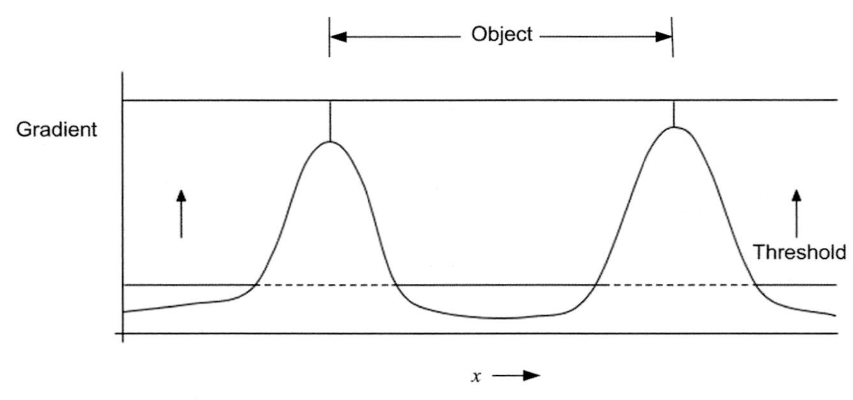

Fig. 7.14 1D illustration of gradient image thresholding using Kirsch's algorithm.

7.3.3 Boundary Finding Based on Gradient Image Thresholding

If we threshold a gradient image at moderate gray level, we find both object and background below threshold and most edge points above (Fig. 7.14). Kirsch's segmentation method makes use of this phenomenon [33]. In this technique, one first thresholds the gradient at a moderately low level to identify the object and the background, which are separated by bands of edge points that are above threshold. Then the threshold is gradually increased. This causes both the object and the background to grow. When they touch, they are not allowed to merge, but rather the points of contact define the boundary. This is essentially an application of the watershed algorithm to the gradient image.

While this method is computationally more expensive than thresholding, it tends to produce maximum gradient boundaries, and it avoids many of the problems of gradient tracking. For multiple object images, the segmentation is correct if and only if it is done accurately by the initial thresholding step. Smoothing the gradient image beforehand may help produce smoother boundaries.

7.3.4 Boundary Finding Based on Laplacian Image Thresholding

The Laplacian is a scalar second derivative operator for 2D images. It is defined as

$$\nabla^2 I(x, y) = \frac{\partial^2}{\partial x^2} I(x, y) + \frac{\partial^2}{\partial y^2} I(x, y) \tag{7.12}$$

and can be implemented digitally by any of the following convolution kernels:

$$l_1 = \begin{bmatrix} 0 & -1 & 0 \\ -1 & 4 & -1 \\ 0 & -1 & 0 \end{bmatrix}, \quad l_2 = \begin{bmatrix} -1 & -1 & -1 \\ -1 & 8 & -1 \\ -1 & -1 & -1 \end{bmatrix}, \quad l_3 = \begin{bmatrix} 1 & -2 & 1 \\ -2 & 4 & -2 \\ 1 & -2 & 1 \end{bmatrix} \tag{7.13}$$

The Laplacian has the advantage that it is an isotropic measure of the second derivative. The edge magnitude is independent of the orientation and can be obtained by convolving the image with only one kernel. As a second derivative operator, the Laplacian produces an abrupt zero crossing at an edge that is easy to find in a noise-free image. Thresholding a Laplacian filtered image at zero gray level may produce closed connected contours at the boundaries of objects. The presence of noise, however, imposes a requirement for a smoothing operation prior to using the Laplacian. Usually, a Gaussian filter is chosen for this purpose. Since convolution is associative, we can combine the Gaussian and Laplacian into a single *Laplacian of Gaussian* (LoG) kernel [34,35]:

$$\mathrm{LoG}(x, y) = -\nabla^2 \frac{1}{2\pi\sigma^2} e^{-\frac{x^2+y^2}{2\sigma^2}} = \frac{1}{\pi\sigma^4}\left[1 - \frac{x^2+y^2}{2\sigma^2}\right] e^{-\frac{x^2+y^2}{2\sigma^2}} \tag{7.14}$$

This LoG filter is also known as the *Mexican hat* filter since it has the shape of a positive peak in a negative dish. It is separable in the x- and y-directions and thus can be implemented efficiently. The parameter, σ, controls the width of the peak, which is related to the amount of smoothing. The edge positions can be determined by the zero crossings in the LoG filtered image.

7.3.5 Boundary Finding Based on Edge Detection and Linking

As discussed earlier, edges are closely associated with the boundaries of objects in an image. Pixels exhibiting the required characteristics are known as edge points. Edge points can be detected and used to establish the boundaries. We can examine each of these pixels and its immediate neighborhood to determine if it is, in fact, on the boundary of an object. Due to noise and shading effects, edge points seldom form the closed connected boundaries that are required for image segmentation. Thus a linking process is usually required to fill in the gaps and associate nearby edge points so as to create a closed connected boundary.

7.3.5.1 Edge Detection

The goal of edge detection is to mark the pixels in an image at which the gray level changes sharply. The two parameters of interest are the slope and direction of the transition. An image in which gray level reflects how strongly each pixel meets the requirements of an edge point is called an "edge map," whereas an image that encodes the direction of the edge instead of the magnitude is known as a "directional edge map." An example pair of edge map and directional edge map is the magnitude and direction of the gradient vector of an image. Edge detection operators examine each pixel neighborhood and quantify the slope, and often the direction as well, of the gray level transition. Most of these operators perform a 2D spatial gradient measurement on an image $I(x, y)$ using convolution with a pair of horizontal and vertical derivative kernels,

g_x and g_y. Each pixel in the image is convolved with both kernels, one estimating the gradient in the x-direction and the other in the y-direction. These kernels are designed to respond maximally to edges running horizontally and vertically relative to the pixel grid. The output of these two convolutions can be combined to form the estimated absolute magnitude of the gradient $|G|$ and its orientation θ at each pixel. This gradient magnitude is computed by taking the square root of the sum of the squares of the outputs from the two orthogonal kernels, that is,

$$|G| = \sqrt{G_x^2 + G_y^2} \tag{7.15}$$

where G_x and G_y are the output of the estimated derivative function in x and y directions, respectively:

$$G_x = I(x, y) * g_x, \quad G_y = I(x, y) * g_y \tag{7.16}$$

Likewise, the gradient direction can be computed from the ratio of G_y and G_x by

$$\theta = \arctan\left(\frac{G_y}{G_x}\right) \tag{7.17}$$

In practical implementations, an approximation to the gradient magnitude is often used instead for faster computation. It is given by

$$|G| = |G_x| + |G_y| \tag{7.18}$$

In the following, we discuss several sets of widely used derivative-based kernels for edge detection.

The Roberts Edge Detector

The Roberts operator represents one of the earliest methods of finding edges in an image using small convolution kernels to approximate the first derivative of the image [36]. It uses the following 2×2 derivative kernels:

$$g_x = \begin{bmatrix} 1 & 0 \\ 0 & -1 \end{bmatrix}, \quad g_y = \begin{bmatrix} 0 & 1 \\ -1 & 0 \end{bmatrix} \tag{7.19}$$

The Sobel Edge Detector

The Sobel operator is characterized by the following pair of 3×3 convolution kernels [37]:

$$g_x = \begin{bmatrix} -1 & 0 & 1 \\ -2 & 0 & 2 \\ -1 & 0 & 1 \end{bmatrix}, \quad g_y = \begin{bmatrix} -1 & -2 & -1 \\ 0 & 0 & 0 \\ 1 & 2 & 1 \end{bmatrix} \tag{7.20}$$

Compared with the Roberts operator, the Sobel operator is a little slower to compute, but its larger convolution kernels smooth the image to a greater extent, making it less sensitive to noise. It also generally produces considerably higher output values for similar edges [37].

The Prewitt edge detector

The Prewitt operator is related to the Sobel operator and uses slightly different kernels [38]:

$$g_x = \begin{bmatrix} 1 & 0 & -1 \\ 1 & 0 & -1 \\ 1 & 0 & -1 \end{bmatrix}, \quad g_y = \begin{bmatrix} -1 & -1 & -1 \\ 0 & 0 & 0 \\ 1 & 1 & 1 \end{bmatrix} \tag{7.21}$$

This operator produces results similar to those of the Sobel operator, but is less isotropic in its response.

The Canny Edge Detector

Generally, edge detection based on the aforementioned derivative-based operators is sensitive to noise. This is because computing the derivatives in the spatial domain corresponds to high-pass filtering in the frequency domain, thereby accentuating the noise. Furthermore, determining edge points by a simple thresholding of the edge map (e.g., the gradient magnitude image) is error prone, since it assumes all the pixels above the threshold fall upon edges. When the threshold is low, more edge points will be detected, and the results become increasingly susceptible to noise. On the other hand, when the threshold is high, subtle edge points may be missed. These problems are addressed by the Canny edge detector, which uses an alternative way to look for and track local maxima in the edge map [39].

The Canny operator is a multistage edge detection algorithm. The image is first smoothed by convolving with a Gaussian kernel. Then a first derivative operator (usually the Sobel operator) is applied to the smoothed image to obtain the spatial gradient measurements, and the pixels with gradient magnitudes that form local maxima in the gradient direction are determined. As local gradient maxima produce ridges in the edge map, the algorithm then performs the so-called "nonmaximum suppression" by tracking along the tops of these ridges and setting to zero all pixels that are not on the ridge top. The tracking process uses a dual-threshold mechanism, known as "thresholding with hysteresis," to determine valid edge points and eliminate noise. The process starts at a point on a ridge higher than the upper threshold. Tracking then proceeds in both directions out from that point until the point on the ridge falls below the lower threshold. The underlying assumption is that important edges are along continuous paths in the image. The dual-threshold mechanism allows one to follow a faint section of a given edge and to discard those noisy pixels that do not form paths but nonetheless produce large gradient magnitudes. The result is a binary image where each pixel is labeled as either an edge point or a nonedge point.

7.3.5.2 Edge Linking and Boundary Refinement

An edge map characterizes the objects in an image with edge points. If the edges are strong enough and the noise level is low, one can threshold an edge map and thin the resulting binary image down to single-pixel-wide closed connected boundaries. Under less-than-ideal conditions, however, the edge points seldom form the closed connected boundaries required for image segmentation. Hence another step is usually required to complete the delineation of object boundaries before object extraction can be performed. "Edge linking" is the process of associating nearby edge points so as to create a closed connected boundary. This process fills in the gaps in the edge map that are often caused by noise and shading in the image.

Generally, edge linking for small gaps can be accomplished by searching a neighborhood around an endpoint for other endpoints, and then filling in boundary pixels as required to connect them. Typically, this neighborhood is a square region of 5×5 or larger. In complex scenes with dense edge points, however, this can oversegment the image. To avoid oversegmentation one can require that the two endpoints agree in gradient magnitude and orientation to within specified tolerances before they are allowed to be connected.

Heuristic Search

Some boundary gaps in an edge map are too wide to be filled accurately with a straight line. For those, one can establish, as a quality measure, a function that can be computed for every connected path between the two endpoints, which we denote as A and B. This edge quality function may, for example, be defined to be the average of the gradient magnitudes of the points, minus some measure of their average disagreement in orientation angles [40,41].

The search starts by evaluating the neighbors of endpoint A as the candidates for taking the first step toward B. Normally only the neighbors that lie in the general direction of endpoint B are considered. The one that maximizes the edge quality function from A to that point is selected. Then it becomes the starting point for the next iteration. When the linking finally reaches B, the edge quality function over the newly created path is compared to a threshold. If the newly created edge is sufficiently strong, it is accepted. Otherwise, it is discarded.

Heuristic search techniques usually perform well in relatively simple images, but they do not necessarily converge to the globally optimal path between endpoints. They become computationally expensive if the gaps to be traversed are numerous and wide, in which case complicated edge quality functions must be used.

Curve Fitting

So far, we discussed edge linking using searching or tracking methods. All of these require the existence of a continuous path of edge points. If the edge points are so sparse that few connected or even nearby edge points are available, it might be desirable to fit a piecewise linear or higher order spline curve through them to establish a boundary suitable for object extraction. For example, one can use a piecewise linear method called "iterative endpoint fitting" [42] for this purpose. Suppose there is a group of edge points lying scattered between two particular edge points A and B, and a subset of these are to be selected to form the nodes of a piecewise linear path from A to B. We start by establishing a straight line from A to B and continue by computing the perpendicular distance from that line to each of the remaining edge points. The furthermost one becomes the next node on the path, which now has two branches. This process is repeated on each new branch of the path until no remaining edge point lies more than a specified distance away from the nearest branch. When this type of fitting is done all around the object, it produces a polygonal approximation to the boundary.

The Hough Transform

The Hough transform [43,44] can detect shapes and establish object boundaries in an image by recognizing evidence in a transformed parameter space. Because the transform requires the data points to be specified in some parametric form, the technique is most commonly used to detect curves of regular shape, such as lines, circles, and ellipses. It is particularly useful when the input image is noisy, and the data points are sparse.

Given an equation for a parameterized 2D curve

$$f(x, y, t_1, \ldots, t_n) = 0 \qquad \text{where } n \geq 2 \tag{7.22}$$

which defines the curve in the (x, y) plane, and $t_1, \ldots, t_n$ are the parameters, one can first select a candidate set of points from the image, such as those points that have high probability of being located on object boundaries. In the n-dimensional parameter space, a histogram is constructed to quantify the strength of evidence with respect to different parameter values. Each edge point that satisfies Eq. (7.22) is transformed into a curve in the parameter space, and the histogram bins that lie along this curve are incremented. In this way, each edge point in the image votes for the parameterized curve it fits best, and the histogram distribution characterizes the relative strength of evidence that the curves with parameter values, $t_1, \ldots, t_n$, are detected in the image.

As an example, a straight line $y = kx + b$ can be expressed in polar coordinates as [42]

$$\rho = x\cos(\theta) + y\sin(\theta) \tag{7.23}$$

where $[\rho, \theta]$ defines a vector from the origin to the nearest point on that line (Fig. 7.15a). This vector will be perpendicular to the line. One can consider a two-dimensional parameter space defined by ρ and θ. Any line in the x, y plane corresponds to a point

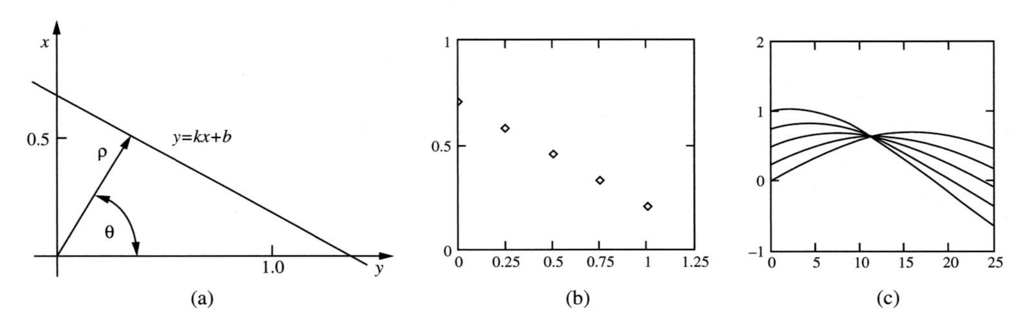

Fig. 7.15 An illustration of the Hough transform: (a) A *straight line* and its polar coordinates; (b) Sample edge points in x, y space; (c) The sinusoids in ρ, θ space that correspond to the edge points.

in the parameter space. Thus the Hough transform of a straight line in image space is a point in the ρ, θ space.

Now consider a particular point x_1, y_1 in the x, y plane. There are many straight lines that pass through the point x_1, y_1, and each of these corresponds to a point in the ρ, θ space. These points, however, must satisfy Eq. (7.23) with x_1 and y_1 as constants. Thus the locus of all such lines in the x, y space is a sinusoid in the parameter space, and any point in the x, y space (Fig. 7.15b) corresponds to a sinusoidal curve in the ρ, θ space (Fig. 7.15c).

If a set of edge points x_i, y_i lies on a straight line with parameters ρ_0, θ_0, then each edge point corresponds to a curve in the ρ, θ space. However, all these curves must intersect at the point ρ_0, θ_0 since this is a line they all have in common (Fig. 7.15c).

Thus to find the straight-line segment that the edge points fall upon, one can set up a two-dimensional histogram in the ρ, θ space. For each edge point x_i, y_i, all the histogram bins in the ρ, θ space that lie on the sinusoid curve for that point are incremented. When this is done for all the edge points, the bin containing ρ_0, θ_0 will be a local maximum in the 2D histogram. One can then search the histogram in ρ, θ space for local maxima and obtain the parameters of linear boundary segments.

Similarly, one can detect and establish boundaries of circular objects using the Hough transform. In this case the parametric equation is

$$(x-a)^2 + (y-b)^2 = r^2 \tag{7.24}$$

where a and b are the coordinates of the center of the circle and r is its radius. Notice that the computational complexity begins to increase, as there are now three variables in the parameter space, and hence a 3D histogram must be constructed. In general, the computational complexity of the transform increases substantially with the number of parameters required to represent the curve. Hence, the Hough transform described here is practical only for curves for which simple analytic parameterizations exist. Nonetheless, the advantages of the technique are a tolerance of gaps in data points, and relatively robust performance when image noise is present.

Active Contours

As the shape of many natural objects cannot be accurately described by rigid graph representations, such as polygons and circles, edge linking based on the techniques discussed so far often results in only coarsely delineated object boundaries. The *active contour*, or "snake," is a boundary refinement technique [45]. The active contour model allows a simultaneous solution for both the segmentation and tracking problems and has been applied successfully in a number of applications [46,47]. The model uses a set of connected points (called a snake), which can move around so as to minimize an energy function formulated for the problem at hand. The curve formed by the connected points delineates the active contour. Properties of the image (e.g., gray level, gradient, etc.) contribute to the energy of the snake, as do constraints on the continuity and curvature of the contour itself. In this way, the snake contour can react to the image and move in a continuous manner, ensuring continuity and smoothness as it converges toward the desired object boundary. Furthermore, the iterative nature of the algorithm allows active adjustment of the weights employed in the energy function to affect the dynamic behavior of the active contour. Processing time is mostly consumed by the first iteration, and subsequent iterations require little additional computation time.

A number of different implementations of active contour have been described in the literature. The first seminal approach was developed using variational calculus and spline models [45]. Other approaches include dynamic programming [48], neural networks [49], and "greedy" algorithms [50]. There are various advantages and disadvantages to each approach. Since dynamic programming and neural network approaches are known to be computationally intensive, variational calculus and greedy algorithms are often preferred. Their main advantages are relative algorithmic simplicity and computational efficiency. The main disadvantage is the extremely local nature of the decision criteria used.

The crucial part of active contour methodology is the formulation of the energy minimization function. Following the notation in [45], given a parametric representation of the snake, $v(s)=(x(s), y(s))$, where $s \in [0, \ 1]$, the energy function can be written as

$$\begin{aligned} E^*_{snake} &= \int_0^1 [E_{snake}(v(s))]ds \\ &= \int_0^1 \left[E_{int}(v(s)) + E_{image}(v(s)) + E_{const}(v(s))\right] ds \end{aligned} \tag{7.25}$$

This is simply an integration of energy along the length of the contour, which in the discrete greedy model would correspond to summing the energy of all the points on the contour after one iteration of the snake. In the greedy implementation, this integration is not actually performed. The greedy nature of the algorithm assumes that local greedy decisions at each contour point automatically result in a tendency toward a global minimum of this function. The energy terms in Eq. (7.25) correspond, respectively, to (1) internal forces between points of the contour (analogous to tension and rigidity), (2) image forces such as gradient magnitude or gray level magnitude, and (3) external constraints. Note that each term in the energy function, E, once computed, must be normalized and weighted in the following manner:

$$E = \omega \cdot \left[\frac{(min - \varepsilon)}{(max - min)} \right] \tag{7.26}$$

where ε is the energy term, ω is a contribution weight, and *min* and *max* are the minimum and maximum energy computations, respectively, in the search neighborhood of a contour point. The internal energy is modeled using two terms, a continuity term (tension) and a curvature term (rigidity), that is,

$$E_{int} = E_{cont} + E_{curv} \tag{7.27}$$

Often, the continuity term is made proportional to the distance between the point being examined and the previous point on the contour. This, however, causes the snake to either contract or expand depending on the sign of the contribution weight. Because the snake boundary is expected to remain close to the original initialized boundary, the continuity term will be made proportional to the difference between that distance and the average interpoint distance for the snake, i.e.,

$$\varepsilon_{cont} = \overline{d} - \| \mathrm{v}_i - \mathrm{v}_{i-1} \| \tag{7.28}$$

where v_i denotes the coordinates of the ith contour point and $\overline{d}$ is the average interpoint distance, calculated at the end of each snake iteration. This will not only encourage equal spacing of points but cause the snake to contract if the interpoint distance is larger than the average and expand if it is smaller. The energy associated with the curvature at a point is approximated by taking the square of the magnitude of the difference between two adjacent unit tangent vectors,

$$\varepsilon_{curv} = \left\| \frac{\vec{\mathrm{u}}_{\mathrm{i}+1}}{\|\vec{\mathrm{u}}_{\mathrm{i}+1}\|} - \frac{\vec{\mathrm{u}}_{\mathrm{i}}}{\|\vec{\mathrm{u}}_{\mathrm{i}}\|} \right\|^2 \tag{7.29}$$

where $\vec{\mathrm{u}}_i = \mathrm{v}_i - \mathrm{v}_{i-1}$ and $\frac{\vec{\mathrm{u}}_{\mathrm{i}}}{\|\vec{\mathrm{u}}_{\mathrm{i}}\|}$ is a discrete approximation to the unit tangent at u_{i}. This gives a quick and reasonable estimate of local curvature and, in general, has the effect of causing the contour to straighten, thus favoring smoother outlines. The contributing measurement for the image energy term can be, for example, the result of a Sobel gradient operator, that is,

$$\varepsilon_{image} = \sqrt{\left[\|\nabla I_x(x, y)\|^2 + \|\nabla I_y(x, y)\|^2\right]} \tag{7.30}$$

Note that the normalization causes the snake to be affected by relative local gradients, regardless of the strength of the gradient. This may be especially effective in localizing the characteristically low contrast cell boundaries that are often found in microscope images.

The term E_{const} corresponds to external constraints that can be modeled with prior knowledge, such as the shape of the objects. For example, one can investigate the use of shape constraints to aid in the segmentation of cells with indistinct overlapping boundaries. A shape bias contributing to the energy of the snake can act as a guiding force to prevent the boundary from taking on an impossible shape.

Fig. 7.16 shows an example of applying the active contour method to the delineation of a cell boundary in a fluorescence microscope image. After 15 iterations of energy function minimization, the improvement in the accuracy of boundary delineation is clearly noticeable.

7.3.6 Encoding Segmented Images

After image segmentation, sometimes it is not necessary to extract the objects from the original image if only gross measurements (e.g., area) of each object are required. In other

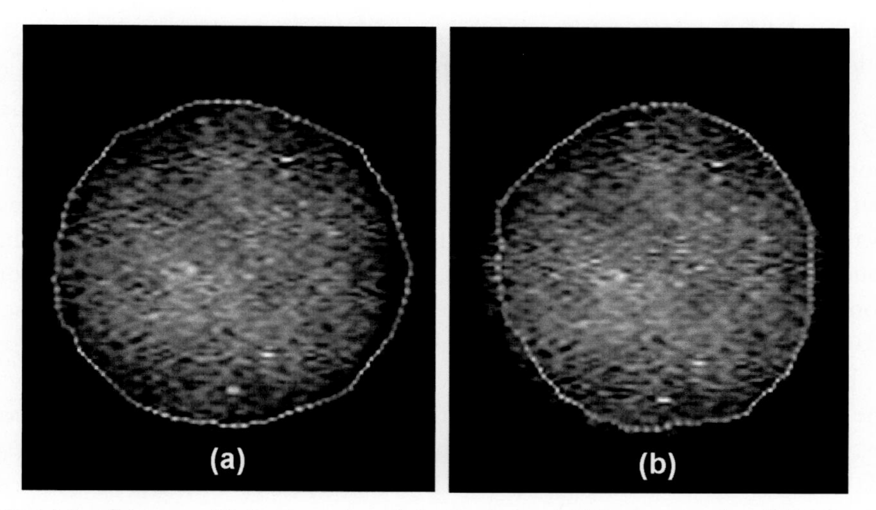

Fig. 7.16 (a) A cell image with the boundary delineated after initial segmentation. (b) The same cell image with improved boundary delineation after 15 iterations of active contour computation.

cases, however, it may be desirable to compose a new image showing the objects extracted or display each object in a separate image. One may also wish to perform further measurement or other processing on each of the individual objects. Hence encoding a segmented image in a convenient format facilitates subsequent measurement and processing of the individual objects. Usually, each object in an image is assigned a sequence number as it is found. This object number can be used to identify and track the individual objects in the image.

7.3.6.1 The Object Label Map

From the segmented regions in an image, one can generate a separate image of size equal to the original, to encode object membership on a pixel-by-pixel basis. In the object label map, the gray level of each pixel encodes the sequence number of the object to which the corresponding pixel in the original image belongs. For instance, all pixels belonging to the 11th object in the image will have a gray level of 11 in the label map.

The object label map is not a particularly compact approach to storing segmentation information as it requires an additional full-size image to encode object membership, even when the image contains only one small object. However, images of this type can be compressed quite significantly since they normally contain large areas of constant gray level. If only object size and shape are of interest for subsequent analysis, the original image may be discarded after segmentation. Further data reduction is possible when there is only one object or if the objects need not be differentiated. In this case the label map becomes a binary image.

In some cases, the computation requirements of an image segmentation algorithm dictate that the process be carried out in several passes over the image data. A binary or multilevel label map is often useful as an intermediate data representation in a multiple-pass image segmentation procedure.

7.3.6.2 The Boundary Chain Code

The boundary chain code (BCC) is another well-known technique often used to encode the results of image segmentation. Compared with the object label map, it is a more compact way to store the image segmentation information [51]. With such an approach, only the boundary is required to define an object, and therefore it is not necessary to store the location of interior pixels. Furthermore, the boundary chain code exploits the fact that boundaries are connected paths; hence highly efficient data encoding, with little redundancy, becomes possible.

The chain code starts by specifying an arbitrarily selected starting point with coordinate (x, y) located on the boundary of the object. This point has eight neighbors. At least one of these must also be a boundary point. The boundary chain code specifies the direction in which a step must be taken to go from the present boundary point to the next. Fig. 7.17 shows the eight directions of the neighboring pixels. The eight possible

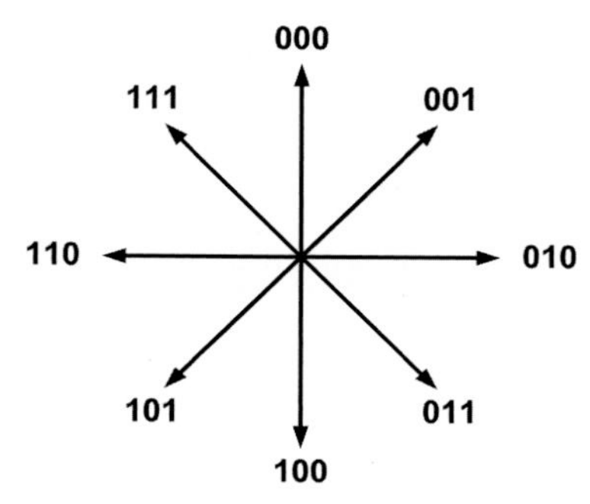

Fig. 7.17 Boundary direction codes.

directions can be represented by three bits. Thus, the boundary chain code consists of the coordinates of the starting point, followed by a sequence of direction codes that specify the path around the boundary.

With the boundary chain code, encoding the segmentation of an object requires only one (x, y) coordinate and then three bits for each boundary point. This is considerably less data than that required for the object label map. When a complex scene is segmented, the program can encode each object boundary as a single record consisting of the object number, the perimeter (number of boundary points), and the chain code. Several size and shape features can also be computed directly from the chain code. If needed, even the object label map can be constructed from the boundary chain code by filling the enclosed region with the object number using a region-filling algorithm [52].

Generation of the boundary chain code usually requires random access to the input image as the boundary must be tracked through the image. The operation is a natural adjunct to the boundary tracking and refinement procedures of image segmentation. When further processing of individual object images is required, the chain code becomes less useful since the interior points of objects must be accessed and computed.

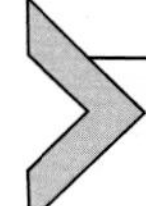

7.4 Summary of Important Points

1. Image segmentation is the process that partitions an image into disjoint regions consisting of connected sets of pixels. These regions correspond to either the background or the objects in the image.
2. Region-based and boundary-based methods are different but complementary approaches to image segmentation.
3. Region-based techniques partition the image into sets of interior and exterior pixels according to similarity of image properties.

4. Boundary-based techniques establish object boundaries by detecting edge pixels that are associated with differences in image properties.
5. Gray level thresholding is a simple region-based segmentation technique.
6. Unless the background gray level and object contrast are relatively constant, it is usually necessary to vary the threshold within the image. This is adaptive thresholding.
7. The selection of the threshold value is crucial and can significantly affect the boundaries and areas of segmented objects.
8. For images of simple objects on a contrasting background, placing the threshold at the dip of the bimodal histogram minimizes the sensitivity of object area measurement to threshold variation.
9. Both the profile function of a concentric circular spot and the average gradient around a contour line can be derived from the histogram or from the perimeter function of its image.
10. Morphological processing can improve the initial segmentation results from thresholding by using procedures such as separation of touching objects and filling of internal holes.
11. Unlike thresholding, region growing and splitting techniques exploit the spatial context in complex scenes.
12. Region growing combines adjacent regions into larger regions, within which the pixels have similar properties.
13. Region splitting partitions larger regions into smaller adjacent regions, within which the pixels have different properties.
14. Edges correspond to the image points where gray level changes abruptly, and usually occur on object boundaries. Edge points can be detected and used to establish the boundaries of objects.
15. Gradient-based methods detect edges by looking for the pixels with large gradient magnitude.
16. Laplacian-based methods search for zero crossings in the second derivative of the image to find edges.
17. Object boundaries can be established by thresholding either the gradient image or the Laplacian image if edges are strong and the noise level is low.
18. The detected edge points seldom form the closed connected boundaries that are required for image segmentation. Therefore edge linking and boundary refinement are usually performed to complete the object boundary delineation process.
19. The Hough transform can fit a parameterized boundary function to a scattered set of edge points.
20. Active contours can be used to refine boundaries that have been found by other methods.
21. The result of image segmentation can be encoded and stored conveniently either as an object label map or as a boundary chain code.

References

[1] E. Davies, Computer Vision, fifth ed., Elsevier, 2018.
[2] K.R. Castleman, Digital Image Processing, Prentice Hall, 1996.
[3] R. Gonzales, R. Woods, Digital Image Processing, fourth ed., Pearson, 2018.
[4] A.K. Jain, Fundamentals of Digital Image Processing, Prentice Hall, 1989.
[5] H.K.D.H. Bhadeshia, Neural networks in material science, ISIJ Int. 39 (10) (1999) 966–979.
[6] C. Charu, Aggarwal, Neural Networks and Deep Learning—A Textbook, Springer, 2018.
[7] I.R.I. Hague, J. Neubert, Deep learning approaches to biomedical image segmentation, Inform. Med. Unlocked (2020), 100297.
[8] Z. Jia, X. Huang, E.I.G. Chang, Y. Xu, Constrained deep weak supervision for histopathology image segmentation, IEEE Trans. Med. Imaging 36 (11) (2017) 2376–2388.
[9] S. Masubuchi, E. Watanabe, Y. Seo, et al., Deep-learning-based image segmentation integrated with optical microscopy for automatically searching for two-dimensional materials, NPJ 2D Mater. Appl. 4 (2020) 3, https://doi.org/10.1038/s41699-020-0137-z.
[10] A. Rosenfield, Connectivity in digital pictures, J. ACM 17 (1970) 146–160.
[11] J.M. Prats-Montalbán, A. Ferrer, Integration of colour and textural information in multivariate image analysis: defect detection and classification issues, J. Chemometr. 21 (1) (2007) 10–23.
[12] E. Bala, A. Ertuzun, A multivariate thresholding technique for image denoising using multiwavelets, EURASIP J. Adv. Signal Process. 8 (2005) 1205–1211.
[13] J.C. Noordam, W.H.A.M. van den Broek, L.M.C. Buydens, Multivariate image segmentation with cluster size insensitive Fuzzy C-means, Chemom. Intel. Lab. Syst. 64 (1) (2002) 65–78.
[14] F. Liu, et al., Adaptive thresholding based on variational background, Electron. Lett. 38 (18) (2002) 1017–1018.
[15] R.J. Wall, The Gray Level Histogram for Threshold Boundary Determination in Image Processing to the Scene Segmentation Problem in Human Chromosome Analysis (PhD thesis), University of California at Los Angeles, 1974.
[16] K.R. Castleman, R. Wall, Automatic systems for chromosome identification, in: T. Caspersson (Ed.), Nobel Symposium 23—Chromosome Identification, Academic Press, 1973.
[17] K.R. Castleman, J. Melnyk, An Automated System for Chromosome Analysis: Final Report, Document No. 5040-30, Jet Propulsion Laboratory, Pasadena, CA, 1976.
[18] P.K. Sahoo, S. Soltani, K.C. Wong, A survey of thresholding techniques, Comput. Vis. Graph. Image Process. 41 (1988) 233–260.
[19] C.A. Glasbey, An analysis of histogram based thresholding operations, Comput. Graphics Image Process. 55 (1993) 532–537.
[20] T.W. Ridler, S. Calvard, Picture thresholding using an iterative selection method, IEEE Trans. Syst. Man Cybern. 8 (8) (1978) 630–632.
[21] I.T. Young, J.J. Gerbrands, L.J. van Vliet, Fundamentals of image processing, in: V.K. Madisetti (Ed.), Video, Speech, and Audio Signal Processing and Associated Standards, Series: The Digital Signal Processing Handbook, second ed., CRC Press, 2009, pp. 1–84.
[22] G.W. Zack, W.E. Rogers, S.A. Latt, Automatic measurement of sister chromatid exchange frequency, J. Histochem. Cytochem. 25 (7) (1977) 741–753.
[23] J. Weszka, A survey of threshold selection techniques, Comput. Graphics Image Process. 7 (1978) 259–265.
[24] M.E. Sieracki, S.E. Reichenbach, K.L. Webb, Evaluation of automated threshold selection methods for accurately sizing microscopic fluorescent cells by image analysis, Appl. Environ. Microbiol. 55 (11) (1989) 2762–2772.
[25] C.L. Viles, M.E. Sieracki, Measurement of marine picoplankton cell size by using a cooled, charge-coupled device camera with image-analyzed fluorescence microscopy, Appl. Environ. Microbiol. 58 (2) (1992) 584–592.
[26] S.R. Sternberg, Parallel architectures for image processing, in: Proc. 3rd International IEEE COMPSAC, 1981.
[27] S.R. Sternberg, Biomedical image processing, IEEE Comput. 16 (1) (1983) 22–34.
[28] R.M. Lougheed, D.L. McCubbrey, The cytocomputer: a practical pipelined image processor, in: Proc. 7th Annual International Symposium on Computer Architecture, 1980.

[29] Y. Tuduki, et al., Automated seeded region growing algorithm for extraction of cerebral blood vessels from magnetic resonance angiographic data, in: Proceedings of the International Conference of the IEEE Engineering in Medicine and Biology Society, vol. 3, 2000, pp. 1756–1759.
[30] T. Pavlidis, Y.T. Liow, Integrating region growing and edge detection, IEEE Trans. Pattern Anal. Mach. Intell. 12 (3) (1990) 225–233.
[31] S. Zucker, Region growing: childhood and adolescence, Comput. Graphics Image Process. 5 (1976) 382–399.
[32] S. Chen, W. Lin, C. Chen, Split-and-merge image segmentation based on localized feature analysis and statistical tests, Comput. Graphics Image Process. 53 (5) (1991) 457–475.
[33] R.A. Kirsch, Computer determination of the constituent structure of biological images, Comput. Biomed. Res. 4 (1971) 315–328.
[34] D. Marr, E. Hildreth, Theory of edge detection, Proc. R. Soc. Lond. B Biol. Sci. 207 (1980) 187–217.
[35] D. Marr, Vision, Freeman, 2010.
[36] J.K. Aggarwal, R.O. Duda, A. Rosenfeld (Eds.), Computer Methods in Image Analysis, IEEE Press, 1977.
[37] L.S. Davis, A survey of edge detection techniques, Comput. Graphics Image Process. 4 (1975) 248–270.
[38] J. Prewitt, Object enhancement and extraction, in: B. Lipkin, A. Rosenfeld (Eds.), Picture Processing and Psychopictorics, Academic Press, 1970.
[39] J. Canny, A computational approach to edge detection, IEEE Trans. Pattern Anal. Mach. Intell. 8 (1986) 679–714.
[40] R. Nevatia, Locating object boundaries in textured environments, IEEE Trans. Comput. 25 (1976) 1170–1180.
[41] J.M. Lester, et al., Two graph searching techniques for boundary finding in white blood cell images, Comput. Biol. Med. 8 (1978) 293–308.
[42] R.O. Duda, P.E. Hart, D. Stork, Pattern Classification, Wiley, 2001.
[43] D. Ballard, C. Brown, Computer Vision, Prentice-Hall, 1982.
[44] D. Ballard, Generalizing the Hough transform to detect arbitrary shapes, Pattern Recogn. 13 (2) (1981) 111–122.
[45] M. Kass, A. Witkin, D. Terzopoulos, Snakes: active contour models, in: Proc. of First International Conf. on Computer Vision, 1987, pp. 259–269.
[46] P. Fua, A.J. Hanson, An optimization framework of feature extraction: applications to semiautomated and automated feature extraction, in: Proc. of the DARPA Image Understanding Workshop, 1989, pp. 676–694.
[47] C. Garbay, Image structure representation and processing: a discussion of some segmentation methods in cytology, IEEE Trans. Pattern Anal. Mach. Intell. 8 (1986) 140–146.
[48] A. Amini, S. Tehrani, T.E. Weymouth, Using dynamic programming for minimizing the energy of active contours in the presence of hard constraints, in: Proc. Second International Conference on Computer Vision, 1988, pp. 95–99.
[49] C.T. Tsai, Y.N. Sun, P.C. Chung, Minimizing the energy of active contour model using a Hopfield network, IEEE Proc. 140 (6) (1993) 297–303.
[50] D.J. Williams, M. Shah, A fast algorithm for active contours, Comput. Graphics Image Process. 55 (1) (1992) 14–26.
[51] H. Freeman, Boundary encoding and processing, in: B. Lipkin, A. Rosenfeld (Eds.), Picture Processing and Psychopictorics, Academic Press, 1970.
[52] T. Pavlidis, Filling algorithms for raster graphics, Comput. Graphics Image Process. 10 (1979) 126–141.

CHAPTER EIGHT

Object Measurement

Fatima A. Merchant, Shishir K. Shah, and Kenneth R. Castleman

8.1 Introduction

Providing objectivity for any image processing task requires quantitative measurement of an area of interest extracted from an image or of the image as a whole. In Chapter 7, we discussed methods for segmenting or extracting objects from an image. In this chapter we discuss the problem of measuring each of the segmented objects so that a quantitative measurement can be associated with the extracted image region. Measuring object properties has been a subject of study since the early 1970s and is the culmination of considerable development [1–6].

Specific object measurement to be performed for an extracted image region is application dependent. It can be used simply to provide a measure of the object morphology or structure by defining its properties in terms of area, perimeter, intensity, color, shape, etc. It can also be used to discriminate between objects by measuring and comparing their properties. In this chapter, we introduce the basic concepts of object measurement. For a more detailed treatment of the subject matter, the reader should consult the broader image analysis literature [7–11].

An image that has undergone segmentation and perhaps morphological post-processing will clearly define objects from which measurements can be computed. The extracted objects can be treated either as binary objects or as gray level objects. In either case, the object of interest is presented with an object label map (Section 7.3.6). Binary objects are typically represented such that pixels belonging to the object take a value of "1" and the remaining pixels are "0."

Object measurements can be broadly classified as (1) geometric measures, (2) those based on the histogram of the object image, and (3) those based on the intensity of the object. Geometric measures include those that quantify object structure and can be computed for both binary and grayscale objects. In contrast, histogram- and intensity-based measures are applicable to grayscale objects. Another category of measures, which are distance-based, can be used for computing the distance between objects, or between two or more components of objects. In the rest of this chapter, we discuss some common measurements for both binary and gray level objects.

Microscope Image Processing
https://doi.org/10.1016/B978-0-12-821049-9.00017-4

Copyright © 2023 Elsevier Inc.
All rights reserved.

8.2 Measures for Binary Objects

A binary object can be described in terms of its size, shape, or distance to other objects. Some common measures are presented in this section.

8.2.1 Size Measures

The size of an object can be defined in terms of its area and perimeter. Area is a convenient measure of overall size. Perimeter is particularly useful for discriminating between objects with simple and complex shapes. Compared to irregular objects that have complex structures, a regular object with a simple shape uses less perimeter to enclose its area. Both area and perimeter measurements can be performed during the extraction of an object from a segmented image.

8.2.1.1 Area

Consider the function $I_n(i,j)$ described for an object label map of an $N \times N$ image (i.e., the result of segmentation as described in Chapter 7):

$$I_n(i,j) = \left\{ \begin{matrix} 1 & \text{if } I(i,j) = n^{\text{th}} \text{ object number} \\ 0 & \text{otherwise} \end{matrix} \right\} \tag{8.1}$$

The area in pixels of the nth object is then given by

$$A_n = \sum_{i=0}^{N-1} \sum_{j=0}^{N-1} I_n(i,j) \tag{8.2}$$

8.2.1.2 Perimeter

The simplest measure of perimeter is obtained by counting the number of boundary pixels that belong to an object. This can be obtained by counting the number of pixels that take a value of 1 and have at least one neighboring pixel with a value of 0. The neighborhood of a pixel is normally defined in terms of either its 4-connectivity or 8-connectivity (Section 7.1.1). Due to the discrete spatial arrangement of pixels, counting boundary pixels is normally biased since small changes in curvature of the boundary will result in a number of 45° or 90° turns. This produces an exaggerated estimate of the perimeter. Unbiased estimators of perimeter based on boundary pixel count using either 4-connectivity or 8-connectivity have been formulated assuming that there is a uniform distribution of orientation changes that occur in the boundary [7,12]. This is given as [12]:

$$\frac{N_4}{1.273} \text{ and } \frac{N_8}{0.900} \tag{8.3}$$

where N_4 and N_8 are the boundary pixel counts using 4-connectivity and 8-connectivity, respectively.

Another concern in computing the perimeter is differentiating between the internal and external perimeter of an object. It is generally understood that the true vertex point of a boundary pixel lies at the center of that pixel. Using the location of the boundary pixels for measuring the perimeter yields the internal perimeter, while using the boundary of the background pixels surrounding the object yields the external perimeter. A simple solution to this is to add π to the internal perimeter measurements [12]. More complicated methods resulting in better estimates of perimeter measurement have been developed as well [13–15].

8.2.1.3 Area and Perimeter of a Polygon

Without loss of generality, we can assume any object to be a polygon. Mathematically, there is a simple way to compute both the area and perimeter of a polygon in a single traversal of its boundary [11]. The area of a polygon can be measured as the sum of areas of all triangles formed by lines that connect the vertices of the polygon to an arbitrary point (x_0, y_0). Let us assume that point to be the origin, as shown in Fig. 8.1. Consider Fig. 8.2, which shows a single triangle having a vertex at the origin. As seen in the figure, the region is divided into rectangles by the horizontal and vertical lines, such that some of the rectangles have sides of the triangle as their diagonals. Thus half of each such rectangle is outside the triangle. By inspection of Fig. 8.2 we can write [11]

$$dA = x_1 y_2 - \frac{1}{2} x_2 y_2 - \frac{1}{2} x_1 y_1 - \frac{1}{2}(x_1 - x_2)(y_2 - y_1) \tag{8.4}$$

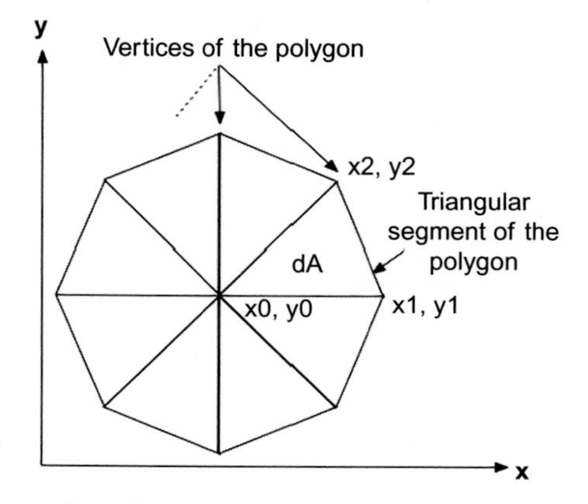

Fig. 8.1 Computing the area of a polygon.

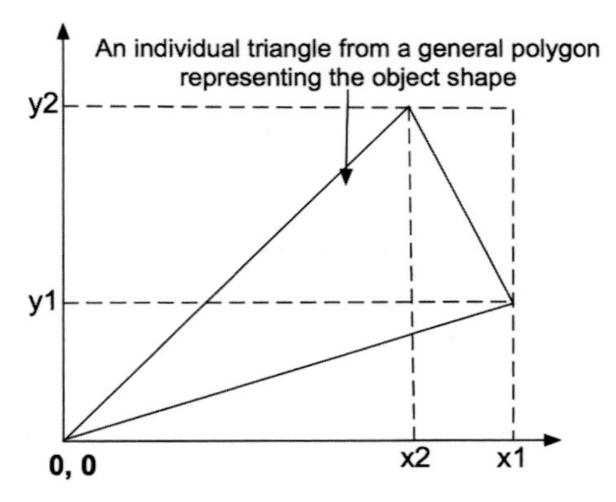

Fig. 8.2 Area measurement for a triangle.

This expression simplifies to

$$dA = \frac{1}{2}(x_1 y_2 - x_2 y_1) \tag{8.5}$$

and the total area becomes

$$A = \frac{1}{2} \sum_{i=1}^{N_b} [x_i y_{i+1} - x_{i+1} y_i] \tag{8.6}$$

where N_b is the total number of boundary points.

Note that if the origin falls outside the object, an area that is not within the polygon can be included in any particular triangle. Note also that depending on the direction in which the boundary is being traversed, the area of a particular triangle can be either positive or negative. By the time a complete traversal around the boundary is complete, the area that falls outside the object has been subtracted out.

Another approach is to use Green's theorem. This says that the area enclosed by a closed curve in the x-, y-plane is given by the contour integral [11]

$$A = \frac{1}{2} \int (x dy - y dx) \tag{8.7}$$

where the integration is carried out around the closed curve. We can discretize Eq. (8.7), [11] yielding

$$A = \frac{1}{2} \sum_{i=1}^{N_b} [x_i (y_{i+1} - y_i) - y_i (x_{i+1} - x_i)] \tag{8.8}$$

and manipulate this expression into the form of Eq. (8.6). The corresponding perimeter is the sum of side lengths of the polygon. Side lengths can be easily calculated from the boundary chain code (Section 7.3.6). If all boundary points are used as vertices, the perimeter will simply be the sum of lateral and diagonal steps written as [11]

$$P = N_e + \sqrt{2}N_o \tag{8.9}$$

where N_e is the number of even and N_o is the number of odd steps in the boundary chain code.

8.2.2 Pose Measures

The pose of an object is typically defined by its location and orientation. Measuring its centroid can indicate the location of an object. Object orientation is normally measured by computing the angle subtended by its major axis.

8.2.2.1 Centroid

Following the definition of Eq. (8.1), the center of the nth object, (i_n^c, j_n^c), can be given as

$$\begin{aligned} i_n^c &= \frac{1}{A_n}\sum_{i=0}^{N-1}\sum_{j=0}^{N-1} iI_n(i,j) \\ j_n^c &= \frac{1}{A_n}\sum_{i=0}^{N-1}\sum_{j=0}^{N-1} jI_n(i,j) \end{aligned} \tag{8.10}$$

where A_n is the area of that object. Since i and j index the image space, the center so computed will be relative to the image coordinate space. Thus the location of the object determined is within the two-dimensional image plane. The same measurement can also be obtained using moments, described in Section 8.2.3.5.

8.2.2.2 Orientation

One method of measuring the orientation of an object is based on computing the axis of the least second moment [16]. Intuitively, this defines the axis of least inertia that can be rotated to align it with the x-axis. Consider the arbitrary shape shown in Fig. 8.3. Defining the orientation of least inertia for this n^{th} object as θ_n, it can be calculated as [11]

$$\tan(2\theta_n) = \frac{2\sum_{i=0}^{N-1}\sum_{j=0}^{N-1} ijI_n(i,j)}{\sum_{i=0}^{N-1}\sum_{j=0}^{N-1} i^2 I_n(i,j) - \sum_{i=0}^{N-1}\sum_{j=0}^{N-1} j^2 I_n(i,j)} \tag{8.11}$$

8.2.3 Shape Measures

Shape measures are increasingly used as features in object recognition and classification applications to distinguish objects of one class from other objects. Shape features are

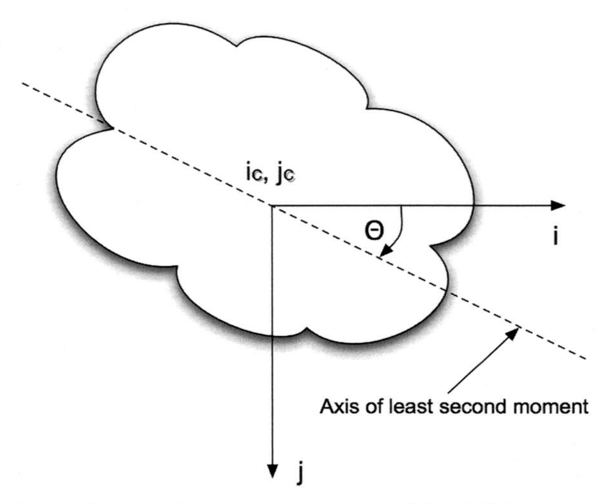

Fig. 8.3 Axis of least inertia for an arbitrary shaped object. *(After [17].)*

generally invariant to translation, rotation, and scaling. These features can be used independently of, or in conjunction with, area and perimeter measurements. In this section, we consider some commonly used shape parameters.

8.2.3.1 Thinness Ratio

Thinness is typically used to define the regularity of an object. Having computed the area, A, and perimeter, P, of an object, the thinness ratio is defined by

$$T = 4\pi\left(\frac{A}{P^2}\right) \tag{8.12}$$

This measure takes a maximum value of 1 for a circle. The same measure can also be used to measure the roundness of an object and is referred to as the *compactness ratio*. Objects of regular shape have a higher thinness ratio than similar irregular ones.

8.2.3.2 Rectangularity

Rectangularity of an object can be measured using the rectangle fit factor [11],

$$R = \frac{A}{A_R} \tag{8.13}$$

This is simply the ratio of the object's area to the area of its *minimum enclosing rectangle* (MER), A_R. The MER for an object is defined as its bounding box aligned such that it encloses all the points in the object with the area minimized. To determine the MER, the object boundary is rotated through 90° in steps of ~3°. Following each stepwise rotation, a horizontally aligned bounding box is fit to the object boundary, and the minimum and maximum x and y values of the rotated boundary points are recorded. At a particular angle, the area of the bounding box is minimized, and this defines the MER. The rectangle fit factor represents how well an object fills its minimum enclosing rectangle. This parameter takes on a maximum value of 1 for rectangular objects. It is bounded between 0 and 1, taking the value $\pi/4$ for circular objects and smaller values for slender, curved objects.

Another related shape measure is the aspect ratio, computed as [18]

$$Aspect = \frac{W}{L} \tag{8.14}$$

which is the ratio of width to length of the minimum enclosing rectangle and is used to distinguish slender objects from roughly square or circular objects.

8.2.3.3 Circularity

Circularity is a shape feature, which is minimized by the circular shape. It is typically used to reflect the complexity of the object boundary, and can be written as [11]

$$C = \frac{P^2}{4\pi A} \tag{8.15}$$

which is the ratio of perimeter squared to area. Circular shapes yield the minimum circularity value of 1.0, and the value increases for complex shapes. Notice that it is the reciprocal of the thinness ratio, defined previously.

Another shape measure that is related to circularity is the boundary energy [4]. Consider an object with perimeter P. We can measure the distance around the boundary starting at some point, p. Then at any instance, the radius of the circle tangent to the boundary at that point defines its radius of curvature, $r(p)$, as shown in Fig. 8.4. The curvature function, $K(p)$, which is periodic with period P at point p is written as [11]

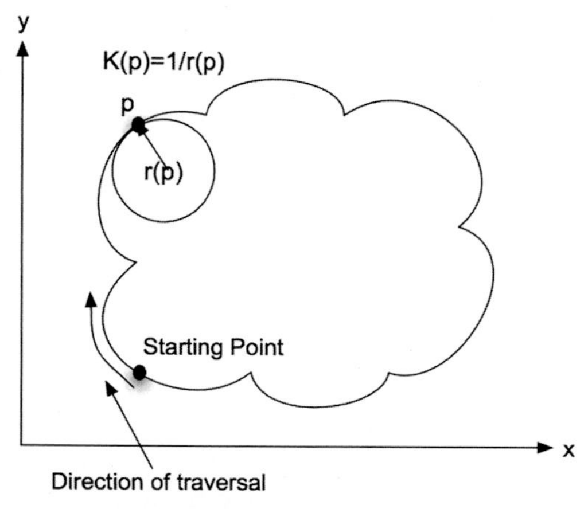

Fig. 8.4 Circularity computation and radius of curvature.

$$K(p) = \frac{1}{r(p)} \tag{8.16}$$

The average energy per unit length of the boundary is given by [11]

$$E = \frac{1}{P} \int_0^P |K(p)|^2 dp \tag{8.17}$$

The circle has, for fixed area, minimum boundary energy given by [11]

$$E_0 = \left(\frac{2\pi}{P}\right)^2 = \left(\frac{1}{R}\right)^2 \tag{8.18}$$

where R is the radius of the circle. Curvature and boundary energy can be computed from the chain code (see Section 7.3.6) [4,5]. It has been shown that the boundary energy reflects the boundary complexity better than the circularity measure of Eq. (8.15) [4].

8.2.3.4 Euler Number

For an image, the Euler number is defined as the number of objects minus the number of holes [19]. For an extracted object, the Euler number is used to represent the completeness of that object, and it corresponds to the number of closed curves contained within the object [20]. The determination of the number of objects and holes in an image has been described earlier, in Chapters 6 and 7.

8.2.3.5 Moments

The moments of a function, which originate from probability theory [21,22], can be used to define a group of shape features that have several desirable properties [23,24]. For a bounded function, $f(x, y)$, the set of moments of two variables is defined by [11]

$$M_{jk} = \int_{-\infty}^{\infty} \int_{-\infty}^{\infty} x^j y^k f(x, y)\ dxdy \tag{8.19}$$

where j and k take on all nonnegative integer values, and they generate an infinite set of moments. The set of moments, $\{M_{jk}\}$, is sufficient to specify the function $f(x, y)$ completely, and is unique for the function, such that only $f(x, y)$ has that particular set of moments.

If $f(x, y)$ takes on the value 1 inside the object and 0 elsewhere, it represents a silhouette function, which ignores internal gray level details and only reflects the shape of the object. This function can be used as a shape descriptor, such that every unique shape corresponds to a unique silhouette, and the corresponding unique set of moments. The order of the moment is given by $j+k$. There is only one zero-order moment, and it gives the area of the object as

$$M_{00}=\int_{-\infty}^{\infty}\int_{-\infty}^{\infty}f(x,y)\ dxdy \tag{8.20}$$

There are two first-order moments and correspondingly more moments of higher orders. All the higher-order moments can be normalized to be invariant to object size, by dividing them by M_{00}.

Central Moments

The so-called *central moments* are computed using the center of gravity as the origin and are therefore location invariant. The coordinates of the center of gravity of the object are given by [11]

$$\mu_{10}=\frac{M_{10}}{M_{00}} \quad \text{and} \quad \mu_{01}=\frac{M_{01}}{M_{00}} \tag{8.21}$$

Moments of higher order, where $j+k>1$, are normally defined in terms of the object's location. This leads to a more generalized mathematical definition for central moments, given as

$$\mu_{jk}=\int_{-\infty}^{\infty}\int_{-\infty}^{\infty}(x-\mu_{10})^j(y-\mu_{01})^k f(x,y)\ dxdy \tag{8.22}$$

Object Dispersion

The three second-order moments provide a measure of how dispersed the pixels in an object are with respect to its centroid. They correspond to $j+k=2$, and are written as

$$\mu_{20}=M_{20}-\frac{M_{10}^2}{M_{00}},\quad \mu_{02}=M_{02}-\frac{M_{01}^2}{M_{00}},\quad \text{and} \quad \mu_{11}=M_{11}-\frac{M_{10}M_{01}}{M_{00}} \tag{8.23}$$

These are proportional to the object's spread over the x-axis, the y-axis, and in both orientations, respectively.

Rotationally Invariant Moments

While central moments are location invariant, they are not rotationally invariant. This means that, given a change in an object's orientation, the central moments will yield different measures. The angle of rotation, θ, that causes the second-order central moment to vanish can be obtained from Eq. (8.11). Then, the principal axes x', y' of an object are at an angle θ from the x-, y-axes. If the moments are computed relative to the principal axes, or if the object is rotated through θ before the moments are computed, then the moments are rotation invariant. In general, it is desirable that object measures be invariant under simple transformations such as translation, rotation, and scale. The area-normalized central moments computed relative to the principal axis are translation, rotation, and scale invariant. Rotationally invariant moments, which are functions of the second-order moments [25], are given by

$$\mu_{20} + \mu_{02} \text{ and } (\mu_{20} - \mu_{02})^2 + 4\mu_{11}^2 \tag{8.24}$$

The magnitude of the invariant moments can be used as shape features for object matching or discrimination.

Zernike Moments

Zernike moments are derived from a complex polynomial that forms an orthogonal set from a unit circle as its domain [26]. The set of these polynomials $\{Z_{mn}(x, y)\}$ are of the form [26]

$$Z_{nm}(x, y) = Z_{nm}(\rho, \theta) = R_{nm} \exp(jm\theta) \tag{8.25}$$

where $n \geq 0$, m is an integer subject to the constraint that $n - |m|$ must be even and $|m| \leq n$, ρ is the length of the vector from the origin to (x, y), and θ is the angle between the vector ρ and the x-axis in the counterclockwise direction. $R_{nm}(\rho)$ in the preceding equation is a radial polynomial defined as [26]

$$R_{nm}(\rho) = \sum_{s=0}^{n-|m|/2} (-1)^s \cdot \frac{(n-s)!}{s!\left(\frac{n+|m|}{2} - s\right)!\left(\frac{n-|m|}{2} - s\right)!}\rho^{n-2s} \tag{8.26}$$

The Zernike moment of order n, with repetition m, for a continuous image function $I(x,y)$ that vanishes outside the unit circle is [26]

$$A_{nm} = \frac{n+1}{\pi} \iint_{x^2+y^2 \leq 1} I(x, y) Z_{nm}^*(\rho, \theta)\, dxdy \tag{8.27}$$

which is the projection of the image function onto the orthogonal basis functions given in Eq. (8.25). Discretizing the integrals in Eq. (8.27), we have [26]

$$A_{nm} = \frac{n+1}{\pi} \sum_x \sum_y I(x, y) V_{nm}^*(\rho, \theta) \quad x^2 + y^2 \leq 1 \tag{8.28}$$

If A_{nm} represents the moments associated with an image, and A_{nm}^r represents the moments of the same image rotated by an angle ϕ, then

$$A_{nm}^r = A_{nm} \cdot \exp(-jm\phi) \tag{8.29}$$

The magnitude of the Zernike moments is therefore invariant to rotation. On the other hand, they are not translation and scale invariant. Nonetheless, translation invariance can be achieved by moving the origin of the image to the centroid of the object. In typical applications requiring object shape measurements, up to eight orders of Zernike moments are used. Only moments of order 2 to 8 are relevant for discrimination purposes since, after object normalization for scale and translation invariance, moment $|A_{00}|$ is constant and moment $|A_{11}|$ is zero [27]. Since the Zernike moments themselves are complex numbers and are sensitive to rotation of the image, typically the magnitudes of the moments (i.e., $|A_{nm}|$) are used as shape features [28].

8.2.3.6 Elongation

Another measure used frequently to describe the shape of an object is elongation. One way of calculating this measure is by taking the ratio of an object's length to its breadth, that is

$$El = \frac{length}{breadth} \tag{8.30}$$

Another way of defining the same measure is based on computing the bounding rectangle for the object and taking the ratio of the long side to the short side. An easy way to approximate this is by scanning the object image and finding the maximum and minimum values along the spatial indices. The ratio can then be written as

$$El = \frac{j_{\max} - j_{\min} + 1}{i_{\max} - i_{\min} + 1} \tag{8.31}$$

The same measure can be calculated as the ratio of second-order moments of the object defining its major and minor axis. Elongation in this case is given by

$$El = \frac{\lambda_1}{\lambda_2} \tag{8.32}$$

where

$$\begin{aligned} \lambda_1 &= \mu_{20}\sin^2\theta + \mu_{02}\cos^2\theta + 2\mu_{11}\sin\theta\cos\theta \\ \lambda_2 &= \mu_{20}\cos^2\theta + \mu_{02}\sin^2\theta - 2\mu_{11}\sin\theta\cos\theta \end{aligned} \tag{8.33}$$

and where θ is the angle of rotation, described in Eq. (8.11). Each of these measures of elongation provide similar but not necessarily identical results.

8.2.4 Shape Descriptors

A shape descriptor is another way of describing an object's shape. It provides a more detailed description of shape than that offered by the single-parameter shape measures

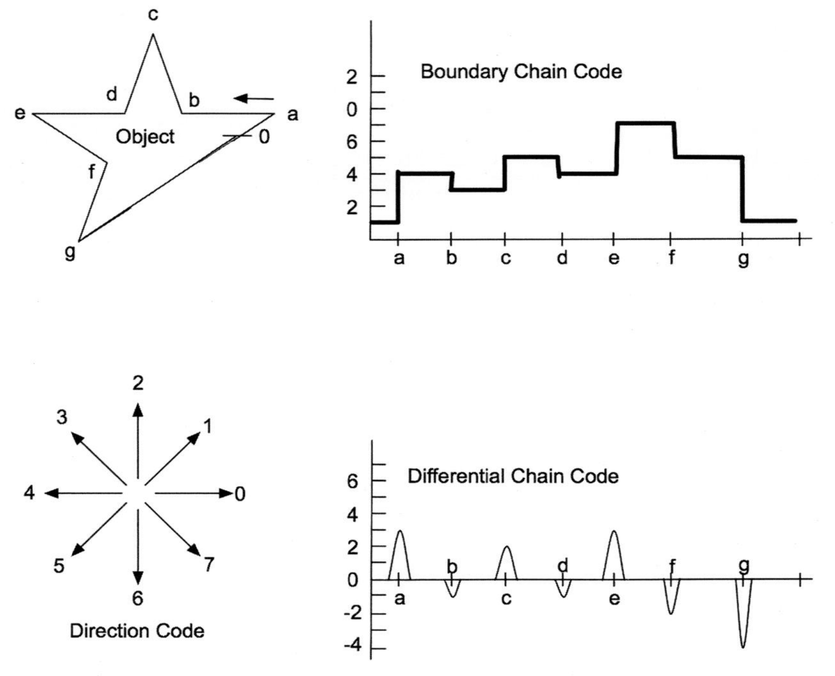

Fig. 8.5 The chain code and its derivative for an arbitrary shape.

described earlier. Shape descriptors also allow a more compact representation of shape than what is reflected by the object image itself.

8.2.4.1 The Differential Chain Code

The most common shape descriptors include the *boundary chain code* (BCC) and its derivative, known as the *differential chain code* (DCC) [14,29,30]. As discussed in Section 7.3.6, the BCC represents the boundary tangent angles as a function of distance around the object. For a simple polygon, Fig. 8.5 shows the associated BCC and DCC. The DCC reflects the curvature of the object's boundary. Convexities and concavities in the boundary show up as peaks and valleys, respectively, in the differential chain code. Both of these functions can be further analyzed to obtain shape measures.

8.2.4.2 Fourier Descriptors

Fourier descriptors exploit the periodicity in the BCC representation of the boundary [31]. The complex boundary function is defined as $B(p)=x(p)+jy(p)$ where $x(p)$ and $y(p)$ are the coordinates of the pth boundary point, measured from an arbitrary starting point, and j is the imaginary unit. $B(p)$ is a complex-valued periodic function with period P, and thus its Fourier series can be computed [11]. It is the low-frequency components in this Fourier series expansion that represent the basic shape of the object. These components are inherently shift invariant, and their complex magnitudes are rotation invariant and independent of the starting point as well. Thus they can be used as shape descriptors.

8.2.4.3 The Medial Axis Transform

Medial axis transformation (MAT) is another data reduction technique that is used as a shape descriptor (Section 6.2.5) [11,32]. The medial axis of an object is a set of points inside the object such that each point is the center of a circle that is tangent to the boundary at two nonadjacent points. Normally a value is associated with each point on the medial axis, and it is the minimum distance to the object boundary from that point.

The simplest technique to find the medial axis is by erosion (Section 6.2). By successively removing the outer perimeter of points one can detect the point whose removal would disconnect the object. That point is then considered to be on the medial axis. Its associated value is simply the number of layers removed or the number of erosion iterations required. The MAT is a useful descriptor for long, narrow, and curved objects. In some applications, the medial axis is used only as a graph, ignoring the associated values. In others, the graph is used to derive additional shape measures, such as the number of branches and the total length [33].

8.2.4.4 Graph Representations

Graphs have been used as a tool for translation and rotation invariant representation of object shapes [34]. These are descriptors that define the structure among a set of points located on the boundary of an object [35]. The two graphs used most often are the minimum spanning tree (MST) and the Delaunay triangulation (DT).

Minimum Spanning Tree

Consider an arbitrary shaped object as shown in Fig. 8.6a, with a set of points, n, given by $P=\{p_1,p_2,\ldots,p_n\}$ located on the boundary. A tree is constructed by connecting pairs of points from the set so as to form a tree structure that "spans" the set of points. There are many ways to draw this tree, but if the sum of branch lengths for a particular tree is less than the sum of branch lengths for any other spanning tree, then that tree is called the *minimum spanning tree* (MST), as shown in Fig. 8.6b. The MST is a type of skeleton of the object, and it can give rise to a number of shape descriptors, such as total, average, and standard deviation of branch length, average branching angle, number of nodes, etc. These descriptors have been used in numerous applications [36,37].

Delaunay Triangulation

In representing an object using Delaunay triangulation (DT), edges are formed by joining pairs of points from the set, $P=\{p_1,p_2,\ldots,p_n\}$, in such a way that as many triangles as

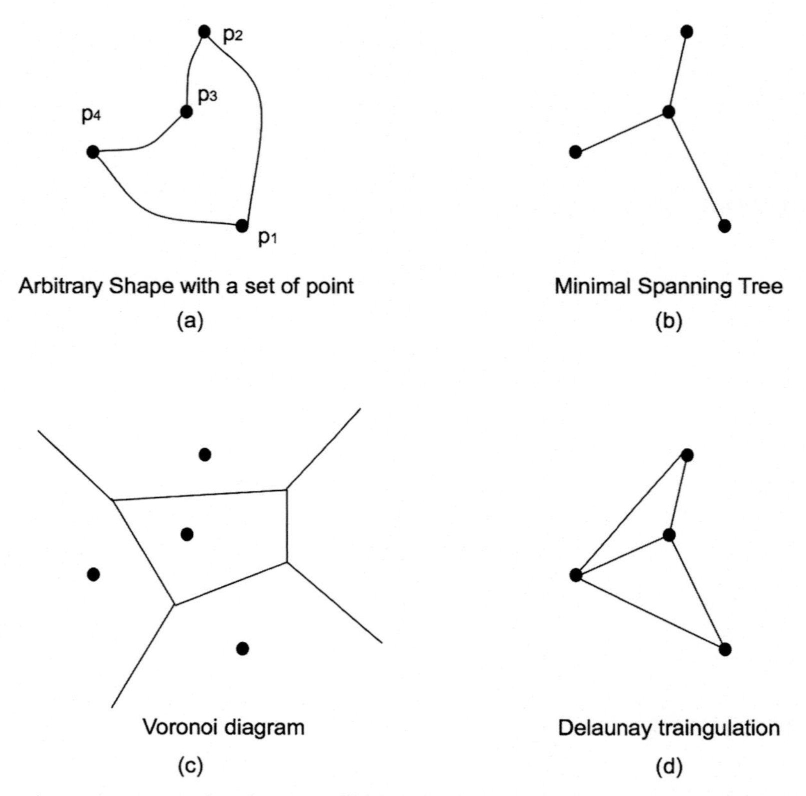

Fig. 8.6 An object represented with a set of boundary points and its associated graphs.

possible are generated, but without any crossing lines. DT is a specific triangulation based on locally equiangular triangles [38] and is normally derived from the Voronoi partitioning of the object shape [39,40]. The Voronoi diagram partitions the object into disjoint regions such that each region R_i is composed of a subset of points p_i defined as

$$R_i = \left\{x : E_d(x, p_i) < E_d\left(x, p_j\right)\right\} \quad \text{for all } j \neq i \tag{8.34}$$

where E_d is the Euclidean distance (described in Section 8.3.1). The partitioning of the example object is shown in Fig. 8.6c. The DT is now defined by joining two points p_i and p_j if and only if their corresponding regions share a side. The resulting triangulation is shown in Fig. 8.6d. Once the triangulation is defined for the object, it can be characterized by the same measures as the MST.

8.3 Distance Measures

Measures of distance provide a way to compute the separation between two points in an image [3,41]. These can be two points within the same object (such as points on the major and minor axes), or points on two different objects. The three most common ways of measuring distance are presented here.

8.3.1 Euclidean Distance

Euclidean distance is by far the most commonly used measure of distance and is given by

$$D_e = \sqrt{(i-k)^2 + (j-l)^2} \tag{8.35}$$

where the two points have spatial indices (i,j) and (k,l), respectively.

8.3.2 City-Block Distance

City-block distance is an approximation to the Euclidean measure that is computationally faster. It is also called "Manhattan distance" or the "absolute value metric." It is written as [12]

$$D_m = |i-k| + |j-l| \tag{8.36}$$

8.3.3 Chessboard Distance

The chessboard distance measure is the maximum separation in either the x- or y-direction between the two points. It is also known as the "maximum value metric" and is written as [12]

$$D_c = \max(|i-k|,\ |j-l|) \tag{8.37}$$

8.4 Gray Level Object Measures

Object measurements derived as a function of the intensity distribution of the object are called gray level object measures. Most of the measures defined previously for binary objects can also be used for gray level objects. There are three main categories of gray level object measurements. Intensity and histogram measures are normally defined as first-order measures of the gray level distribution, while texture measures quantify second- or higher-order relationships among gray level values.

8.4.1 Intensity Measures

Images most often contain regions that show heterogeneous intensity distributions. Intensity-based measures can be used to quantify intensity variations across and between objects. Some of the commonly used measures are described in the following sections.

8.4.1.1 Integrated Optical Density

The integrated optical density (IOD) is the sum of the gray levels of all pixels in the object [8–11]. It reflects the "mass" or "weight" of the object and is numerically equal to the area multiplied by the mean interior gray level. Consider for an object, if (i,j) are the spatial indices, $I(i,j)$ represents the gray level, and A is the area of the object, then

$$IOD = \sum_{i,j \in A} I(i,j) \tag{8.38}$$

8.4.1.2 Average Optical Intensity

The average optical density (AOD) is merely IOD divided by area [8–11]. The total number of object pixels is the simplest measure of an object's area. Thus the AOD of an $M \times N$ image can be calculated by

$$AOD = \frac{1}{A}\sum_{i=1}^{M}\sum_{j=1}^{N} I(i,j) \tag{8.39}$$

where $A = M \times N$ is the area of the image. For an object, the summations are taken over all pixels inside the object, and A is the number of pixels in the object.

8.4.1.3 Contrast

A measure of contrast of an object is the brightness (AOD) difference between the object and the surrounding background.

8.4.2 Histogram Measures

The histogram of the image of an object provides a description of the distribution of intensity values within the object. When normalized by the size of the object, the histogram is the probability density function (pdf) of the gray levels. Thus measures derived from the normalized histogram of the object image provide statistical descriptors characterizing the gray level distribution of the object [8–11]. Common first-order measures calculated on the histogram include mode, mean, standard deviation, skew, energy, and entropy. Second-order measures are based on joint distribution functions and are representative of the texture of the object [42]. Consider the gray level probability density function (pdf) given as

$$P(g) = \frac{h(g)}{M} \tag{8.40}$$

where $h(g)$ is the number of pixels with gray level g, and M is the total number of pixels in the image. Each of the first-order measures can be calculated from the pdf.

8.4.2.1 Mean Gray Level

Mean gray level measures the average intensity of the image. It can be calculated as

$$\bar{g} = \sum_{g=0}^{L-1} P(g) \cdot g \qquad (8.41)$$

where L is the number of gray levels present in the image. It should be noted that this is the same as AOD.

8.4.2.2 Standard Deviation of Gray Levels

Standard deviation is a measure that provides an understanding about the spread of intensities across the image. This is also an indicator of contrast in the image. Standard deviation is measured by

$$\sigma_g = \sqrt{\sum_{g=0}^{L-1} (g - \bar{g})^2 \cdot P(g)} \qquad (8.42)$$

8.4.2.3 Skew

Skew measures the asymmetry in the image's intensity distribution. We can calculate skew by

$$\kappa = \frac{1}{\sigma_g^3} \sum_{g=0}^{L-1} (g - \bar{g})^3 \cdot P(g) \qquad (8.43)$$

8.4.2.4 Entropy

Entropy provides a measure of an image's smoothness in terms of gray level values. The more gray levels there are in the image, the higher the entropy will be. Entropy can be calculated by

$$Entropy = -\sum_{g=0}^{L-1} P(g) \cdot \log_2[P(g)] \qquad (8.44)$$

8.4.2.5 Energy

Energy is another measure that shows how the gray level values are distributed within the image. It has an inverse relation to entropy in that the energy of an image is highest if it has only one gray level value. The more gray levels there are in an object, the lower its energy will be. We can calculate energy as

$$Energy = \sum_{g=0}^{L-1} [P(g)]^2 \qquad (8.45)$$

8.4.3 Texture Measures

The word "texture" was originally used to refer to the appearance of fabric. A general definition is "the arrangement or characteristics of the constituent elements of anything, especially as regards to surface appearance or tactile qualities" [43]. A more relevant definition for image analysis is "an attribute representing the spatial arrangement of the gray levels of pixels in a local region" [44]. In the current context, we are specifically concerned with the measurement of the texture of an object in an image. Perception of texture is scale dependent. For example, in viewing an image of a tiled floor from a distance, texture would be perceived as the repetition observed in tile placement. By contrast, observing an individual tile on the tiled floor may lead to perceiving the texture within that tile. Broadly speaking, we can define texture as patterns of local variations in image intensity, which are too fine to be distinguished as separate objects at the observed resolution [17].

Electronic noise induced by a camera is an example of a *random texture*. Here the gray level variation exhibits no recognizable repeating pattern. Crosshatching, by contrast, is a *pattern texture* that does exhibit a visible regularity. Statistical properties such as standard deviation of gray level (i.e., texture amplitude) and autocorrelation width (i.e., texture size) are commonly used to characterize random textures. Similarly, pattern textures can be characterized by measurements that quantify the nature and directionality of the pattern, if it has any.

A *texture feature* quantifies some characteristic of the gray level variation within an object. It is normally independent of object position, orientation, size, shape, and average brightness. Presented here are some of the more common methods for computing *texture features*.

8.4.3.1 Statistical Texture Measures

Statistical measures of intensity variation include standard deviation, variance, and skew. These can be computed as moments of the gray level histogram, H, of the object. Similarly, a feature referred to as the "module" can be computed as [45]

$$I = \sum_{i=1}^{N} \frac{H_i - {}^{M}\!/_{N}}{\sqrt{\dfrac{H_i\left(1 - {}^{H_i}\!/_{M}\right) + M\left(1 - {}^{1}\!/_{N}\right)}{N}}} \tag{8.46}$$

where M is the number of pixels in the object, and N is the number of gray levels in the grayscale. Although the human eye is insensitive to textural differences of order higher than the second (i.e., the variance), texture features such as the "module" often rely on quantifiable differences, where they exist.

The Gray Level Co-Occurrence Matrix

The gray level co-occurrence matrix (GLCM) provides a number of second-order statistics relating to the gray level relationships in a neighborhood around a pixel [42,46,47]. Computation of GLCM features is a two-step process. The GLCM is created as the first step; then it is used to compute a number of statistics, and those are the texture features.

The GLCM, $\mathbf{P}_d$, is a 2D histogram that specifies how often two gray levels occur in pairs of pixels separated by a particular offset distance. First, one must pick the offset distance and direction. Then, each entry, (i, j), in $\mathbf{P}_d$ corresponds to the number of occurrences of the gray levels i and j, in pairs of pixels that are separated, in the image, by the chosen distance and direction. Once the GLCM is formed, one can compute specific statistical values from it (see the following). Selecting a different offset direction and distance gives rise to a new GLCM.

There are several widely used statistical and probabilistic features that can be derived from the GLCM [48,49]. Examples include "entropy," given by

$$H = -\sum_{i,j} \mathbf{P}_d(i,j)\log(\mathbf{P}_d(i,j)) \tag{8.47}$$

"inertia," which is

$$I = \sum_{i,j} (i-j)^2\, \mathbf{P}_d(i,j) \tag{8.48}$$

"energy," defined as

$$E = \sum_{i,j} [\mathbf{P}_d(i,j)]^2 \tag{8.49}$$

"maximum probability," given as

$$P = \max_{i,j} \mathbf{P}_d(i,j) \tag{8.50}$$

"inverse differential moment," defined by

$$IDM = \sum_{i,j(i \neq j)} \frac{\mathbf{P}_d(i,j)}{(i-j)^2} \tag{8.51}$$

and "correlation," denoted by

$$C = \frac{1}{\sigma_i \sigma_j} \sum_{i.j} (i-\mu_i)\left(j-\mu_j\right) \mathbf{P}_d(i,j) \tag{8.52}$$

Some co-occurrence matrix-based texture features correspond to characteristics that can be recognized by the eye [50], but many do not. In general, one must determine experimentally which of these features have discriminating power.

8.4.3.2 Power Spectrum Features

Power spectrum features are measures of texture that are derived from the Fourier transform of the object image. The power spectrum, defined as the magnitude squared of the 2D Fourier spectrum, gives rise to a set of texture measures. These measures can be defined by averaging the power spectrum in annular rings to produce a 1D function that ignores directionality, or by averaging along radial lines to produce a 1D function that shows only the directionality. These one-dimensional functions of frequency (or angle) can be further reduced to single measurement values that reflect salient characteristics of the texture pattern.

8.5 Object Measurement Considerations

Image analysis can provide several measures of an object's structure by defining its characteristics in terms of area, perimeter, elongation, compactness, contrast, and texture, as shown in Table 8.1. This table shows object measures computed for different particle types from urinary sediment, including red blood cells (RBCs), calcium oxalate crystals (CAOXs), white blood cells (WBCs), bacteria (BACT), granular casts (GRANs), cellular cast (CCST), squamous epithelial cells (SQEPs), sperm (SPRM), and hyphae yeast (HYST). It is clear that size measures such as area and perimeter can be used to distinguish smaller particles such as RBCs from the larger cell types, such as SQEPs. Similarly, the elongation and compactness measures readily differentiate the circular shaped cells (e.g., RBCs, WBCs, and SQEPs) from the elongated cells (e.g., SPRM and HYST). Finally, intensity-based measures such as contrast and texture measures can be used to differentiate between low contrast objects (e.g., SPRM, SQEPs, and HYST) and high-contrast objects such as crystals (e.g., CAOXs, RBCs, and WBCs), and between objects with textured interiors (e.g., SPRM, SQEPs, and HYST) and relatively untextured objects (e.g., CAOXs, RBCs, and WBCs).

In computing measurements of an object, it is important to keep in mind the specific application and its requirements. A critical factor in deciding on an object measurement to be used is its robustness. The robustness of a measurement is its ability to provide consistent results in different applications. For example, if we wish to design a system that can differentiate between types of cells under different illumination conditions, we may not want to use an intensity measure, such as average optical density, as the only measurement made on the object. This would provide inconsistent results due to lighting changes that will alter the measured AOD of cells. Instead, we may wish to measure cell area. Another important consideration is the invariance of the measurement under rotation, translation, and scale. When deciding on the set of object measures to use, these considerations should guide one in identifying a suitable choice.

Table 8.1 Measurements of object structure for a variety of particle types.

Cell type		Object measures					
		A (μm^2)	P (μm)	Elongation	Compactness	Contrast	Texture
RBC	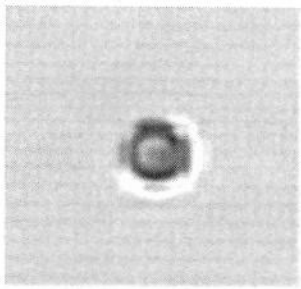	56.16	22.8	1.123	1.142	0.4874	127.4
CAOX	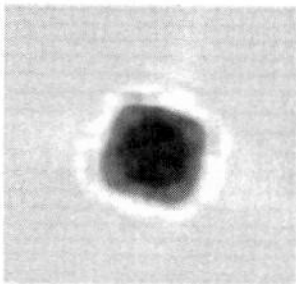	236.16	50.4	1.034	1.065	0.8083	106.5
WBC	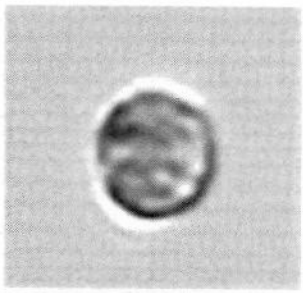	390.24	63.6	1.119	1.242	0.5081	73.8
BACT	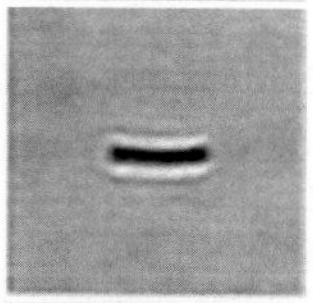	146.88	49.2	2.429	1.17	0.1189	122.9
GRAN	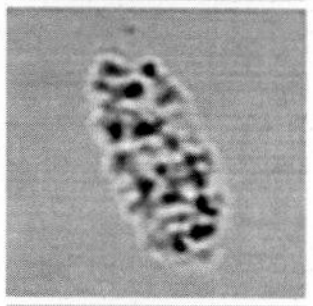	1121.76	145.2	2.373	1.451	0.3202	192.6
CCST	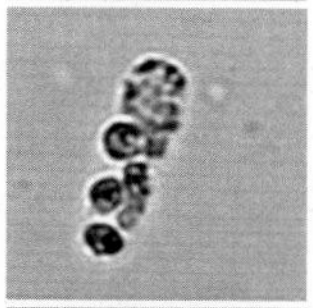	1882.08	205.2	3.158	1.411	0.4852	183.6
SQEP	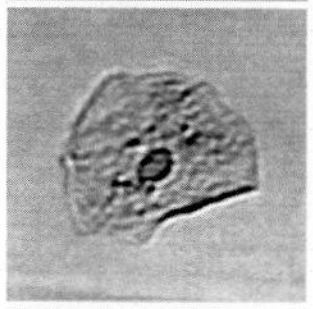	5127.84	256.8	1.139	1.355	0.1422	275.2
SPRM	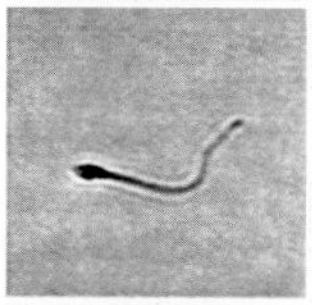	416.16	146.4	2.629	3.680	0.0275	346.0
HYST	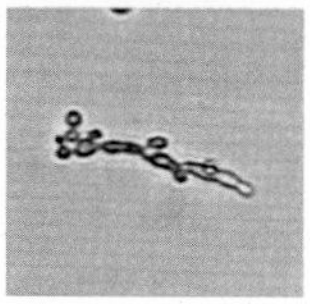	918.72	225.6	3.933	2.670	0.2900	344.8

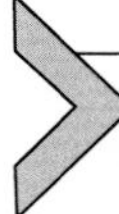

8.6 Summary of Important Points

1. Object measurements are normally computed from the binary representation of a segmented object or the gray level intensity distribution within the object boundary.
2. Measurements of an object can be based on its size, shape, or intensity values.
3. Area, length, width, and perimeter are common measures of object size.
4. Object shape can be captured by measures of circularity, rectangularity, moments, and Euler number, among other features.
5. Histogram measures capture the statistics of an object's gray levels.
6. Texture measures capture the statistics of an object's gray level structure.
7. Objects can be described using compact descriptors such as chain codes, the medial axis transform, or graphs.

References

[1] K. Fukunaga, Introduction to Statistical Pattern Recognition, Academic Press, 1972.
[2] H. Blum, Biological science and visual shape (part I), J. Theor. Biol. 38 (1973) 205–287.
[3] R.O. Duda, P.E. Hart, Pattern Classification and Scene Analysis, Wiley-Interscience Publication, 1973.
[4] I.T. Young, J.E. Walker, J.E. Bowie, An analysis technique for biological shape (I), Inf. Control 25 (1974) 357–370.
[5] J.E. Bowie, I.T. Young, An analysis technique for biological shape-II, Acta Cytol. 21 (3) (1977) 455–464.
[6] J.E. Bowie, I.T. Young, An analysis technique for biological shape-III, Acta Cytol. 21 (6) (1977) 739–746.
[7] L. Dorst, Discrete Straight Lines: Parameters, Primitives and Properties, Delft University of Technology, 1986.
[8] T.Y. Young, K. Fu, Handbook of Pattern Recognition and Image Processing, Academic Press Inc, 1986.
[9] A.K. Jain, Fundamentals of Digital Image Processing, Prentice-Hall, Englewood Cliffs, 1989.
[10] R. Gonzales, R. Woods, Digital Image Processing, Addison-Wesley Publishing Company, 1992.
[11] K.R. Castleman, Digital Image Processing, Prentice Hall, 1996.
[12] C.A. Glasbey, G.W. Horgan, Image Analysis for Biological Sciences, John Wiley & Sons, Inc, 1995.
[13] L. Dorst, A.M. Smeulders, Best linear unbiased estimator for properties of digitized straight lines, IEEE Trans. Pattern Anal. Mach. Intell. 8 (1986) 276–282.
[14] J. Koplowitz, A.M. Bruchstein, Design of perimeter estimators for digitized planar shapes, IEEE Trans. Pattern Anal. Mach. Intell. 11 (6) (1989) 611–622.
[15] L. Dorst, A.M. Smeulders, Length estimators for digitized contours, Comput. Vis. Graph. Image Process. 40 (1987) 311–333.
[16] S.E. Umbaugh, Computer Vision and Image Processing: A Practical Approach Using CVIPtools, first ed., Prentice Hall, 1998.
[17] R. Jain, R. Kasturi, B.G. Schunk, Machine Vision, McGraw-Hill, 1995.
[18] S. Petitjean, S. Ponce, D.J. Kriegman, Computing exact aspect graphs of curved objects: algebraic surfaces, Int. J. Comput. Vis. 9 (3) (1992) 231–255.
[19] S.E. Umbaugh, Computer Imaging: Digital Image Analysis and Processing, CRC Press, 2005.
[20] J.R. Munkers, Element of Algebraic Topology, Perseus Press, 1993.
[21] E. Kreyzig, Introductory Mathematical Statistics, John Wiley & Sons, 1970.

[22] A. Papoulis, Probability, Random Variables, and Stochastic Processes, McGraw-Hill, 1965.
[23] C.H. Teh, R.T. Chen, On image analysis by the method of moments, IEEE Trans. Pattern Anal. Mach. Intell. 10 (1988) 496–513.
[24] F.L. Alt, Digital pattern recognition by moments, J. ACM 9 (1962) 240–258.
[25] M.K. Hu, Visual pattern recognition by moment invariants, IRE Trans. Inf. Theory (1962) 179–187.
[26] A. Khotanzad, Y.H. Hong, Invariant image recognition by Zernike moments, IEEE Trans. Pattern Anal. Mach. Intell. 12 (1990) 489–497.
[27] Y. Abu-mostafa, D. Psaltis, Image normalization by complex moments, IEEE Trans. Pattern Anal. Mach. Intell. 7 (1) (1985) 46–55.
[28] M.V. Boland, M.K. Markey, R.F. Murphy, Automated recognition of patterns characteristic of subcellular structures in fluorescence microscopy images, Cytometry 33 (1998) 366–375.
[29] A. Rosenfeld, Survey: image analysis and computer vision, Comput. Vis. Image Underst. 66 (1) (1977) 33–93.
[30] D.H. Ballard, C.M. Brown, Computer Vision, Prentice Hall, 1982.
[31] C.T. Zahn, R.Z. Roskies, Fourier descriptors for plane closed curves, IEEE Trans. Comput. C-21 (1972) 269–281.
[32] R.J. Wall, A. Kilnger, S. Harami, An algorithm for computing the medial axis transform and its inverse, in: Workshop on Picture Data Description and Management, IEEE Computer Society, 1977.
[33] R.M. Haralick, S.R. Sternberg, X. Zhuang, Image analysis using mathematical morphology, IEEE Trans. Pattern Anal. Mach. Intell. 9 (1987).
[34] G.T. Toussaint, The relative neighborhood graph of a finite planar set, Pattern Recogn. 12 (1980) 261–268.
[35] C.T. Zahn, Graph-theoretical methods for detecting and describing Gestalt clusters, IEEE Trans. Comput. 20 (1971) 68–86.
[36] B. Rosenberg, D.J. Langridge, A computational view of perception, Perception 2 (4) (1973) 415–424.
[37] B. Weyn, et al., Computer assisted differential diagnosis of malignant mesothelioma based on syntactic structure analysis, Cytometry 35 (1) (1999) 23–29.
[38] R. Sibson, Locally equiangular triangulation, Comput. J. 21 (1978) 243–245.
[39] B.A. Lewis, J.S. Robinson, Triangulation of planar regions with applications, Comput. J. 21 (1978) 324–332.
[40] P.J. Green, R. Sibson, Computing Dirichlet tessellations in the plane, Comput. J. 21 (2) (1978) 168–173.
[41] R.O. Duda, P.E. Hart, D.G. Stork, Pattern classification, John Wiley & Sons, 2001.
[42] R.M. Haralick, Statistical and structural approaches to texture, Proc. IEEE 67 (5) (1979) 786–804.
[43] S.I. Landau, Webster Illustrated Contemporary Dictionary, Doubleday, NY, 1982.
[44] IEEE Standard 601.4-1990, IEEE Standard Glossary of Image Processing and Pattern Recognition Terminology, IEEE Press, 1990.
[45] G.E. Lowitz, Can a local histogram really map texture information? Pattern Recogn. 16 (2) (1983) 141–147.
[46] S.H. Peckinpaugh, An improved method for computing gray-level co-occurrence matrix based texture measures, Comput. Vis. Graph. Image Process. 53 (6) (1991) 574–580.
[47] P. Kruzinga, N. Petkov, Nonlinear operator for oriented texture, IEEE Trans. Image Process. 8 (10) (1999) 1395–1407.
[48] M. Tuceryan, A.K. Jain, Texture analysis, in: C.H. Chen, L.F. Pau, P.P. Wang (Eds.), Handbook of Pattern Recognition and Computer Vision, World Scientific Publishing Company, 1993.
[49] R.M. Haralick, L.G. Shapiro, Computer and Robot Vision, Addison-Wesley, 1992.
[50] H. Tamura, S. Mori, T. Yamawaki, Texture features corresponding to visual perception, IEEE Trans. Syst. Man Cybern. 8 (1978) 460–473.

CHAPTER NINE

Object Classification

Kenneth R. Castleman and Qiang Wu

9.1 Introduction

Classification is the step that tells us what is in the image. Assuming the objects in the image have been segmented and measured, classification identifies them by assigning each of them to one of several previously established types, categories, or classes. There are several mathematical approaches that can be taken to address the classification problem, and a complete coverage is beyond our scope. Here we illustrate the process of classification with the very useful maximum likelihood method. This technique is widely used because it minimizes the probability of making an incorrect assignment. More specifically, we present the minimum Bayes risk classifier, assuming Gaussian statistics, along with several of its interesting special cases. We also address a few other classification strategies.

9.2 The Classification Process

When we encounter an unknown object in a microscopic image, we know three things about it. First, we know something about the objects in general. We know the *a priori probability* that an object belongs to each of the classes. For example, if we are attempting to separate abnormal from normal cells, we might know from past experience that 90% of all cells encountered are normal. Thus the a priori probability for class 1 (normal) is 0.9, while for class 2 (abnormal) it is 0.1.

$$P(C_1) = 0.9 \qquad \text{and} \qquad P(C_2) = 0.1 \tag{9.1}$$

This knowledge applies to all of the objects. Quite a large sample of cells might be required to estimate these prior probabilities [1]. Second, we know something about each class. We know the probability density function (pdf) of the features for each of the classes. Third, we know something about each particular object. We know its measured feature values. This is the quantitative data we have that is unique to that particular object. Given these three pieces of information, we seek to make an optimal assignment of that object to a class. For the moment we take the probability of error as the performance criterion, and we seek to minimize it.

Microscope Image Processing
https://doi.org/10.1016/B978-0-12-821049-9.00010-1

Copyright © 2023 Elsevier Inc.
All rights reserved.

9.2.1 Bayes' Rule

We now consider how to combine the three things we know about an object to find its most likely class. After an object has been measured, we should be able to use the measurement data and the class-conditional pdfs to improve our knowledge of the object's most likely class membership. The *a posteriori probability* that the object belongs to class i is given by Bayes' theorem; that is,

$$P(C_i|\, x) = \frac{P(C_i)p(x|\, C_i)}{p(x)} \tag{9.2}$$

where $P(C_i)$ is the a priori probability of class i, $p(x\,|\,C_i)$ is the pdf of the feature, x, for class i, and

$$p(x) = \sum_{i=1}^{N} p(x|\, C_i)P(C_i) \tag{9.3}$$

is the normalization factor required to make the set of a posteriori probabilities sum to unity. Bayes' theorem, then, allows us to combine the a priori probabilities of class membership with the measurement data and the class-specific pdf, to compute, for each class, the probability that the measured object belongs to that class. Given this information, we might choose to assign each object to its most likely class.

9.3 The Single-Feature, Two-Class Case

To illustrate the classification process, we first consider the simple case where two types of objects must be sorted on the basis of a single measurement. For this example, assume we are attempting to separate abnormal from normal cells on the basis of nuclear diameter alone. This means that the cells encountered belong either to class 1 (normal) or to class 2 (abnormal). For each cell, we measure one property, nuclear diameter, and this is the feature we call x.

It may be that the pdf of the diameter measurement, x, is already known for one or both classes of cells. If not, we would have to estimate it by measuring a large number of normal and abnormal cells and plotting histograms of their nuclear diameters. After normalization to unit area, and perhaps some smoothing, these histograms can be taken as estimates of the corresponding pdfs. If the histogram fits the Gaussian form, to a reasonable approximation, we can compute the mean, μ, and variance, σ, and use the parametric representation for the normal distribution

$$p(x) = \frac{1}{\sqrt{2\pi}} e^{-\frac{(x-\mu)^2}{2\sigma^2}} \tag{9.4}$$

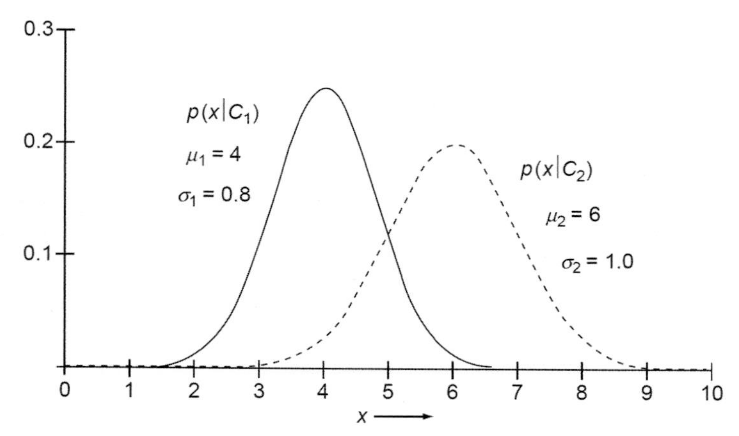

Fig. 9.1 Probability density functions for a two-class problem.

There are other standard statistical distributions for the pdf that might fit the histograms if the Gaussian form does not. If we use the Gaussian, then only the mean and variance are required to completely specify the pdf for a class.

9.3.1 A Priori Probabilities

The *a priori* probabilities represent our knowledge about an object before it has been measured. In this example, we assume that an unmeasured cell has a nine-to-one chance of being normal (Eq. 9.1).

9.3.2 Conditional Probabilities

Fig. 9.1 shows what the two pdfs might look like. We denote the conditional pdf for normal cell diameter as $p(x \mid C_1)$, which can be read as "the probability that diameter x will occur, given that the object belongs to class 1." Similarly, $p(x \mid C_2)$ is the probability of diameter x occurring, given class 2 (abnormal).

If we scale each of the pdfs by the a priori probability of its class, as in Fig. 9.2, we get a better picture of the error situation. We could establish a decision rule by setting a threshold value, T, on the nuclear diameter and classifying cells normal if they fall below that value and abnormal if they fall above it. The area under the dotted curve, to the left of the threshold, is proportional to the probability of calling an abnormal cell normal. Similarly, the area under the solid curve, to the right of the threshold, is proportional to the probability of misclassifying a normal cell.

9.3.3 Bayes' Theorem

Before an object has been measured, our knowledge of it consists merely of the a priori probabilities of class membership. After measurement, however, we can combine the

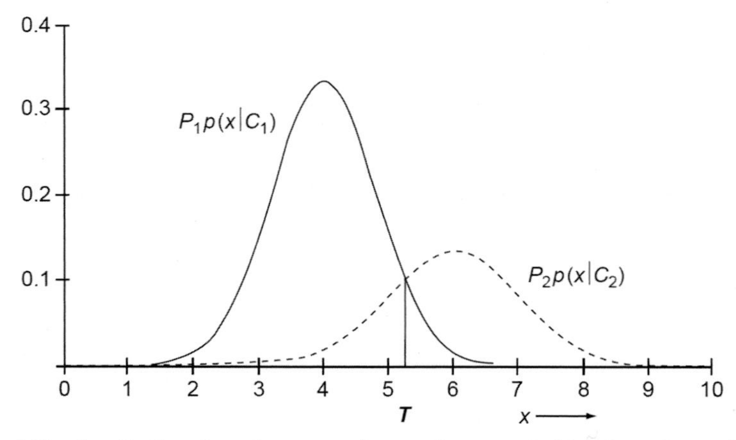

Fig. 9.2 Probability density functions for a two-class problem, scaled by the prior probabilities.

measurement with the conditional pdfs to improve our knowledge of the object's class membership. After measurement, the a posteriori probability that the object belongs to class i is given by Bayes' theorem [2–7], that is,

$$P(C_i|\,x) = \frac{P(C_i)p(x|C_i)}{p(x)} \tag{9.5}$$

where $P(C_i)$ is the a priori probability of class i, $p(x\,|\,C_i)$ is the pdf of the feature, x, for class i, and

$$p(x) = \sum_{i=1}^{N} p(x|C_i)P(C_i) \tag{9.6}$$

is the required normalization factor.

Bayes' theorem, then, allows us to combine the a priori probabilities of class membership with the measurement and the class-specific pdf, to compute, for each class, the probability that the measured object belongs to that class. Fig. 9.3 shows the *a posteriori* probabilities for this example. For any nuclear diameter, x, the solid curve gives the probability that a cell having that diameter belongs to class 1. The dotted curve gives the probability that the cell belongs to class 2.

In our cell-sorting example, we would assign the object to class 1 (i.e., we would call it normal) if

$$P(C_1|\,x) > P(C_2|\,x) \tag{9.7}$$

and assign it to class 2 (abnormal) otherwise. At the decision threshold, T, where equality holds in Eq. (9.7), we may assign arbitrarily. The classifier defined by this decision rule is

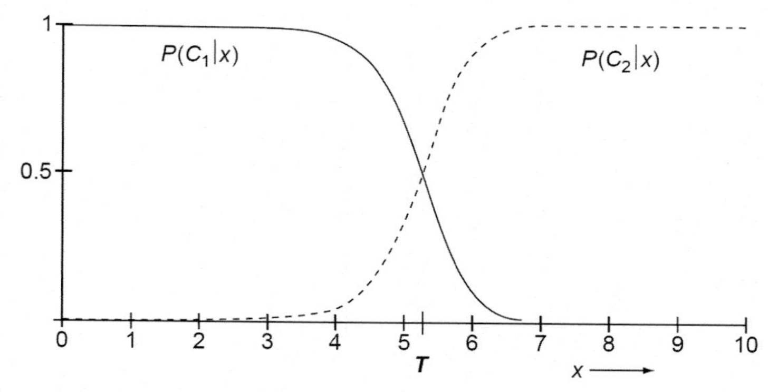

Fig. 9.3 A posteriori probabilities for the two-class problem.

called a maximum-likelihood classifier because it maximizes the probability of a correct assignment.

Note that, in Fig. 9.3, cells with nuclear diameter less than 4 μm can be confidently assigned to class 1, while cells larger than 6 μm easily can be called abnormal. It is for cells with nuclear diameter near 5 μm that one would expect misclassification errors to occur.

9.4 The Three-Feature, Three-Class Case

We next consider the case where there are three types of objects, and three measurements are made on each. The particular example we use here is the classification of pixels in a color image. The three measurements made on each pixel are the red, green, and blue intensity values. We assume that each pixel belongs to one of three classes: the interior of a normal cell, the interior of an abnormal cell, or the background.

Each pixel can be considered to represent a point in a three-dimensional color space. Thus each of the different-colored objects in the image will correspond to a "cloud" of points in color space, and classification becomes the task of isolating these clusters. More specifically, we wish to define a set of decision surfaces that carve up the space into three disjoint regions, one for each class.

9.4.1 The Bayes Classifier

One straightforward and quite powerful approach is the use of the Bayes maximum-likelihood classifier. It generates second-order surfaces that partition the color space into disjoint regions, one for each object type. Assuming Gaussian distributions for the clusters of points in color space, the Bayes classifier maximizes the probability that each pixel will be assigned correctly.

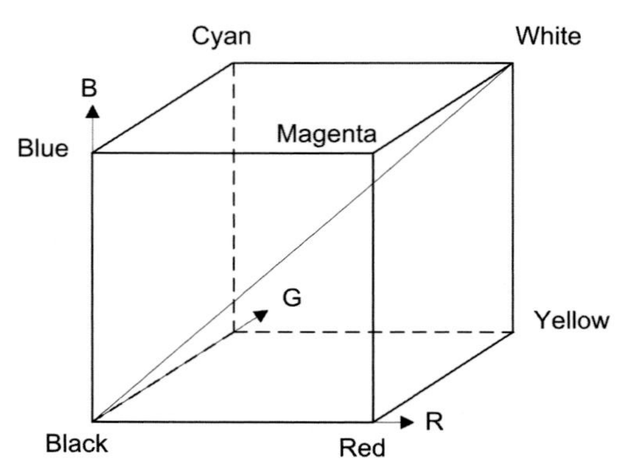

Fig. 9.4 The three-dimensional RGB color space.

We illustrate the use of the Bayes classifier with a simple example. We assume that a three-color RGB system (Red, Green, Blue) is used to digitize a fluorescent microscope image. The vector of gray levels at a single pixel location is

$$\mathbf{x} = [x_j] = \begin{bmatrix} x_1 \\ x_2 \\ x_3 \end{bmatrix}$$

where

$$j = \begin{cases} 1 \Rightarrow & \text{red} \\ 2 \Rightarrow & \text{green} \\ 3 \Rightarrow & \text{blue} \end{cases} \tag{9.8}$$

We further assume the images contain two types of objects, normal and abnormal cells. Both types of cells bind the blue fluor; the normals also bind the green fluor, but the abnormals pick up the red fluor instead. One would expect a three-dimensional histogram (scatterplot) of the color space to show three clusters of points, one each for background, normal cells, and abnormal cells. Since the background is dark, its cluster will fall near the origin of the color space. The normal cells will give rise to a cloud of points near the cyan corner, while the abnormals will fall near the magenta corner, as in Fig. 9.4.

9.4.1.1 Prior Probabilities

Let us assume that the area of a typical image is 90% occupied by background and 10% by cells. Further assume that, overall, only 10% of the cells are abnormal. Thus the vector of prior probabilities, where $P_i = P\{\text{pixel belongs to class } i\}$, is

$$\mathbf{P} = \begin{bmatrix} 0.90 \\ 0.09 \\ 0.01 \end{bmatrix}$$

where

$$i = \begin{cases} 1 & \Rightarrow \text{background} \\ 2 & \Rightarrow \text{normal} \\ 3 & \Rightarrow \text{abnormal} \end{cases} \tag{9.9}$$

9.4.1.2 Classifier Training

The first step is to train the classifier to recognize the three types of pixels. For this we require a training set containing pixels that are known to fall in the background, inside normal, and inside abnormal cells. It is the statistics of these training set pixels that constitute the knowledge that the classifier has about the problem. Estimating these statistics is the process of classifier training.

9.4.1.3 The Mean Vector

Using the training set, we calculate, for each class, i, the mean pixel brightness in each color, j. That is,

$$\mu_{ij} = \frac{1}{N_i} \sum_{k=1}^{N_i} x_{ijk} \tag{9.10}$$

where N_i is the number of pixels in class i, and x_{ijk} is the value, in color j, of pixel k in class i of the training set. The mean vector for class i is

$$\boldsymbol{\mu}_i = \begin{bmatrix} \mu_{i1} \\ \mu_{i2} \\ \mu_{i3} \end{bmatrix} \tag{9.11}$$

This vector, the mean vector of the training set, is an estimate of the mean of the entire population of class i pixels. If the training set is both adequately large and representative of the population, it will be a good estimate. Otherwise, it will be a poor estimate, and classifier accuracy will suffer, as will our ability to predict how well it will work.

9.4.1.4 Covariance

We calculate the covariance matrix for each class [2,5,8] as

$$\mathbf{S}_{ij_1j_2} = \frac{1}{N_i - 1} \left[\sum_{k=1}^{N_i} x_{ij_1k} x_{ij_2k} - N_i \mu_{ij_1} \mu_{ij_2} \right] \tag{9.12}$$

9.4.1.5 Variance and Standard Deviation

The diagonal elements of a covariance matrix are the variances of the features, for that class. The variance is the square of the standard deviation. That is,

$$\sigma_{ij}^2 = S_{ijj} \quad \text{and} \quad \sigma_{ij} = \sqrt{S_{ijj}} \tag{9.13}$$

9.4.1.6 Correlation

From the covariance matrix we can compute the correlation matrix for each class [2,5,8]. For class i, this is

$$\mathbf{C}_{ij_1j_2} = \frac{\mathbf{S}_{ij_1j_2}}{\sigma_{ij_1}\sigma_{ij_2}} \tag{9.14}$$

The elements of the correlation matrix are bounded by ± 1. A correlation of 1.0 means the corresponding two features are proportional to each other. A correlation of -1 means each is proportional to the negative of the other. A correlation of 0 means the two features are uncorrelated. Using highly correlated features is not only redundant, but it can actually degrade classifier accuracy. One can either combine highly correlated features (by averaging, for example), or simply discard all but one of them. In this example we assume the three features are not highly correlated.

9.4.1.7 The pdf

The probability density function for class i is

$$p_i(\mathbf{x}) = \frac{1}{\sqrt{(2\pi)^n|\mathbf{S}_i|}} \exp\left[-\frac{1}{2}\left|(\mathbf{x}-\boldsymbol{\mu}_i)^T\mathbf{S}_i^{-1}(\mathbf{x}-\boldsymbol{\mu}_i)\right|\right] \tag{9.15}$$

where $\mathbf{x}$ is the vector of RGB values for a pixel, as in Eq. (9.8), and $n=3$ is the number of features in use. The superscript T indicates the matrix transpose, and $\mathbf{S}_i^{-1}$ is the inverse of the covariance matrix for class i. Eq. (9.15) is the multidimensional generalization of Eq. (9.3).

9.4.1.8 Classification

Each class of pixels is now characterized by its prior probability, mean vector, and covariance matrix. The classifier has been trained. We now have enough statistical information about the problem to begin classifying pixels. By Bayes' rule, the likelihood that an unknown pixel having color vector $\mathbf{x}$ belongs to class i is

$$L_i = P_i\frac{pdf_i(\mathbf{x})}{p(\mathbf{x})} \quad \text{where} \quad p(\mathbf{x}) = pdf_1(\mathbf{x}) + pdf_2(\mathbf{x}) + pdf_3(\mathbf{x}) \tag{9.16}$$

or

$$L_i = \frac{P_i}{p(\mathbf{x})\sqrt{(2\pi)^3|\mathbf{S}_i|}} \exp\left[-\frac{1}{2}\left|(\mathbf{x}-\boldsymbol{\mu}_i)^T\mathbf{S}_i^{-1}(\mathbf{x}-\boldsymbol{\mu}_i)\right|\right] \tag{9.17}$$

Thus we can compute the three likelihoods (one for each class) and assign the pixel to the most likely (largest likelihood) class. If the pdfs are, as we have assumed, Gaussian (normal) density functions, then no other partitioning of color space will result in lower overall error rates [2].

9.4.1.9 Log Likelihoods

Since the logarithm is a monotonic function, we can take the log of both sides of Eq. (9.17) and use the resulting value for classification purposes. Eq. (9.17) then becomes

$$\ln(L_i) = \ln(P_i) - \frac{1}{2}\left|(\mathbf{x}-\boldsymbol{\mu}_i)^T S_i^{-1}(\mathbf{x}-\boldsymbol{\mu}_i)\right| - \frac{3}{2}\ln(2\pi) - \frac{1}{2}\ln(|\mathbf{S}_i|) - \ln(p(\mathbf{x})) \quad (9.18)$$

The third and fifth terms are constants and, for classification purposes, can be omitted, leaving

$$LL_i = \ln(P_i) - \frac{1}{2}\left|(\mathbf{x}-\boldsymbol{\mu}_i)^T S_i^{-1}(\mathbf{x}-\boldsymbol{\mu}_i)\right| - \frac{1}{2}\ln(|\mathbf{S}_i|) \quad (9.19)$$

The first term accounts for the prior probabilities, while the third term accounts for the within-class scattering of the features. The larger this variation, the less confidently one can assign the pixel to that class. The second term is the square of the *Mahalanobis distance.* It represents the variance-normalized distance, in feature space, from the unknown color to the class mean.

9.4.1.10 The Mahalanobis Distance Classifier

We can simplify the Bayes' classifier further by computing only the Mahalanobis distance from an unknown pixel to each class mean

$$D_i = \sqrt{\frac{1}{2}\left|(\mathbf{x}-\boldsymbol{\mu}_i)^T S_i^{-1}(\mathbf{x}-\boldsymbol{\mu}_i)\right|} \quad (9.20)$$

and assigning each pixel to the class having the nearest mean. This results in what is called a *Mahalanobis distance classifier.* It corresponds to the special case where the prior probabilities are equal among the classes, and likewise for the within-class variations. This distance classifier is sometimes used when the prior probabilities are unknown and the covariance matrix cannot be estimated accurately, due to limited training set size. Distance in feature space can be computed in other ways as well (e.g., Euclidean distance), and this gives rise to other types of distance classifiers.

9.4.1.11 Uncorrelated features

While 30 or so pixels per class in the training set might be sufficient to estimate the feature means and variances, considerably more might be required to estimate the off-diagonal elements of the covariance matrix. If the training set is necessarily small, one solution is to set the off-diagonal elements to zero. This is equivalent to assuming that the features are uncorrelated. Under pressure of limited training set size, this can yield a more stable and better performing classifier than one designed around inadequately estimated covariances. Using a distance classifier, which automatically assumes uncorrelated features, results in an even simpler classifier. Since there are so many pixels in an image, however, it is possible to accumulate quite large training sets, and covariances could be estimated quite accurately in this three-feature, three-class example.

9.4.2 A Numerical Example

To illustrate the operation of the three-class Bayes classifier we include a numerical example using six pixels from each class in the training set. While this is hopelessly inadequate for any real case, it serves to illustrate the calculations. This example will permit readers who choose to implement a Bayes classifier to check their implementation for numerical accuracy.

Assume the training set is

$$\begin{bmatrix} 33 & 31 & 46 & 57 & 18 \\ 42 & 10 & 24 & 38 & 56 \\ 62 & 50 & 34 & 21 & 33 \end{bmatrix} \begin{bmatrix} 6 & 28 & 47 & 21 & 58 \\ 96 & 116 & 126 & 84 & 73 \\ 70 & 82 & 96 & 117 & 90 \end{bmatrix} \begin{bmatrix} 78 & 115 & 122 & 76 & 134 \\ 12 & 52 & 34 & 70 & 22 \\ 81 & 100 & 146 & 78 & 70 \end{bmatrix} \tag{9.21}$$

for the RGB color values of five pixels each of background, normal cells, and abnormal cells, respectively. The mean vectors for the three classes are then

$$\boldsymbol{\mu}_1 = \begin{bmatrix} 37 \\ 34 \\ 40 \end{bmatrix} \quad \boldsymbol{\mu}_2 = \begin{bmatrix} 32 \\ 99 \\ 91 \end{bmatrix} \quad \boldsymbol{\mu}_3 = \begin{bmatrix} 105 \\ 38 \\ 95 \end{bmatrix} \tag{9.22}$$

The covariance matrices are.

$$\mathbf{S}_1 = \begin{bmatrix} 223.5 & -79 & -112.25 \\ -79 & 310 & -58.5 \\ -112.25 & -58.5 & 257.5 \end{bmatrix} \quad \mathbf{S}_2 = \begin{bmatrix} 428.5 & -24 & 86.25 \\ -24 & 482 & -79.75 \\ 86.25 & -79.75 & 306 \end{bmatrix}$$
$$\mathbf{S}_3 = \begin{bmatrix} 700 & -154.5 & 265.75 \\ -154.5 & 542 & 21.5 \\ 265.75 & 21.5 & 934 \end{bmatrix} \tag{9.23}$$

from which the standard deviations (square roots of diagonal elements) are

$$\boldsymbol{\sigma}_1 = \begin{bmatrix} 15.0 \\ 17.6 \\ 16.0 \end{bmatrix} \quad \boldsymbol{\sigma}_2 = \begin{bmatrix} 20.7 \\ 22.0 \\ 17.5 \end{bmatrix} \quad \boldsymbol{\sigma}_3 = \begin{bmatrix} 26.5 \\ 23.3 \\ 30.6 \end{bmatrix} \tag{9.24}$$

the correlation matrices (Eq. 9.14) are

$$\mathbf{C}_1 = \begin{bmatrix} 1 & -.300 & -.468 \\ -.300 & 1 & -.207 \\ -.468 & -.207 & 1 \end{bmatrix} \qquad \mathbf{C}_2 = \begin{bmatrix} 1 & -.053 & .238 \\ -.053 & 1 & -.208 \\ .238 & -.208 & 1 \end{bmatrix}$$

$$\mathbf{C}_3 = \begin{bmatrix} 1 & -.251 & .329 \\ -.251 & 1 & .030 \\ .329 & .030 & 1 \end{bmatrix} \tag{9.25}$$

and the determinants of the covariance matrices are

$$|\mathbf{S}_1| = 1.053 \times 10^7 \qquad |\mathbf{S}_2| = 5.704 \times 10^7 \qquad |\mathbf{S}_3| = 2.917 \times 10^8 \tag{9.26}$$

Suppose we have an unknown pixel having color vector

$$\mathbf{x} = \begin{bmatrix} 38 \\ 80 \\ 78 \end{bmatrix} \tag{9.27}$$

The three likelihoods (from Eq. 9.17) are

$$L = \begin{bmatrix} 0.000015 \\ 0.088083 \\ 0.000213 \end{bmatrix} \tag{9.28}$$

and the pixel would be assigned to class 2. The log likelihoods, with constant terms dropped (Eq. 9.19), are

$$\mathbf{LL} = \begin{bmatrix} -20.9 \\ -12.3 \\ -18.3 \end{bmatrix} \tag{9.29}$$

Again, the pixel would be assigned to class 2. The Mahalanobis distances to the class means (Eq. 9.20) are

$$\mathbf{D} = \begin{bmatrix} 3.57 \\ 0.97 \\ 1.99 \end{bmatrix} \tag{9.30}$$

This pixel would be assigned to class 2 by a distance classifier as well, since it is the closest. Thus all three classifiers would assign this pixel to class 2.

9.5 Classifier Performance

Once a classifier has been designed and trained, it is necessary to test it to establish its accuracy. This is usually done by classifying a test set of known objects and tabulating the number of errors. If the test set is the same as the training set, the performance estimates will be optimistically biased. If it includes none of the training data, they will be pessimistically biased. If the test set is large, the effects of this bias will be slight. If the number of available preclassified objects is small, one can use the "round robin" or "leave one out" method. Here the classifier is trained on all but one of the objects and tested on the remaining object. This process is repeated until every object has been used for testing. The results of the various experiments are then averaged together to estimate the error rates.

9.5.1 The Confusion Matrix

A very handy tool for specifying the accuracy of a multiclass classifier is the confusion matrix. This is an $N \times N$ matrix, $\mathbf{C}$, where N is the number of classes. The columns of $\mathbf{C}$ correspond to the classes that objects actually belong to, while the rows of $\mathbf{C}$ correspond to the classes to which objects can be assigned. Thus the element c_{ij} corresponds to the situation of an object that belongs to class j being assigned to class i. That is, true class $= j$, assigned class $= i$.

There are several ways one can set up the confusion matrix to summarize the results of a classifier test. A *raw confusion matrix* results when the value of each element is simply set to the number of times the corresponding situation occurred in a particular test of the classifier. Other values, however, may be more useful. For example, *sensitivity* is defined, for each class, as the probability that an object belonging to that class will be correctly assigned. We obtain an estimate of the sensitivity matrix by dividing the raw confusion matrix elements by the total number of objects in the true class. Each element, then, shows what fraction of the objects that actually belong to that class are assigned to that class. The columns of the sensitivity matrix sum to unity.

Specificity, for a particular class, is defined as 1 minus the ratio of the number of objects incorrectly assigned to that class divided by the total number of objects that are not in that class. Specificity is seldom a very useful parameter because it almost always takes on values quite close to unity. A more useful specification is *positive predictive value* (PPV). This is the probability that an object assigned to a class actually belongs to that class. The PPV matrix is estimated by dividing the elements in each row of the raw confusion matrix by the total number of objects assigned to that class. In this case the rows sum to unity.

When analyzing the performance of a classifier, one finds the sensitivity matrix and the PPV matrix to be very useful. In short, the sensitivity matrix tells you where each type of object is going, while the PPV matrix tells you what is going into each of the classes. Studying these two matrices can yield considerable insight into the strengths and weaknesses of a particular classifier.

As an example, consider the sensitivity matrix and PPV matrix shown in Fig. 9.5. They correspond to the three-class pixel classifier example discussed previously. We see from Fig. 9.5 that this classifier has two problems. First, 9% of the pixels in abnormal

Sens	True Class		
Assigned Class	1	2	3
1	98%	2%	1%
2	1%	96%	9%
3	1%	2%	90%

PPV	True Class		
Assigned Class	1	2	3
1	98%	1%	1%
2	2%	96%	2%
3	2%	7%	91%

Fig. 9.5 Confusion matrices.

cells are being called normal. This could lead to abnormal cells being missed. Second, 7% of the pixels that are called abnormal are actually normal. This could lead to false positive errors. We would conclude that this classifier needs to be improved in its ability to discriminate between pixels in the normal and abnormal classes.

9.6 Bayes Risk

A useful generalization of the maximum likelihood Bayes classifier allows one to bias the classifier so as to reduce the occurrence of certain costly types of misclassification errors, in exchange for making more of other, less serious, errors [2,5]. In our three-class example, there are nine elements in the confusion matrix. Three correspond to correct decisions, while the remaining six represent different types of errors. Suppose that it is considered to be more serious to confuse a pixel from an abnormal cell with one from a normal cell, or to call a normal pixel abnormal, than it is to make any of the four other possible errors. The minimum Bayes risk classifier allows us to account for this difference in the severity of errors.

9.6.1 The Minimum-Risk Classifier

We begin by setting up a *cost matrix*. It has the same format as the confusion matrix, except that its elements represent the "cost" of having that situation occur. Specifically, C_{ij} represents the cost of assigning to class j a pixel that actually belongs to class i. If $i=j$, this corresponds to a correct classification, and a cost of zero might be assigned to those elements. If all misclassification errors are equally unfortunate, then ones could be placed in all of the off-diagonal elements. In this case a maximum likelihood classifier results. However, larger values can be assigned to cost matrix elements that correspond to the more serious errors. One possible cost matrix for our three-class pixel classifier example is

$$\mathbf{C} = \begin{bmatrix} 0 & 1 & 1 \\ 1 & 0 & 4 \\ 1 & 2 & 0 \end{bmatrix} \tag{9.31}$$

Here we have said (1) that correct classifications cost nothing, (2) that calling an abnormal pixel normal has a cost of 4, and (3) that calling a normal pixel abnormal has a cost of 2, while (4) the four remaining errors have unit cost. Note that the actual cost values are unitless, and their values are relevant only in relation to one another.

Given a cost matrix, we can set up the Bayes classifier to minimize its long-term cost of operation. The Bayes risk for assignment to a particular class is the cost of each outcome times the likelihood of that outcome, summed over all possible assignments to that class. It can be computed, for the unknown pixel of Eq. (9.27), as

$$R_i = \sum_{j=1}^{3} C_{i,j} L_j \quad \mathbf{R} = \begin{bmatrix} 0.0883 \\ 0.0009 \\ 0.1762 \end{bmatrix} \tag{9.32}$$

In this example the unknown pixel would be assigned to Class 2 because the Bayes risk is lowest there. We would expect this classifier to perform better in practice than the maximum likelihood classifier since the two most serious error rates have been reduced (but at the expense of more of the less serious errors).

9.7 Relationships Among Bayes Classifiers

Note that the minimum-risk classifier is the most general Bayes classifier. The maximum-likelihood classifier is a special case of that, namely when all costs are set to be equal. Further reductions in generality result when the *a priori* probabilities are assumed to be equal, or the off-diagonal covariances are assumed to be 0. The minimum-distance classifier is a further restricted special case that results when both the prior probabilities and the within-class variation are ignored.

In this numerical example all three forms of the Bayes classifier, the minimum-risk, maximum likelihood, and minimum distance classifiers, assigned the unknown pixel to the same class. This will not be the case in general, as objects that fall near the decision boundaries will be assigned differently by the different classifiers. Objects that fall near their class mean will be classified correctly by any of the classifiers. It is the outliers, those that defy the assumption of Gaussian statistics, that make classification difficult.

9.8 The Choice of a Classifier

If a considerable amount of training data is available, one can simply estimate the required pdfs and use those estimates in the classification process. Such classifiers are called "nonparametric." Often, however, it is difficult to obtain large numbers of preclassified objects. In this case one can assume a particular functional form for the pdf (the Gaussian, for example) and use the training data only to estimate the parameters of the pdf. This gives rise to a "parametric classifier." Considerably less training data is then required for training, and one benefits from the powerful mathematics that have been developed for those cases.

It is often useful to begin a classifier design effort with a classical Bayes classifier, as described previously. At the very least, this establishes a baseline of performance against which other types of classifiers can be tested and evaluated. Further, if the underlying assumptions are met, the Bayes classifier, assuming Gaussian statistics, may well perform as well as or better than any other classifier.

Problems arise when the underlying pdfs do not fit the assumed functional form. The classifier's performance, and one's predictions of its accuracy, are only as good as the underlying assumptions. It is rather difficult to prove that a population of objects actually fits, for example, a Gaussian (normal) distribution. As a rule of thumb, if the marginal distributions (one-dimensional histograms) are unimodal and symmetrical, one can often assume multivariate Gaussian statistics (although there are no guarantees, and notable exceptions exist). Even so, the Gaussian pdf decays quite rapidly for values distant from the mean. Any occurrence of data beyond 4σ, for example, is essentially impossible. For actual data, however, this may not match reality. In such cases one can use a "heavy-tailed" pdf that decays to zero more slowly than the Gaussian. Such distributions include the log-normal, Cauchy, Levy, power-law, and student's *t*-distributions [9].

9.8.1 Subclassing

Even if the marginal distributions are not unimodal and symmetrical, one can do things to make them unimodal and symmetrical. If the feature histograms of a class are multimodal (i.e., they have two or more peaks) one would suspect that two or more distinct subclasses exist within the class. By subdividing the class, one can often achieve unimodal pdfs, but with a larger number of classes. This is a fair trade if it justifies the assumption of Gaussian statistics. An example is shown in Fig. 9.6. This is the single-feature, two-class example used earlier in this chapter. Here the nuclear diameter histogram of normal cells is unimodal, but that of the abnormals is bimodal.

If we reexamine the training set we may find that two distinct populations of cells exist within the abnormal class. In this case we can establish a new class called "atypical" and assign some of the previously "abnormal" cells to it, based on their morphology. The result is a three-class problem where the feature histograms are unimodal (Fig. 9.7). This is a very profitable trade if it permits the assumption of Gaussian statistics.

9.8.2 Feature Normalization

An asymmetric or non-Gaussian pdf often can be corrected by a suitably designed nonlinear transformation of the feature values [5]. A feature histogram, normalized to

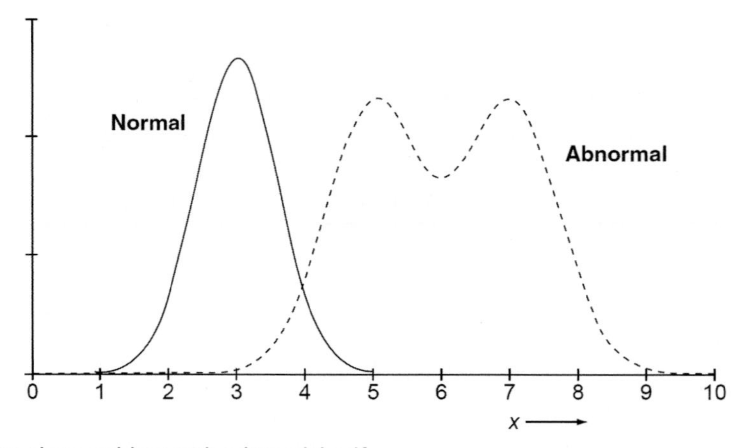

Fig. 9.6 A two-class problem with a bimodal pdf.

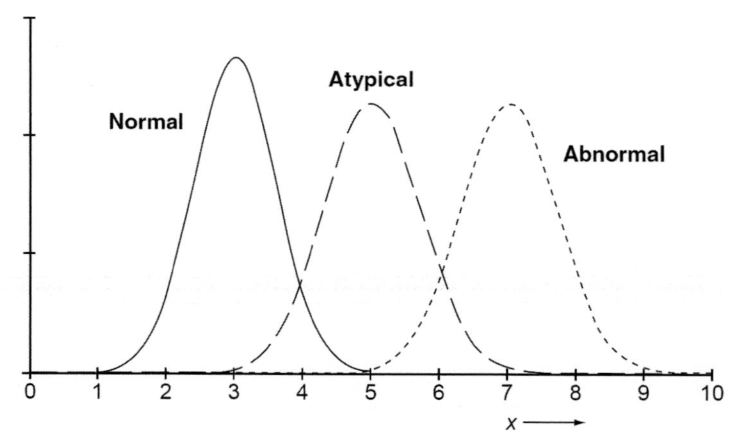

Fig. 9.7 The use of subclassing to eliminate a bimodal pdf. The result is a three-class problem with unimodal pdfs.

unit area, is an estimate of the pdf of that feature. The *cumulative distribution function* (CDF) is the integral of the pdf, that is,

$$P(x) = \int_0^x p(u)du = \frac{1}{A_o}\int_0^x H(u)du \tag{9.33}$$

where $H(x)$ is the histogram, $p(x)$ is the pdf, $P(x)$ is the CDF, and A_0 is the area under the histogram. The CDF is quite a well-behaved function, increasing monotonically from 0 to 1. If it is used to transform feature x into a new feature y, that is, $y = P(x)$, then the feature y will have a flat histogram (uniform distribution). As a special case, the CDF corresponding to a Gaussian pdf will transform a feature with a Gaussian pdf into one with a flat histogram. It follows that the inverse function of the Gaussian CDF will transform a feature with a flat histogram into one with a Gaussian histogram. Thus we can transform a feature so that it has a Gaussian histogram by concatenating two nonlinear transformations:

$$y = P_2(P_1(x)) \tag{9.34}$$

where $P_1(x)$ is the CDF of the feature, and $P_2(x)$ is the inverse of the CDF of a Gaussian. $P_1(x)$ makes the pdf uniform, and $P_2(x)$ makes it Gaussian.

Note that, in a multifeature classification problem, transforming the individual features to have Gaussian pdfs does not guarantee that the overall multivariate pdf will be Gaussian. As a practical matter, however, such a transformation can make the assumption of Gaussian statistics much less of an approximation. Feature normalization works best in the commonly occurring case where the raw feature histograms are not radically different from a Gaussian to begin with.

9.9 Nonparametric Classifiers

If the functional form of the pdfs of the classes is unknown, then the parametric approach cannot be used. In this case one must estimate the pdfs directly from the training data [2]. This generally requires a much larger training set. However, the maximum likelihood and minimum risk formulations still apply.

The basic problem of nonparametric pdf estimation is straightforward: given a set of training samples, model the pdf of the data without making any assumptions about the functional form of the distribution. Suppose we have N_j training samples from class j. To estimate the pdf, the L-dimensional feature space can be partitioned into small regions that are L-dimensional hypercubes, with volume $V = h^L$, where h is the bin size. Let $\mathbf{R}$ be such a region and k_j be the number of samples from class j falling into $\mathbf{R}$, with $k_j \leq N_j$. A straightforward estimate of the pdf can be expressed as:

$$\hat{p}(\mathbf{x}|C_j) = \frac{k_j/N_j}{V} \tag{9.35}$$

This basic estimator corresponds to an L-dimensional histogram. Essentially, the feature space is divided into a finite number of hypercube bins and the probability density at the center of each hypercube is estimated by the fraction of samples in the training set that fall into that hypercube bin. The bin size, h, and the starting position of the first bin are two "parameters" that determine the shape of the histogram.

The histogram is a simple and effective form of pdf estimation, but it has several drawbacks. The shape of the estimate is affected by both the bin size and the starting point of the bins. The discontinuities of the estimate are not due to the underlying probability density, but are caused by the particular choice of bin locations. A more serious problem is the curse of dimensionality, since the number of bins grows exponentially with the number of dimensions. In high dimensions we would require a very large number of training samples, or else most of the bins would be empty.

A more advanced nonparametric pdf estimation method makes use of the so-called "Parzen window" for a unit hypercube centered at the origin:

$$\psi(\mathbf{v}) = \begin{cases} 1, & |v_q| \leq \frac{1}{2}, \; q = 1, 2, \ldots, L \\ 0, & otherwise \end{cases} \tag{9.36}$$

Hence $\psi((\mathbf{x}-\mathbf{x}_i)/h)$ equals to unity if $\mathbf{x}_i$ is within the hypercube at $\mathbf{x}$ and is zero elsewhere. The number of training samples that fall into this hypercube can be written as

$$k = \sum_{i=1}^{N_j} \psi((\mathbf{x}-\mathbf{x}_i)/h)$$

Substituting it in Eq. (9.35), we have

$$\hat{p}(\mathbf{x}|C_j) = \frac{1}{N_j} \sum_{i=1}^{N_j} \frac{1}{V} \psi((\mathbf{x}-\mathbf{x}_i)/h) \tag{9.37}$$

Notice that the Parzen window density estimate resembles the histogram, except that the hypercube locations are determined by the training sample points rather than by the histogram bins. The expression in Eq. (9.37) shows that the estimate $\hat{p}(\mathbf{x}|C_j)$ is made of an average of functions of x and the samples $\mathbf{x}_i$. Based on this formulation, we can adopt two basic approaches. We can choose a fixed value of k and determine the corresponding volume V from the training samples. This gives rise to the so-called "k nearest-neighbor" (kNN) approach. Alternatively, we can also choose a fixed value of the volume V and determine k from the samples. This leads to the methods commonly referred to as "kernel density estimation" (KDE).

9.9.1 Nearest-Neighbor Classifiers

The kNN is a very intuitive nonparametric approach that classifies unknown objects based on their similarity to the samples in the training set. For an unknown object $\mathbf{x}$, it finds the k "nearest" samples x_i in the training set and assigns $\mathbf{x}$ to the class that appears most frequently among them. A great advantage of the kNN approach is that no estimation of the pdf is required since the function is only approximated locally, and all computation is deferred until the classification stage. However, the disadvantages are the memory requirement to store training samples and the computational complexity required to search for the k nearest samples during the classification of each unknown object.

On the other hand, with the KDE methods one can generalize the hypercube Parzen window with a smooth nonnegative kernel function $\psi(\mathbf{x})$ that satisfies the condition $\int \psi(\mathbf{x})\, d\mathbf{x} = 1$. Just as the Parzen window estimate can be considered a sum of boxes centered at the samples, the smooth kernel estimate is a sum of "bumps" placed at the samples, and the kernel function determines the shape of the bumps. Usually, $\psi(\mathbf{x})$ is chosen to be a radially symmetric, unimodal pdf, such as the multivariate Gaussian. The kernel function is used essentially for interpolation, and each sample contributes to the estimate according to its distance from $\mathbf{x}$. It can be shown [2] that both of these approaches converge to the true pdf as $N_j \to \infty$, that is, $\lim_{N_j \to \infty} \hat{p}(\mathbf{x}|C_j) = p(\mathbf{x}|C_j)$, provided that V shrinks with N_j, and k grows with N_j properly.

For applications with high dimensional feature space, the curse of dimensionality affects all classifiers without exception. The available training samples are usually inadequate to obtain an accurate estimation in these cases. One solution to the problem is to choose independent features so that $p(\mathbf{x}|C_j) = \prod_{i=1}^{L} p(x_i \mid C_j)$, by mapping the original features using a proper subspace transformation such as the independent component analysis (ICA) [10]. Thus the problem of estimating an L-dimensional multivariate pdf $p(\mathbf{x} \mid C_j)$ is collapsed to that of estimating multiple one-dimensional univariate pdfs $p(x_i \mid C_j)$, $i = 1, 2, \ldots, L$. This way, the training set size requirement becomes much easier to meet.

9.10 Feature Selection

Ideally one would prefer to use a rather small number of highly discriminating, uncorrelated features. Increasing the number of features increases the dimensionality, and hence the volume of the feature space [11–13]. This, in turn, increases the requirements for training set and test set size [2–6]. Adding features that have poor discrimination or are highly correlated with the other features can actually degrade classifier performance [2].

9.10.1 Feature Reduction

There are well-developed mathematical procedures for reducing a large number of features down to a smaller number without severely limiting the discriminating power of the set. *Principal component analysis* (PCA) [8] and *linear discriminant analysis* (LDA), also known as *Fisher discriminant analysis* (FDA) [14], discussed shortly, are among the best-known subspace methods that can be used to reduce the dimensionality of the feature space. Both generate a new set of features, each of which is a linear combination of the original features. In both cases the new features are ranked so that one can select only a few of the most useful ones, thereby reducing the number of features.

9.10.1.1 Principal Component Analysis

In general, suppose $\mathbf{x}$ is an L-dimensional feature vector and $\mathbf{W}$ is a $Q \times L$ matrix; then

$$y_i = \sum_{j=1}^{L} w_{i,j} x_j, \quad i = 1, 2, \ldots, Q \qquad \text{or} \qquad \mathbf{y} = \mathbf{W}\mathbf{x} \tag{9.38}$$

defines a linear transformation of the vector $\mathbf{x}$. The result is a $Q \times 1$ vector $\mathbf{y}$, which is a projection of $\mathbf{x}$ onto a linear subspace defined by the transform matrix $\mathbf{W}$. Each element y_i is the inner product of a basis vector, which is the ith row of $\mathbf{W}$, with the input vector $\mathbf{x}$.

Consider a set of L-dimensional sample feature vectors $\mathbf{x}_1$, $\mathbf{x}_2$, $\ldots\mathbf{x}_M$. Without loss of generality, we can assume these are zero-mean vectors since we can always redefine $\mathbf{x} = \mathbf{x}' - \boldsymbol{\mu}$, where $\boldsymbol{\mu}$ is the mean vector of all these samples. Then $\mathbf{X}$ is an $L \times M$ data matrix whose columns comprise the M sample vectors $\mathbf{x}_1$, $\mathbf{x}_2$, $\ldots\mathbf{x}_M$, and $\mathbf{S}_t = \mathbf{X}\mathbf{X}^T$ is defined as the total scatter matrix of the sample vectors. The aim of PCA is to find the transform matrix of a subspace whose basis vectors correspond to the maximum-scatter directions in the original L-dimensional feature space. Therefore the PCA transform matrix, $\mathbf{W}_{\text{PCA}}$, is chosen to maximize the determinant of the total scatter matrix of the projected samples

$$\mathbf{W}_{PCA} = \arg\max_{\mathbf{W}} |\widetilde{\mathbf{S}}_t| \tag{9.39}$$

where $\widetilde{\mathbf{S}}_t = \mathbf{W}\mathbf{S}_t\mathbf{W}^T$. The solution to this equation is the transformation matrix, $\mathbf{W}$, constructed so that its row vectors are the eigenvectors, $\mathbf{w}_j$, of the scatter matrix, $\mathbf{S}_t$, arranged in the order of decreasing magnitude of the corresponding eigenvalues λ_j, that is,

$$\mathbf{S}_t\mathbf{w}_j = \lambda_j \mathbf{w}_j, \qquad j = 1, 2, \ldots, Q \tag{9.40}$$

where the λ_js are nonzero eigenvalues associated with the eigenvectors $\mathbf{w}_j$; Q denotes the rank of $\mathbf{S}_t$ and it cannot exceed the lesser of L and M.

Because of the maximum-scatter projection, PCA provides an optimal transformation for representing the original data vector, $\mathbf{x}$, from a lower-dimensional subspace in terms of minimum mean square error (MSE) [2]. Let $\hat{\mathbf{W}}_{PCA}$ be the $R \times L$ matrix ($R < L$) formed by discarding the lower L-R rows of $\mathbf{W}_{\text{PCA}}$. Then the transformed $R \times 1$ vector $\hat{\mathbf{y}}$ is given by $\hat{\mathbf{y}} = \hat{\mathbf{W}}_{PCA}\mathbf{x}$. The $\mathbf{x}$ vector can still be reconstructed as $\hat{\mathbf{x}} = \hat{\mathbf{W}}^T{}_{PCA}\hat{\mathbf{y}}$ with approximation error given by $MSE = \sum_{k=R+1}^{L} \lambda_k$. Overall, PCA decorrelates the new features and maximizes their variance. The number of new PCA features is equal to the number of original features, but one can decide how many of them to use.

9.10.1.2 Linear Discriminant Analysis

Unlike PCA, LDA seeks a linear subspace that best discriminates among object classes, rather than the one that represents samples with least MSE. Specifically, LDA selects the transform matrix $\mathbf{W}_{\mathrm{LDA}}$ in such a way that the ratio of the between-class scatter to the within-class scatter is maximized [14]. If we define the between-class scatter matrix as

$$\mathbf{S}_b = \sum_{i=1}^{c} M_i(\boldsymbol{\mu}_i - \boldsymbol{\mu})(\boldsymbol{\mu}_i - \boldsymbol{\mu})^T \tag{9.41}$$

and the within-class scatter matrix as

$$\mathbf{S}_w = \sum_{i=1}^{c}\sum_{j=1}^{M_i} \left(\mathbf{x}_j - \boldsymbol{\mu}_i\right)\left(\mathbf{x}_j - \boldsymbol{\mu}_i\right)^T \tag{9.42}$$

where M_i is the number of samples in class i, c is the number of object classes, $\boldsymbol{\mu}_i$ is the mean of type i sample vectors, and $\boldsymbol{\mu}$ is the total mean of sample vectors of all classes. The optimization criterion here is to maximize the determinant ratio of between-class and within-class scatters of the projected samples

$$\mathbf{W}_{LDA} = \underset{\mathbf{W}}{\arg\max} \left\{ \frac{|\widetilde{\mathbf{S}}_b|}{|\widetilde{\mathbf{S}}_w|} \right\} \tag{9.43}$$

where $\widetilde{\mathbf{S}}_b = \mathbf{W}\mathbf{S}_b\mathbf{W}^T$ and $\widetilde{\mathbf{S}}_w = \mathbf{W}\mathbf{S}_w\mathbf{W}^T$. It has been proven [14] that if $\mathbf{S}_w$ is non-singular, the determinant ratio in Eq. (9.43) is maximized when the row vectors of the transform matrix, $\mathbf{W}$, are the generalized eigenvectors of $\mathbf{S}_w^{-1}\mathbf{S}_b$ corresponding to

$$\mathbf{S}_w^{-1}\mathbf{S}_b\mathbf{w}_i = \lambda_i\mathbf{w}_i, \qquad i = 1, 2, \ldots, m \tag{9.44}$$

where λ_i, $i = 1, 2, \ldots, m$ are the generalized eigenvalues, and m is the number of nonzero generalized eigenvectors, $m \le c - 1$. Notice that the dimensionality of the LDA subspace is upper-bounded by $c-1$, meaning that the total number of LDA features is 1 less than the number of classes. This is because $\mathbf{S}_b$ is of rank $c-1$ or less. Also, since the rank of $\mathbf{S}_w$ is, at most, $M-c$, M must be greater than or equal to $L+c$ in order to ensure that $\mathbf{S}_w$ does not become singular.

In summary, since LDA maximizes the ability of the new features to discriminate among the classes, it is generally considered to be more effective than PCA for feature reduction prior to classification.

9.11 Neural Networks

A completely different approach to classification is the use of artificial neural networks (ANNs) [7]. Here a network is composed of one or more layers of interconnected processing elements (PEs). Each PE creates its output as a weighted sum of its inputs (see Fig. 9.8). The feature values are the inputs to the first layer, and the output values of the final layer are used to assign the object to a class.

The ANN is trained by adjusting the weighting factors in each of its PEs. A large training set of preclassified objects is presented to the network repeatedly, and in random order. Each time, the weights are adjusted to bring the output value toward its correct value. The training process is continued until the error rate stops declining.

The computation performed by such a PE is a function of a dot product, namely

$$O = g(\mathbf{X} \cdot \mathbf{W}) = g\left[\sum_{i=1}^{N} x_i w_i\right] = g(S) \tag{9.45}$$

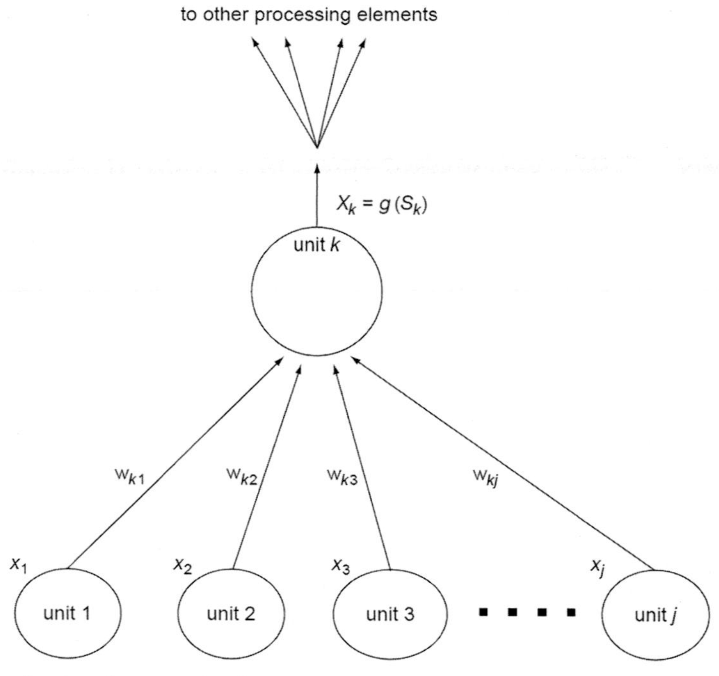

Fig. 9.8 A neural network processing layer.

where O is the (scalar) output, $\mathbf{X}$ is the input vector, and $\mathbf{W}$ is the weight vector associated with that processing element. The weights are adjusted during the training process, and they remain fixed during routine usage.

The weighted sum is subjected to a nonlinear transformation by the *activation function*, $g(S)$, which has a sigmoid (S-curve) shape. It is monotonically increasing, differentiable, and it asymptotically approaches 0 and 1 at large negative and positive values of its argument, respectively. An example is

$$g(S) = \frac{1}{1 + e^{-S}} \tag{9.46}$$

The primary purpose of the activation function is to restrict the output of the PE to the range [0,1]. By convention, outputs are all positive, but interconnection weights can be either positive or negative.

One advantage of the ANN is that it is not necessary to know the statistics (i.e., pdfs) of the features in order to develop a functioning classifier. Further, the decision surfaces that the ANN can implement in feature space are more complex than the second-order surfaces that the parametric Bayes classifier, for example, generates. This can be helpful when the pdfs are multimodal.

A disadvantage of the ANN, as compared to the statistical classifiers previously discussed, is that it is a "black box," and one is hard pressed to understand or explain its behavior. It also lacks the rich analytical underpinning of the classical approach that provides guidance in the design and development process. This makes it difficult to prove optimality or to predict error rates. Further, if the training is not done properly, on representative training sets of sufficient size, then the net can "overfit" or "memorize" the training set, that is, perform well on the training set but not generalize to objects previously unseen.

Artificial neural networks are discussed in much greater detail in Chapter 15.

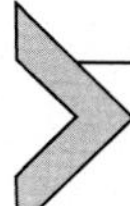

9.12 Summary of Important Points

1. A well-trained Bayes classifier can be quite effective at multiclass, multifeature classification, even in the presence of considerable noise.
2. One should pay particular attention to numerical precision issues since some of the parameters in the probability calculations can become quite large or quite small.
3. If the marginal distributions are unimodal and symmetrical, it may be useful to assume Gaussian statistics (multivariate normal pdfs).
4. A nonlinear transformation can make a feature's pdf symmetrical.
5. A multimodal pdf suggests the presence of subclasses. Judicious use of subclassing and feature transformations can often make the Gaussian assumption work.

6. When a particular functional form for the pdf (the Gaussian, for example) is known, less training data is required since it is used only to estimate the parameters. This gives rise to a parametric classifier.
7. If the functional form of the pdf is not given or it is known to be non-Gaussian, one must estimate the pdfs directly from the training data. Such classifiers are nonparametric, and they usually require considerably more training data.
8. When available training samples are inadequate to estimate the pdfs accurately, one can choose independent features by mapping the original features using a proper transformation such as independent component analysis. With this method the problem of estimating a multivariate pdf is simplified to that of estimating multiple univariate pdfs, thereby considerably reducing the size requirements of the training set.
9. PCA and LDA are two well-known techniques for reducing a large number of features down to a smaller number without losing their discriminating power. PCA decorrelates the new features and maximizes their variance, whereas LDA maximizes the ability of the features to discriminate among the classes.
10. An ANN classifier has the advantages that it is not necessary to know the statistics of the features in order to function, and the decision surfaces it can implement in feature space are more complex than the second-order surfaces that the parametric Bayes classifier generates. However, its disadvantages are that it is a "black box" and it is difficult to prove optimality or to predict error rates. It also lacks the rich analytical underpinning that supports the design of statistical classifiers.

References

[1] B.S. White, K.R. Castleman, Estimating cell populations, Pattern Recogn. 13 (5) (1981) 365–370.
[2] R.O. Duda, Pattern Classification, second ed., Wiley, 2007.
[3] S. Theodoridis, Machine Learning: A Bayesian and Optimization Perspective, second ed., Elsevier Academic Press, 2020.
[4] C.M. Bishop, Pattern Recognition and Machine Learning, Springer, 2011.
[5] K.R. Castleman, Digital Image Processing, Prentice-Hall, 1996.
[6] W. Meisel, Computer-Oriented Approaches to Pattern Recognition, Academic Press, New York, 1972.
[7] R. Schalkoff, Pattern Recognition—Statistical, Structural and Neural Approaches, John Wiley & Sons, New York, 1992.
[8] L. Ott, W. Mendenhall, Understanding Statistics, fifth ed., PWS-KENT, Boston, 1990.
[9] C. Forbes, M. Evans, M. Hastings, B. Peacock, Statistical Distributions, fourth ed., John Wiley and Sons, 2011.
[10] A. Hyvärinen, J. Karhunen, E. Oja, Independent Component Analysis, John Wiley & Sons, 2001.
[11] R. Fisher, The statistical utilization of multiple measurements, Ann. Eugen. 8 (1938) 376–386.
[12] I.T. Young, Further considerations of sample size and feature size, IEEE Trans. IT-24 (6) (1978) 773–775.
[13] A.K. Jain, B. Chandrasekaran, Dimensionality and sample size considerations in pattern recognition practice, in: Handbook of Statistics, vol. 2, North Holland Publishing Company, 1982, pp. 835–855.
[14] L. Kanal, B. Chandrasekaran, On dimensionality and sample size in statistical pattern recognition, Pattern Recogn. **3** (1971) 225–234.

CHAPTER TEN

Multispectral Fluorescence Imaging

Fatima A. Merchant and Ammasi Periasamy

10.1 Introduction

Fluorescence microscopy is one of the most basic tools used in biological sciences for the visualization of cells and tissues. The popularity of fluorescence microscopy for the examination of biological specimens, both fixed and live, stems from its inherent ability to target fluorescent probes at molecules at low concentrations with high selectivity and specificity, and relatively high signal-to-noise ratios (SNRs) due to separation of the excitation light from the recorded fluorescence image.

Modernization of imaging techniques, robotic instrumentation, development of new fluorescent tagging proteins and synthetic fluorophores, and the rapid growth in computer and informatics technology have only compounded its utilization in the observation of the temporal and spatial dynamics of cellular components and activity. The past few decades have witnessed a renaissance of fluorescence microscopy, with digital imaging playing a pivotal role in automated detection and analysis of molecular and cellular processes, resulting in a shift of paradigm from qualitative to quantitative biology. The current emphasis in biology is now on quantitative analysis of information, so that observations can be integrated, and their significance understood. Digital image processing can provide numerical data to quantify and substantiate biological processes observed by fluorescence microscopy. This chapter covers the principles of fluorescence, highlighting problems inherent to fluorescence microscopy and methods to digitally correct them. In subsequent sections, various fluorescence microscopy techniques are introduced, with an emphasis on the image processing and analysis algorithms used for enhancing and analyzing fluorescence images.

10.2 Basics of Fluorescence Imaging

Electrons in certain types of molecules can absorb light, reach excited higher energy states, and then decay back to their ground state by losing energy in the form of heat and emitted light. If the electron's spin is unchanged, the excited state is called the "singlet state," whereas if the spin is altered by the excitation, the electron enters the "triplet state." Decay from the singlet excited state results in *fluorescence* emission, whereas decay from the triplet state is known as *phosphorescence*. The phenomenon of

Microscope Image Processing
https://doi.org/10.1016/B978-0-12-821049-9.00007-1
Copyright © 2023 Elsevier Inc.
All rights reserved.

fluorescence occurs when certain molecules (called fluorophores, fluorochromes, or fluorescent dyes) absorb light and reach an excited, unstable electronic singlet state (S_1). Under normal conditions, an unexcited molecule typically resides in the stable ground state S_0, at its lowest vibrational or rotational energy level. The absorption of a photon moves a molecule to one of the vibrational or rotational energy levels of a higher energy state (S_1). Internal energy conversions (time on the order of ~1 ps) then force the molecule to relax back to the lowest energy level of S_1. From here, they transition back to the singlet ground state (S_0), following the emission of fluorescent light at a characteristic wavelength (time order of ~1–10 ns). Internal conversion again relaxes the molecule back to the lowest energy level of S_0.

Fluorescence emission always occurs due to decay from the lowest energy level of S_1 to some level of the ground state, regardless of the initial state of excitation. Thus the energy of excitation is greater than the energy of emission, and the emission spectrum is independent of the energy of the exciting photon. The energy of the emitted photon is the difference between the energy levels of the two states, and it determines the wavelength of the emitted light λ_{EM} as follows:

$$\lambda_{EM} = \frac{hc}{E_{EM}} \tag{10.1}$$

where E_{EM} is the difference between the energy levels of the two states during emission (EM) of light, h is Planck's constant, and c is the speed of light. The wavelength of emission is always longer than that of excitation, and the difference between the two is known as the Stokes shift.

The emitted fluorescence can be expressed as

$$I_{EM} = I_{EX} \cdot \varepsilon \cdot c \cdot x \cdot \phi \tag{10.2}$$

where I_{EX} is the intensity of the illuminating light, ε is the extinction coefficient of the fluorophore, c is the concentration of the fluorophore, x is the optical path length, and ϕ is the quantum efficiency, which reflects the fluorophore's ability to convert absorbed light into emitted fluorescence (i.e., the ratio of the number of photons emitted to the number of photons absorbed). The level of intensity or brightness of the emission produced by a fluorophore depends on its ability to absorb light at a particular wavelength (i.e., its extinction coefficient, ε), and the quantum efficiency. Typical values of ε for fluorophores at their characteristic absorption wavelength are in the range of tens of thousands, whereas the ϕ may range from 0 (no fluorescence) to 1 (100% efficiency). Fluorescence intensity is also influenced by the intensity of the incident illumination, that is, with an increase in illumination intensity, a larger number of fluorophore molecules are excited, and the number of emitted photons increases. Under conditions of constant illumination wavelength and intensity and low fluorophore concentrations, the emitted fluorescence is a linear function of the number of fluorophore molecules present.

Nonlinearity typically results at very high fluorophore concentrations partly due to reabsorption of the emitted light.

In fluorescence imaging, specimens are labeled with fluorophores. The distribution of fluorescence is then observed under exciting illumination and captured by photosensitive detectors that measure the intensity of the emitted light and create a digital image of the sample. Conventional wide-field microscopy (epifluorescence mode) is most commonly used for fluorescence microscopy of thin fixed samples, whereas confocal and two-photon fluorescence microscopy are more appropriate for thicker samples and for dynamic live cell imaging.

10.2.1 Image Formation in Fluorescence Imaging

Fluorescence microscopy is an incoherent imaging process. Each point in the specimen contributes independently to the light intensity distribution in the observed image. Each fluorophore molecule in the sample acts as a light source, and image formation occurs by the integration of these secondary light sources in the specimen. Thus the principle of linear superposition applies in fluorescence microscopy, such that the combination of multiple sources generates an image that is the sum of the individual responses of the point sources. Most fluorescence microscopes are *epifluorescence* systems [1], wherein both sample illumination and the collection of emitted light from the sample occur through the same objective lens. Imaging in a fluorescence microscope can be modeled as a space-invariant linear system (Chapter 2), where the intensity distribution is given by [1]

$$I(x,y,z) \propto \int_{R^3} \left| h_{\lambda_{em}}\left(\frac{x}{M}-u, \frac{y}{M}-v, \frac{z}{M^2}-w\right)\right|^2 \chi(u,v,w)\ du\ dv\ dw \tag{10.3}$$

where M is the magnification of the objective, and χ is an object-dependent function related to the fluorophore concentration that describes the specimen's ability to emit light at λ_{em}. The incoherent point spread function (Chapter 2), denoted by $|h_{\lambda_{em}}|^2$, is the impulse response that defines the image of an ideal point object and is given by the two-dimensional (2D) Fourier transform of the pupil function [1]

$$h_\lambda(x,y,z) = \int_{R^2} P(u,v)\ exp\left(i2\pi z\frac{u^2+v^2}{2\lambda f^2}\right)\ exp\left(-i2\pi\frac{xu+yv}{\lambda f}\right)\ du\ dv \tag{10.4}$$

where f is the focal length of the objective, and P is the pupil function representing the circular aperture of the objective. The focal length, f, is related to the radius of the circular aperture of the objective as $f = r/NA$, where NA is the numerical aperture. NA is a measure of the angular dimension of the light cone emerging from the sample that is collected by the objective and affects both resolution and depth of field. Fluorescence intensity typically increases with an increase in the NA. The light intensity at the detector is approximately proportional to NA^4 [2]. Detailed descriptions of image formation, imaging resolution, and sampling concepts can be found in Chapters 1–3.

10.3 Optics in Fluorescence Imaging

Wide-field, confocal, and two-photon microscopes can be used to perform single and/or multidimensional fluorescence microscopy. The key requirement is that the microscope must be appropriately equipped to allow (1) illumination at the required excitation wavelength, (2) separation and effective removal of the excitation light from the emitted fluorescence, and (3) detection of the emitted light. Mercury or xenon lamps that produce "white light" (i.e., visible spectrum, with peaks at certain characteristic wavelengths) are used as the light source in wide-field microscopy, whereas lasers are used for illumination in confocal and two-photon imaging. The selection of light of a particular wavelength is achieved by inserting an excitation filter in the illumination path and an emission filter (also known as a barrier filter) in the fluorescence emission path. Separation of reflected excitation light from the emitted light is accomplished by a dichroic mirror, which reflects or transmits light depending on wavelength (e.g., reflects excitation and transmits emission wavelengths). Information regarding fluorescence microscopes, filters, and objectives can be found in other publications [3,4].

Finally, it is critical to choose an appropriate detector. In fluorescence microscopy, CCD (charge-coupled device) cameras and image-intensified systems are often used. For detailed information on detector selection and performance evaluation, see [4]. There is also a community-driven initiative, the Quality Assessment and Reproducibility for Instruments & Images in Light Microscopy (QUAREP-LiMi) working on improving reproducibility for light microscopy image data through quality control management of instruments and images [5].

10.4 Limitations in Fluorescence Imaging

Most modern microscopy equipment is constructed to minimize phase distortion and optical aberrations. Nevertheless, no optical system is completely free of distortion, and, in practice, aberrations are always present to some extent. For a detailed description of the sources of aberrations in fluorescence microscopy, see [6]. The sources of aberrations in fluorescence microscopy broadly fall into two categories: instrumentation-based and sample-based. In addition, sample preparation and microscope handling also constitute an important and frequent source of inhomogeneity observed in fluorescence images.

10.4.1 Instrumentation-Based Aberrations

The choice and configuration of a fluorescence microscopy system has a profound effect on the quality of the acquired image. The illumination source, type of microscope, objectives, excitation and emission filters, and the photosensitive detector must all be carefully and appropriately chosen and adjusted to achieve sharp images with high SNR [4]. Illumination sources should be stable and should generate reliable, reproducible, and uniform field illumination with proper spectral separation and registration. The quality and performance of the objective lens is directly related to the system resolution, and it must be chosen to fit the application requirements. Similarly, detectors should be chosen to satisfy the Nyquist sampling criteria, linearity of photometric response, high SNR, and sensitivity (Chapters 2 and 3). Photomultiplier tubes (PMTs) and CCDs, when used properly, are linear detectors, and are thus used in most fluorescence microscopy applications. Despite appropriately chosen instrumentation for performance optimization, certain limitations inherent to the equipment may result in measurement noise in images [7–9]. Common sources of noise and related image processing algorithms for correction are described in the following sections.

10.4.1.1 Photon Shot Noise

Apart from the diffraction-limited spatial resolution inherent in light microscopy, the major source of aberration introduced by the imaging process is intrinsic photon noise. In fluorescence imaging, the quantum nature of light gives rise to a fundamental limitation of any photodetector, known as "photon shot noise," which results from the random nature of photon emission. The absorption of a photon by the CCD creates a photoelectron in the CCD well, and, during read-out, the accumulated photoelectrons are counted. The number of photons collected by the detector determines the image amplitude at any given point in the image. Photon shot noise is random, with a Poisson distribution of the number of detected photons given by [10]

$$p(N) = \prod_{i=1}^{M} \frac{\mu^N exp(-\mu)}{N!} \tag{10.5}$$

where N is the number of detected photons and μ is the mean of the Poisson process. Although other factors, such as the integration time and quantum efficiency of the detector, influence photon shot noise, its intrinsically random nature makes it impossible to avoid. Thus, even in the absence of other noise sources, photon noise leads to the finite SNR of detectors. This limiting SNR is given by [7]

$$SNR_{photon} = 10\, log_{10}(\mu) \tag{10.6}$$

Thus photon shot noise is independent of the detector electronics and can be reduced only by increasing the light intensity or the exposure time. Consequently, the SNR can only be improved by increasing the exposure of specimens to light.

10.4.1.2 Dark Current

Dark current is another intrinsic source of noise present in photodetectors. It is the number of induced electrons per pixel that arise from sources such as thermal agitation. Thermal energy is mostly responsible for the generation of dark current, with higher temperatures resulting in higher kinetic energy of the electrons and stronger dark current. For CCD detectors, dark current tends to charge the CCD pixel wells when the integration time or the temperature is too high. For PMTs, thermal energy often results in spontaneous electron emissions, and consequently dark current as well. Dark current also follows the Poisson distribution for the number of thermal electrons produced over a given time interval. The presence of dark current introduces an offset in the pixel value, resulting in noise and a reduced dynamic range. The most effective method to alleviate dark current noise is to cool the detector. Consequently, most modern high-sensitivity scientific grade cameras are available as cooled device detectors that have reduced thermal noise, even at long integration times.

10.4.1.3 Auxiliary Noise Sources

Electronics associated with CCD cameras are also responsible for generating noise during readout. Similarly, fluctuations in the internal gain of PMT devices also result in noise. This type of noise is called readout noise, and its amplitude depends on the camera's readout rate [7,10]. Readout noise is inversely related to the pixel readout frequency, with the power spectral density of the noise decreasing as $1/f$. Scientific grade CCD cameras that operate in the frequency range of 20–500 kHz typically have low readout noise that can be ignored. For higher readout frequencies, the readout noise appears as additive, Gaussian distributed noise that can be problematic.

10.4.1.4 Quantization Noise

All digital detectors produce quantization noise. The digitization of the CCD image into a collection of integer values (i.e., the digital image) introduces a form of noise that is not band-limited. Quantization noise is inherent in the amplitude quantization process that occurs in the analog-to-digital converter (ADC). It is independent of the signal and can be modeled as additive noise, when the dynamic range has intensity levels $\geq 2^4$ (equivalent to bit depth, B, of 4). For a digitized image, if the ADC is adjusted so that 0 corresponds to the minimum video signal value, and 2B − 1 corresponds to the maximum video value, then the SNR for quantization, S_Q, is given by [10]

$$S_Q = 6 \times B + 11 \tag{10.7}$$

measured in decibels. It is good practice to keep the quantization noise level to one-half the root mean square noise level due to other sources.

10.4.1.5 Other Noise Sources

Finally, the ambient radiation, especially in the infrared domain, can also be a source of background noise in images. With recent advances in imaging hardware, most detectors are manufactured such that noise from other sources is negligible, and detector noise is essentially limited to only photon shot noise. Most of the state-of-the-art photodetectors have high quantum efficiency, low dark current (especially for cooled devices), high sensitivity, and excellent linearity.

10.4.2 Sample-Based Aberrations

The optical characteristics of a sample are not only important for image formation, but also play a role in introducing image aberrations. The sample-based aberrations most often observed in fluorescence images include effects of photobleaching, autofluorescence, absorption, and scattering.

10.4.2.1 Photobleaching

Photobleaching is an inherent phenomenon in fluorophores, and it has a damaging effect in fluorescence microscopy. Specifically, it causes an effective reduction in and, ultimately, a complete elimination of fluorescence emission. The term "bleaching" covers all of the processes that cause the fluorescent signal to fade permanently, due to photon-induced chemical damage and covalent modification. This is different from the process of *quenching* (reduction in the excited state lifetime and quantum yield), which is the reversible loss of fluorescence that occurs due to noncovalent interactions between a fluorophore and its molecular environment.

At the molecular level, photobleaching occurs when an excited electron reaches the triplet state. This state is long-lived, allowing more time for the fluorophores to react with another molecule to produce an irreversible covalent chemical modification. For example, in the presence of oxygen, a fluorophore molecule can transfer its energy to oxygen molecules, exciting them to a reactive singlet state. Chemical reactions with singlet oxygen molecules then covalently alter the fluorophore. This leads to permanent changes by which the molecule loses its capability to fluoresce altogether, or it becomes nonabsorbent at the excitation wavelength. The kinetics of bleaching vary among fluorophores since the number of excitation and emission cycles for each fluorophore depend on its molecular structure and the local milieu. Overall, bleaching is an irreversible process that has a collective effect: that is, a reduction in exposure time or excitation intensity does not prevent bleaching, but only reduces the rate at which it occurs.

Another detrimental side effect of the process is phototoxicity, that may result from the interaction of the radical singlet oxygen species with other organic molecules. In fluorescence imaging, bleaching limits the total intensity of light and the exposure time in all samples, and in dynamic studies it may decrease viability of living tissue over time. On a positive note, the photobleaching phenomenon has been the basis of many fluorescence measurement techniques (described in Section 10.7). Sample preparation and experimental methods to reduce photobleaching are described in detail elsewhere [11], whereas image processing-based correction is described in Section 10.5.7.

10.4.2.2 Autofluorescence

Autofluorescence is the term used to describe the emission of fluorescence from organic or inorganic molecules that are naturally fluorescent. In fluorescence imaging, molecules that emit fluorescence in the visible domain can be visualized, even in unstained samples. Autofluorescence is a significant source of background noise in fluorescence images. In quantitative fluorescence imaging, background signals due to autofluorescence become problematic when there is an overlap with the emission of a target fluorophore, and more so when the latter is sparsely expressed or exhibits weak fluorescence. Autofluorescence is observed in both plant and animal tissue. For example, the fluorescent pigment lipofuscin, is found in the cytoplasm of mammalian cells, and in chloroplasts of plant cells. The mounting and embedding material may also exhibit low amounts of autofluorescence. Although sample preparation methods are available to reduce autofluorescence [12], computational background subtraction, described in Section 10.5.3.2, can prove to be a very effective method for correction.

10.4.2.3 Absorption and Scattering of the Medium

Absorption and scattering of light are the most common causes of optical aberrations observed in fluorescence imaging. Light at both the excitation and emission wavelengths can be scattered due to particles, refracted due to interfaces, or absorbed by the specimen material itself. This typically occurs due to (1) the presence of particles in the specimen with sizes comparable to the wavelength of light, and (2) a mismatch in the refractive indices of the specimen and the immersion layers. The optical aberration introduced is observed as a reduction in the intensity of light with increasing depth into the specimen. In light microscopy, the effect is manifested both as a reduction in the intensity of illumination with increasing depth, and a limitation on the depth within the specimen from which an emitted light signal can be detected. These effects are apparent as a reduction in image contrast when imaging deep within the specimen, and they are most obvious in wide-field and confocal microscopy. The problems are somewhat alleviated in two-photon imaging wherein excitation at longer wavelength allows imaging deeper within the sample.

10.4.3 Sample and Instrumentation Handling-Based Aberrations

Variability in sample preparation and improper illumination alignment are the two most common and frequently observed sources of light intensity inhomogeneity in fluorescence images. Sample preparation can result in variation of light intensity across the image due to inconsistent staining and nonspecific fluorescence probe binding. Besides sample handling problems, uneven illumination of the specimen due to improper centering of the lamp contributes to further degradation of the image signal. Both halogen (transmitted light) and mercury (fluorescence light) lamps must be adjusted for uniform illumination of the field of view prior to use. Moreover, microscope optics and cameras can also introduce vignetting, in which the corners of the image are darker than the center. Depending on the nature of the intensity variation observed, digital image correction can be implemented to remove brightness variations as described in the following section.

10.5 Image Corrections in Fluorescence Microscopy

In modern biological microscopy, fluorescence imaging is used to record and quantify location, functional status, and abundance of a tagged target molecule. This requires the application of image correction techniques and calibration methods for both image visualization and quantitative analysis. Digital image processing algorithms for correction and calibration strategies are discussed in the following sections.

10.5.1 Background Shading Correction

The process of eliminating nonuniformity of image background intensity by application of image processing to facilitate visualization, segmentation, or to obtain accurate quantitative intensity measurements is known as background correction, background flattening, flat-field correction, or shading correction. These brightness variations in images are typically observed as intensity shadings across the field of view. In microscopy, image shading may occur due to nonuniform illumination, inhomogeneous detector sensitivity, dirt particles in the optics, nonspecific sample staining, or autofluorescence. Some of the most frequently used methods for correction or elimination of background shading are described here.

For fluorescence imaging, intensity shading can be described by the following model [10]:

$$b(x, y) = I_{ill}(x, y) \cdot a(x, y) \tag{10.8}$$

where $b(x, y)$ is the image produced by the interaction of the illumination $I_{ill}(x, y)$ with the sample $a(x, y)$, and (x, y) represents the spatial coordinates. Assuming low fluorophore concentrations, $c(x, y)$, for fluorescence imaging, the sample can be denoted as $a(x, y) = c(x, y)$. Incorporation of the effects of the detector gain and offset gives the following discrete expression in integer coordinates [10]:

$$c[m, n] = gain[m, n] \cdot b[m, n] + offset[m, n] \tag{10.9}$$

where $b[m, n]$ is the digital image that would have been recorded if there were no shading (Eq. 10.8) in the image. Eq. (10.9) can then be expressed as

$$c[m, n] = gain[m, n] \cdot I_{ill}[m, n] \cdot a[m, n] + offset[m, n] \quad (10.10)$$

The goal of all correction algorithms is to determine $a[m, n]$ given $c[m, n]$. Algorithms used to correct shading effects are described in the following.

10.5.2 Correction Using the Recorded Image

In the first approach, the recorded digital image $c[m, n]$ is used to estimate the background shading pattern. Three different methods can be used to estimate the shading pattern. In the first, low-pass filtering is applied to smooth $c[m, n]$, where the smoothing effect is larger than the size of objects in the image. The choice of the low-pass filter requires that the spatial frequencies where the shading persists be known. The corrected image $\hat{a}[m, n]$ is then estimated as follows [10]:

$$\hat{a}[m, n] = c[m, n] - LowPass\{c[m, n]\} + \text{constant} \quad (10.11)$$

Second, instead of low-pass filtering $c[m, n]$, smoothing can be performed with morphological filtering (Chapter 6). Morphological filtering is used when the background variation is irregular and cannot be estimated by surface fitting (Section 10.5.4). The assumption behind this method is that foreground objects are limited in size and are smaller than the scale of background variations, and the intensity of the background differs from that of the features. The approach is to use an appropriate structuring element to describe the foreground objects. Neighborhood operations are used to compare each pixel to its neighbors. Regions larger than the structuring element are taken as background. This operation is performed for each pixel in the image, producing a new image. The result of applying this operation to the entire image is to shrink the foreground objects by the radius of the structuring element, and to extend the local background brightness values into the area previously occupied by the objects. The choice of the appropriate structuring element for smoothing depends on the size of the largest object of interest in the image. For example, an opening operation can be used to estimate the shape of the background. The corrected image $\hat{a}[m, n]$ is then estimated as [10]

$$\hat{a}[m, n] = c[m, n] - Sm\{c[m, n]\} + \text{constant} \quad (10.12)$$

where $Sm\{c[m,n]\}$ is a morphological smoothing operation.

Finally, homomorphic filtering can be used to estimate the shading. The logarithm of Eq. (10.10) is taken, assuming the $offset[m, n]$ to be zero, the term $\{gain[m, n] \cdot I_{ill}[m, n]\}$ to be slowly varying (low frequency), and $a[m, n]$ to be rapidly changing (high frequency). Thus high-pass filtering can effectively attenuate any shading. The background is characterized as low intensity and as having low spatial frequency content. The corrected image can be computed by taking the exponent (i.e., the inverse logarithm) [10]

$$\hat{a}[m, n] = exp\{HighPass\{ln\{c[m, n]\}\}\} \quad (10.13)$$

This approach is also known as frequency domain filtering. It assumes that the background variation in the image is a low-frequency signal and can be separated in frequency space from the higher frequencies that define the objects of interest in the image. The high-pass filter removes the low-frequency background components. This approach is typically used to remove shading due to nonuniform illumination or staining, and it is most appropriate when no information is available about the imaging system used to acquire the data. When implemented by convolution, the kernels tend to be quite large.

10.5.3 Correction Using Calibration Images

In this approach calibration images are acquired in advance using the imaging system. Although several different algorithms are used, the underlying principle is to use previously acquired calibration or test images. A few such algorithms are described here.

10.5.3.1 Two-Image Calibration

One approach is to record two calibration images using the microscope. First, a black image is acquired by blocking all light to the detector (i.e., $b[m,n]=0$ in Eq. 10.9). The recorded image is then $BLACK[m,n]=offset[m,n]$. A second calibration image is then recorded using a white reflecting surface or uniformly fluorescing glass, with no specimen in the light path, such that $a[m,n]=1$, giving $WHITE[m,n]=gain[m,n] \cdot I_{ill}[m,n] + offset[m,n]$. The corrected image is then computed as follows [10]:

$$\hat{a}[m,n] = \text{constant} \cdot \frac{c[\text{m},\text{n}] - BLACK[m,n]}{WHITE[m,n] - BLACK[m,n]} \tag{10.14}$$

where the term *constant* determines the dynamic range. The choice of the parameter *constant* is important, since the corrected image will have integer values obtained by rounding real numbers. If small values are chosen for *constant*, the dynamic range will be limited, whereas if large values are chosen, image display can become problematic. The exposure time (or integration time) must be the same when acquiring the images in Eq. (10.14). The black image is also known as the dark current image. This approach corrects images for shading due to uneven illumination and dark current effects.

10.5.3.2 Background Subtraction

Another approach is to use a single calibration image. Background subtraction can produce a flat background and compensate for nonuniform lighting, nonuniform camera response, or minor optic artifacts (such as dust specks that mar the background of images captured from a microscope). One background image is acquired in which a uniform reference surface or specimen is inserted in place of actual samples, and a background image is recorded. This image represents the intensity variations that occur without a specimen in the light path, leaving only those due to inhomogeneity in the illumination source, system optics, or camera. It can then be used to correct all subsequently recorded images. When the background image is subtracted from a given image, areas that are similar to the background will be replaced with values close to the mean background intensity. The process of background subtraction is applied to even out the background intensity variations in a microscope image. If the camera is logarithmic with a gamma of 1.0, then the background image should be subtracted. However, if the camera is linear, then the acquired image should be divided by the background image. In the process of subtracting (or dividing) one image by another, the dynamic range of the original data will be reduced.

10.5.4 Correction Using Surface Fitting

This approach also uses a recorded image, but rather than a smoothed version of the recorded image, it uses the process of surface fitting to estimate a background image devoid of any objects. This method is especially useful when a reference specimen or the imaging system is not available to experimentally acquire a background image. Typically, a polynomial function is used to estimate variations of background brightness as a function of location. The process involves an initial determination of an appropriate grid of background sample points. To be accurate, the fit must be done through background pixels only. It is critical that the points selected for surface fitting represent true background areas in the image, and that no foreground (object) pixels are included. If a foreground pixel is included in the fitting process, the surface fit will be biased, resulting in an overestimation of the background. In some cases it is practical to locate the points automatically for background fitting. This is feasible when working with images that have distinct objects that are well separated throughout the image area and contain the darkest (or lightest) pixels present. The image can then be subdivided into a grid of smaller squares or rectangles. The darkest (or lightest) pixels in each subregion are located, and these points are used for the fitting.

Another issue is the spatial distribution and the number of sample points. The greater the number of valid points that are uniformly spread over the entire image, the greater will be the accuracy of the estimated surface fit. A least-squares fitting approach may then be used to determine the coefficients of the polynomial function. For a third-order (cubic) polynomial, the functional form of the fitted background is

$$B(x, y) = a_0 + a_1 \cdot x + a_2 \cdot y + a_3 \cdot xy + a_4 \cdot x^2 + a_5 \cdot y^2 + a_6 \cdot x^2 y + a_7 \cdot xy^2 + a_8 \cdot x^3 + a_9 \cdot y^3 \tag{10.15}$$

This polynomial has 10 fitted constants, $a_0 - a_9$. A good fit with diminished sensitivity to minor fluctuations in individual pixels requires fitting the surface through at least three times as many pixels as the total number of coefficients to be estimated. Fig. 10.1 demonstrates the process of background fitting.

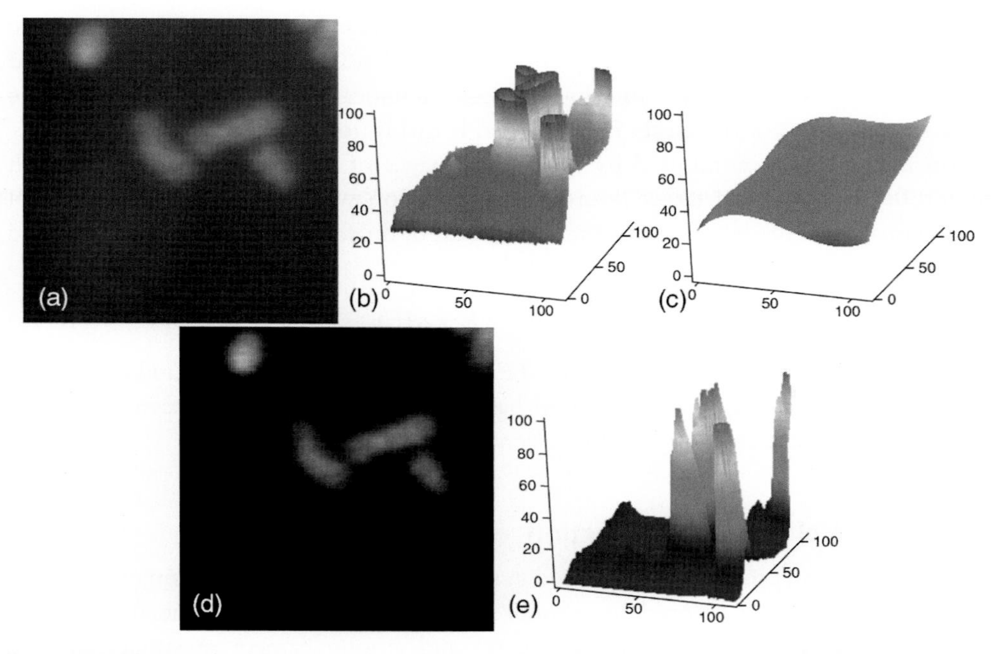

Fig. 10.1 Background subtraction via surface fitting. Panel A shows the original image, Panel B presents its 2D intensity distribution as a surface plot, Panel C shows the background surface estimated via the surface-fitting algorithm, Panel D shows the background subtracted image, and Panel E presents its 2D intensity distribution as a surface plot. *(Reproduced with permission from [13])*

Several other approaches for shading correction have been published. Reducing brightness variations by subtracting a background image, whether obtained by experimental measurement, mathematical fitting, or image processing, is not a cost-free process. Subtraction reduces the dynamic range of the image, and clipping must be avoided in the subtraction process, or it might interfere with subsequent analysis of the image. Given the diversity of biological samples being imaged with fluorescence microscopy, no single background estimation algorithm is suitable for all shading correction. The choice of the correction technique is application dependent.

10.5.5 Autofluorescence Correction

The background correction methods described earlier are also effective in attenuating shading effects due to autofluorescence. In another approach, two images are recorded, one illuminated at the excitation wavelength of the fluorophore of interest and the other at the excitation wavelength of the autofluorescence [14]. Subtraction of the latter autofluorescence only image from the total fluorescence image (the specific fluorophore plus the autofluorescence) results in an autofluorescence-free image. This technique assumes the autofluorescence image can be captured using appropriate narrow-band excitation filters such that the autofluorescence excitation wavelength lies outside the excitation spectrum of the target fluorophore. If this is not the case, an additional correction factor can be computed by recording images of an unstained sample at both the autofluorescence and target fluorophore excitation wavelengths. The ratio, R, of the two images at the desired emission wavelength of the target fluorophore is then used to compute the specific fluorescence as [14]

$$SF = TF - (R \times AF) \tag{10.16}$$

where SF is the specific fluorescence, TF is the total fluorescence, and AF is the autofluorescence. An alternative method based on time-delayed fluorescence imaging can also be used to eliminate autofluorescence effects [15].

10.5.6 Spectral Overlap Correction

The most common problem encountered in multicolor fluorescence imaging (i.e., when two or more different-colored fluorophores are used) results from the unavoidable overlap among fluorophore emission spectra and among camera sensitivity spectra. The individual colors, rather than being confined to one color channel, are smeared across all the channels. Intensity overlap between various color channels is problematic both for visualization and quantification. Examples include studies involving measuring concentrations of tagged molecules, colocalization of multiple target molecules, and segmentation of components in color images. A method called "color compensation" effectively isolates three fluorophores by separating them into three color channels (RGB) of the digitized color image [16].

The color compensation algorithm for fluorescence assumes that the measured signal at each pixel is a linear combination of the overlapping spectra at that point. Further, these algorithms also assume that the measured signal is linearly proportional to the concentration of the fluorophore or dye at that point. This assumption holds when the absorption and fluorophore concentrations are low, but it may be disrupted by energy transfer between colocalized fluorophores (Section 10.7.5). For an N-color system, each $N \times 1$ pixel vector is premultiplied by an $N \times N$ compensation matrix. The elements of this matrix must be determined experimentally for a particular combination of camera, color filters, and fluorophores.

The color compensation algorithm processes the RGB image so that each fluorophore is isolated to a single channel. A 3×3 matrix, C, specifies how the fluorophore brightnesses are spread among the three color channels. Each element c_{ij} is the proportion of the brightness from fluorophore i that appears in color channel j of the digitized image. The columns of this matrix sum to unity. If x is the 3×1 vector of actual fluorophore brightness values at a particular pixel, then $y = Cx + b$ is the vector of RGB gray levels recorded by the digitizer at that pixel. The column vector b accounts for the black level offset of the digitizer: that is, b is the gray level that corresponds to zero brightness in channel i. The true brightness values can then be determined as

$$x = C^{-1}[y - b] \tag{10.17}$$

The color compensation matrix is thus the inverse of the color spread matrix. It operates on the RGB pixel vector after the black level has been subtracted from each channel. The values in the color smear matrix can be determined experimentally from digitized images of specimens stained with single fluorophores. This technique can also account for unequal integration times [17]. Fig. 10.2 demonstrates the process of spectral unmixing via the color compensation algorithm [18].

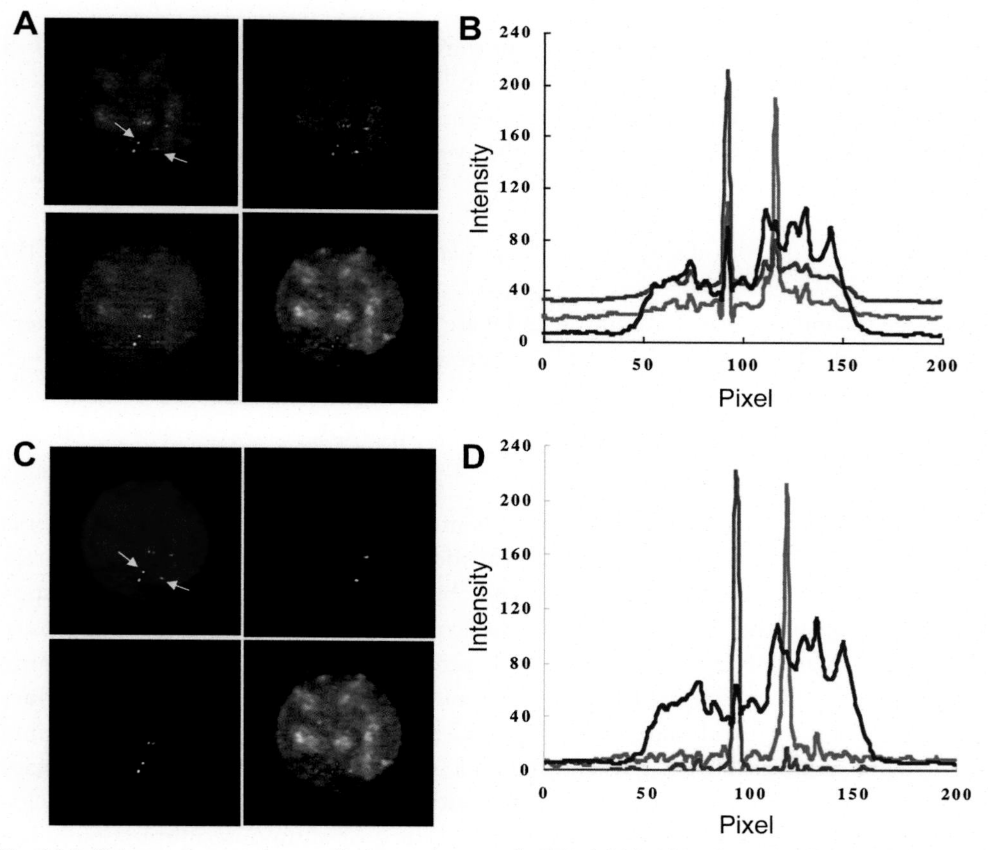

Fig. 10.2 Color compensation applied to an image of a FISH-labeled lymphocyte. The nuclei are counterstained with DAPI, and dots are labeled with *red* (chromosome 21) and *green* (chromosome 13). (A) Original image and individual *red* (R), *green* (G), and *blue* (B) channels. The *blue fluorophore* (nuclear stain) is seen in both the *red* and *green* channels. Similarly, *green dots* are also seen in the *red channel* and vice versa. (B) Intensity profile of a line segment through a *red* and *green dot* (*white arrows* in original image). (C, D) Color-compensated images. The smearing of colors across the individual channels is effectively removed. Original magnification = × 630. *(Reproduced with permission from [18])*

10.5.7 Photobleaching Correction

Photobleaching causes an overall decrease in intensity during intermittent or continuous illumination over a period of time, and it can interfere with both visualization and quantification in fluorescence imaging. Thus it is often necessary to analyze the kinetics of photobleaching of fluorophores and make corrections. In fluorescence imaging, photobleaching manifests itself either as a first-order, second-order, or higher-order exponential decay of the average intensity of the image series as a function of time.

For a first-order bleaching process, the average intensity of an image at time, t, is given by [19]

$$\left\langle i(x, y, t) \right\rangle_t = \left\langle i(x, y, t_0) \right\rangle_0 exp\,[-kt] \tag{10.18}$$

where $\langle i(x, y, t_0)\rangle_0$ is the average intensity of the first image in the time series, and k is the bleaching decay constant having reciprocal time units. The angular brackets in Eq. (10.18) indicate spatial averaging over the entire image or a region of interest.

For a second-order bleaching process, the average intensity of an image at time, t, is given by

$$\left\langle i(x, y, t) \right\rangle_t = A\,exp[-kt] + B\,exp[-jt] \tag{10.19}$$

where j is a second bleaching rate, and A and B are amplitude constants. Similarly, higher-order exponential decay functions can be expressed as

$$\left\langle i(x, y, t) \right\rangle_t = \sum_i D_i \exp\,[-k_i t] \tag{10.20}$$

Using the appropriate photobleaching process model (Eq. 10.18, 10.19, or 10.20), it is possible to empirically characterize the intensity decay from the image (or region of interest) time series data. The decay curves can be obtained by fitting the appropriate exponential function to the data. To perform the exponential fitting, the average intensity values of the image or region of interest is determined for all the images of a bleaching series. The photobleaching decay curves are generated by plotting these intensity values as a function of time (in arbitrary units). Fig. 10.3 presents an example of photobleaching observed in a time series of images captured for a triple stained slide of bovine pulmonary artery endothelial cells. The cells were probed with antibovine-tubulin mouse monoclonal antibody and visualized with bodipy fluorescein goat antimouse immunoglobulin. The actin filaments are labeled with Texas Red-X phalloidin, and the nuclei are counterstained with DAPI (4′,6-diamidino-2-phenyl indole dihydrochloride). The images were acquired using a triple bandpass filter appropriate for Texas Red dye, fluorescein, and DAPI. The images in Fig. 10.3 were taken at 2-min intervals to excite the three fluorophores simultaneously while also recording the combined emission signals.

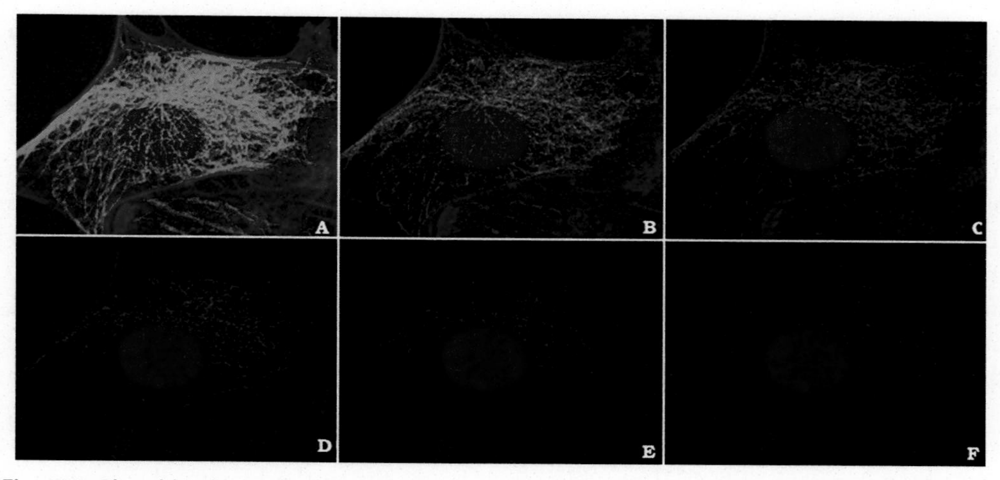

Fig. 10.3 Photobleaching of DAPI, fluorescein, and Texas Red. Time points were taken in 2-min intervals over a 10-min period. (A–F) Time = 0, 2, 4, 6, 8, and 10 min, respectively. The integration time for each image was 2.5 s. Note that all three fluorophores have a relatively high intensity in (A), but the Texas Red intensity has dropped rapidly at 2 min and is almost gone at 4 min (C). Similarly, the intensity of the *green fluorescence* drops dramatically over the course of 10 min.

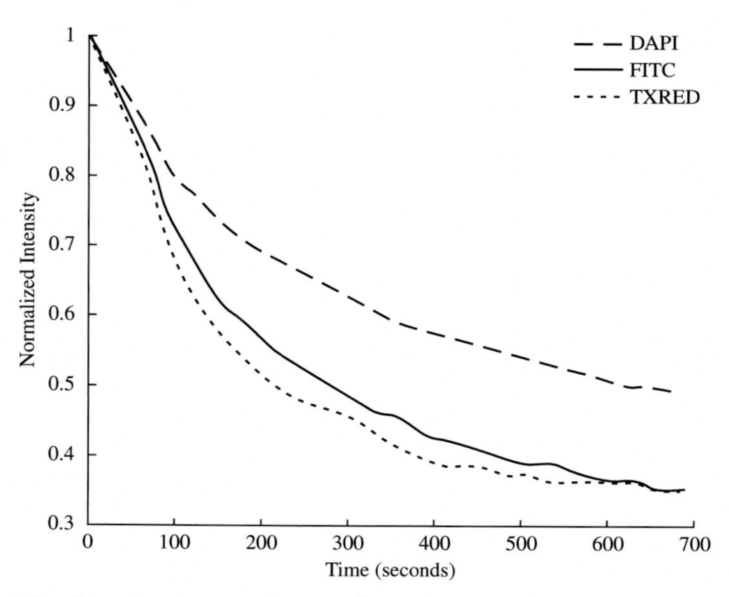

Fig. 10.4 Photobleaching decay curves for DAPI, fluorescein, and Texas Red.

The photobleaching decay curves were generated by plotting these intensity values as a function of time (Fig. 10.4).

Nonlinear fitting methods such as the Levenberg-Marquardt algorithm can be used to determine the best-fit parameters. Typically, the fitting procedure minimizes a merit function with nonlinear dependencies to find the best-fit parameters. The minimization proceeds iteratively by giving trial values for the parameters, and the algorithm improves the trial solution until the value for the merit function stops (or effectively stops) decreasing. Generally, the estimated parameters are constrained to be positive so that the exponential function does not become negative. Finally, for correction, every pixel is multiplied by the ratio of the fitted value at time zero to the fitted value at the time the image was recorded.

Alternatively, several other functions have been defined to model the photobleaching process. For example, a stretched exponential decay [20] and a single exponential function plus a constant term [21] provide much better fits to the kinetics than a monoexponential function.

10.6 Quantifying Fluorescence

In modern biological microscopy, fluorescence imaging of live and fixed tissue is now used routinely. Quantitative analysis of fluorescence images allows measurements of (1) amounts or concentrations of cellular components, and their interactions, and (2) dynamics of cellular processes in space and time, at the subcellular, cellular, and tissue levels. The use of sophisticated imaging systems, in conjunction with the image correction methods described earlier, allows a direct correlation between the distribution of fluorescence and the digitally measured fluorescence signal, such that quantitative analysis can be performed to obtain statistically significant and meaningful mathematical simulations of the underlying processes. A key to quantitative analysis is (1) to take into account critical details of the entire protocol, starting with the optical properties of both the microscope and specimen, (2) apply the appropriate corrections as needed, and (3) compute numerical values based on the properties of the fluorophores and the imaging parameters.

10.6.1 Fluorescence Intensity and Fluorophore Concentration

The amount of fluorescence emitted, I_{EM}, is proportional to the amount of light absorbed by the fluorophore. Given that I_{EX} is the intensity of the light that illuminates a sample, and I, is the amount of light that passes through the sample, then the portion of light absorbed is $I_{EX} - I$. The intensity of the emitted fluorescence can then be denoted by

$$I_{EM} = \phi(I_{EX} - I) \qquad \text{where} \qquad (I_{EX} - I) \approx I_{EX} \cdot \varepsilon \cdot c \cdot x \tag{10.21}$$

as defined earlier in Eq. (10.2). According to the Beer-Lambert law, the relationship between the exciting, emitted, and absorbed light can be expressed as

$$I_A = -\,log\left(\frac{I}{I_{EX}}\right) \quad \text{or} \quad I = I_{EX}e^{-I_A} \tag{10.22}$$

Thus Eq. (10.21) can be rewritten as [22]

$$I_{EM} = \phi I_{EX}\left(1 - e^{-I_A}\right) \tag{10.23}$$

From Eq. (10.22), it can be seen that when the absorption, I_A is zero, I_{EM} is zero. When I_A approaches infinity, $I_{EM}=\phi I_{EX}$. Thus when I_A is small, the term $(1-e^{-I_A})$ approaches I_A, and $I_{EM}=I_A\phi I_{EX}$. That is, the amount of light absorbed by the fluorophore is related to the concentration of the fluorophore, with $I_A \propto kc$, where k is a constant, and c is the fluorophore concentration. Thus, at low concentrations, the fluorescence intensity is directly proportional to the concentration of the fluorophore with $I_{EM}=kc\phi I_{EX}$, and at high concentrations $I_{EM}=\phi I_{EX}$, which is independent of the concentration. The linear relationship between absorbance and fluorescence holds at low absorbance values ($\sim\leq 0.2$) [22]. This linear relationship of fluorescence to excitation intensity enables quantitative measurements of concentration or amounts using fluorescence microscopy. It should be noted that fluorescence is a relative quantity, and a large number of factors can affect fluorescence measurements (discussed in Section 10.4). Thus careful calibration of the instrumentation, standards, and image correction methods are required for accurate quantification. Several quantitative microscopy techniques based on fluorescence imaging and the associated image analysis approaches are described in the following section.

10.7 Fluorescence Imaging Techniques

The choice of the fluorescence imaging method depends on the application. While the simplest form of fluorescence imaging involves the use of single-color fluorescence microscopy to measure the amounts and localization of cellular components, dual color imaging is used for colocalization analysis, and more sophisticated approaches, such as fluorescence resonance energy transfer, are used to study protein interactions and for the dynamic investigation of molecular processes.

10.7.1 Immunofluorescence

Immunofluorescence is extensively used for the visualization and quantitation of the distribution of specific cellular components (such as proteins) in cells or tissue. There are two major types of immunofluorescence techniques. In direct immunofluorescence the primary antibody is labeled with a fluorescent dye, and directly binds to the appropriate antigen. The specificity of this method is high, but the overall fluorescence signal is weak. In indirect immunofluorescence a secondary antibody labeled with a fluorophore is used to recognize a primary antibody. This approach allows an increase of the fluorescence intensity due to the larger number of antigenic sites available for binding the fluorescently labeled antibody.

Immunofluorescence microscopy is used to image the expression of proteins such as receptors, ion channels, and enzymes. This technique is useful for visualization of proteins but has two inherent limitations. The processing of tissue sections for mounting and staining may introduce artifacts, and, more importantly, structure and function cannot be studied in real time.

The most commonly observed problems in immunofluorescence microscopy are nonspecific fluorescence due to cross-reactivity and background fluorescence, and photobleaching of the labeled target. Image preprocessing procedures typically involve the use of background correction, minimization of autofluorescence, and estimation of photobleaching characteristics. Analysis procedures include specific algorithms for localization of the spatial distribution of the target molecules and quantitative estimation of the relative amounts.

Fig. 10.5 shows an image of ovarian cancer cells cultured on a substrate with circular micropatterns to produce cellular patterning. The cytoskeletal architecture of the cells was made visible by direct immunofluorescence labeling. The cytoskeletal morphology shows actin stress fibers in green along the periphery of the circle patterns.

Fig. 10.6 shows fluorescently labeled molecules of a genetically modified bacterial protein, H5, interacting with NIH/3T3 mouse fibroblasts. The interaction of H5 with cell membranes was studied by visualization and quantitation of the toxin molecules via indirect immunofluorescence labeling and confocal imaging [23]. The relative amounts and spatial distribution of the interacting protein was determined by digital image analysis. The 3D spatial distribution of the immunofluorescently labeled H5 was generated by determining the distance of each nonzero voxel from the center of mass of the cell. The total amount of H5 interacting with cell membranes was determined in terms of the volume of immunofluorescently labeled H5 normalized with respect to the cell volume. Image analysis algorithms included preprocessing of the acquired images using median filtering followed by segmentation using adaptive gray level thresholding, and 3D region labeling. Surface modeling estimation using superquadric surfaces was used to identify the localization of the toxin molecules.

10.7.2 Fluorescence In Situ Hybridization (FISH)

Fluorescence in situ hybridization is a molecular cytogenetic technique that is very similar to immunofluorescence. It is specifically used for the visualization and localization of deoxyribonucleic acid (DNA) and ribonucleic acid (RNA) sequences [24]. It has been widely applied in many areas of diagnosis and research, including prenatal and postnatal screening of chromosomal aberrations, preimplantation genetic diagnosis, cancer cytogenetics, gene mapping, molecular pathology, and developmental molecular biology.

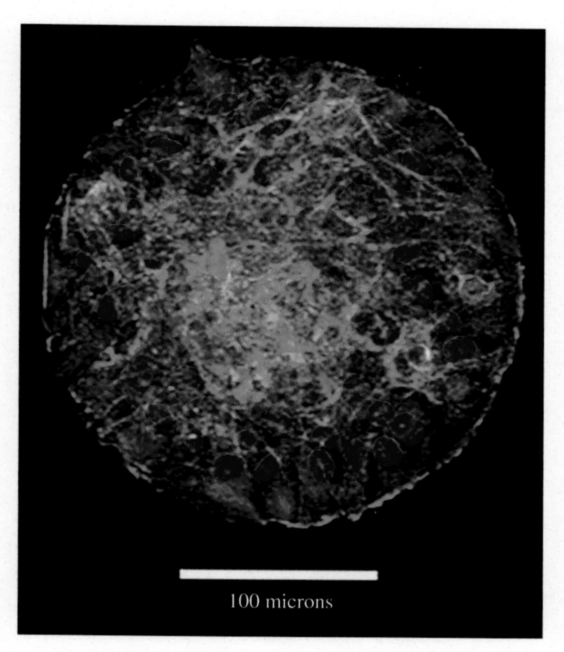

Fig. 10.5 Image of the cytoskeletal architecture of ovarian cancer cells determined via immunofluorescence labeling and imaging of cytoskeletal components. The cytoskeletal architecture of the cells was visualized via direct immunofluorescence labeling and imaging of the fluorescein isothiocyanate (FITC) phalloidin stained filamentous actin network and indirect immunofluorescence labeling of vimentin, whereas the nucleus was counterstained with 4′,6-diamidino-2-phenylindole (DAPI). Images were acquired using a laser scanning confocal microscope. Evaluation of cytoskeletal morphology shows actin stress fibers (*green*) aligned along the periphery of the circle patterns suggesting contact guidance behavior, and characteristic spindle-shaped stretching and anchoring of the F-actin cytoskeleton (*green*) seen closer to the periphery of the circle patterns, versus rounded clumping of the cytoskeleton in the interior cells.

FISH allows the microscopic analysis of chromosomal abnormalities such as an increase or reduction in the number of chromosomes, or a translocation of part of one chromosome onto another.

The basic principle of FISH is that a DNA probe for a specific chromosomal region will recognize and hybridize to its complementary sequence. Similar to immunofluorescence, two approaches are used for staining, direct and indirect. For direct labeling, the probe is tagged with a fluorescent dye, whereas in indirect labeling it is chemically modified by the addition of hapten molecules (biotin or digoxigenin) and then fluorescently labeled. The target DNA is counterstained with another fluorophore of a complementary color. The probe DNA can be observed on its target by using a fluorescent microscope with filters specific for the fluorophore label and the counterstain. Normally, in interphase cells, the nucleus is counterstained using DAPI, and biotin- and digoxigenin-labeled

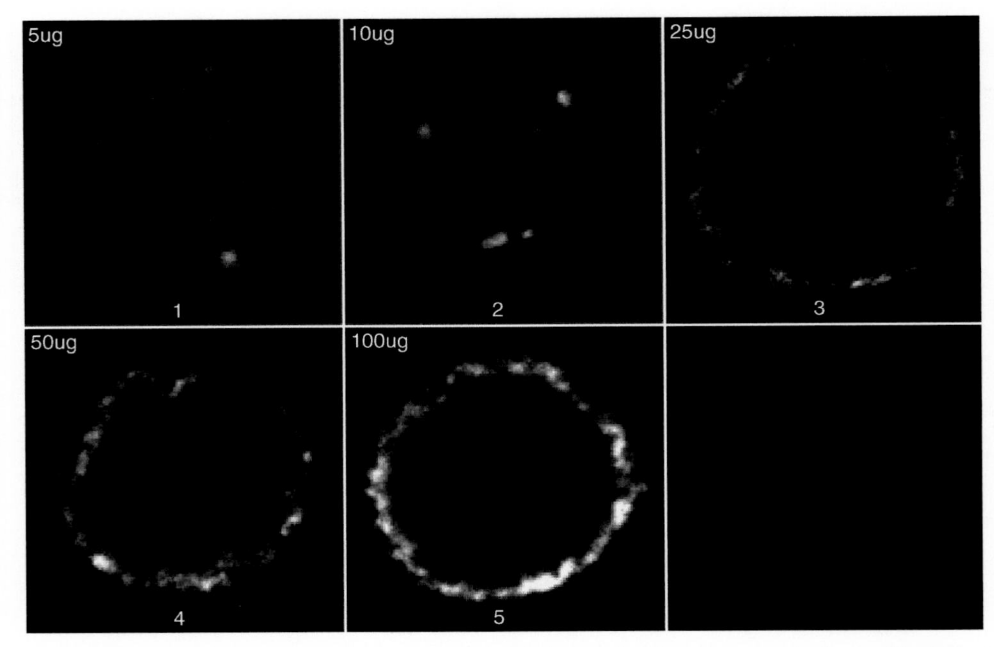

Fig. 10.6 Montage showing immunofluorescently labeled molecules of a genetically modified *Staphylococcus aureus* alpha toxin protein interacting with mouse fibroblasts. Panels labeled 1, 2, 3, 4, and 5 show single optical section confocal micrographs of cells treated with 5, 10, 25, 50, and 100 μg/ml of H5, respectively, for 20 min.

chromosome-specific DNA probes are used in conjunction with secondary reagents (e.g., avidin or antidigoxigenin conjugated to green (e.g., fluorescein) or red (e.g., Texas Red) dyes) to visualize blue nuclei with colored dots (red, green, aqua, orange, etc.). Typically, multilabel (two-, three-, or four-color) FISH analysis is performed for cytogenetic analysis. The number of detected chromosomes or genes can be increased using combinatorial labeling, where the number of targets is 2^{n-1}, with n colors used for labeling.

Analysis of an interphase FISH specimen consists of determining the number of dots of each color, per cell. The enumeration of chromosomes is directly influenced by hybridization and detection efficiency, as well as the geometry or chromosomal location within the nucleus. Automated image analysis has been used to address most of these issues. Effective and efficient algorithms address the problem of out-of-focus dots by performing optical section deblurring and image fusion, and for differentiating between overlapping dots, split dots, and duplicated dots [25].

An example algorithm for automated image analysis is described here. Preprocessing is usually performed using background subtraction (Section 10.5.3.2) and color compensation (Section 10.5.6). Automated gray level thresholding (generally, blue for the DAPI counterstain) is used to obtain binary images of cells. The cells are then uniquely identified using a region labeling procedure (Chapter 7). The 8-connected pixel neighborhood is used to determine the pixel belonging to a certain object. Each pixel in the connected neighborhood is then assigned a unique number so that all the pixels belonging to an object will have the same unique label. The number of pixels in each object is computed and used as a measure of cell size. Subsequently, shape analysis is used to discard large cell clusters and noncircular objects. Further, a morphological technique is used for automatically cutting touching cells apart (Chapter 6). The morphological algorithm shrinks the objects until they separate, and then thins the background to define cutting lines. An exclusive OR operation then separates the cells. Cell boundaries are smoothed by a series of erosions and dilations, and the smoothed boundary is used to obtain an estimate of the cellular perimeter. Performing a binary AND operation on the thresholded image and the morphologically processed mask with the other two red and green planes of the color compensated image yields grayscale images containing only dots that lie within the cells. Objects are then located by thresholding in the probe color channels, using smoothed boundaries as masks. A minimum size criterion is used to eliminate noise spikes, and shape analysis is used to flag noncompact dots. The remaining objects are counted. The spatial location of each isolated dot is compared with the cell masks to associate each chromosomal dot with its corresponding cell [18].

Statistical modeling approaches may be applied to determine unbiased estimates of the proportion of cells having a given number of dots. For example, the befuddlement theory provides guidelines for dot-counting algorithm development by establishing the point at which further reduction of dot-counting errors will not materially improve the estimate [26]. This occurs when statistical sampling error outweighs dot-counting error.

Automated image analysis of FISH labeled interphase cells is illustrated in Fig. 10.7. The image shows six female (XX) cells counterstained blue (DAPI). The X chromosomes are labeled in red (Texas Red). In Fig. 10.7, Panel A is the original image, panel B shows the background subtracted image, panel C shows the color-compensated image, and panel D presents results of automated cell and dot finding. As seen in the figure, automated image analysis correctly finds single cells, separates touching cells, and detects the red dots in individual cells.

10.7.3 Quantitative Colocalization Analysis

Colocalization studies are used to identify functionally related molecules. They involve the simultaneous analysis of the location and expression of multiple target molecules. The most common application is to determine the spatial colocalization between two fluorescently labeled proteins. Quantitative colocalization analysis allows estimation of the extent to which two or more proteins routinely occur at the same physical location in a cell or tissue region. This enables the mapping of potential protein-to-protein interactions with subcellular precision, thereby providing a better understanding of how intracellular mechanisms are regulated. In terms of image analysis this involves the quantification of overlapping signals in multiple channels.

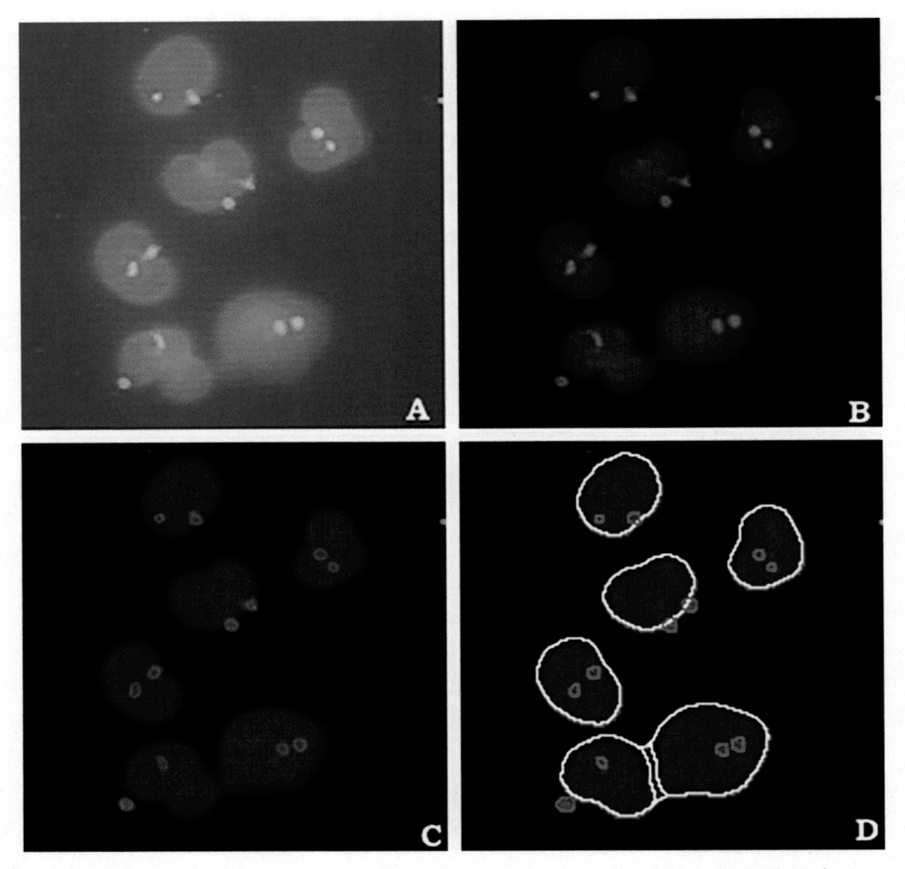

Fig. 10.7 FISH image of six female (XX) cells. Cells are counterstained blue (DAPI); X chromosomes are labeled in *red* (Texas Red). Results of automated image analysis, (A) original image, (B) background subtracted image, (C) color-compensated image, and (D) results of automated cell and dot finding.

As described earlier, it is essential to correct the fluorescence images for background staining, spectral bleed-through, and other influences such as lamp alignment and camera exposure settings, before colocalization can be quantified. Several approaches have been proposed for colocalization analysis. Techniques such as cross-correlation analysis [27] and cluster analysis of the 2D histogram [28] have been applied to prove the existence of colocalization. Generally, given two proteins labeled using antibody (Ab) 1, colored red, and Ab 2, colored green, colocalization is based on the fact that the superimposition of the two proteins appears yellow.

Fig. 10.8 shows colocalization images of nascent DNA labeled with CldUrd (green) and IdUrd (red). Optical sections through the center of double-labeled nuclei of Chinese Hamster V79 cells show two early S-phase DNA replication patterns. Nascent DNA was

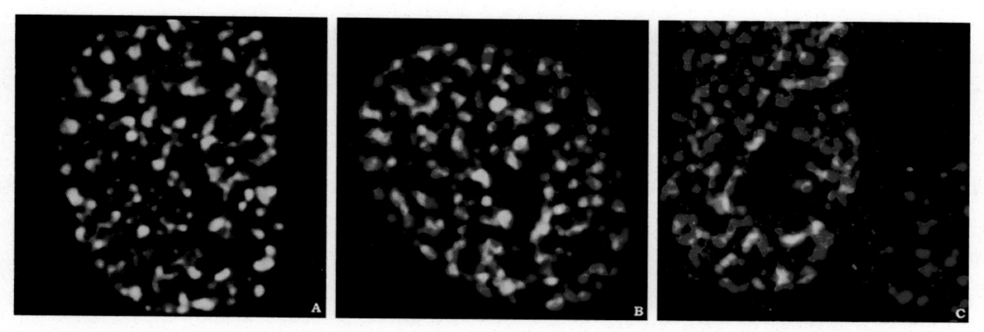

Fig. 10.8 Images of double-labeled Chinese Hamster V79 cell nuclei. Nascent DNA is labeled with CldUrd (*green*) and IdUrd (*red*) at two different times. Time between the labels was 0 min in the control experiment (A), showing virtually complete colocalization, 25 min in (B) and 45 min in (C). Progressively less colocalization is seen with increasing time as the *red* and *green* labeled DNA separates. *(Images courtesy of Erik Manders)*

labeled with CldUrd (green) and IdUrd (red) at two different times. Increasing the time between labelings shows decreasing colocalization, indicating that the DNA replication machinery moves through the nucleus.

The Pearson's correlation coefficient, r_p, provides a nonlinear estimate of the number of colocalized signals in the red and green channels as

$$r_p = \frac{\sum_i (R_i - R_{av}) \cdot (G_i - G_{av})}{\sqrt{\sum_i (R_i - R_{av})^2 \cdot \sum_i (G_i - G_{av})^2}} \quad (10.24)$$

where R_i and G_i are the values of pixel i of the red and green component of a dual-color image, respectively, and R_{av} and G_{av} are the average values of R_i and G_i, respectively; r_p provides information on the similarity of shape independent of the average intensity of the signals. Its value ranges from -1 to 1, and interpretation of the degree of overlap of the two signals may be ambiguous when negative values are encountered.

Alternatively, an overlap coefficient can be computed as follows [29]:

$$r = \frac{\sum_i R_i \cdot G_i}{\sqrt{\sum_i (R_i)^2 \cdot \sum_i (G_i)^2}} \quad (10.25)$$

The value of r ranges from 0 to 1, and it is independent of differences in signal intensities, when compared to r_p. The fraction of colocalizing regions in each component of dual color images can be computed by dividing r into two different coefficients [30], as

$$r_1 = \frac{\sum_i R_i \cdot G_i}{\sum_i R_i^2} \quad \text{and} \quad r_2 = \frac{\sum_i R_i \cdot G_i}{\sum_i G_i^2} \tag{10.26}$$

The coefficients r_1 and r_2 are dependent on the intensities of the red and green signals, respectively. Two other coefficients, known as the Mander's colocalization coefficients, can be defined to be independent of the signal intensities as follows [30]:

$$M_1 = \frac{\sum_i R_{i,\text{coloc}}}{\sum_i R_i}, \quad \text{where} \quad R_{i,\text{coloc}} = R_i \quad \text{if} \quad G_i > 0 \quad \text{and} \quad R_{i,\text{coloc}} = 0 \quad \text{if} \quad G_i = 0$$

$$M_2 = \frac{\sum_i G_{i,\text{coloc}}}{\sum_i G_i}, \quad \text{where} \quad G_{i,\text{coloc}} = G_i \quad \text{if} \quad R_i > 0 \quad \text{and} \quad G_{i,\text{coloc}} = 0 \quad \text{if} \quad R_i = 0 \tag{10.27}$$

Mander's colocalization coefficients are proportional to the amount of fluorescence of the colocalizing objects in each component of the image, relative to the total fluorescence in that component. The significance of the Mander's coefficients can be assessed by a comparison with an expected random pattern obtained by repeatedly randomizing the pixel distribution in one of the channels [31]. Although these coefficients provide an indication of the coexistence of two proteins, they provide no information about whether the intensity of staining for the two proteins varies in synchrony (i.e., whether the two target proteins are structural elements of a common complex).

Intensity correlation analysis can be used to determine whether the staining intensities in the dual color images are associated in a random, a dependent, or a segregated manner [32]. If N is the number of pixels, and R_i and G_i are the intensities in the red and green channels, and R_{av} and G_{av} the mean values of the red and green distributions, respectively, then we compute the sum of products of the differences between the pixel values and their means as follows [32]:

$$\sum_i^N (R_i - R_{av})(G_i - G_{av}) \tag{10.28}$$

The intensity correlation analysis technique involves generating scatterplots of the red dye or green dye against the product of the differences of each pixel in red and green intensities from their respective means (Eq. 10.28). The resulting plots emphasize the high intensity-stained pixels, allowing the identification of protein pairs based on the variations in protein concentrations across the cell and not simply their locations. If the two proteins are randomly distributed, then the value of Eq. (10.28) will tend toward zero, whereas if they are dependent, the value will be positive, and if they are segregated the value tends to be negative. The polarity of each $(R_i - R_{av})(G_i - G_{av})$ value can be used to compute the intensity correlation quotient (ICQ), which provides a statistically testable, single-value assessment of the relationship between the stained protein pairs. The ratio of the number of positive values to the total number of pixel pairs is first computed, and then the ICQ is determined by subtracting 0.5 from this value to distribute the quotients in the −0.5 to +0.5 range. With random (or mixed) staining $ICQ \approx 0$; with dependent staining $0 < ICQ \leq +0.5$; and for segregated staining $0 > ICQ \geq -0.5$ [32]. The intensity correlation ICQ analysis allows identification of potential low-affinity protein complexes while retaining information on the cellular and subcellular location.

10.7.4 Fluorescence Ratio Imaging (RI)

In fluorescence imaging, there are three major applications of ratio imaging, namely, combinatorial ratio labeling, comparative genome hybridization (CGH), and ion ratio imaging. Combinatorial ratio labeling is used in multispectral fluorescence in-situ hybridization (MFISH), wherein the ratio between the intensities of different fluorophores is used to expand the number of colored labels. This number is typically limited by the number of fluorophores that can be spectrally separated. In CGH, an estimate of the DNA sequence copy number as a function of position on the chromosome is obtained by measuring the ratio between sample DNA and the reference/control DNA to detect gene amplifications and deletions. Finally, ion ratio imaging is also used to measure either absolute or relative changes in spatial and temporal ion concentrations within living cells. This is achieved by measuring the fluorescence emission of special dyes that have been designed to change their spectral properties or emission intensities upon binding to the ion of interest.

The principle underlying fluorescence ratio imaging is that the ratio of intensities at two or more wavelengths, computed for a single pixel or a region of interest, avoids the major problems associated with intensity variations in single wavelength images. Gray level changes in fluorescence images are very difficult to interpret, because of variations in (1) the intensity of the peak absorption or emission signals, (2) local dye concentration, (3) photon shot noise, (4) fluctuations in the excitation light intensity, and (5) variations in detector gain. These intensity variations can be misinterpreted as a change in the concentration of the target. Moreover, specimen thickness may be problematic in single-wavelength images that display the amount of fluorescence only, with thicker portions of the sample looking brighter than smaller regions. The calculation of a ratio between the two channels corrects the result for intensity fluctuations and specimen thickness [33].

The protocol for ratio imaging is as follows. The ion-sensitive fluorescent dye is introduced into the cell, and two images are captured. The first image is acquired at the characteristic emission wavelength for binding, and the second at a reference wavelength (unbound). A ratio is then taken of the gray level changes at the same location at the two different wavelengths. This normalizes changes in the cell that are independent of a change in target concentration, such as the distance through a cell. Ratio imaging thus allows detection of ion transport and binding sites in living cells with relatively short optical path lengths (a few micrometers). It requires that the sources of measurement error (Section 10.4) be understood and appropriately corrected (Section 10.5) before quantitative measurement of concentration.

There is a wide range of fluorophores, with different spectral properties for numerous different ions. Intracellular calcium is the most widely imaged ion since it is involved in many different physiological processes, including muscle contraction, the release of neurotransmitters, ion channel gating, second messenger pathways, etc. Two types of fluorescence indicators [34] are available for the measurement of intracellular calcium: (1) fluorescent dyes (Fluo-4, Fura-2, calcium green, etc.), and (2) fluorescent proteins (aequorin, derivatives of the green fluorescence protein (GFP) such as yellow camaleons, etc.). An advantage of the fluorescent proteins is that they can be targeted to different cell compartments while most dyes cannot.

The procedure for ratio imaging is relatively straightforward. For example, for analysis of Ca^{+2} concentration using Fura-2 fluorescence dye, two excitation wavelengths are used, 340 nm and 380 nm. In the absence of any stimulus, in Fura-2 loaded cells, the Ca^{+2} is bound in cell compartments. When excited at 380 nm, the Fura-2 molecules show strong fluorescence at an emission of 510 nm, whereas an excitation at 340 nm produces only weak fluorescence. On stimulation, the cell releases Ca^{+2} from storage compartments and the Fura-2 molecules form complexes with the released Ca^{+2} ions. The emitted signal now increases when excited with 340 nm and decreases when excited with 380 nm. The images are corrected for background variations, and the ratio between the signals of the two excitation channels is used to quantify the change of intensity. The ratio images can also be calibrated so that ratio values correspond to concentrations as [35]

$$\left[Ca^{+2}\right] = \frac{K_D \beta (R - R_{min})}{(R_{max} - R)} \tag{10.29}$$

where R is the ratio of Fura-2 fluorescence with 340 nm excitation divided by the fluorescence with 380 nm excitation at a given point in time. R_{max} is the ratio when all the Fura-2 is bound to Ca^{+2}, R_{min} is the ratio when all the Fura-2 is in the free acid form, β is the ratio of fluorescence of free Fura-2 divided by the fluorescence of Ca^{+2}-bound Fura-2 with 380 nm excitation, and K_D is the dissociation constant for Fura-2 and Ca^{+2} binding. Typically, the values of R_{max}, R_{min}, and β for intracellular Ca^{+2} concentration calibration are determined as follows. Following the recording of Ca^{+2} dynamics during an experiment, a calcium ionophore, such as ionomycin, is introduced into a cellular buffer solution that has a high free $[Ca^{+2}]$ to raise the intracellular $[Ca^{+2}]$. Recording images after the cytosol reaches equilibrium allows the determination of R_{max}, as the 340 nm excitation signal is at its highest and the 380 nm excitation signal is at its lowest.

Subsequently, a calcium chelator (e.g., ethylene glycol tetraacetic acid), is added to drop the intracellular [Ca^{+2}], and, at equilibrium, images are acquired to measure R_{min}. The value of K_D is fluorophore-specific and can be obtained from the manufacturer. Ratio imaging can be performed using wide-field, confocal, or two-photon microscopy, with the most critical requirement being the simultaneous or near-simultaneous acquisition of two images at different excitation or emission wavelengths.

10.7.5 Fluorescence Resonance Energy Transfer (FRET)

Fluorescence resonance energy transfer (FRET) can be used to obtain information on the immediate environment (in nanometer ranges, 1–10 nm) of a labeled molecule, to detect macromolecular interactions, and to determine the intra- and intermolecular proximity of two appropriately paired fluorophores. Protein-to-protein interactions mediate most cellular processes. Identification of a protein's interacting partners is critical in understanding its function, placing it in a biochemical pathway, and thereby establishing its relationship to important disease processes.

Fluorescence resonance energy transfer is a process involving the radiation-less transfer of energy from a donor fluorophore to an appropriately positioned acceptor fluorophore [36]. FRET can occur when the emission spectrum of a donor fluorophore significantly overlaps (>30%) the absorption spectrum of an acceptor. In the absence of spectrum overlap, FRET cannot occur. For FRET, the emission dipole of the donor and the acceptor absorption dipole should be oriented to each other. However, there is no FRET if the dipoles are oriented perpendicular to each other. A dipole is an electromagnetic field that exists in a molecule with two oppositely charged regions. With overlapping spectra of donor emission to the acceptor absorption, the donor's oscillating emission dipole looks for a matching absorption dipole of the acceptor to oscillate in synchrony. The magnitude of the relative orientation of the dipole-dipole coupling values ranges between 1 and 4, and the efficiency of energy transfer varies inversely with the sixth power of the distance separating the donor and acceptor fluorophores. Thus the distance over which FRET can occur is limited to a range of 1–10 nm. When the spectral overlap, dipole orientation, and distance criteria are satisfied, excitation of the donor fluorophore results in sensitized fluorescence emission of the acceptor, indicating that the tagged proteins are separated by <10 nm. Proximity between two labeled cellular components considerably surpasses the resolution of normal light microscopy, which can resolve distances of ~200 nm. The most commonly used fluorophore pairs for FRET can be found in the literature along with their respective spectra and filter combinations [37]. A new generation of "Super-Resolution" techniques have been spawned with a number of new commercial instruments, improving the resolution to 25–140 nm, depending on fixed vs. live specimens and the tailored selection of fluorophores [38].

FRET has been implemented using wide-field, confocal, and two-photon fluorescence microscopy. Wide-field FRET (W-FRET) imaging provides the 2D spatial distribution of steady-state protein-to-protein interactions [39]. An advantage of W-FRET is that it allows the use of any excitation wavelength using interference filters for various fluorophore pairs. A disadvantage of W-FRET is that it contains out-of-focus information in the FRET signal. However, this low-cost system is used widely to monitor protein associations in living specimens where confocality is not an issue, such as following events in the nucleus. In confocal FRET (C-FRET) and two-photon (2p) excitation FRET (2p-FRET) microscopy, one can discriminate the out-of-focus information to obtain the FRET signal at the selected focal plane [40]. Moreover, 2p-FRET microscopy uses infrared (IR) laser light as an excitation wavelength and has the ability to excite most of the selected fluorophore pairs. Confocal systems use fixed laser lines for a number of wavelengths or white-light lasers with flexible excitation wavelengths. It is important to choose 2p-FRET fluorophore pairs with a different 2p absorption cross-section to avoid simultaneous excitation by one wavelength. The use of infrared laser light excitation instead of ultraviolet laser light also reduces phototoxicity in living cells. In general, 2p-FRET microscopy is superior for deep tissue FRET imaging. Also, spectral imaging, either by confocal or two-photon systems, is an intensity-based imaging technique that provides an excellent way to obtain a FRET signal [41] (see Figs. 10.9 and 10.10).

Fig. 10.9 presents images of GHFT1 cells expressing alpha enhancer binding protein (C/EBPα) either with Venus FP (acceptor) alone or Cerulean FP (donor) or with both Cerulean and Venus. The same optical settings were used for imaging both single- and double-labeled cells. These images were unmixed based on the reference spectra of donor and acceptor. The experimental details regarding data acquisition and processing are described in the literature [41,42], and the methodology discussed in this section is applicable to any commercially available confocal spectral imaging system.

In Fig. 10.9, Spect I, panel DA_DS shows the spectral image from the double-labeled specimen under donor excitation. Images e_s and f_s are unmixed from DA_DS. Spect II, panel DA_AS shows the spectral image from the same double-labeled specimen but under acceptor excitation, and g_s is unmixed from DA_AS. Spect III, panel A_AS is the spectral image from a single-labeled acceptor specimen under the same acceptor excitation as that from the double-labeled specimen, and image d_s is unmixed from A_AS. Spect IV, panel A_DS is the spectral image from a single-labeled acceptor specimen under the same donor excitation as that from the double-labeled specimen, and c_s is unmixed from A_DS, and it is only acceptor bleed-through. The three blank images in the middle column are donor components from unmixing for DA_AS, A_DS, and A_AS, and are not required for the data analysis.

Fig. 10.10 illustrates results of spectral unmixing on the f_s image in panel Spect I of Fig. 10.9, which contains both the FRET signal and the acceptor spectral bleed-through signal. The bleed-through signal was removed using an unmixing algorithm [41,42] as

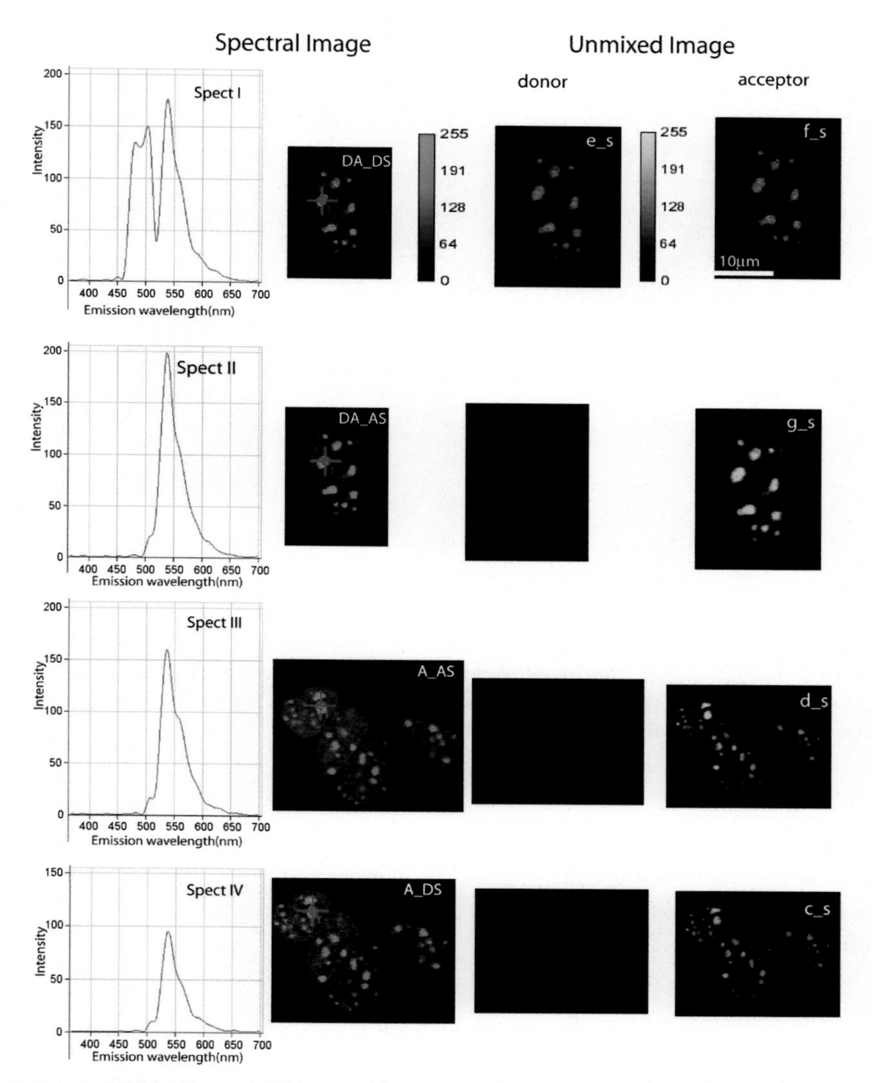

Fig. 10.9 Spectral FRET data acquisition and linear unmixing. GHFT1 cells expressing alpha enhancer binding protein (C/EBPα) labeled with either Venus FP (acceptor) or Cerulean FP (donor) alone or with both Cerulean and Venus are shown. The top two rows show a double-labeled specimen, and the bottom two rows a single-labeled specimen. Rows one and four were illuminated with donor excitation, and rows two and three with acceptor excitation. A Carl Zeiss laser scanning 510 confocal spectral imaging system collected the data, and the same optical settings were used for imaging both labeled cells. Cerulean FP and Venus FP fingerprints were obtained from single-labeled donor and single-labeled acceptor to use for unmixing. Linear unmixing was implemented using Cerulean FP, Venus FP, and background spectra in a spectral unmixing algorithm provided by the Carl Zeiss Company.

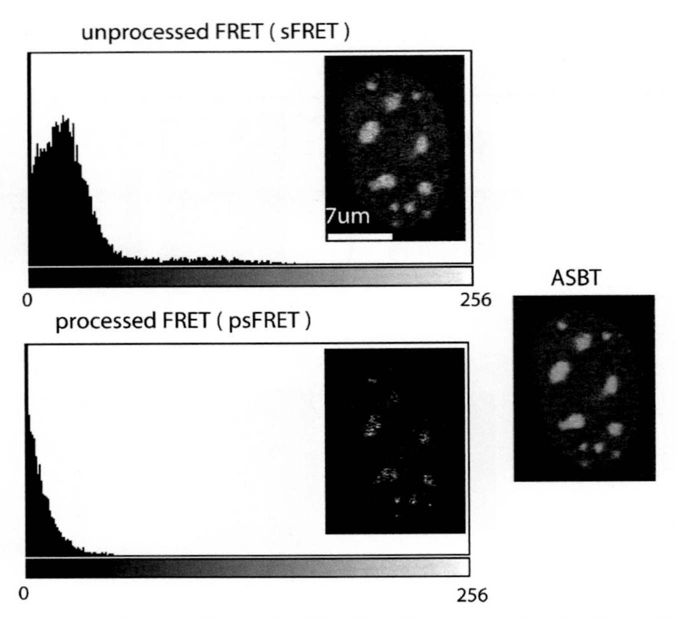

Fig. 10.10 The f_s image in the panel Spect I of Fig. 10.9 shows both the FRET signal and the acceptor spectral bleed-through signal. The bleed-through signal was removed by an unmixing algorithm [41,42]. The localization of C/EBPα protein in the living cell nucleus is visible in the unmixed image. The image improvement is also evident in the two histograms.

shown in Fig. 10.10. The localization of C/EBPα protein in the living cell nucleus shows up clearly in the processed image.

The Förster's basic rate equation for a donor and acceptor pair at a distance, *d*, from each other, is

$$k_\tau = \frac{1}{\tau_D}\left(\frac{R_0}{d}\right)^6 \tag{10.30}$$

where k_τ is the rate of energy transfer, τ_D is the donor-excited state lifetime in the absence of the acceptor and R_0 is the Förster distance (a distance at which coupling efficiency reaches 50%).

The apparent energy transfer efficiency (E) is represented by the equation

$$E = \frac{R_0}{\left(R_0^6 + d^6\right)} \tag{10.31}$$

Typically, the efficiency drops if the distance between donor (D) and acceptor (A) molecules changes from the Förster distance. The energy transfer efficiency

(E) and the distance (d) measurement between donor and acceptor can also be calculated using the following equations; however, the often-referred-to statement of FRET being a "spectroscopic ruler" is not supportable, but the great value of FRET is confirmation of proximity and the relative distances between molecules

$$d = R_0\left\{\left(\frac{1}{E}\right) - 1\right\}^{1/6} \tag{10.32}$$

$$R_0 = 0.211\left\{\kappa^2 n^{-4} Q_D J(\lambda)\right\}^{1/6} \tag{10.33}$$

where Q_D is the quantum efficiency of the donor molecule, J is the spectral overlap between the donor emission and acceptor absorption power spectrum ($cm^2\ s^4/mol$), n is the refractive index of the energy transfer medium, and κ^2 is orientation factor, which varies between 0 and 4. The overlap integral J, which expresses the degree of spectral overlap between the donor emission fluorescence and the acceptor absorption, is given as

$$J(\lambda) = \int_0^{\infty} f_D(\lambda)\ \varepsilon_A(\lambda)\ \lambda^4\ d\lambda \tag{10.34}$$

where λ is the wavelength of the light, $\varepsilon_A(\lambda)$ is the molar extinction coefficient of the acceptor at that wavelength, and $f_D(\lambda)$ is the normalized fluorescence intensity at that wavelength. The energy transfer efficiency may also be calculated by ratioing the donor image in the presence and absence of the acceptor

$$E = 1 - \left[\frac{I_{DA}}{I_D}\right] \tag{10.35}$$

where I_{DA} and I_D are the fluorescence intensities in the presence and the absence of acceptor, respectively.

As described earlier, all these techniques require the removal of certain signal contaminations, such as detector dark current noise and background fluorescence [41,42]. One of the important conditions for FRET to occur is the overlap of the emission spectrum of the donor with the absorption spectrum of the acceptor. As a result of spectral overlap, the FRET signal is always contaminated by donor emission into the acceptor channel and by the excitation of acceptor molecules by the donor excitation wavelength. Both of these signals are termed spectral bleed-through (SBT) of signal into the acceptor channel. There are various methods to assess the SBT contamination in FRET image acquisition [43–45]. One approach eliminates both the donor and acceptor SBT problems and corrects the variation in fluorophore expression level (FEL), using the same cell as used for FRET imaging [40,46,47]. The letters D, A, and DA are used to denote donor, acceptor, and donor-acceptor pairs. Seven images (denoted by a, b, c, d, e, f, g) are acquired as follows: double-labeled (three images: donor excitation/donor "e" and acceptor "f" channel; acceptor excitation/acceptor channel "g"), single-labeled donor (two images:

donor excitation/donor "a" and acceptor "b" channel), and single-labeled acceptor (two images: donor excitation/acceptor "c" channel; acceptor excitation/acceptor channel "d"). This approach works on the assumption that the double-labeled cells and single-labeled donor and acceptor cells, imaged under the same conditions, exhibit the same SBT dynamics [47]. A practical impediment to implementation arises from the fact that there are three different cells (D, A, and D + A), where individual pixel locations cannot be compared. Instead, comparison of pixels with matching fluorescence levels is performed. The algorithm follows fluorescence levels pixel-by-pixel to establish the level of SBT in the single-labeled cells, and then applies these values as a correction factor to the appropriate matching pixels of the double-labeled cell. The following equations are used to remove the spectral bleed-through signal from the FRET channel image. The corrected FRET signal image, denoted as PFRET (processed FRET) is computed as [46–48].

$$PFRET = UFRET - DSBT - ASBT \tag{10.36}$$

where UFRET (image "f") is uncorrected FRET, ASBT is the acceptor spectral bleed-through signal, and DSBT is the donor spectral bleed-through signal, computed as follows. To correct the DSBT, three images are required (one double-labeled "e" and two single-labeled donor images "a" and "b"). To obtain the DSBT values, the following equations are used [46–48]:

$$rd_{(j)} = \frac{\sum_{i=1}^{m} b_i}{\sum_{i=1}^{m} a_i}, \quad DSBT_{(j)} = \sum_{p=1}^{n} \left(e_p * rd_{(j)}\right), \quad DSBT = \sum_{j=1}^{k} DSBT_{(j)} \tag{10.37}$$

where j is the jth range of intensity, $rd_{(j)}$ is the donor bleed-through ratio for the jth intensity range, m is the number of pixel in "a" for the jth range, a_i is the intensity of pixel i, $DSBT_{(j)}$ is the donor bleed-through factor for the range j, n is the number of pixels in "e" for the jth range, e_p is the intensity of pixel p, k is the number of range, and $DSBT$ is the total donor bleed-through.

The $ASBT$ correction follows the same approach as $DSBT$ using three images (one double-labeled "g" and two single-labeled acceptor images "c" and "d"). To obtain the $ASBT$ values, the following equations are used [46–48]:

$$ra_{(j)} = \frac{\sum_{i=1}^{m} c_i}{\sum_{i=1}^{m} d_i}, \quad ASBT_{(j)} = \sum_{p=1}^{n} \left(g_p * ra_{(j)}\right), \quad ASBT = \sum_{j=1}^{k} ASBT_{(j)} \tag{10.38}$$

where j is the jth range of intensity, $ra_{(j)}$ is the acceptor bleed-through ratio for the jth intensity range, m is the number of pixels in "d" for the jth range, d_i is the intensity of pixel i, $ASBT_{(j)}$ is the acceptor bleed-through factor for the range j, n is the number of pixels in "g" for the jth range, g_p is the intensity of pixel p, k is the number of range and $ASBT$ is the total acceptor bleed-through.

The *PFRET* image is then used for further data analysis, such as estimation of distance between donor and acceptor molecule and the energy transfer efficiency E%, as follows. The sensitized emission in the acceptor channel is due to the quenching of the donor or energy transferred signal from the donor molecule in the presence of the acceptor. Adding the *PFRET* value to the DSBT to the intensity of the (quenched) donor in the presence of the acceptor gives I_D (unquenched donor) [49]. This I_D is from the same cell used to obtain the I_{DA}. Then from Eq. (10.36), we have

$$E_n = 1 - \left[\frac{I_{DA}}{I_{DA} + PFRET}\right], \quad \text{where } I_D = I_{DA} + PFRET + DSBT \tag{10.39}$$

It is important to note that several processes are involved in the excited state during energy transfer. The equation for energy transfer efficiency E_n (see [47] for details) is thus calculated by generating a new I_D image and by including the detector spectral sensitivity for donor and acceptor channel images and the donor quantum yield, with the *PFRET* signal as follows [47]:

$$E_n = 1 - \left\{\frac{I_{DA}}{\left[I_{DA} + PFRET \times \left(\frac{\psi_{dd}}{\psi_{aa}}\right) \times Q_d/Q_a\right]}\right\} \tag{10.40}$$

where

$$\frac{\psi_{dd}}{\psi_{aa}} = \left[\left(\frac{\text{PMT gain of donor channel}}{\text{PMT gain of acceptor channel}}\right) \times \left(\frac{\text{spectral sensitivity of donor channel}}{\text{spectral sensitivity of acceptor channel}}\right)\right] \tag{10.41}$$

and Q_d is donor quantum yield. Using R_0 defined in Eq. (10.33), the distance, d_n, between donor and acceptor is computed as [46–48].

$$d_n = R_0\left\{\left(\frac{1}{E_n}\right) - 1\right\}^{1/6} \tag{10.42}$$

An alternative approach to measure FRET is to determine the fluorescent lifetime of the donor in the presence and absence of the acceptor. The advantages of this approach are that it is directly dependent upon excited state reactions, and independent of the fluorophore concentration and light path length (and consequently photobleaching), thus allowing the relative donor-acceptor distance to be mapped more accurately.

10.7.6 Fluorescence Lifetime Imaging (FLIM) FRET

Fluorescence lifetime is the average amount of time a molecule spends in the excited state before returning to ground state. The fluorescence lifetime of a fluorophore carries information about events in its local microenvironment, which affects its photophysical processes. FRET adds a nonradiative dissipation pathway for the excited state energy of the donor and thus shortens its fluorescence lifetime. Using FLIM, E can be quantified by measuring the reduction of the donor fluorescence lifetime, resulting from quenching in the presence of an acceptor [50]. The FLIM-FRET approach has several advantages over intensity-based FRET imaging: (i) FLIM-FRET measurements typically do not require corrections for spectral bleedthrough, necessary for intensity-based measurements of acceptor sensitized emission; (ii) fluorescence lifetime measurements are insensitive to the change in fluorophore concentration, excitation intensity, or light scattering and to some extent of photobleaching – all these factors potentially induce artifacts in intensity-based imaging; and (iii) FLIM-FRET methods have the capability to estimate the percentage of "FRETing" and "non-FRETing" donor populations [51], which cannot be distinguished by most intensity-based FRET methods; the E measured in most intensity-based FRET methods is often called *apparent* E, which inter alia includes "non-FRETing" donors in the calculation.

FLIM-FRET measurements require advanced instrumentation as well as understanding of the basic physics for data analysis and interpretation. In the past 22 years, rapid developments in FLIM have greatly advanced and simplified the technique, and various FLIM methods have been developed for biological and clinical applications [52–54]. More importantly, commercial stand-alone FLIM systems or those integrated with existing multiphoton, confocal, or wide-field microscopes have become available from Picoquant, Becker & Hickl, Lambert Instruments, ISS, Intelligent Imaging Innovations, and others. FLIM therefore has become more of a routine tool for many laboratories for FRET studies [52,55]. Leray et al. [56] applied a 3D polar plot analysis to spectrally resolved FLIM data, demonstrating that this multimodel fitting-free approach yields more accurate FRET measurements.

FLIM techniques are generally subdivided into the time domain (TD) and the frequency domain (FD), although the basic physics for both are essentially identical [50,52–54]. TD FLIM uses a pulsed light source synchronized to high-speed detectors and electronics to directly measure the fluorescence decay profile to estimate the fluorescence lifetime. FD FLIM employs a modulated excitation light source and measures the phase shift(s) and amplitude attenuation(s) of the emission relative to the excitation to estimate the fluorescence lifetime. The repetition rate of the excitation source in TD FLIM or the fundamental modulation frequency of the excitation source in FD FLIM are chosen according to the fluorescence lifetime to be measured, e.g., megahertz for measuring nanosecond lifetimes.

10.7.6.1 *Time Correlated Single Photon Counting (TCSPC) FLIM-FRET*

FLIM can provide a robust verification of intensity-based FRET measurements. In Fig. 10.11 the two-photon excitation TCSPC FLIM method was used to measure the quenched donor (Cerulean) fluorescence lifetimes due to the energy transfer from bZip-Cerulean to bZip-Venus [57]. TCSPC FLIM datasets were acquired from cells coexpressing bZip-Cerulean and bZip-Venus and cells that only express bZip-Cerulean as the donor-alone control. The fluorescent lifetime decay kinetics for the bZip-Cerulean (FRET donor) in the absence and presence of bZip-Venus (acceptor) were determined by fitting the decay data into a single or double exponential decay model, respectively, with the estimated instrument response function (IRF). By applying a suitable intensity threshold, fitting was only applied to pixels in the centromeric

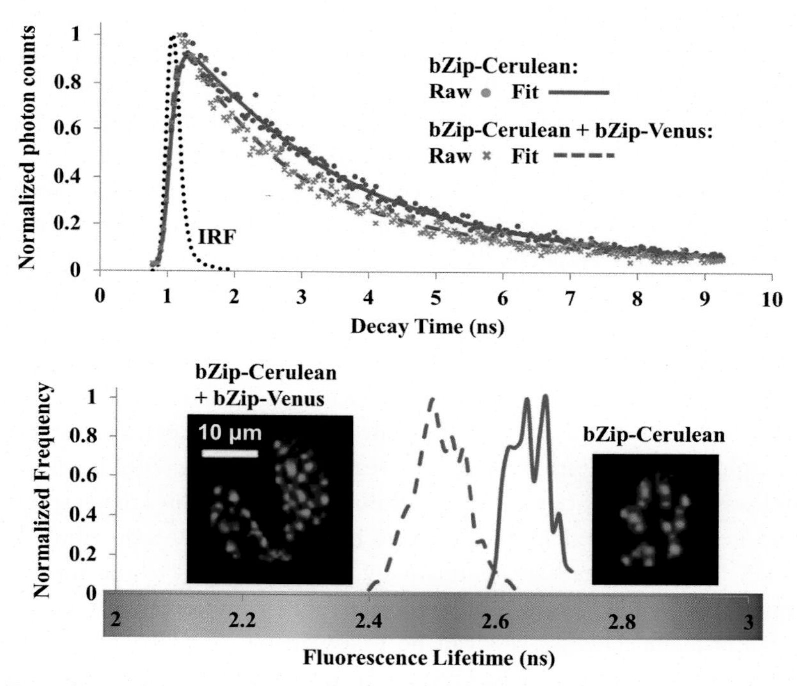

Fig. 10.11 Two-photon excitation (TPE) TCSPC FLIM-FRET microscopy. The average unquenched Cerulean lifetime obtained from 10 cells only expressing bZip-Cerulean was 2.75 ns, and the quenched bZip-Cerulean lifetimes measured from 10 cells coexpressing bZip-Cerulean and bZip-Venus ranged from 2.0 to 2.6 ns, producing an E% range of 5.5–27.3% [52]. (Biorad Radiance 2100 Confocal/Multiphoton imaging system and Nikon 60 ×/1.2NA water immersion objective; a Coherent Mira-900 ultrafast (repetition rate of ~78 MHz) pulsed (pulse width of ~150 fs) laser was tuned to 820 nm for the TPE wavelength; a Becker & Hickl (BH) SPC-150 board was used to synchronize the laser pulse to the scanning clock; photons were counted by using a BH PMH-100-0 photomultiplier tube detector; all decay data sets were analyzed using the BH SPCImage software Version 3.8.9). *(Adapted from [57])*

heterochromatin regions of the cell nucleus. The comparison between the representative decay data points, fitting curves, and fluorescence lifetime images and distributions of the two cases clearly shows bZip-Cerulean in the presence of bZip-Venus decays faster (i.e., has a shorter lifetime) than that in the donor-alone control cells, demonstrating bZip-Cerulean was quenched by bZip-Venus due to FRET. FLIM results clearly demonstrated FRET between bZip-Cerulean and bZip-Venus, confirming the ranges produced by intensity-based FRET.

In summary, FLIM-FRET is an important technique for investigating a variety of phenomena that produce changes in molecular proximity, and for monitoring intermolecular interactions and localization of proteins in cells and tissues [39,55].

10.7.7 Fluorescence Recovery After Photobleaching (FRAP)

Fluorescence recovery after photobleaching is used for studying the dynamic behavior of labeled molecules, most specifically the behavior of proteins in living cells. The process involves photobleaching a region of interest, thereby allowing the temporal study of the consequent fluorescent recovery in that bleached region because of the movement of nonbleached fluorescent molecules from the surrounding area. The extent and the speed at which this recovery occurs are measures for the fraction of mobile molecules and the speed at which they move, respectively [58]. The basic FRAP experiment is straightforward. First, a region of limited dimensions within a larger volume is illuminated with a short pulse of an intense laser beam at the excitation wavelength of the dye to be bleached. Subsequently, the molecules in the exposed region are no longer fluorescent. If the target labeled molecule is fixed, the region will remain dark. However, if the target molecules are mobile, they diffuse, with new fluorescent molecules from the surrounding unbleached regions moving into the bleached region and mixing with the bleached molecules. This leads to a continuous increase of fluorescence in the bleached region until the bleached and new fluorescent molecules have been completely redistributed over the entire volume. If the bleached area is relatively large compared to the total volume in which the target molecules reside, the final recovery of fluorescence will be less than the prebleaching level. This process can be followed on a microscope by visualizing the fluorescence either in the bleached region or in the total volume. After a statistically relevant number of cells or regions have been sampled, the average mobility of the fluorescent molecules can be determined by averaging the normalized fluorescence intensity of the individual regions.

One approach to normalizing FRAP data is to express the data relative to the prebleaching value

$$I_{norm,t} = \frac{\left(I_t - I_{background}\right)}{\left(I_{prebleach} - I_{background}\right)} \tag{10.43}$$

where $I_{prebleach}$ is the measurement before bleaching (or the average of a number of recordings before bleaching), and $I_{background}$ is the signal level in the absence of any fluorescence [58]. Alternatively, normalization can be performed by fitting the data to the analytically derived descriptions of diffusion, as follows [58]:

$$I_{norm,t} = \frac{(I_t - I_0)}{\left(I_{final} - I_0\right)} \tag{10.44}$$

where I_{final} is the final value at the completion of fluorescence recovery. This process yields a curve that starts at zero immediately after bleaching and reaches unity at recovery. The curve fit can be performed using any equation that represents the diffusion process, and with any fitting algorithm, such as the least-squares method. A variety of analytical functions representing 2D and 3D diffusion models and the Monte Carlo simulation approach to generating FRAP curves are covered in the literature [58–61].

Another application of FRAP is to quantitatively determine the immobile fraction of the labeled molecule under investigation. This is achieved as

$$I_{norm,t} = \frac{(I_t - I_0)}{\left(I_{prebleach} - I_0\right)} \tag{10.45}$$

where I_0 is the intensity immediately after bleaching. This approach yields a curve for which the prebleaching value is unity, and the fluorescence level immediately after bleaching is zero [58,59]. If the fraction of mobile bleached molecules is negligible, compared to the total amount of fluorescence molecules, that is, if the FRAP curve would return to prebleaching levels when no immobile fraction was present, then the immobile fraction can be estimated as $1 - I_{norm,final}$.

The confocal laser scanning microscope is ideal for performing FRAP experiments as the laser illumination can be limited to defined coordinates, allowing any region to be selected for bleaching. Applications of FRAP include the study of exchange between cells or organelles, diffusion of proteins within membranes or organelles, and determination of protein turnover in complexes. To aid in localization of the target molecule post-bleaching, sometimes a second fluorophore that remains visible throughout the time course of the experiment is introduced to the target. This process is termed fluorescence localization after photobleaching (FLAP) [62]. Another complementary technique is fluorescence loss in photobleaching (FLIP), which involves repeated photobleaching of a region of fluorescence within the cell [62]. Over time, this repeated bleaching leads to permanent loss of fluorescent signal throughout the cell. This indicates that free exchange of molecules occurs between the photobleached region and the rest of the cellular compartments. Thus nonbleaching of molecules within the cell indicates where molecules are isolated and specifically localized in distinct cellular compartments.

10.7.8 Total Internal Reflectance Fluorescence Microscopy (TIRFM)

Total internal reflectance fluorescence microscopy (TIRFM) is used to monitor the behavior of biomolecules directly at the single-molecule level, both in vitro and in living cells. It is most suited to image events occurring at, or close to, the plasma membrane of live cells. This technique is based on the principle that when the excitation light is incident above a critical angle upon the glass/water interface, the light is totally reflected internally, and it generates a thin electromagnetic field (called the evanescent field) with the same wavelength as the incident light [63]. The evanescent field that decreases exponentially is created at a depth of ~150 nm below the glass/water interface [64,65]. The intensity of the evanescent field decays exponentially with distance from the glass surface, so it is only able to excite molecules near the glass surface. This has the advantage that fluorophores further from the surface are not excited, resulting in a reduction of background noise. The constant of the exponential spatial decay of the field is the penetration depth, d_p, which depends on the physical parameters of the total internal reflection setup and the incident light.

In TIRF microscopy the field penetrates from the cover glass to the liquid in which the specimen is mounted. The penetrating energy is used as excitation to image a thin slice in the liquid substrate. The image intensity of a pixel depends on its depth $z(x, y)$ above the interface, and is given by [64]

$$I_z(x, y) = I_{\max} \exp\left(-(z(x,y) - z_{\min})/d_p\right) \tag{10.46}$$

where $I_{\max}$ is the maximum intensity, and $z_{\min}$ is the corresponding height, whose value is an experimental constant depending on the thickness of the region under observation.

TIRFM is used to measure the time course, trajectory, and distribution of molecular properties for individual biomolecules in vitro and in living cells. Specific applications include examination of plasma membrane events, such as endocytosis and exocytosis, spatial and quantitative analyses of receptor-ligand binding interactions, mobility of plasma membrane components, dynamics of cytoskeletal filaments, and the monitoring of chemical reactions. Image processing and analysis algorithms are typically dictated by the application, and most are customized to analyze quantitatively the parameters of interest.

10.7.9 Fluorescence Correlation Spectroscopy (FCS)

Fluorescence correlation spectroscopy (FCS) imaging is a highly sensitive method that provides fast temporal and high spatial resolution for single molecule imaging. The basis of this technique is monitoring the fluorescence fluctuations that arise from molecule diffusion within a small optically defined volume (on the order of femtoliters), following excitation by a focused laser beam. These fluctuations in fluorescence may arise as a result of flow, chemical reaction, and Brownian diffusion. In conventional imaging, fluctuations in intensity are problematic and constitute noise. However, in FCS the fluctuations constitute the signal. The imaging approach is to record the fluctuations in fluorescence intensity as a function of time. FCS is based on correlation analysis, comparing time signals for a series of lag times, τ. The decrease in the autocorrelation function of the time series provides a measure of the diffusion coefficient of the fluorophore.

The most commonly used normalized autocorrelation function, relating the fluorescence intensity $I(t)$ at time t, to that τ seconds later, $I(t+\tau)$, is [66]

$$G(\tau) = 1 + \frac{\langle \delta I(0) \delta I(\tau) \rangle}{\langle I \rangle^2} \tag{10.47}$$

In FCS the following analytical expression is used for the evaluation of experimental data [66]:

$$G(\tau) = 1 + \frac{1}{N}\left(1 + \frac{\tau}{\tau_D}\right)^{-1}\left(1 + \frac{\tau}{R^2 \tau_D}\right)^{\frac{-1}{2}}, \quad \text{where} \quad R = \frac{\omega_z}{\omega_{xy}} \tag{10.48}$$

Eq. (10.48) gives the correlation function for translational diffusion. The function is time-dependent and represents the fluorescence fluctuation that occurs due to Brownian motion. Assuming that the molecules within the observed volume element follow Poisson statistics, N is the average number of molecules in the sample volume. The number of molecules is inversely proportional to the amplitude of the autocorrelation curve, and τ_D is the average time that the molecules take to move across the observation volume in the radial direction, and R is the ratio of the axial half-axis to the lateral half-axis of the observation volume [66].

The autocorrelation curve resulting from processing FCS data is interpreted by fitting equations to it. Typically, mathematical functions that represent different diffusion models (such as Eq. 10.48) are fitted to the autocorrelation curve. Parameters such as the translational diffusion time, which characterizes the average residence time of the fluorescent molecule in the defined excitation volume, and the average number of fluorescent molecules in the volume, can then be determined. FCS has been used to measure the location, accumulation, and mobility of fluorescently tagged molecules in membranes and the cell wall from very low-bulk concentrations [67].

10.8 Summary of Important Points

1. Fluorescence imaging provides an incomparable degree of flexibility given its ability to maintain a "molecular" resolution with high specificity in a tremendously complex biological background.
2. Fluorescence occurs when a molecule absorbs a photon and then emits a photon as it returns to a lower state.

3. Fluorescence microscopy is an incoherent imaging process, that is, each point of the object or sample contributes independently to the light intensity distribution in the observed image.
4. Wide-field, confocal, and two-photon microscopes can be used to perform single and multidimensional fluorescence microscopy. The key requirements are illumination at the required wavelength and effective blocking of the excitation light during detection of the emitted light.
5. In practice, instrument- and sample-based aberrations are always present in fluorescence microscopy. It is critical that the sources of noise be identified and appropriately corrected.
6. Fluorescence images can be corrected easily for background inhomogeneity, dark current, autofluorescence, photobleaching, and intensity attenuation with depth.
7. The linear relationship between fluorescence and excitation intensity enables quantitative measurements of concentration using fluorescence microscopy.
8. Fluorescence is a relative quantity, and a large number of factors can affect fluorescence measurements, requiring calibration of instrumentation, standards, and image-correction methods for accurate quantification.
9. Several techniques are available for fluorescence microscopy. The choice of the fluorescence microscopy method used depends on the application.
10. Immunofluorescence is extensively used for the visualization and quantitation of the distribution of specific cellular components (such as proteins) in cells or tissue.
11. Fluorescence in situ hybridization is a molecular cytogenetic technique that is specifically used for the visualization and localization of DNA and RNA sequences.
12. Colocalization studies are performed to identify functionally related molecules, and they involve the simultaneous analysis of the location and expression of multiple target molecules.
13. Ratio imaging is most widely used to measure either absolute or relative changes in spatial and temporal ion concentrations within living cells.
14. FRET can be used to obtain information on the immediate environment (in the nanometer range) of a labeled molecule, to detect macromolecular interactions, and to determine the intramolecular and intermolecular proximity between two appropriately paired fluorophores.
15. FLIM can be used for studying FRET by detecting the decrease in the fluorescence lifetime of the donor fluorophore.
16. FRAP, FLIP, and FLAP are used for studying the dynamic behavior of labeled molecules, most specifically the behavior of proteins in living cells.
17. TIRFM is used to monitor the behavior of biomolecules directly at the single-molecule level, both in vitro and in living cells.
18. FCS is a highly sensitive method for single molecule imaging.

References

[1] C. Vonesch, et al., The colored revolution of bioimaging, IEEE Signal Process. Mag. 23 (3) (2006) 20–31.
[2] M. Abramowitz, et al., Basic principles of microscope objectives, Biotechniques 33 (2002) 772–781.
[3] J.W. Dobrucki, U. Kubitscheck, Fluorescence microscopy, Fluoresc. Microsc. 2 (12) (2017) 85–132.
[4] P. Montero Llopis, et al., Best practices and tools for reporting reproducible fluorescence microscopy methods, Nat. Methods 18 (2021) 1463–1476.
[5] U. Boehm, et al., QUAREP-LiMi: a community endeavor to advance quality assessment and reproducibility in light microscopy, Nat. Methods 18 (12) (2021) 1423–1426.
[6] A. Payne-Tobin Jost, J.C. Waters, Designing a rigorous microscopy experiment: validating methods and avoiding bias, J. Cell Biol. 218 (5) (2019) 1452–1466.
[7] I.T. Young, Quantitative microscopy, IEEE Eng. Med. Biol. 15 (1) (1996) 59–66.
[8] L.R. Van Den Doel, et al., Quantitative evaluation of light microscopes based on image processing techniques, Bioimaging 6 (1998) 138–149.
[9] Z. Jericevic, et al., Validation of an imaging system: steps to evaluate and validate a microscope imaging system for quantitative studies, Methods Cell Biol. 30 (1989) 47–83.
[10] I.T. Young, J.J. Gerbrands, L.J. van Vliet, Fundamentals in Image Processing, Delft University Press, 1998.
[11] J.W. Lichtman, J.A. Conchello, Fluorescence microscopy, Nat. Methods 2 (12) (2005) 910–919.
[12] M. Neumann, D. Gabel, Simple method for reduction of autofluorescence in fluorescence microscopy, J. Histochem. Cytochem. 50 (3) (2002) 437–439.
[13] F.A. Merchant, K.R. Castleman, Computer assisted microscopy, in: A.C. Bovik (Ed.), Handbook of Image and Video Processing, Academic Press, 2005.
[14] C.A. van De Lest, et al., Elimination of autofluorescence in immunofluorescence microscopy with digital image processing, J. Histochem. Cytochem. 43 (7) (1995) 727–730.
[15] G. Marriott, et al., Time resolved imaging microscopy: phosphorescence and delayed fluorescence imaging, Biophys. J. 60 (6) (1991) 1374–1387.
[16] K.R. Castleman, Color compensation for digitized FISH images, Bioimaging 1 (1993) 159–163.
[17] K.R. Castleman, Digital image color compensation with unequal integration periods, Bioimaging 2 (1994) 160–162.
[18] F.A. Merchant, K.R. Castleman, Strategies for automated fetal cell screening, Hum. Reprod. Update 8 (6) (2002) 509–521.
[19] D.L. Kolin, S. Costantino, P.W. Wiseman, Sampling effects, noise and photobleaching in temporal image correlation spectroscopy, Biophys. J. 90 (2006) 628–639.
[20] J.M. Zwier, et al., Image calibration in fluorescence microscopy, J. Microsc. 216 (1) (2004) 15–24.
[21] J. Murray, Evaluating the performance of fluorescence microscopes, J. Microsc. 191 (1998) 128–134.
[22] B. Herman, Fluorescence Microscopy, second ed., BIOS Scientific Publishers, 1998.
[23] F.A. Merchant, M. Toner, Spatial and dynamic characterization of the interaction of *Staphylococcal aureus* α-toxin with cell membranes, in: Advances in Heat and Mass Transfer in Biotechnology HTV 355/BED, 37, 1997, pp. 3–8.
[24] A. Alamri, J.Y. Nam, J.K. Blancato, Fluorescence in situ hybridization of cells, chromosomes, and formalin-fixed paraffin-embedded tissues, Methods Mol. Biol. 1606 (2017) 265–279.
[25] F.A. Merchant, Confocal microscopy, in: A.C. Bovik (Ed.), Handbook of Image and Video Processing, Academic Press, 2005.
[26] K.R. Castleman, B.S. White, Dot-count proportion estimation in FISH specimens, Bioimaging 3 (1995) 88–93.
[27] B. van Steensel, et al., Partial colocalization of glucocorticoid and mineralocorticoid receptors in discrete compartments in nuclei of rat hippocampus neurons, J. Cell Sci. 109 (1996) 787–792.
[28] D. Demandolx, J. Davoust, Multicolour analysis and local image correlation in confocal microscopy, J. Microsc. 185 (1997) 21–36.
[29] E.M. Manders, et al., Dynamics of three-dimensional replication patterns during the s-phase, analyzed by double labeling of DNA and confocal microscopy, J. Cell Sci. 103 (1992) 857–862.

[30] E.M. Manders, F.J. Verbeek, J. Aten, Measurement of colocalization of objects in dual-color confocal images, J. Microsc. 169 (1993) 375–382.
[31] S.V. Costes, et al., Automatic and quantitative measurement of protein-protein colocalization in live cells, Biophys. J. 86 (2004) 3993–4003.
[32] Q. Li, et al., A syntaxin 1, Gαo, and N-type calcium channel complex at a presynaptic nerve terminal: analysis by quantitative immunocolocalization, J. Neurosci. 24 (61) (2004) 4070–4081.
[33] R. Wegerhoff, O. Weidlich, M. Kassens, Basics of light microscopy and imaging, in: Imaging & Research, Git Verlag Gmbh & Co. KG, 2009. http://www.gitverlag.com/.
[34] J. Bruton, A.J. Cheng, H. Westerblad, Measuring Ca^{2+} in living cells, Adv. Exp. Med. Biol. 1131 (2020) 7–26.
[35] G. Grynkiewicz, M. Poenie, R.Y. Tsien, A new generation of Ca^{2+} indicators with greatly improved fluorescent properties, J. Biol. Chem. 260 (1985) 3440–3450.
[36] A. Periasamy, Fluorescence resonance energy transfer microscopy: a mini review, J. Biomed. Opt. 6 (3) (2001) 287–291.
[37] L. Mantovanelli, B.F. Gaastra, B. Poolman, Fluorescence-based sensing of the bioenergetic and physicochemical status of the cell, Curr. Top. Membr. 88 (2021) 1–54.
[38] A.M. Szalai, C. Zaza, F.D. Stefani, Super-resolution FRET measurements, Nanoscale 13 (44) (2021) 18421–18433.
[39] H. Wallrabe, A. Periasamy, FRET-FLIM microscopy and spectroscopy in the biomedical sciences, Curr. Opin. Biotechnol. 16 (1) (2005) 19–27.
[40] J.D. Mills, et al., Illuminating protein interactions in tissue using confocal and two-photon excitation fluorescent resonance energy transfer (FRET) microscopy, J. Biomed. Opt. 8 (2003) 347–356.
[41] Y. Chen, A. Periasamy, Localization of protein-protein interactions in live cells using confocal and spectral imaging FRET microscopy, Indian J. Exp. Biol. 45 (1) (2007) 48–57.
[42] Y. Chen, et al., Characterization of spectral FRET imaging microscopy for monitoring the nuclear protein interactions, J. Microsc. 228 (Pt. 2) (2007) 139–152.
[43] G.W. Gordon, et al., Quantitative fluorescence resonance energy transfer measurements using fluorescence microscopy, Biophys. J. 74 (1998) 2702–2713.
[44] V.S. Kraynov, et al., Localized Rac activation dynamics visualized in living cells, Science 290 (2000) 333–337.
[45] Z. Xia, Y. Liu, Reliable and global measurement of fluorescence resonance energy transfer using fluorescence microscopes, Biophys. J. 81 (2001) 2395–2402.
[46] M. Elangovan, et al., Characterization of one- and two-photon excitation energy transfer microscopy, Methods 29 (2003) 58–73.
[47] Y. Chen, J.D. Mills, A. Periasamy, Protein interactions in cells and tissues using FLIM and FRET, Differentiation 71 (2003) 528–541.
[48] Y. Chen, M. Elangovan, A. Periasamy, FRET data analysis-the algorithm, in: A. Periasamy, R.N. Day (Eds.), Molecular Imaging: FRET Microscopy and Spectroscopy, Oxford University Press, 2005.
[49] Y. Sun, A. Periasamy, Additional correction for energy transfer efficiency calculation in filter-based Förster resonance energy transfer microscopy for more accurate result, J. Biomed. Opt. 15 (2) (2010) 020513 (1–3).
[50] J.R. Lakowicz, Principles of Fluorescence Spectroscopy, Springer, New York, 2006.
[51] M.A. Digman, et al., The phasor approach to fluorescence lifetime imaging analysis, Biophys. J. 94 (2) (2008) L14–L16.
[52] Y. Sun, R.N. Day, A. Periasamy, Investigating protein-protein interactions in living cells using fluorescence lifetime imaging microscopy, Nat. Protoc. 6 (2011) 1324–1340.
[53] A. Periasamy, R.M. Clegg, FLIM applications in the biomedical sciences, in: A. Periasamy, R.M. Clegg (Eds.), FLIM Microscopy in Biology and Medicine, CRC Press, London, 2009.
[54] T.W.J. Gadella (Ed.), FRET and FLIM Techniques, Elsevier, Oxford, 2009.
[55] H. Wallrabe, A. Periasamy, Imaging protein molecules using FRET and FLIM microscopy, Curr. Opin. Biotechnol. 16 (2005) 19–27.
[56] A. Leray, et al., Three-dimensional polar representation for multispectral fluorescence lifetime imaging microscopy, Cytometry A 75 (2009) 1007–1014.

[57] Y. Sun, et al., Förster resonance energy transfer microscopy and spectroscopy to localize protein-protein interactions in live cells, Cytometry A 83A (9) (2013) 780–793.
[58] A.B. Houtsmuller, Fluorescence recovery after photobleaching: application to nuclear proteins, Adv. Biochem. Eng. Biotechnol. 95 (2005) 177–199.
[59] A.B. Houtsmuller, et al., Action of DNA repair endonuclease ERCC1/XPF in living cells, Science 284 (1999) 958–961.
[60] G. Carrero, et al., Using FRAP and mathematical modeling to determine the in vivo kinetics of nuclear proteins, Methods 29 (1) (2003) 14–28.
[61] J.G. Blonk, et al., Fluorescence photobleaching recovery in the confocal scanning light microscope, J. Microsc. 169 (1993) 363–374.
[62] T.H. Ward, F. Brandizzi, Dynamics of proteins in Golgi membranes: comparisons between mammalian and plant cells highlighted by photobleaching techniques, Cell. Mol. Life Sci. 61 (2004) 172–185.
[63] T. Wazawa, M. Ueda, Total internal reflection fluorescence microscopy in single molecule nanobioscience, Adv. Biochem. Eng. Biotechnol. 95 (2005) 77–106.
[64] S. Hadjidemetriou, D. Toomre, J.S. Duncan, Segmentation and 3D reconstruction of microtubules in total internal reflection fluorescence microscopy (TIRFM), Med. Image Comput. Comput. Assist. Inter. 8 (1) (2005) 761–769.
[65] D. Toomre, D.J. Manstein, Lighting up the cell surface with evanescent wave microscopy, Trends Cell Biol. 11 (7) (2001) 298–303.
[66] M. Gosch, R. Rigler, Fluorescence correlation spectroscopy of molecular motions and kinetics, Adv. Drug Deliv. Rev. 57 (2005) 169–190.
[67] N. González Bardeci, et al., Dynamics of intracellular processes in live-cell systems unveiled by fluorescence correlation microscopy, IUBMB Life 69 (1) (2017) 8–15.

CHAPTER ELEVEN

Three-Dimensional Imaging

Fatima A. Merchant and Alberto Diaspro

11.1 Introduction

Cells and tissues are inherently three-dimensional (3D), and cellular activities occur in 3D space. Thus 3D imaging techniques are required to study them. Three-dimensional light microscopy offers a noninvasive, minimally destructive option for obtaining spatial and volumetric information about the structure and function of cells and tissues. The last three decades have seen the development of new methods for 3D imaging and a consequent increase in techniques for 3D image processing and analysis. In this chapter we describe commonly used methods for the acquisition of 3D microscopy data and related image processing and analysis algorithms.

11.2 Image Acquisition

The techniques available for 3D imaging include electron microscope tomography and optical sectioning light microscopy. An advantage of optical microscopy is that it is nondestructive, in that it does not involve any physical manipulation of the specimen, thus allowing imaging of intact cells and tissues. The focus of this book is image analysis in optical microscopy, and other microscopy methods, such as electron microscopy, are not covered.

The two basic approaches to 3D optical microscopy are far-field and near-field imaging. Far-field methods, such as conventional light microscopy, rely on light diffraction from specimens for the formation of images, and their resolution is limited by the wavelength of visible light. Near-field methods utilize a solid mechanical probe to examine the specimen surface and are not limited by the Abbé equation (see Chapter 2). The near-field optical microscope, scanning tunneling microscope, and the atomic force microscope all employ near-field optics. Near-field microscopy methods are slower than far-field techniques and pose the danger of mechanical damage to the specimen.

In this chapter we address the most popular and widely commercialized approaches to 3D microscopy. These include the far-field techniques of wide-field, confocal, and multiphoton microscopy. Other high-resolution 3D microscopy methods that are newly developed and not yet commercially available are only briefly discussed.

Microscope Image Processing
https://doi.org/10.1016/B978-0-12-821049-9.00009-5
Copyright © 2023 Elsevier Inc.
All rights reserved.

11.2.1 Wide-Field 3D Microscopy

Conventional wide-field light microscopy can be used to collect 3D information from a specimen in the form of a series of two-dimensional (2D) images taken at different focal planes. When used to collect 3D data, wide-field microscopy is referred to as *computational optical sectioning microscopy* or *deconvolution microscopy* [1]. The drawback of this method is that light emitted from planes above and below the in-focus plane is also captured in each optical section image because the entire specimen is illuminated. All of the transmitted or emitted light is detected by the camera. This reduces both lateral resolution and depth discrimination. Computational deconvolution, based on the physics of image formation and recording, can be used to remove the out-of-focus information from each 2D image in the series. This results in a set of section images that more nearly represent the specimen.

11.2.2 Confocal Microscopy

One of the most important developments in 3D imaging is confocal microscopy [2]. The advent of affordable computers in the mid-1970s pushed the development of digital image processing software, and the popular use of laser sources in the 1980s supported the confocal approach. This contributed to expanded use of 3D optical microscopy in research laboratories. The concept of confocality was initially implemented by Naora and colleagues to reduce background noise in a spectroscopic setup designed for fluorescence studies [3]. Later, Minsky [4] proposed and patented a confocal microscope in line with the concept extensively developed by Petran at Pilsen [5], Davidovits and Egger at Yale [6], Wilson and Sheppard at Oxford [7], and Brakenhoff et al. in Amsterdam [8].

Confocal imaging is a microscopy technique that provides resolution and depth discrimination beyond that of conventional wide-field microscopes. Confocal microscopy achieves a theoretical improvement in axial resolution of approximately $1/\sqrt{2}$ compared to conventional wide-field microscopy. The confocal microscope has three important features that provide its advantages over conventional microscopes. First, the lateral resolution can be as much as one and a half times better than that of a conventional microscope. Second, and most importantly, the confocal microscope can remove out-of-focus information and thus produce an image of a very thin section of a specimen. Third, because of the absence of out-of-focus information, much higher contrast images are obtained.

The confocal microscope, developed in 1957, uses point illumination and a pinhole in front of the detector to eliminate out-of-focus information. Only the light within the plane of focus is detected, so the image quality is much better than that of wide-field images. In contrast to wide-field microscopy, in most confocal microscopes the illumination light is focused to the smallest possible spot in the plane of focus, using a coherent light source such as a collimated laser beam.

Confocal microscopes are categorized into two major types depending on the imaging instrument design. The scanning confocal microscope scans the specimen by physically moving either the stage or the illumination beam. By contrast, the spinning-disk ("Nipkow disk") confocal microscope employs a stationary stage and light source. Generally speaking, confocal laser scanning microscopy yields better image quality, but the imaging frame rate is quite slow (<3 frames/s). In contrast, spinning-disk confocal microscopes can achieve imaging at video rates, which is desirable for dynamic observations. Detailed reviews of the concepts, advantages, and aberrations of confocal microscopy are available [9].

Although live cell imaging is possible with confocal microscopy, phototoxicity and photobleaching from repeated exposures to visible light limit its practical application. In conventional *one-photon* confocal fluorescence microscopy, the total excitation, which depends linearly on incident illumination intensity, is constant in each plane throughout the specimen. Confocal microscopy exposes the entire sample to high-energy photons every time an optical section is generated. Since fluorophores are bleached when excited, photobleaching occurs throughout the thickness of a sample when collecting a series of images. This limits not only the maximum time for image collection, but also the amount of time that living tissues can be observed because of photodamage by toxic by-products.

The integration of confocal imaging with additional fluorescence-related parameters, such as fluorescence lifetime (Chapter 10), allows accessing high-content data, thereby providing new details about the sample being studied. The advent of fast field programmable gate array (FPGA) acquisition technology, which can replace the commonly used time-correlated single-photon counting (TCSPC) approach, enables high efficiency in detecting the time of arrival of photons. Although time resolution cannot reach the performance of TCSPC systems [10], it is adequate for most microscopy applications. It allows simultaneous imaging modalities, such as intensity detection, average arrival time, gating up to 16 time-gates simultaneously, and a powerful lifetime-based component separation algorithm. The technical key to this lies in the use of fast FPGA-based acquisition hardware, photon-counting detectors, hybrid detectors, and a tunable white light laser illumination source [11].

11.2.3 Multiphoton Microscopy

In 3D imaging, photodamage and photobleaching can be minimized by using *multiphoton microscopy* optical sectioning, in which excitation is confined to the optical section being observed by the process of *two-photon absorption*. Two-photon excitation (2PE) microscopy is perhaps the most important development in fluorescence microscopy since the introduction of confocal imaging [12]. When Denk and colleagues demonstrated the capability of 2PE imaging in biology [12], it brought about a revolution in fluorescence microscopy, stimulating original research ranging from tracking individual molecules in living cells to the observation of entire organisms [13,14].

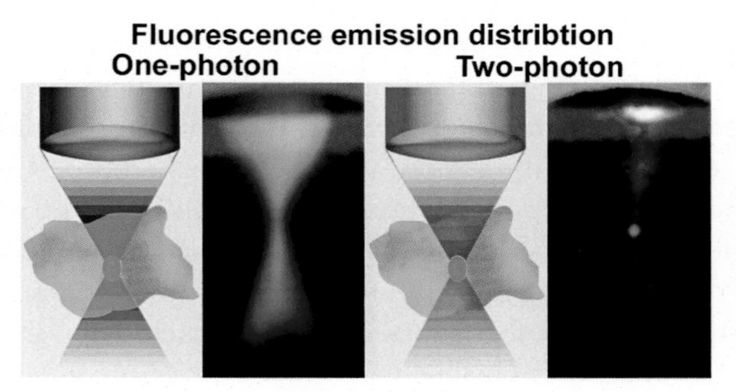

Fig. 11.1 Fluorescence emission distribution under 1PE *(blue)* and 2PE excitation *(red)*. *(Adapted from [15])*

Two-photon laser scanning microscopy is a nonlinear process that retains the optical sectioning ability of confocal microscopy while improving upon its ability to image live cells. It involves exciting the fluorophore by the simultaneous absorption of two photons. The combined effect of two low-energy photons is to raise the fluorophore from the ground state to the same excited state as would be caused by a single photon of half the wavelength (i.e., twice the energy). The fluorophore then relaxes back to the ground state, emitting the absorbed energy as fluorescence, just as if it had been excited by a single photon. Unlike linear fluorescence excitation, the emission wavelength is shorter than the excitation wavelength.

Fig. 11.1 shows the spatial distribution of fluorescence emission obtained by two-photon selection of the excited photon. The main difference between two-photon excitation and confocal microscopy is that in the former case none of the emitted photons upon nonlinear excitation are lost. This fact is self-evident when considering the parameters governing the process, which can be summarized in the following relationship that defines the probability, n_a, that a certain fluorophore simultaneously absorbs two photons during a single excitation pulse, in the paraxial approximation [12,15]:

$$n_a = \frac{(\delta_2 \cdot P_{ave}^2)}{\tau_p \cdot f_p^2} \left(\frac{NA^2}{2hc\lambda} \right)^2 \tag{11.1}$$

where P_{ave} is the average power of the illumination beam, δ_2 is the two-photon cross section of the fluorescent molecule, and λ is the excitation wavelength. Introducing 1 GM (Goppert-Mayer) = 10^{-58} [m^4 s], for a δ_2 of approximately 10 GM, focusing through an objective, NA >1, an average incident laser power of ≈1–50 mW, operating at a wavelength ranging from 680 to 1100 nm with 80–150 fs pulse width and 80–100 MHz repetition rate, one should get fluorescence without saturation. The repetition time of pulses is on the order of the typical excited-state lifetime, which is a few nanoseconds for commonly used fluorescent molecules. For this reason, the typical repetition rate is around 100 MHz, i.e., one order of magnitude slower than typical fluorescence lifetime. During the pulse time (10^{-13} s of duration for a typical lifetime in the 10^{-9} s range) the molecule has no time to relax to the ground state, which is a prerequisite for absorption of another photon pair. Therefore, whenever n_a approaches unity, saturation effects start occurring. Such a consideration allows one to optimize optical and laser parameters in order to maximize the probability of the two-photon excitation process [15].

Because a fluorophore must absorb two photons for excitation, fluorescence depends on the square of the incident beam intensity. Moreover, the intensity of the exciting light falls off as $1/z^2$ (where z is the distance from the focal plane) above and below the focal plane. Thus, the probability of exciting a fluorophore falls off as $1/z^4$. This highly nonlinear behavior limits the excitation to a high-intensity region near the focal plane of the focused laser beam (Fig. 11.1). The extremely high intensities near the focal plane confine 80% of the fluorescence excitation to a 10^{-10} μl volume when a high numerical aperture (*NA*) objective is used. This process uniquely localizes the excitation to the diffraction-limited spot of the focused beam, giving rise to the intrinsic optical sectioning ability of two-photon microscopy. Its high 3D resolution is due to the confinement of absorption, and consequently excitation, to the focal volume. Therefore out-of-focus photobleaching and photodamage and the attenuation of the excitation beam by out-of-focus absorption are avoided. Photodamage at the focal plane does occur, as with conventional confocal microscopy, but damage above and below the plane of focus is greatly reduced [16].

Two-photon imaging is also useful with ultraviolet excitable dyes in live cells, as the excitation is achieved with infrared light and the cells are never exposed to the more damaging ultraviolet excitation. Moreover, the infrared excitation used in two-photon imaging penetrates into tissue more efficiently than shorter wavelengths, allowing for imaging of thicker specimens. Since the probability of absorbing two photons depends on the square of the illumination intensity, infrared lasers that compress all of their output into very short ($\sim 10^{-13}$ s) high energy pulses (~2 kW) are used. It is possible to produce these very short, intense light pulses with a mode-locked laser light source. Mode-locked lasers generate pulses relatively far apart ($\sim 10^{-8}$ s), so that a peak power in the kilowatt range is reduced to a mean power of only a few tens of milliwatts at the specimen, and these moderate mean power levels do not damage the specimen. The laser most commonly employed to date is a tunable titanium-doped sapphire (Ti:sapphire) laser that operates in the 690–1000 nm range. This allows two-photon excitation of fluorophores that are normally excited by single ultraviolet, blue or green light photons.

The theoretical resolution of two-photon microscopy is typically up to 1.3 times better (smaller resolution) than that of conventional fluorescence microscopy because of the longer excitation wavelength used. When maximum resolution is required, two-photon microscopy is often coupled with confocal detection (by a pinhole) [17], or two-beam interference illumination is combined with confocal detection [18].

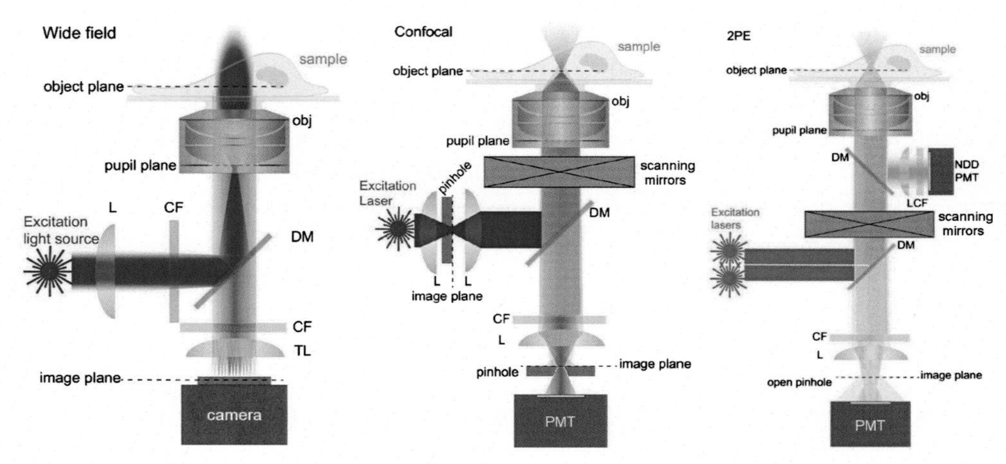

Fig. 11.2 Wide-field (left), confocal (center), and 2-photon (right) imaging configurations.

11.2.4 Microscope Configuration

Optical sectioning using a classical wide-field optical microscope, can be implemented under defocused conditions realized by moving the focus of the lens or the sample along the optical axis. Fig. 11.2 shows the microscope configuration for wide-field (left), confocal (center), and 2-photon (right) imaging. In a wide-field microscope, the entire specimen is bathed in light from a mercury or xenon source, and the image is projected directly onto an image-capture device. Fluorescence emitted by the specimen from out-of-focus regions often interferes with the resolution of those features that are in focus. In the confocal microscopy, light rays coming from out-of-focus positions are blocked at the photodetector by means of point-by-point scanning through pinholes in both the illumination and detection pathways. One characteristic of 2PE is that there is no need of rescan since photons can come only and uniquely from the focal region. For this reason, one can select either a descanned or nondescanned solution for collecting the emitted photons. The nondescanned approach allows maximizing the number of photons collected and is particularly useful when dealing with strongly scattering samples in the range of 50 μm to 1.3 mm thick [19].

11.2.5 Other 3D Microscopy Techniques

Recent years have seen the development of several other methods for 3D microscopy. Optical coherence tomography (OCT) is a 3D microscopy technique wherein axial resolution is improved by interferometric measurement of the time-of-flight of short-coherence light [20]. Typically, a Michelson-type interferometer is illuminated by a femtosecond laser pulse or superluminescent light emitting diode (LED) light, and the reference arm is dithered to generate a heterodyne signal by the interference with the back-scattered light from the sample point. The 3D image is constructed by mechanical scanning over the sample volume.

Another technique is optical projection tomography (OPT), which is based on the transmission and detection of visible light through a 360-degree step-rotated specimen. Data are collected at each angular position, and the 3D image is reconstructed by a back-projection algorithm. OPT has the advantage over confocal microscopy and OCT of increased depth penetration [21]. An alternative is multiple imaging axis microscopy, where wide-field images (2D projections) are taken of a stationary sample using several objective lenses mounted at different angles, and the recorded images are then combined to reconstruct a 3D image [22].

Another noteworthy approach for removing undesired information from optically sectioned images of a thick sample (>50 μm) is the light-sheet based microscope, also known as the single-plane illumination microscope (SPIM) [23]. Here the sample is illuminated, rotated about a vertical axis, and observed in a direction perpendicular to the illumination plane. An advantage of SPIM is that only those parts of the sample that are being observed are illuminated, so out-of-focus light is not generated. The sample is attached to a stage that can be rotated and translated. This allows 3D data stacks to be recorded along different directions. SPIM complements wide-field, confocal, and two-photon fluorescence microscopy, but is comparatively fast and reduces the effects of photobleaching and phototoxicity, while facilitating 3D superresolution of thick objects [24].

Other methods that offer improved 3D resolution include interference and structured illumination methods such as (1) 4Pi-confocal [18], (2) I^nM (incoherent interference illumination microscopy) [25], and (3) HELM (harmonic excitation light microscopy) [26]. In interference microscopy, two or more light sources are used to generate a periodic pattern of light at the sample plane. When these patterns are used to excite fluorescence they interact with the sample structure, and the recorded emission carries higher resolution information than what can be achieved by conventional microscopy. Typically, two objective lenses are used at the front and rear of the sample, and the emitted fluorescence is collected from both objectives and combined to interfere at the image plane. Structured illumination microscopy is covered in Chapter 12.

11.3 3D Image Data

Three-dimensional data obtained from microscopes consists of a stack of optical sections, referred to as the "z-series" (see Fig. 11.3). The optical sections are obtained at fixed intervals along the z-axis. Each 2D image is called an optical slice or optical section, and all the slices together comprise a volume dataset. Building up the z-series in depth allows the 3D specimen to be reconstructed.

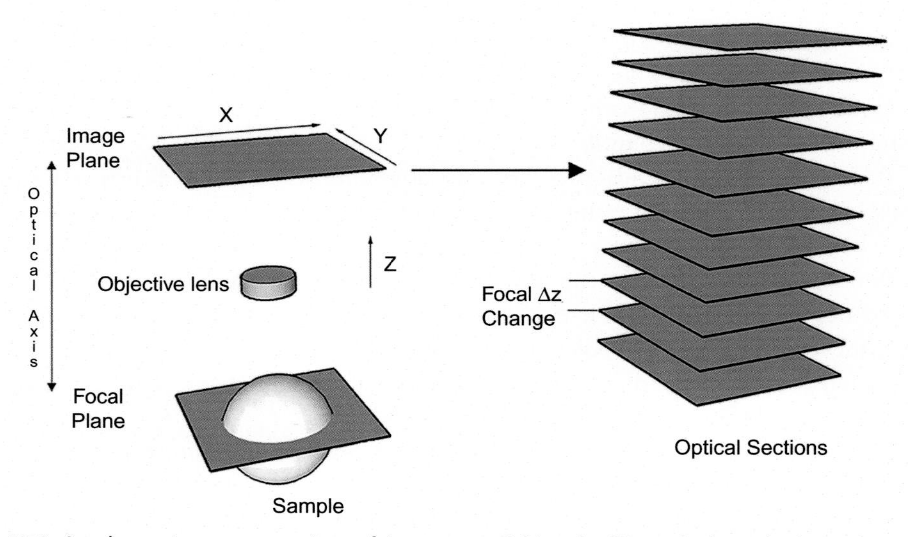

Fig. 11.3 A schematic representation of image acquisition in 3D optical sectioning microscopy, wherein the focal plane moves down through the specimen, and images are recorded at equally spaced levels along the *z*-axis.

11.3.1 3D Image Representation

For image processing and analysis purposes, a 3D digital image is represented as a 3D array whose elements are called "volume elements" or "voxels." Other terms used to refer to 3D images include 3D dataset, 3D volume, volumetric image, and *z*-stack. Many 3D image processing techniques are simply extensions of the corresponding 2D image processing algorithms. A few of the truly 3D algorithms that operate on volume datasets are discussed.

11.3.1.1 3D Image Notation

A 3D image, $f(x, y, z)$, is represented as a 3D matrix of dimensions $L \times M \times N$, where x, y, z, denote column, row, and slice coordinates, respectively. Each voxel has a physical size $D_x \times D_y \times D_z$ in units that are typically millimeters or micrometers. Ideally, a voxel is a cube with the same dimension in depth as in lateral spacing, but in practice (especially in 3D microscopy), this is rarely achieved, and the z-dimension is much thicker. In the following sections, we describe the methods for deblurring and restoration of images obtained by optical sectioning. Then we describe some 3D image processing, analysis, and display techniques.

11.4 Image Restoration and Deblurring

The aim of image restoration is to bring an image toward what it would have been if it had been recorded without degradation by the microscope imaging process and by noise. In other words, the goal is to recover the function $f(x, y, z)$ from a series of images $g(x, y, z)$ taken at different focal plane levels z. While this approach faces theoretical limitations, it can be done well enough to make it an important tool in biological research, particularly in microscopy. Optical microscopy had its key transition to 3D imaging when Agard and Sedat published their paper dealing with the 3D topography of an intact nucleus using fluorescently stained *Drosophila* polytene chromosomes [27]. Their approach was an extension of one earlier proposed by Castleman [28,29], and it led to the development of other computational approaches based on processing optical slices of biological samples.

Typically, restoration encompasses the processes of noise removal and image deblurring via deconvolution. These topics are treated for 2D images in Chapter 5. The primary task of 3D deblurring is to reduce the out-of-focus haze arising from objects located above and below the plane of focus. In order to achieve this, the deconvolution process estimates the amount of out-of-focus light characterizing the particular microscope optics in use, and then attempts to either subtract it out or redistribute it back to its point of origin in the specimen volume.

11.4.1 The Point Spread Function

The imaging process can be characterized by the 3D *point spread function* (psf), which is the 3D image of a point source of light. The psf is the basic function that describes the microscope's ability to form an image (see Chapter 2). For example, single fluorescent molecules in a specimen are essentially point sources of light. A copy of the 3D psf is produced in the image by each of these point sources, and the 3D summation of these copies forms the 3D image. Thus the out-of-focus light in the image arises from the summed contributions of many psfs. This process can be modeled mathematically by the convolution operation, whereby light emitted from out-of-focus points in the specimen is convolved with the defocus psf.

Ideally the psf would be a single point in 3D space, and each point in the specimen would appear as a single point in the 3D image. But the actual psf is not a point. Therefore blurring by the psf causes the image of a point object in the specimen to be expanded to fill a small space in the 3D image. It appears not as a point but as a blurred region in the image. Deconvolution can be used to reverse this process and deblur the image. It is a process that reduces the size of the image of a point source in the specimen. Typically, the 2D images in a stack of optical sections can be processed individually to deblur each optical slice in the series. The 3D psf is considered to be a collection of 2D psfs that vary in size and shape with distance along the optical axis. Thus the 3D deblurring process can be implemented either as a single 3D deconvolution or as a series of 2D deconvolutions, one for each recorded image.

The first step in any 3D deconvolution algorithm design is to determine the 3D psf. Three methods are commonly used for estimating the psf of a microscope: experimental, analytical, and theoretical. Experimentally, images of one or more point like objects (typically fluorescent beads smaller than the resolution of the microscope) are collected and used to estimate the psf [30]. The recommended bead diameter is one-third of the Rayleigh resolution value [31], or $0.41\lambda/NA$ (Chapter 2). For an $NA = 1.4$ objective operating with green fluorescence (500 nm), a bead diameter of 0.15 μm is recommended. It is important to emulate the microscopy conditions of the planned experiment as closely as possible. For example, the same mounting medium should be used, and the optical system should have the same objective lens, relay lenses, and detectors. Typically, an experimentally measured psf closely matches the actual psf for that experimental setup. An experimentally measured psf, however, may exhibit low signal-to-noise ratio (SNR) with symmetric features in radial planes, but asymmetry along the axial direction due to spherical aberration. Typically, high-quality optical components that are precisely aligned produce empirical psfs that are symmetrical.

The second method for determining the psf is analytical. In this case, the psf is determined by computationally extracting it directly from recorded 3D image data. Section 11.4.3.6 discusses this being done by blind deconvolution.

11.4.1.1 Theoretical Model of the Psf

In the theoretical method, the psf is computed using a mathematical model from diffraction theory (see Eq. 11.2). A theoretically determined psf will have axial and radial symmetry. Fig. 11.4 presents cross sections through theoretically determined psfs for wide-field, confocal, and 2PE microscope systems [15]. Both 2PE and confocal shapes exhibit a better signal-to-noise ratio than the wide-field case. 2PE distribution is larger due to the fact that a wavelength twice that in the wide-field and confocal case is responsible for the intensity distribution. Optical conditions are as follows: excitation wavelengths are 488 nm and 900 nm for 1PE and 2PE, respectively; emission wavelength is 520 nm; numerical aperture is 1.3 for an oil immersion objective with oil refractive index value set at 1.515. As seen in the figure, the psf for the wide-field system, which exhibits an Airy disk pattern, clearly shows the central disk and side lobes. The confocal psf, which is equal to the square of the Airy disk, exhibits a narrower central disk with very weak side-lobes.

The psf can be calculated using a diffraction-based theoretical model [32], as

$$h(x, y, z; \boldsymbol{\psi}) = \left| \int_0^1 J_0 \left(\frac{K N_a \rho \sqrt{x^2 + y^2}}{M} \right) exp\{ jK\varphi(z, \boldsymbol{\psi})\} \ \rho \ d\rho \right|^2 \tag{11.2}$$

where J_0 is the Bessel function of the first kind, $K = 1/\lambda$ is the wave number, N_a is the numerical aperture of the microscope, M is the magnification of the lens, and $j^2 = -1$ (see also Section 2.4.1). The vector $\boldsymbol{\psi}$ denotes the measurement setup parameters (i.e., the

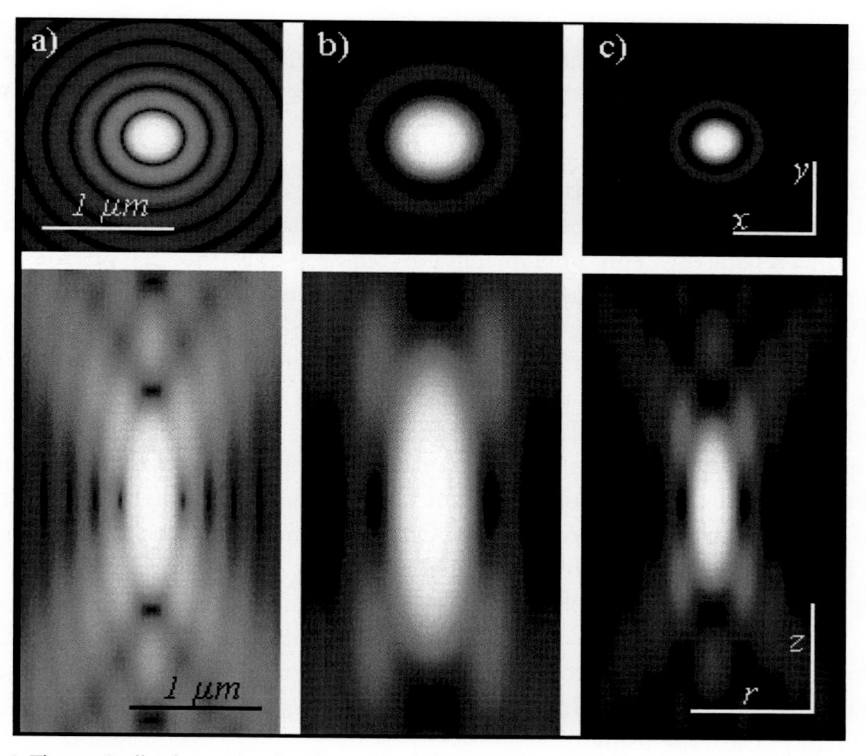

Fig. 11.4 Theoretically determined point spread function for wide-field, confocal, and two-photon microscopy. *(Reproduced with permission from [15])*

thickness and refractive index of the immersion oil, the coverslip, the specimen, and the tube length of the microscope), and ρ denotes normalized radius in the back focal plane. The distance from the in-focus plane to the point of evaluation is z, and $\varphi(\cdot)$ denotes the phase aberration. When more than one or two physical parameters are unknown in Eq. (11.2), the psf can be modeled in terms of purely mathematical parameters by replacing the phase term $exp\{jK\varphi(z,\boldsymbol{\psi})\}$ with $A(\rho)\,exp\{jW(\rho,z)\}$, where [32]

$$W(\rho,z) = zC(\rho)+B(\rho)$$

$$A(\rho) = \sum_{k=0}^{k_a} a_k \chi_k(\rho)$$

$$B(\rho) = \sum_{k=0}^{k_b} b_k \vartheta_k(\rho)$$

$$C(\rho) = \sum_{k=0}^{k_c} c_k v_k(\rho)$$

The unknown parameters a_k, b_k, c_k, and $\chi_k(\cdot)$, $\vartheta_k(\cdot)$, $\nu_k(\cdot)$, are suitably chosen bases that need not be the same. It is possible to represent $A(\rho)$, $B(\rho)$, and $C(\rho)$ either as power series expansions or in terms of radial Zernike polynomials [33]. Tutorials on determining the psf are available [34,35].

11.4.1.2 Approximate Methods

Optical section deblurring can be done as a 2D convolution operation by treating each image plane separately. This implementation requires only the 2D psf for various amounts of defocus. It is often useful to specify the resolution of a microscope by using the 2D optical transfer function (OTF) (Section 2.4.2). The OTF is a property of the optical system alone, and using it facilitates comparing different optical systems in terms of their resolution. The OTF is the Fourier transform of the psf, and it is also the autocorrelation function of the pupil function of the imaging lens. Once calculated or measured, it can be used to model the imaging process in the frequency domain. Further, convolution in the spatial domain is equivalent to multiplication in the frequency domain. These relationships among the psf, OTF, pupil function, and Fourier transform greatly facilitate the analysis of imaging systems.

Hopkins [36] derived an expression for the image plane OTF of a defocused optical system. Stokseth [37] developed a simplified approximation to that expression, and Castleman [28,29] modified Stokseth's approximation to make it more accurate for the small values of defocus normally encountered in optical section deblurring. This approximation takes the form

$$H(w, q) \approx \frac{1}{\pi}[2\beta - \sin(2\beta)]\, jinc\left[4k\,w\left(1 - \frac{|q|}{f_c}\right)\frac{q}{f_c}\right] \tag{11.3}$$

where q is spatial frequency in the radial direction, λ is the wavelength of the narrowband incoherent illumination, $k=1/\lambda$, and f_c is the optical cutoff frequency, $2NA/\lambda$. The amount of defocus is specified by w, the maximum path length error (path length difference between axial and marginal rays). It is given by

$$w = -d_i - \delta z \cos\alpha + \left(d_i^2 + 2d_i\delta z + \delta z^2 \cos^2\alpha\right)^{\frac{1}{2}} \quad \alpha = \arctan\frac{A}{d_i} \tag{11.4}$$

where d_i is the distance from the lens to the image (see Fig. 2.1), δz is the defocus distance (the distance from the focal plane to the current specimen plane), A is the aperture diameter, and

$$jinc(x) = 2\frac{J_1(x)}{x} \quad \beta = \cos^{-1}\frac{q}{f_c} \quad q^2 = u^2 + v^2 \quad f_c = \frac{2NA}{\lambda} \tag{11.5}$$

Here f_c is the optical cutoff frequency in incoherent light, and q, u, and v are frequency variables in the radial, x, and y directions, respectively. The numerical aperture is

$NA = n \cdot \sin(\alpha)$, where α is the acceptance semiangle of the objective lens and n is the refractive index of the medium of the specimen.

When deblurring optical sections, we are primarily interested in the adjacent planes, where defocus is relatively small. This approximation is accurate at zero defocus and differs less than 1% from Stokseth's approximation [37] at $w = 5\lambda$. The circularly symmetric OTF can be inverse Fourier transformed to produce the defocus psf required for 2D optical section deblurring (Chapter 5).

11.4.2 Models for Microscope Image Formation

Most microscopes can be assumed to be shift invariant linear systems with a position-independent psf, where point sources located anywhere in the sample create psfs of constant shape at the detector. In its simplest form, then, the formula for microscope image formation incorporates two known quantities, namely the psf, $h(x, y, z)$, and the recorded 3D image, $g(x, y, z)$, and one unknown quantity, the actual distribution of light in the 3D specimen, $f(x, y, z)$. These terms are related by the imaging equation

$$g(x, y, z) = f(x, y, z) * h(x, y, z) \tag{11.6}$$

where * indicates 3D convolution. In practice, however, image recording is corrupted by intrinsic and extrinsic noise. Intrinsic noise obeys a Poisson model, and is introduced when each photon hits the detector, thus creating a random number of photoelectrons. Other sources introduce random extrinsic noise that can be modeled as additive Gaussian noise. Mathematical models for 3D fluorescence microscopy imaging based on the Poisson and Gaussian noise statistics are commonly used [32].

11.4.2.1 Poisson Noise

Poisson modeling [32] of both the signal emitted by the specimen and the background noise can be represented mathematically as follows:

$$og(x, y, z) = P(o[f(x, y, z) * h(x, y, z)]) + P(o[b(x, y, z)]) \quad x, y, z \in R \tag{11.7}$$

where o is the reciprocal of the photon-conversion factor, $og(x, y, z)$ is the number of photons measured in the detector, P is a Poisson process, and $b(x, y, z)$ is the background noise. For fluorescence microscopy, the photon-conversion factor is proportional to several physical parameters, such as the integration time and the quantum efficiency of the detector. Both $P(o[f(x, y, z) * h(x, y, z)])$ and $P(o[b(x, y, z)])$ are independent Poisson random variables and, hence, the measured output is a Poisson random variable.

11.4.2.2 Gaussian Noise

The Gaussian noise model [32] is given by

$$g(x, y, z) = f(x, y, z) * h(x, y, z) + w(x, y, z) \quad x, y, z \in R \tag{11.8}$$

where $w(x, y, z)$ represents the additive Gaussian noise. A background term is not included here because it can be estimated and removed from the acquired image before starting the deconvolution operation. For the Poisson noise model, the background term cannot be incorporated in a term that is independent of $(f(x, y, z) \otimes h(x, y, z))$ and thus must be explicitly defined. As explained earlier, deconvolution recovers an estimate of $f(x, y, z)$ from $g(x, y, z)$, given a knowledge of $h(x, y, z)$ and $w(x, y, z)$. The following section describes some of the algorithms commonly used for implementing deconvolution.

11.4.3 Algorithms for Deblurring and Restoration

Deconvolution algorithms can be divided into two classes, "deblurring" and "image restoration," based on their mode of implementation (2D or 3D). Image *restoration* algorithms operate simultaneously on every voxel in a 3D image stack and are truly implemented in 3D to restore the actual distribution of light in the specimen by determining $f(x, y, z)$. In contrast, algorithms that are applied sequentially to each 2D plane of a 3D image stack, one at a time, are classified as *deblurring* procedures. Appropriately, they are not actual deconvolution methods because they do not use the 3D imaging formula (Eq. 11.6). In this chapter, the term deconvolution is used to refer to both optical section deblurring and image restoration.

Different deconvolution methods solve for $f(x, y, z)$ in different ways, and six major categories of deconvolution algorithms are discussed here. These approaches differ in the type of mathematical model used to emulate image formation and recording and in the underlying assumptions used in the simulation to reduce complexity. This section describes the following approaches to deconvolution: (1) the no-neighbors methods, (2) neighboring plane methods, (3) linear methods, (4) nonlinear methods, (5) statistical methods, and (6) blind deconvolution. Prior to deconvolution, the 3D image data recorded from a microscope should be preprocessed (or corrected) to remove background inhomogeneities such as those due to spatially nonuniform illumination, uneven sensitivity of the detector, and intensity attenuation with depth (Chapter 10). Other image-preprocessing algorithms and enhancement methods are described in Chapter 5.

11.4.3.1 No-Neighbor Methods

The no-neighbor processing scheme is a 2D method for deblurring an individual section through the object from a single image [38]. It is based on the principle that in-focus structures in the image are "sharper" than out-of-focus ones, which tend to blur or flatten out. While sharpness is typically represented by higher spatial frequencies, the out-of-focus components tend to be composed of lower spatial frequencies (i.e., the light intensity varies more slowly over the field-of-view). The idea is that attenuating lower spatial frequencies will remove mainly out-of-focus structures and leave behind the in-focus objects of interest. Retaining the higher frequencies also tends to improve the picture by sharpening edges of structures. This approach, then, is simply a sharpening (high-pass) filter (Chapter 5) that can be implemented by convolution. This approach is applicable for specimens that tend to be composed mostly of small structures such as point-like objects or filaments [38]. A significant advantage of this approach is its computational simplicity and the resulting speed of processing. A disadvantage is that most specimens are actually a complex mixture of low and high spatial frequency content, and there is the danger of filtering out components of interest.

11.4.3.2 Nearest-Neighbor Method

This method also falls under the category of deblurring algorithms, which were first introduced in 1971 for deblurring light microscope image stacks [39]. Each 2D image in a 3D stack contains information from the corresponding in-focus specimen plane plus a sum of defocused adjacent specimen planes. The diffraction-limited optical transfer function (Chapter 2) tends to discriminate against high spatial frequencies but passes low-frequency content. It is thus possible to partially remove defocused structures by subtracting adjacent plane images that have been blurred by convolution with the appropriate defocus psf. In the nearest-neighbor method only two immediately adjacent planes are used, one above and the other below the plane of focus. While the image taken at any one focal plane actually contains out-of-focus light from all specimen planes, the nearest-neighbor method assumes that the strongest contributions come from the nearest two adjacent planes. For example, as shown in Eq. (11.9), the nearest-neighbor algorithm operates on an image at plane k by blurring its neighboring plane images $k \pm 1$, using a digital blurring filter, and then subtracting the blurred plane images from image k [29].

$$\hat{f}_k(x, y) = g_k(x, y) - c[g_{k-1}(x, y) * h_{k-1}(x, y) + g_{k+1}(x, y) * h_{k+1}(x, y)] \tag{11.9}$$

where the subscript k indicates the optical section slice number. One 2D image is sharpened at a time, with the neighboring images immediately above and below it blurred, and a fraction c of them subtracted out. Fig. 11.5 shows the effect of deblurring using the nearest-neighbor method [40]. Panels a–c show three optical slices taken from

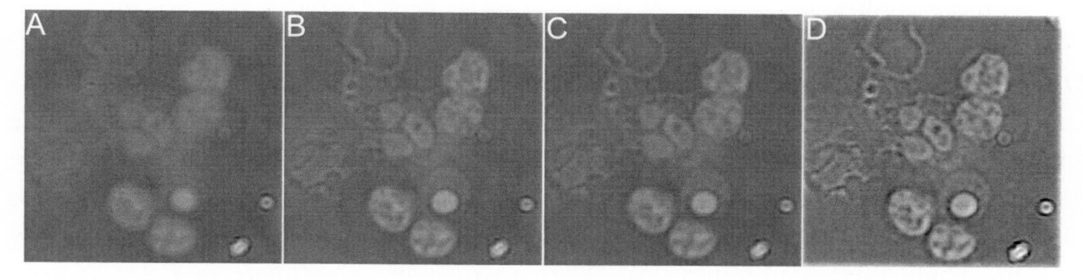

Fig. 11.5 Deblurring using the nearest-neighbor method. Panels a–c show three optical slices taken from May-Giemsa stained blood cells. The images were obtained at a z-interval of 0.5 μm. The optical section in (b) was deblurred, using neighboring slices, immediately above (a), and below (c). The resulting deblurred image is shown in (d). *(Reproduced with permission from [40])*

May-Giemsa stained blood cells. The images were obtained at a z-interval of 0.5 μm. The centermost slice (Panel b) was deblurred, using neighboring slices, immediately above (Panel a), and below (Panel c). The deblurred image is shown in Panel d.

Multineighbor methods extend this concept to a user-selectable number of planes. The entire 3D stack is processed by applying the algorithm to each image in the stack to remove blurring from that plane. Not only does increasing the number of planes used for deblurring increase the computational load of the technique, but the amount of improvement falls off rapidly as more planes are used. The deblurred images can be further sharpened with a Wiener deconvolution filter (Chapter 5) to reduce the remaining blur and noise. After processing, the content of each image corresponds predominantly to the in-focus information from the corresponding specimen plane.

Nearest-neighbor methods work best when the amount of blur from one plane to the next is significant, that is, when the planes are relatively far apart. They are most applicable to specimens containing sparse structures distributed through a transparent or nonfluorescing tissue, such as isolated cells or thin filaments. Since the algorithms involve relatively simple calculations on single image planes, they are useful in situations when quick image sharpening is needed and computing power is limited. The advantages of this approach include computational speed, improved contrast, and sharpening of features in each optical slice.

There are also several disadvantages to this approach. These methods are sensitive to noise and may produce noisier images, since noise from several planes tends to accumulate. They also tend to introduce structural artifacts because each optical slice can contain diffraction rings or light from other structures that may be sharpened as if it were in that focal plane. Contrast is also reduced by the deblurring process, as intensities are removed rather than redistributed. Therefore, it is better to use these methods for visualization, rather than prior to quantitative analysis, such as morphometric measurements, quantitative fluorescence measurements, and intensity ratio calculations. For specimens containing structures that fluoresce over large areas or volumes these simple methods do not perform well, and some of the approaches described in the following sections may be more applicable.

11.4.3.3 Linear Methods

Linear methods fall under the category of image restoration techniques and represent the simplest of the true 3D methods for deconvolution. With linear methods, the deconvolution algorithm is implemented in the frequency domain. This is useful because convolution in the spatial domain corresponds to multiplication in the frequency domain. These methods employ Eq. (11.6) to recover an estimate of $f(x, y, z)$ from $g(x, y, z)$, and require knowledge of the 3D psf. Transforming Eq. (11.6) into the frequency domain yields

$$G(u, v, w) = F(u, v, w)H(u, v, w) \tag{11.10}$$

where u, v, and w are frequency variables in the x-, y- and z-directions, respectively. The spectrum of the specimen function is

$$F(u, v, w) = G(u, v, w)H'(u, v, w) \tag{11.11}$$

where $H'(u, v, w)$ is the inverse 3D optical transfer function (OTF), given by

$$H'(u, v, w) = \frac{1}{H(u, v, w)} \tag{11.12}$$

The OTF, $H(u, v, w)$, which is the Fourier transform of the psf of the microscope, describes mathematically how the system treats periodic structures (Chapter 2). Essentially, the OTF drops off at higher frequencies and goes to zero at, $f_c = 2NA/\lambda$, the optical cutoff frequency. Frequencies above the cutoff are not recorded in the microscope image. As seen in Eqs. (11.6)–(11.12), division in the Fourier domain is equivalent to deconvolution in the spatial domain. This inverse filtering is the simplest way to remove the effects of the psf, and it forms the basis of all linear methods of deconvolution. Most linear methods thus involve Fourier transforming an image and then dividing by the Fourier transform of the psf.

In linear methods noise in the measured data becomes a problem, particularly at frequencies where the denominator $H(u, v, w)$ is small, or even zero. As discussed in Chapter 2, the OTF is almost always small at high spatial frequencies. Further, the specimen typically contains little energy at the high frequencies, so noise will usually dominate the upper part of the frequency spectrum. To address this problem, linear deconvolution methods often adopt some noise reduction strategy. The goal is to strike a balance by reducing the contribution of high spatial frequency noise while retaining sharpness in the image. Several algorithms for linear deconvolution have been described, including (1) inverse filtering, (2) Wiener deconvolution, (3) regularized least squares, (4) linear least squares restoration, and (5) Tikhonov-Miller regularization.

Inverse Filtering

An inverse filter uses an approximate direct linear inversion of the imaging equation, Eq. (11.11), as

$$\hat{f}(x, y, z) = F^{-1}\left[\frac{G(u, v, w)}{H(u, v, w)}\right] \tag{11.13}$$

The main drawback of this method is that $H(u, v, w)$ is usually a low-pass filter and therefore $1/H(u, v, w)$ is a high-pass filter that takes on large values at the higher frequencies. Thus Eq. (11.13) becomes numerically unstable for small values of $H(u, v, w)$, and this greatly increases the high-frequency noise contribution. This makes simple inverse filtering very sensitive to noise. One method to combat this is to limit the inverse OTF as follows [32]:

$$\hat{f}(x,y,z) = \begin{cases} F^{-1}\left\{\dfrac{G(u,v,w)}{H(u,v,w)}\right\} & \text{if} \quad |H(u,v,w)| \geq \varepsilon \\ 0 & \text{if} \quad |H(u,v,w)| \leq \varepsilon \end{cases} \tag{11.14}$$

where ε is a small positive constant. The choice of ε controls the tradeoff between resolution and the amount of noise in the resulting image estimate. For example, small ε results in sharper images with finer resolution, but more noise.

Another approach for reducing noise is to apply an adjustable smoothing operation. The inverse filtering algorithm is most useful when a known low-pass filter has blurred the image since it is theoretically possible to recover that image by inverse filtering. In microscopy, deconvolution algorithms use either a theoretical or a measured psf to remove image blurring rapidly and effectively. The results are usually qualitative and generally better than the no- and nearest-neighbor methods. However, as discussed previously, inverse filtering is very sensitive to additive noise. The Wiener filtering approach, described next, implements an optimal tradeoff between inverse filtering and noise smoothing.

Wiener Deconvolution

One of the most widely used image restoration techniques is Wiener deconvolution. Unlike simple inverse filtering, this method attempts to reduce noise while restoring the original signal. It implements a balance between inverse filtering and noise smoothing that is optimal in the mean-square-error (MSE) sense. Assuming the white Gaussian noise imaging situation described in Eq. (11.8), the orthogonality principle implies that the Wiener filter can be expressed in the Fourier domain as [29,32]

$$\hat{H}(u,v,w) = \frac{H^{*}(u,v,w)}{\left\{|H(u,v,w)|^2 + \left(\frac{P_w(u,v,w)}{P_f(u,v,w)}\right)\right\}} \tag{11.15}$$

where $\hat{H}(u,v,w)$ is the 3D Fourier transform of $\hat{h}(x,y,z)$, "*" is the complex conjugation operation, and $P_w(u,v,w)$ and $P_f(u,v,w)$ are the power spectral densities of the noise and the specimen, respectively (Chapter 5). Eq. (11.15) can be rewritten as

$$\hat{H}(u,v,w) = \frac{1}{H(u,v,w)}\left[\frac{|H(u,v,w)|^2}{|H(u,v,w)|^2 + \frac{1}{SNR(u,v,w)}}\right] \tag{11.16}$$

where

$$SNR(u,v,w) = \frac{P_f(u,v,w)}{P_w(u,v,w)}$$

Eq. (11.16) can be interpreted as two filters in cascade in the frequency domain, where $1/H(u, v, w)$ is the inverse filter, and the term in brackets is the Wiener filter. The term $SNR(u, v, w)$ is the signal-to-noise ratio as a function of frequency. In the absence of noise, (i.e., infinite SNR) the Wiener deconvolution filter reduces to the standard inverse filter. However, as noise is added, the SNR decreases, and the term in square brackets decreases as well. The Wiener filter attenuates certain frequencies according to their SNR. It not only performs deconvolution, which is a high-pass filtering operation, but it also reduces noise with low-pass filtering. Wiener deconvolution performs well in the presence of noise, but it has other problems that limit its effectiveness. First, the MSE criterion of optimality is not a particularly good one for human observation. It tends to smooth the image more than the eye would like. Second, it weighs all errors equally, regardless of their location in the image. Finally, classical Wiener deconvolution cannot handle a spatially variant microscope psf.

Linear Least Squares

The linear least squares (LLS) method is also used to restore images corrupted with additive white Gaussian noise. This approach is based on linear algebra. The Gaussian noise imaging model in Eq. (11.8) can be rewritten in discrete form as a matrix-vector equation

$$\boldsymbol{G} = \boldsymbol{HF} + \boldsymbol{W} \tag{11.17}$$

where **G** is the recorded image, **F** is the specimen object, **W** is the noise, and **H** is the matrix representing the psf. In this formulation **F**, **G**, and **W** are vectors formed by stacking the columns of the respective images. If $\mathbf{W}=0$, or if we know nothing about the noise, we can set up the restoration as a least squares minimization problem in the following way. We wish to select $\hat{\boldsymbol{F}}$ so that, if it is blurred by **H**, the result will differ from the observed image **G**, in the mean square sense, by as little as possible. Since **G** itself is simply **F** blurred by **H**, this is a satisfying approach. If **F** and $\hat{\boldsymbol{F}}$, both having been blurred by **H**, are nearly equal, then hopefully $\hat{\boldsymbol{F}}$ is a good approximation to **F**. This formulation is distinctly different from that used in the Wiener filter. There we sought to minimize the difference between the restored signal and the original. Here we are satisfied to minimize the difference between the blurred original and a similarly blurred estimate of the original. We cannot expect the results of these two formulations to be the same.

Let $e(\hat{\boldsymbol{F}})$ be a vector of residual errors that results from using $\hat{\boldsymbol{F}}$ as an approximation to $\boldsymbol{F}$. Eq. (11.17) then becomes

$$\boldsymbol{G} = \boldsymbol{HF} = \boldsymbol{H}\hat{\boldsymbol{F}} + e(\hat{\boldsymbol{F}}) \quad \text{or} \quad e(\hat{\boldsymbol{F}}) = \boldsymbol{G} - \boldsymbol{H}\hat{\boldsymbol{F}} \tag{11.18}$$

and we seek to minimize the function

$$\varphi(\hat{\boldsymbol{F}}) = \| e(\hat{\boldsymbol{F}}) \|^2 = \| \boldsymbol{G} - \boldsymbol{H}\hat{\boldsymbol{F}} \|^2 = (\boldsymbol{G} - \boldsymbol{H}\hat{\boldsymbol{F}})^T (\boldsymbol{G} - \boldsymbol{H}\hat{\boldsymbol{F}}) \tag{11.19}$$

where $\|a\| = \sqrt{a^T a}$ denotes the Euclidean norm of a vector, that is, the square root of the sum of the squares of its elements. Then, setting to zero the derivative of $\varphi(\hat{\boldsymbol{F}})$ with respect to $\hat{\boldsymbol{F}}$, and solving for $\hat{\boldsymbol{F}}$ gives [41]

$$\hat{\boldsymbol{F}} = \left[\boldsymbol{H}^T \boldsymbol{H}\right]^{-1} \boldsymbol{H}^T \mathbf{G} \tag{11.20}$$

where the matrix **H** can be constructed from the discrete psf and has a Toeplitz structure. The variance and the error are linked to the eigenvalues, which are used for the matrix inversion operation. The algorithm searches the optimal number of eigenvalues to be used in the inversion process by discarding the lowest eigenvalues. The estimate $\hat{\boldsymbol{F}}$ of the 3D image is obtained in a single pass. One drawback is that, since the OTF of a microscope gives rise to a singular or quasisingular matrix, the inversion of such a matrix system is an ill-posed problem [41].

Regularization

A "well-posed" estimation problem is one in which (1) a solution exists, (2) that solution is unique, and (3) it depends on the input data in a continuous fashion. The deconvolution of **G**, i.e., Eq. (11.17) is often an ill-posed problem. This means that a large, uncontrolled amplification of the noise can be expected, and the resulting solution of the inverse problem is useless. Regularization is used to find a useful solution. This is a procedure that seeks a solution that approaches the true input distribution as the amount of noise is reduced, and the image is sampled finely over a large volume [31]. It provides additional information for solving the ill-posed problem. For example, additional constraints upon the image to be restored are utilized, such as nonnegativity, statistical properties of the image, or even information about the degree of smoothness of the image. Mostly this results in a tradeoff between the smoothness of the reconstruction (with less noise) and the degree of the deblurring. A regularized filter typically imposes certain constraints on the estimate, allowing the algorithm to select the most reasonable estimate from the large number of solutions that might arise because of the noise. Moreover, the result is usually smoothed by the elimination of higher frequencies that are well beyond the resolution limit of the microscope [42].

Tikhonov Regularization Tikhonov regularization is the most used method for regularization of ill-posed problems. Here the same image distortion model is assumed as in the linear least squares method (discussed previously). The approach involves minimizing the Tikhonov functional, which is given by [42]

$$\varphi(\hat{\boldsymbol{F}}) = \left\|\boldsymbol{H}\hat{\boldsymbol{F}} - \mathbf{G}\right\|^2 + \lambda \left\|\boldsymbol{C}\hat{\boldsymbol{F}}\right\|^2 \tag{11.21}$$

where λ is the regularization parameter and **C** is the regularization matrix. The matrix **C** penalizes the solution of $\hat{\boldsymbol{F}}$ in the regions where it oscillates due to the noise. In matrix notation $\hat{\boldsymbol{F}}$ is given as:

$$\hat{\boldsymbol{F}} = \left(\boldsymbol{H}^T\boldsymbol{H} + \lambda\boldsymbol{C}^T\boldsymbol{C}\right)^{-1}\boldsymbol{H}^T\boldsymbol{G} \quad (11.22)$$

Regularization can be applied in one step within an inverse filter (discussed earlier), or it can be applied iteratively as discussed in the following section.

11.4.3.4 Nonlinear Methods

The linear methods described previously are quick to compute, but they have several drawbacks. These include the inability to incorporate prior knowledge about the true image, such as ringing artifacts that may be created near edges and negative intensities that might occur in the deconvolved image. Artifacts result from an inability to estimate the high-frequency components that are cut off by the diffraction-limited objective. Because these frequency components are not in the recorded image, it is difficult to obtain a correct estimate of the specimen. One method to address this problem is to use algorithms that incorporate a priori information about the specimen, such as non-negativity, finite support, and smoothness. This may be used to enforce regularization constraints that the specimen estimate must satisfy. Nonlinear methods are usually iterative, involving operations that are performed repetitively until certain criteria are satisfied. Thus these methods are also called "constrained iterative algorithms." The constraints minimize noise or other distortions, consequently improving the restoration of the blurred image. The drawback is that iterative methods require more computational time. Some of the most widely used nonlinear methods are described in the following sections. They provide a way to impose constraints on the restored image, after each iteration step, to avoid convergence to an infeasible solution.

Jansson-van Cittert Method

The Jansson-van Cittert (JVC) method of repeated convolution has been applied to digital microscope images [1,43]. It is an iterative spatial domain transform that generates successive approximations of the specimen image. The JVC method is most useful for image reconstruction in microscopy, where a consistent blurring function, (i.e., the psf) has degraded an image. The effects of blurring are attenuated using the following iterative process

$$\hat{f}_{k+1} = \hat{f}_k + r\left(g - h * \hat{f}_k\right) \quad (11.23)$$

where g is the recorded image (typically a version that has been smoothed to reduce noise), $\hat{f}_k$ is the kth iteration image estimate, and h is the microscope psf. The relaxation function, r, controls, during the iteration, voxel specific constraints and image convergence constraints. The method proceeds by repeatedly adding a high-pass filtered version, $\left(g - h * \hat{f}_k\right)$, scaled by r, of the current image iteration, k, to itself. Typically, r is a finite weight function that is defined over a positive intensity range. It is used to prevent unusually bright intensities in the estimated image and negative intensities, because a specimen cannot have negative fluorescence. Thus any voxel value in the estimate that becomes negative during the computation is automatically set to zero. Convergence is achieved when the difference between the image estimate at iteration k, $\hat{f}_k$, and the smoothed image, g, approaches zero.

A critical factor in reconstruction quality is the mitigation of noise. As iterations proceed, noise is amplified. Most implementations suppress this with a smoothing filter (e.g., Gaussian), which simultaneously attenuates both signal and noise. Residual structures are then amplified by a high-pass filter. Most often, the smoothing operation does not work well for low SNR images.

The Nonlinear Constrained Least Squares Method

In the nonlinear least squares (NLS) approach, the sum of the squared difference between the distorted recorded image and the estimated image is minimized. That is, the NLS method aims to iteratively find the specimen function $\hat{f}(x, y, z)$ that minimizes

$$\sum_{i,j,k} \left| g_{i,j,k} - \left[\hat{f} * h\right]_{i,j,k} \right|^2 \tag{11.24}$$

where i, j, and k are the voxel coordinates of the recorded 3D image, and the $*$ represents 3D convolution. In contrast to the linear least squares method, the function $\hat{f}(x, y, z)$ is refined iteratively until the MSE is minimized. High-frequency noise imposes the same challenges described previously. The recorded image may contain components that do not correspond to any (blurred) physically possible component of the specimen. This situation can force the iterative process to include artificial high-frequency components in the reconstructed specimen. Impulse noise in $g(x, y, z)$, for example, might correspond to physically impossible high-frequency components in $\hat{f}(x, y, z)$. Combined with truncation (and often undersampling) particularly in the z-direction, least squares reconstruction can lead to inaccurate results. Remedies for these problems include smoothing $\hat{f}(x, y, z)$ between iterations [43] and terminating the reconstruction process before the high-frequency artifacts build up.

The Carrington Algorithm

Carrington proposed a regularization method, based on a minimization with constraints in the least-squares sense. The Carrington algorithm seeks the nonnegative function $\hat{f}(x, y, z)$ that minimizes [41,44]

$$\min_{\{\hat{f}\geq 0\}} \sum \left| g(x,y,z) - \iiint h(x,y,z)\hat{f}(x,y,z)\,dx\,dy\,dz \right|^2 + \alpha \iiint \left| \hat{f}(x,y,z) \right|^2 dx\,dy\,dz \tag{11.25}$$

where α is a constant. The first term in the equation represents the difference between the original and restored images. The second term enforces smoothness on $\hat{f}(x, y, z)$ to prevent noise in $g(x, y, z)$ from introducing unwarranted oscillations. The value of α determines the amount of smoothing that is enforced upon $\hat{f}(x, y, z)$. If α is too small, we face the same problems as with least squares deconvolution. If α too large, $\hat{f}(x, y, z)$ will be too smooth to show details of interest. The factor α permits us to establish a compromise between precision and robustness.

The Iterative Constrained Tikhonov-Miller Algorithm

The Tikhonov-Miller algorithm is the linear restoration filter that minimizes the Tikhonov functional from Eq. (11.21). The iterative constrained Tikhonov-Miller (ICTM) algorithm [45,46] is a nonlinear algorithm that iteratively minimizes the Tikhonov functional. In this method, regularization is achieved by imposing the nonnegativity constraint by clipping to zero the negative intensities at each iteration step. This algorithm finds the minimum of the Tikhonov functional using the method of conjugate gradients [47]. The conjugate gradient direction is given by [42]

$$d_k = r_k + \xi_k d_{k-1} \tag{11.26}$$

where for the kth iteration $\xi_k = \|r_k\|^2/\|r_{k-1}\|^2$ and r_k is the steepest descent direction, given by

$$r_k = \left(-\frac{1}{2}\right)\nabla_{\hat{f}}\,\phi(\hat{f}) = \left(\boldsymbol{H}^T\boldsymbol{H} + \lambda\boldsymbol{C}^T\boldsymbol{C}\right)\hat{f}_k - \boldsymbol{H}^T\boldsymbol{G}.$$

The next new conjugate gradient estimate is found by [42]

$$\hat{f}_{k+1} = \begin{cases} \hat{f}_k + \beta_k d_k & \hat{f}_k + \beta_k d_k \geq 0 \\ 0 & \text{otherwise} \end{cases} \tag{11.27}$$

where β_k is the optimal step size, and it can also be analytically determined as $\beta = d^T r / d^T A d$. In the presence of this constraint, it is not always possible to produce correct values for β, and this results in poor performance of the ICTM algorithm [45]. Thus values for β are generally found using an iterative one-dimensional minimization algorithm such as the golden section rule [47], or by using a first-order Taylor series expansion of Eq. (11.27) with respect to β. Although the ICTM algorithm used with a nonnegativity constraint is less sensitive to errors than noniterative TM restoration, it is more computationally expensive.

11.4.3.5 Maximum Likelihood Restoration

Most of the constrained deconvolution methods described previously assume that the noise statistics follow a Gaussian distribution and can be modeled as additive noise. However, this assumption is not always valid, especially in the case of photon-limited imaging where the noise is mainly due to the photon counting nature of the image-detection process. In these situations, when the noise component is either large or dominated by Poisson noise, statistical processing methods are typically used to improve the deconvolution process. They incorporate information about the statistics of the noise and impose necessary constraints. Statistical methods are usually based on the premise that, given the probability distribution of the observed image for a known specimen and a known microscope psf, one can statistically estimate the true specimen image that best satisfies not only the mathematical description of image formation and recording, but also the original recorded image [48]. These methods are also capable of recovering certain information that is not passed by the objective lens [48]. The algorithms are iterative and computationally more expensive than the constrained iterative methods.

Maximum likelihood (ML) restoration has been applied to optical section deblurring. If the psf, h, of the microscope is known, the *probability density function* (pdf) $Pr(g|f)$, which is the likelihood of the recorded image g, is a function of the true specimen image f. Then the problem of image restoration is to estimate the unknown parameters $f(X)$, $\forall X \in S_f$, where S_f is the support of f. Using the log-likelihood function, the ML solution for the image estimate $\hat{f}$ that is most likely to give rise to the observed image g is found by solving [49]

$$\hat{f} = \arg\min_{f} - \log Pr(g|f,h) \tag{11.28}$$

In the case of Gaussian noise, the probability density function is given by [49]

$$Pr(g|f) = \frac{1}{(2\pi)^{N/2\sigma N}} \exp\left(-\frac{\|g - Hf\|^2}{2\sigma^2}\right) \tag{11.29}$$

where N is the number of voxels in the image and σ^2 is the noise variance. Eq. (11.28) can now be written as

$$\hat{f} = \arg\min_{f} \frac{\|g - Hf\|^2}{2\sigma^2}$$

which is similar to the least squares solution. Iterative techniques, such as the steepest descent method, may then be used for minimization [49]

$$\hat{f}_{k+1} = \hat{f}_k + \eta \boldsymbol{H}^T(g - \boldsymbol{H}f) \tag{11.30}$$

where T denotes matrix transpose, subscript k denotes the kth iteration, and η is a predetermined parameter such that $0 < \eta < 2/\sigma_1^2$, where σ_1 is the largest singular value of the matrix **H**. This method is usually referred to as the Landweber method [50]. For large values of η, convergence to the optimal image is rapid, but may be unstable, whereas smaller values for η, although slower, afford more stability. It is important to have a criterion for stopping the iterations. For example, the discrepancy error, $e^k = \|Hf^k - g\|_2^2$, is widely used for stopping the iterations when it is less than a predetermined threshold [50]. The number of iterations also allows regularization, and the process can be stopped to avoid overfitting the noise. Other constraints, like nonnegativity and bandlimitedness, can also be applied after each iteration step.

The EM-ML Algorithm

In the case of Poisson noise, if A represents the mean of the number of photons counted at all the image voxels, then the probability of counting exactly n photons during the exposure time of one voxel is Poisson distributed with density [49]:

$$Pr(n) = \frac{A^n}{n!} \, exp\,(-A) \tag{11.31}$$

This uncertainty in the number of photons appears as Poisson noise in the observed image, and it is correlated with intensity. The pdf can be written as [49]

$$Pr\,(g|f,h) = \frac{(\boldsymbol{H}f)^g}{g!} \, exp\,(-\boldsymbol{H}f) \tag{11.32}$$

The maximum likelihood solution is then found by setting $\partial \, log\,(Pr(g\,|f))/\partial f$ to zero, which is the *expectation maximization-maximum likelihood* (EM-ML) algorithm [48,51].

The expectation maximization (EM) algorithm is an iterative procedure to compute the maximum likelihood (ML) estimate in the presence of missing or hidden data. The aim is to estimate the model parameter(s) (specimen function, $\hat{f}$) for which the observed data are the most likely. In microscopy, the recorded (diffraction-limited) image $g\,(x, y, z)$ represents an incomplete dataset, whereas the specimen image $f(x, y, z)$ is the image to be estimated. Details of the implementation are given in the references [48,51]. The EM algorithm has a slow convergence rate and is quite computationally intensive.

The Richardson-Lucy Algorithm

The Richardson-Lucy (RL) technique is similar to the maximum likelihood algorithm for Poisson noise [52]. The maximization of $Pr(g|f,h)$ with respect to f leads to the iterative form [49]

$$\hat{f}_{k+1} = \left(\boldsymbol{H}^T \frac{g}{\boldsymbol{H}f}\right)\hat{f}_k \tag{11.33}$$

where k is the iteration number and $\boldsymbol{H}^T$ denotes the transpose of the convolution matrix corresponding to the psf. If the initial guess for the estimated image $\hat{f}_0$ is nonnegative, the estimate at the kth iteration will remain nonnegative. This makes the algorithm sensitive to the initial guess and influences its performance. Typically, a smooth initial solution is used to avoid the amplification of high-frequency noise. The Richardson-Lucy algorithm is constrained but not regularized. The number of iterations is used to stop the computations. The MSE (between the estimated image and the true solution) decreases with the number of iterations until it reaches a minimum and then increases again when noise overfitting begins. At this point, the algorithm is usually stopped to avoid noise amplification.

Maximum Penalized Likelihood Method

Typically, in ML restoration, it is difficult to determine the optimal number of iterations. Regularization can be applied as a penalty function added to the likelihood function. This is known as the maximum penalized likelihood (MPL) method [48]. In MPL, the likelihood function is modified such that it decreases when the noise increases. That is, the log-likelihood function can be modified by subtracting a penalty term that increases with unwanted characteristics, such as high-frequency content or intensity saturation. For example, an "intensity penalty" is applied to address problems associated with very bright spots in the images, and a "roughness penalty" is applied to prevent high-frequency noise, wherein large changes between neighboring voxels are encountered [48,53].

Maximum A Posteriori Method

Alternatively, Bayesian statistics can be used for regularization in the form of a prior probability distribution, known as the maximum a posteriori (MAP) methods. In this approach, prior knowledge about the true image, i.e., the image to be estimated, is taken in the form of a prior pdf. For example, if the pdf $Pr(f)$ represents prior knowledge about the true image, then using Bayes' theorem, this prior distribution can be modified, based on the recorded image, into the a posteriori distribution [29]. According to Bayes' theorem, the posterior probability can be calculated as (see Chapter 9)

$$Pr(f\,|g) = \frac{Pr(g|f)Pr(f)}{Pr(g)} \tag{11.34}$$

where $Pr(g)$ depends on the observed image only and can be regarded as a normalizing constant, and the likelihood $Pr(g|f)$ denotes the conditional pdf of g given f. The mode of the posterior distribution is often selected to be the estimated true image. In this case, it is known as the MAP solution and is obtained by maximizing $Pr(g|f)\,Pr(f)$. The $Pr(f)$ can be regarded as a penalty function that penalizes undesired features of the solution.

The maximization of the penalized likelihood can be interpreted as the maximization of the posterior probability [49]

$$\hat{f} = arg \max_{f} Pr(g|f)Pr(f) \tag{11.35}$$

The selection of the prior probability distribution is difficult. A prior distribution that performs well for one class of images might not be suitable for another. Several forms of the prior distribution have been published, including the Gibbs distribution [49], the Good's roughness penalty [54], and the total variations penalty function [55]. Alternatively, when information about the true image (other than nonnegativity) is not available, the probability $Pr(f)$ can be based on the entropy of f [56].

The difference between MAP and MPL is that, in MPL methods, only the undesired characteristics of the true image need to be known, whereas in MAP the probability distribution of the specimen must be known a priori [48]. MPL reduces to MAP if the penalty function is taken to be the prior pdf.

11.4.3.6 Blind Deconvolution

Given an inaccurate psf, all the image restoration methods described previously are ineffective in estimating the true specimen image. In practice, a truly accurate determination of the psf is difficult to obtain. Noise is always present in an experimentally measured psf, and a theoretical psf cannot completely account for the aberrations present in the microscope optics. Blind deconvolution algorithms do not require knowledge of the psf. Instead, they estimate the microscope psf and the original 3D specimen image simultaneously from the acquired image.

Several authors have described blind deconvolution algorithms [53,57,58]. One approach is to constrain the psf to be circularly symmetric and bandlimited [57]. Another approach is to apply a quadratic parameterization to enforce nonnegativity on f and use a psf parameterization based on phase aberrations in the pupil plane for h [58].

A parametric blind deconvolution algorithm based on a mathematical model of the psf is described here. In general, an iterative estimate of the object is computed as [32,53]

$$\hat{f}_{i+1}^{k+1}(x,y,z) = \left\{ \left[\frac{g(x,y,z)}{\hat{f}_i^{k}(x,y,z) * \hat{h}^{k-1}(x,y,z)} \right] * \hat{h}^{k-1}(-x,-y,-z) \right\} \hat{f}_i^{k}(x,y,z) \tag{11.36}$$

and the psf is estimated as

$$\hat{h}_{i+1}^{k}(x,y,z) = \left\{ \left[\frac{g(x,y,z)}{\hat{h}_i^{k}(x,y,z) * \hat{f}^{k-1}(x,y,z)} \right] * \hat{f}^{k-1}(-x,-y,-z) \right\} \hat{h}_i^{k}(x,y,z) \tag{11.37}$$

In the first iteration step, the object estimate is simply the recorded image, which is convolved with a theoretical psf calculated from the optical parameters of the imaging system. The resulting blurred specimen estimate is compared with the raw image, and a correction is computed. This correction is used to generate the next estimate. The same correction is also applied to the psf, generating a new psf estimate. In further iterations, the psf estimate and the object estimate are updated together using Eqs. (11.36), (11.37). Constraints, such as axial circular symmetry and bandlimitedness, can be imposed on the psf. These constraints are imposed at each iteration step, after the computation of each new estimate. Circular symmetry can be enforced on the psf by averaging the values equidistant from the optical axis. The frequency bandlimit constraint is imposed by setting to zero all values of the Fourier transform of $\hat{h}_{k+1}(x, y, z)$ that lie above the cutoff frequency.

In 3D microscopy, lateral and axial resolutions are usually different, and the psf can be adjusted accordingly. Blind deconvolution methods take twice as long per iteration as the ML technique because two functions (i.e., object and psf) are being estimated. They can produce better results than using a theoretical psf, especially if unpredicted aberrations are present. The algorithm adjusts the psf to fit the data and can thus partially correct for spherical aberration. Blind deconvolution is typically used when the psf is unknown.

11.4.3.7 Space-Variant Deconvolution

Although both deconvolution approaches, those that employ experimental or theoretical psfs and the ones that implement blind deconvolution, are largely effective at deblurring images, they fail in the reconstruction of perfectly spherical objects. This is because they incorporate depth-invariant symmetric estimates of the psf that fail to represent the actual shape of the psf, which includes nonnegligible radial asymmetry, depth variance, axial asymmetry, and specimen dependence. Additionally, most deconvolution methods assume that the microscopy system is spatially invariant, introducing considerable errors. Thus, to improve performance of 3D deconvolution, approaches have been proposed to address space-variant deconvolution [59,60], depth-related variation of the psf [61,62], and depth variant asymmetric deconvolution [63].

11.4.3.8 Interpretation of Deconvolved Images

Deconvolution techniques can be applied to wide-field and even confocal and two-photon microscope images. Generally, of all the deconvolution algorithms described here, the methods that most precisely model the image formation characteristics lead to the best results. Depending on the imaging modality being used (wide-field, confocal, two-photon, etc.), and the system optics, the psf can have a different shape. The results obtained from deconvolution procedures also depend heavily on image quality, specifically signal-to-noise ratio and sampling density. Finely sampled, low-noise images yield the best results. Artifacts often arise due to the noise sensitivity of the algorithm, a poor

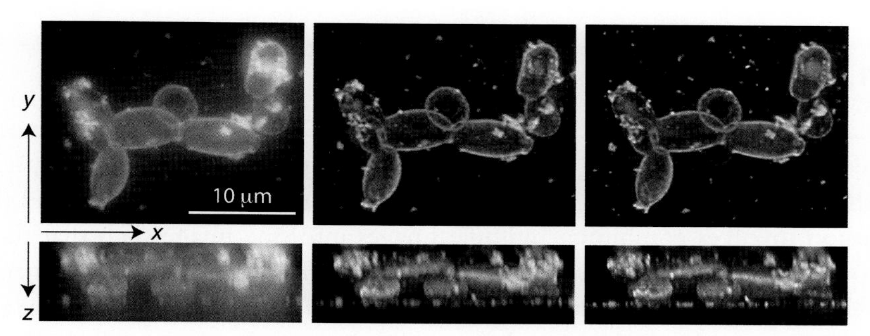

Fig. 11.6 Linear versus nonlinear deconvolution. Deconvolution using a Wiener filter is compared with a constrained maximum-likelihood iterative method. The results are presented using maximum intensity projection. Top row presents the *xy*-projection (horizontal), and bottom row presents the *xz*-projection (vertical). The leftmost column shows an image of hystoplasma capsulatum collected with a 100×/1.3 NA oil-immersion objective at a fluorescent wavelength of 570 nm. The center column presents deconvolution results obtained with a Wiener filter, and the right column shows results obtained with a constrained maximum-likelihood iterative method. Scale bar = 10 μm, with a 0.1 μm pixel size along all axes. *(Courtesy of José-Angel Conchello. Original image contributed by W. Goldman.)*

choice of algorithm regularization parameters, inaccurate psf estimation, and coarse data sampling. Interpretation of deconvolved images typically requires some knowledge of the processing methods so that the user can recognize artifacts and identify real features. For a discussion of interpreting deconvolved images, refer to [35,64]. Fig. 11.6 presents deconvolution results using linear and nonlinear methods. Deconvolution using a Wiener filter (linear, noniterative method) is compared with a constrained maximum-likelihood iterative (nonlinear) method.

11.4.3.9 Commercial and Free Deconvolution Packages

Commercial deconvolution software packages are available from Media Cybernetics (AutoDeblur) and Scientific Volume Imaging (HuygensPro), and from most major microscope manufacturers such as Zeiss, Olympus, Leica, and Nikon. Most of these packages include features for 3D psf generation in addition to image-deblurring algorithms. The key commercial vendors in this field market different implementations of the nonlinear algorithms, such as the maximum likelihood estimation (MLE) and expectation maximization (EM) algorithms. Freely available software includes ImageJ [65] or Fiji [66] with deconvolution packages such as DeconvolutionLab and DeconvolutionLab2 [64], Iterative Deconvolve 3D [67], Parallel Iterative Deconvolution [68], and BatchDeconvolution [69]. An open-source toolbox DVDeconv [70] for implementing depth-variant asymmetric deconvolution is available for Matlab (Mathworks, Natick, MA). A comparison of the commercial and free deconvolution software can be found in [71].

11.5 Image Fusion

Image fusion is an approach frequently used in 3D microscopy to combine a set of optical section images into a single 2D image containing the detail from each optical section in the stack. This simulates a microscope with much larger depth-of-field (DOF). Image fusion involves combining multiple images into one image such that the fused image contains the interesting components collected from all of the input images. An implicit assumption used by most image fusion techniques is that all input images are registered, which means that each physical component is aligned to the same location in all of the images. For extended DOF microscope imaging, since all optical section images are captured from same field of view, albeit at different focal planes, such an assumption is easily met.

The central idea of image fusion methods for extended DOF microscope imaging is to select, from different optical section images, those pixels or regions that contain the most in-focus information, as in-focus components contain more structural information than out-of-focus components. The amount of information in a region or component of an image is usually estimated by a quantitative focus measure, which is a 2D function of (x, y) location in the image. A 2D fusion map is generated to indicate, at each (x, y) position, which of the optical section images has the highest focus measure. This fusion map is used to determine from which optical section image that component in the composite image should be gathered. Finally, the selected components from all corresponding optical section images are assembled, according to the fusion map, to form the extended DOF image.

Given a set of N optical section images, as $I_1(x, y), I_2(x, y), \ldots, I_N(x, y)$, the component focus measure for each of these images at location (x, y) is $R_n(x, y)$, $n=1, 2, \ldots, N$. The fusion map is then defined as

$$M(x, y) = \arg\max_n |R_n(x, y)|, \quad n = 1, 2, \ldots, N \tag{11.38}$$

which is an index function with integer value ranging from 1 to N. The composite image resulting from image fusion is given by [72]

$$I'(x, y) = \bigcup_{n=M(x, y)} \{I_n(x, y)\} \tag{11.39}$$

The preceding discussion assumes single-channel grayscale images. For RGB images a color space transformation should first be performed to convert the images from RGB to another space such as the YUV or principal component analysis (PCA) space. Then the component focus measure $R_n(x, y)$ is computed using only one grayscale image from the transformed space (such as the luminance or the first principal component), since this image encodes most focus-relevant information. Subsequently, the fusion map is generated based on this grayscale image, and it is used to assemble the fused image from the stack of RGB section images. A detailed description of three different image fusion schemes for extended DOF microscope imaging, including (1) pixel-based, (2) neighborhood-based, and (3) multiresolution (wavelet transform) based is described in detail in [72].

11.6 Three-Dimensional Image Processing

Basic 2D imaging processing was discussed in earlier chapters. Three-dimensional image processing is an extension of those techniques, and it can be performed using three different approaches. The first approach involves performing 2D image processing operations on the individual optical sections of the image stack. This approach does not take into account the 3D nature of the dataset and may be suboptimal in some situations. The second approach is to perform image processing using the entire 3D dataset, treating the voxel (volume element) as the basic unit of brightness in three dimensions. In this approach, all operations are performed on cubic voxel arrays. Finally, the fused image of a stack can be processed by 2D techniques. For most image processing techniques, the 2D counterpart can be easily extended to 3D. The following sections discuss several image processing algorithms that are useful in 3D microscopy.

11.7 Geometric Transformations

Transformations for 3D images, such as translation and reflection, are direct extensions of the corresponding 2D image transforms (Chapter 4). Image translation by a vector (dx, dy, dz) is given by:

$$\begin{bmatrix} x' \\ y' \\ z' \end{bmatrix} = \begin{bmatrix} x \\ y \\ z \end{bmatrix} + \begin{bmatrix} dx \\ dy \\ dz \end{bmatrix} \tag{11.40}$$

where x, y, and z are the original coordinates, and x', y', and z' are the corresponding translated voxel locations. The translation operation can be formulated as [73]

$$g(x, y, z) = f(x + dx, y + dy, z + dz) \tag{11.41}$$

Similarly, reflection about the z-axis can be formulated as

$$g(x, y, z) = f(x, y, D_z - z) \tag{11.42}$$

where D_z is the image dimension in the z-direction, that is, the total number of optical slices. Reflection along the x- and y-directions can be computed in a similar fashion.

Rotation is another operation that is often applied to 3D images to enable visualization of the dataset from different viewpoints. This operation requires each voxel, (x, y, z), in the original volume to be mapped into a new position, (xx, yy, zz). The computation can be performed as [73]

$$\begin{bmatrix} x - x_c \\ y - y_c \\ z - z_c \end{bmatrix} = \begin{bmatrix} \cos\theta & -\sin\theta\cos\varphi & \sin\theta\sin\varphi \\ \sin\theta & \cos\theta\cos\varphi & -\cos\theta\sin\varphi \\ 0 & \sin\varphi & \cos\varphi \end{bmatrix} \begin{bmatrix} xx - xx_c \\ yy - yy_c \\ zz - zz_c \end{bmatrix} \quad (11.43)$$

where (x_c, y_c, z_c), (xx_c, yy_c, zz_c) are the rotation centers in the input and output images, and θ and ϕ are the rotation angles about the z- and x-axis, respectively. Typically, such computations result in noninteger values for the coordinates that need to be truncated or rounded. This may result in more than one voxel being mapped to the same position, which leaves holes in the rotated image (see Chapter 4). This can be avoided by using an inverse implementation where, for each voxel (xx, yy, zz) of the output image, the corresponding input voxel (x, y, z) value is evaluated. Both translation and rotation are rigid body transforms wherein lengths and angles are preserved. Scaling is an affine transformation, where lengths are changed. It can be computed as

$$\begin{bmatrix} x' \\ y' \\ z' \end{bmatrix} = \begin{bmatrix} s_x & 0 & 0 \\ 0 & s_y & 0 \\ 0 & 0 & s_z \end{bmatrix} \begin{bmatrix} x \\ y \\ z \end{bmatrix} \quad (11.44)$$

where s_x, s_y, s_z are the scaling factors in the x-, y-, and z-directions, respectively.

11.8 Pointwise Operations

Pointwise operations that are performed in 2D (Chapter 5) can also be applied in 3D on a voxel-by-voxel basis. For example, gray level mathematical operations such as subtraction and addition and binary operations such as OR, XOR, and AND can be expressed for 3D images as follows [73]:

$$\begin{aligned} &\textit{addition}: c(x, y, z) = a(x, y, z) + b(x, y, z) \\ &\textit{subtraction}: c(x, y, z) = a(x, y, z) - b(x, y, z) \\ &\textit{binary OR}: c(x, y, z) = a(x, y, z) \,||\, b(x, y, z) \end{aligned} \quad (11.45)$$

11.9 Histogram Operations

The gray level histogram, introduced for 2D images in Chapter 5, is a function showing, for each gray level, the number of voxels in the image that have that level. In other words, the pdf of the voxel intensities is called the *histogram*. For a 3D image of dimensions $D_x \times D_y \times D_z$ having K discrete intensity levels $i_1, \ldots, i_K$, the histogram value at a certain intensity is given by the frequency of occurrence of that intensity [73]

$$p(i_k) = \frac{n_k}{D_x D_y D_z} \tag{11.46}$$

where n_k is the number of voxels having intensity i_k. A histogram is plotted as a display of the frequencies at each intensity value, where the gray level intensity is the abscissa and the ordinate is the frequency of occurrence. The cumulative histogram is a variation of the histogram in which the vertical axis represents not just the counts for a single gray level, but rather denotes the counts for the intensity in consideration, plus all values less than that intensity. The cumulative density function (cdf) is given by [73]

$$P(i_k) = \frac{1}{D_x D_y D_z} \sum_{l=1}^{K} n_l \tag{11.47}$$

The histogram contains important information about image quality and can be used for image enhancement (Chapter 5). A popular histogram-based enhancement operation is *histogram equalization*, wherein the histogram of the output image is made to be uniform (flat) such that it has equal numbers of voxels at every gray level. Histogram equalization for an input image $f(x, y, z)$ to generate the output $g(x, y, z)$ can be achieved using a pointwise intensity transformation function as follows [73]:

$$g(x, y, z) = T(f(x, y, z)) \text{ such that } g = T(f) = \int_0^f p_f(w)\, dw \tag{11.48}$$

where p_f is the histogram of the input image and image intensities are normalized to the [0,1] range. Although image contrast is improved, histogram equalization tends to enhance noise, and image smoothing is often applied prior to histogram equalization for improved performance. Histogram equalization does not lend itself to interactive image enhancement applications because it generates only one result, an approximation to a uniform histogram. In situations where a particular histogram shape, capable of highlighting certain gray level ranges, is desired, another method known as *histogram specification* can be used. This transforms an image so that its histogram more closely resembles a given histogram. An image, $f(x, y, z)$, with a histogram function, p_f, can be transformed into a new image, $g(x, y, z)$, having a given histogram, p_g, by [73]

$$g(x, y, z) = P_g^{-1}\left[P_f(f(x, y, z))\right] \tag{11.49}$$

where P_f and P_g are the cumulative histograms for the 3D images $f(x, y, z)$ and $g(x, y, z)$, respectively, and P_g^{-1} is the inverse function of P_g.

11.10 Filtering

Image filtering operations are typically used either to reduce noise by smoothing or to emphasize edges and small detail. Analogous to their 2D counterparts (Chapter 5), 3D filters can be defined as operators that map one 3D image into another 3D image as $b(x,y,z) = f(a(x,y,z))$, where $b(x,y,z)$ is the output (filtered) image that results from processing the input image $a(x,y,z)$. Each voxel in the output image is computed as a function of one or several voxels located in its neighborhood in the input image. Depending on the nature of the filter function f, the operation can be classified as linear or nonlinear. Digital convolution filters are linear since the output voxel values are linear combinations of the input image voxels, but other types of filters are nonlinear.

11.10.1 Linear Filters

Linear filters can be implemented either as convolution operations in real space or as multiplication operations in Fourier space. A linear filter can be specified either by its convolution kernel, analogous to the psf, or by its transfer function. The kernel and the transfer function form a Fourier transform pair.

11.10.1.1 Finite Impulse Response Filters

A finite impulse response (FIR) filter has a convolution kernel (matrix of filter coefficients) that is zero outside a relatively small region near the origin. It is implemented as a convolution of the kernel with the input image $a(x,y,z)$, as [73]

$$\begin{aligned} b(x,y,z) &= h(x,y,z) * a(x,y,z) \\ &= \sum_{i=0}^{L-1} \sum_{j=0}^{M-1} \sum_{k=0}^{N-1} h(i,j,k)\ a(x-i, y-j, z-k) \end{aligned} \tag{11.50}$$

where $h(x,y,z)$ is the filter kernel, defined over $0 \le x < L$, $0 \le y < M$, and $0 \le z < N$.

The moving average filter is also known as the arithmetic mean filter, where the values of all the filter coefficients in the kernel are equal to $1/LMN$ for a kernel of dimensions $L \times M \times N$. From this definition, one can see that if a voxel value is noise, taking the average gray level of the voxels surrounding it tends to normalize its gray level. Thus an averaging filter smooths areas containing high-frequency content and only subtly changes areas of relatively constant gray level. While it is useful for reducing random noise, it is not as effective against impulse noise as the nonlinear filters described in the following text. With a less restricted kernel, Eq. (11.50) can be used to implement a variety of linear filters for low-pass or high-pass filtering, simply by selecting an appropriate kernel or transfer function (see Chapter 5).

11.10.2 Nonlinear Filters

Nonlinear filters can also be implemented by $b(x, y, z) = f(a(x, y, z))$ as earlier, except that f is no longer a linear function. They are more general and can be designed to preserve edge information and remove impulse noise. Some representative nonlinear 3D filters are discussed in the following sections.

11.10.2.1 Median Filter

A median filter belongs to the class of order statistics filters, i.e., filters based on a rank ordering of the input voxel values. Most implementations use a square kernel of neighborhood pixels (e.g. 3×3, 5×5) as the filter window. Consider the voxels $X_1, X_2, \ldots, X_N$ where N is an odd number, within a filter window of size $N = L \times M \times N$. Then the rank ordered voxel values can be expressed as $X_1 < X_2 < \ldots < X_i \cdots < X_N$, and the median is defined as the middle voxel value, given by $X_{med} = X_{N+1/2}$, and the median filter is defined as the value that minimizes [73]

$$\sum_{i=1}^{N} |X_i - X_{med}| \tag{11.51}$$

The median filter is implemented by moving the filter window over the 3D dataset. At each window position, the central voxel is replaced by the value of X_{med} inside the window neighborhood. This process is slow, especially in 3D, due to the requirement of sorting all the voxels in each neighborhood based on their gray level. One approach to reducing the computational complexity is to apply successive 1D median filtering along the rows, columns and sections (image planes). This process is called a "separable median filter" and the resulting output image differs from that of a true 3D median filter.

11.10.2.2 Weighted Median Filter

The weighted median filter is a variation of the median filter that incorporates spatial information of the voxels when computing the median value. A weighted 3D median filter is implemented as follows [73]:

$$b(x, y, z) = median\{w_1 \square X_1, \ldots, w_N \square X_N\} \tag{11.52}$$

where $X_1 \ldots X_N$ are the voxel values inside a window centered at (x, y, z) and $w \square X$ denotes replication of X, w times. The advantage of this approach is that, by applying larger weight values on certain voxels, spatial or structural information can be incorporated into the filtering process.

11.10.2.3 Minimum and Maximum Filters

Another set of widely used nonlinear filters are the minimum and maximum filters. Like the median filter, these two are also order statistics filters. The maximum filter fills the central voxel of the 3D window with the maximum X_N among the ordered set of voxel intensities, whereas the minimum filter replaces it with the minimum value, X_1. Unlike the median filter, these filters have much reduced computational complexity. The implementation of these filters is also speeded up by performing three successive 1D filtering operations along the three axis directions. Maximum filters are used to suppress negative impulses and to brighten up the image, whereas minimum filters reduce positive impulses and tend to darken the image.

11.10.2.4 *α-Trimmed Mean Filters*

The α-trimmed mean filters are widely used for the restoration of signals and images corrupted by additive symmetric noise. The filter rejects a certain percentage of outlying voxels (extremely high and low values), and then averages the remaining voxels within the filter window. The filter is implemented as follows [73]:

$$b(x, y, z) = \frac{1}{(1-2\alpha)N} \sum_{i=\alpha N+1}^{(1-\alpha)N} X_i \qquad 0 \leq \alpha \leq 0.5 \tag{11.53}$$

where N is the number of voxels within a filter window centered at (x, y, z). The approach involves ordering the voxels within the filter window and then rejecting a percentage (specified by α), of the lower and upper rank voxels. The filter output is the average of the remaining voxels. The rejection parameter, α, is selected in relation to the noise statistics. The filters are simple to implement and can be applied iteratively for optimizing α and thereby controlling the filtering effects. These filters are generally used to eliminate spike noise without smearing the image.

An extension of the trimmed filter is the modified trimmed mean filter. This filter sets a threshold T, and averages only those voxel values where the difference from the mean or median value is less than the threshold. This process is written as [73]

$$b(x, y, z) = \frac{\sum_i a_i X_i}{\sum_i a_i} \quad \text{where} \quad a_i = \begin{cases} 1 & \text{if } |X_i - \overline{X}| \leq T \\ 0 & \textit{otherwise} \end{cases} \tag{11.54}$$

where $\overline{X}$ is either the mean or the median value within the filter window. The threshold is determined based on the noise statistics.

11.10.3 Edge Detection Filters

Similar to boundaries in 2D images, the surfaces of different 3D regions are represented by intensity changes in the data volume and can be detected using specialized linear filters that highlight edges. Three-dimensional edge operators are direct extensions of the 2D edge-based operators (Chapter 7), and 3D edges (or, more appropriately, surfaces) represent discontinuities in image intensity. The intensity gradient can be expressed in 3D as [73]

$$\nabla f(x, y, z) = \left(\frac{\partial f}{\partial x}, \frac{\partial f}{\partial y}, \frac{\partial f}{\partial z} \right) \tag{11.55}$$

Edge activity can then be detected by computing the gradient magnitude or the L_1 norm as follows [73]:

$$\text{edge}(x, y, z) = \sqrt{\left(\frac{\partial f}{\partial x}\right)^2 + \left(\frac{\partial f}{\partial y}\right)^2 + \left(\frac{\partial f}{\partial z}\right)^2}$$
$$\text{or} \qquad \text{edge}(x, y, z) = \left|\frac{\partial f}{\partial x}\right| + \left|\frac{\partial f}{\partial y}\right| + \left|\frac{\partial f}{\partial z}\right| \tag{11.56}$$

The algorithm for edge detection involves convolution with a kernel that creates an approximation of the partial derivative in a given direction. Thus, for 3D edge detection, three masks are used, one for each direction. For example, the Sobel edge detector (Chapter 7) can be generalized for 3D processing as a set of three 3D masks, one for each of three directions [73]. For the y-direction (vertical edges), this $3 \times 3 \times 3$ matrix is

$$\begin{bmatrix} -1 & 0 & 1 \\ -2 & 0 & 2 \\ -1 & 0 & 1 \end{bmatrix} \begin{bmatrix} -2 & 0 & 2 \\ -3 & 0 & 3 \\ -2 & 0 & 2 \end{bmatrix} \begin{bmatrix} -1 & 0 & 1 \\ -2 & 0 & 2 \\ -1 & 0 & 1 \end{bmatrix} \tag{11.57}$$

The other two operators are obtained by 90-degree rotations of those shown in Eq. (11.57). Edge detection filters are commonly used for image segmentation. For the most part, 3D filtering is an extension of the 2D filtering process described in Chapter 5, but with somewhat increased computational complexity due to the added third dimension. For a more complete description of filters, the reader should consult a textbook on image processing [29,73].

11.11 Morphological Operators

Mathematical morphological operators are a subclass of nonlinear filters that are used for shape and structure analysis, and filtering in binary and grayscale images. Three-dimensional morphological operators are straightforward extensions of their 2D counterparts (as described in Chapter 6) with sets and functions defined in the 3D Euclidean grid $\boldsymbol{Z}^3$. In this section we extend the concepts introduced in Chapter 6 to three dimensions. An in-depth description of the mathematical basis of morphological operators is presented in [73–75].

11.11.1 Binary Morphology

Mathematical morphological operators are based on set theory. An object O in a binary 3D image can be denoted as

$$O = \{\boldsymbol{v} : f(\boldsymbol{v}) = 1, \boldsymbol{v} = (x, y, z) \in \boldsymbol{Z}^3\} \tag{11.58}$$

where f is called the characteristic function of O. Similarly, the object background O^c can be defined as follows [73]:

$$O^c = \{\boldsymbol{v} : f(\boldsymbol{v}) = 0, \boldsymbol{v} = (x, y, z) \in \boldsymbol{Z}^3\} \tag{11.59}$$

All morphological operations utilize a structuring element (also known as the kernel), which determines the precise details of the effect that the operator has on the input image. In 3D morphology, the structuring element is a small cluster of voxels, arranged in a geometric pattern (sphere, cube, octahedron, etc.) relative to some origin. Normally Cartesian coordinates are used to represent each voxel, and the origin is typically at the center of the cluster. Similarly to convolution, the origin of the structuring element is typically translated to each voxel position in the input image in turn. There the points within the translated structuring element are compared with the underlying image voxel values. The details of the operation and the effect of the outcome depend on which morphological operator is being used. The two basic mathematical operations of dilation and erosion are denoted as [73]

$$\begin{aligned} \textit{dilation}: \; O \oplus B^s &= \{\boldsymbol{v} \in \boldsymbol{Z}^3 : B_v \cap O \neq 0\} \\ \textit{erosion}: \; O \ominus B^s &= \{\boldsymbol{v} \in \boldsymbol{Z}^3 : B_v \subset O\} \end{aligned} \tag{11.60}$$

where B^S is the symmetric of B with respect to the origin (0,0,0). Dilation is an expanding operation, whereas erosion has a shrinking effect. The successive application of erosion and dilation is opening, and dilation followed by erosion is closing [73]:

$$\begin{aligned} \textit{opening}: &\; (X \ominus B^s) \oplus B \\ \textit{closing}: &\; (X \oplus B^S) \ominus B \end{aligned} \tag{11.61}$$

Opening smooths the surface and typically smears sharp spurs on the object boundary, whereas closing fills small holes and typically unites objects that are close to each other. Note that applying either opening or closing more than once produces no further effect.

11.11.2 Grayscale Morphology

Grayscale images can be denoted as functions, f, whose domain D is a subset of the Euclidean grid $\boldsymbol{Z}^3$ [73]:

$$f(\boldsymbol{v}), \boldsymbol{v} = (x, y, z) \in D \subset \boldsymbol{Z}^3 \tag{11.62}$$

The structuring element can also be denoted as a function g within domain G [73]

$$g(\boldsymbol{v}), \boldsymbol{v} = (x, y, z) \in G \subset \boldsymbol{Z}^3 \quad \text{and the symmetric of } g \text{ is} \quad g^s(\boldsymbol{v}) = g(-\boldsymbol{v}) \tag{11.63}$$

The operations of dilation, erosion, opening, and closing are then defined as follows [73]:

$$\begin{aligned} dilation&: \ [f \oplus g^s](\boldsymbol{v}) = \max_{\boldsymbol{v} \in D, \boldsymbol{v}-\boldsymbol{d} \in G} \{ f(\boldsymbol{v}) + g(\boldsymbol{v} - \boldsymbol{d}) \} \\ erosion&: \ [f \ominus g^s](\boldsymbol{v}) = \min_{\boldsymbol{v} \in D, \boldsymbol{v}-\boldsymbol{d} \in G} \{ f(\boldsymbol{v}) - g(\boldsymbol{v} - \boldsymbol{d}) \} \\ opening&: \ [(f \ominus g^s) \oplus g](\boldsymbol{v}) \\ closing&: \ [(f \oplus g^s) \ominus g](\boldsymbol{v}) \end{aligned} \tag{11.64}$$

where $\boldsymbol{d}$ is a vector that defines the translation. Grayscale opening suppresses positive impulses but enhances negative ones, whereas closing does the converse. Additionally, opening and closing can be combined to create filters such as the close-open and open-close filters, which suppress both negative and positive impulses. Another morphological operator of interest is the top-hat transform, which is very similar to a high-pass filter. It can be applied to 3D images to produce peaks of the object features. The top-hat transform is denoted as [73]

$$g(\boldsymbol{v}) = f(\boldsymbol{v}) - f_{kB}(\boldsymbol{v}) \tag{11.65}$$

where $kB = B \oplus B \oplus B \oplus \ldots \oplus B$ *k times.*

Finally, mathematical operators can also be used as edge detectors, as follows [73]:

$$e(\boldsymbol{v}) = f(\boldsymbol{v}) - [f \ominus kB](\boldsymbol{v}) \tag{11.66}$$

where e is the output edge image.

Three-dimensional image stacks are usually anisotropic, that is, the sampling interval along the axial dimension is larger than in the radial dimension. For this reason, the structuring element should be chosen as an anisotropic 3D kernel. For example, the quasispherical structuring element of size $3 \times 7 \times 7$ shown in Eq. (11.67) has been used for a 3D morphological opening operation [76].

0	0	0	0	0	0	0
0	0	1	1	1	0	0
0	1	1	1	1	1	0
0	1	1	1	1	1	0
0	1	1	1	1	1	0
0	0	1	1	1	0	0
0	0	0	0	0	0	0

(a)

0	0	1	1	1	0	0
0	1	1	1	1	1	0
1	1	1	1	1	1	1
1	1	1	1	1	1	1
1	1	1	1	1	1	1
0	1	1	1	1	1	0
0	0	1	1	1	0	0

(b)

0	0	0	0	0	0	0
0	0	1	1	1	0	0
0	1	1	1	1	1	0
0	1	1	1	1	1	0
0	1	1	1	1	1	0
0	0	1	1	1	0	0
0	0	0	0	0	0	0

(c)

(11.67)

From Eq. (11.67), the kernel (b) is applied to the optical slice in the middle, whereas (a) and (c) are applied to slices above and below.

Other widely used 3D morphological techniques include the watershed algorithm and skeletonization (see Chapter 6). The interested reader is directed to other publications for detailed information on these techniques [73–76].

11.12 Segmentation

Segmentation is a procedure that classifies all the voxels in an image to ultimately divide the image into regions, each of which corresponds to a different object. Two-dimensional segmentation is discussed in Chapter 7. Here we generalize those concepts to 3D. The segmentation process is based on the notion that voxels belonging to a certain region share some similar characteristics, such as intensity, texture, or spatial position. Segmentation algorithms may be applied either to unprocessed images or after the application of certain transformations or filters, and they may be automated or interactive, that is, requiring human input.

Segmentation is often the most challenging step in image analysis. If segmentation is done improperly, then all subsequent stages of image analysis are incorrect. The challenges of segmentation are further confounded when processing 3D microscopy images. The difficulty in segmenting regions of volumetric images arises from several factors. Regions are often touching each other or overlapping and irregularly arranged, with no definite shape. Illumination variations are also common in thick specimens, with intensity falling off with increasing depth due to factors such as absorption, scattering, and diffraction of the light by structures located above and below the focal plane.

There are two methods to implement segmentation of 3D images. The first approach is slice-by-slice segmentation, wherein the 2D images in the 3D stack are individually processed, and the 3D regions are extracted when the processed 2D slices are stacked. The disadvantage of this implementation is that discontinuities of volume between the slices may occur, and thus it is not possible to derive accurate 3D-volumetric morphology. Object boundaries that lie parallel to the focal plane cannot be extracted accurately.

The second approach is the volume-oriented approach, which is based on processing the complete set of consecutive slices, in its entirety, as a single 3D image. The trade-off here is that volumetric information is retained, but the computational complexity is increased. Regardless of the implementation chosen, there are three general approaches to segmentation: point-based, edge-based, and region-based methods (Chapter 7).

11.12.1 Point-Based Segmentation

In thresholding, which is also known as point-based segmentation, voxels are allocated to categories according to the range of intensity in which a voxel value lies. For example, if voxels that form a certain object fall within a specific intensity range that is different from the intensity of the background voxels and from the intensity range of other objects in the image, then the object can be segmented using a pair of intensity thresholds. Given a pair of thresholds, t_1 and t_2, the voxel located at position (x, y, z) with grayscale value $f(x, y, z)$ is allocated to a category C_1 if $t_1 < f(x, y, z) \leq t_2$. Otherwise, the voxel is allocated to a different category.

The success of thresholding algorithms depends heavily upon the selected threshold, the selection of which is challenging and often subjective. Several automatic threshold selection methods have been developed, but very often the procedure requires some user interaction. Reviews of threshold selection methods for 2D images are available [77–79]. While most algorithms simply use the histogram, others make use of contextual information such as gray level occurrences in adjacent pixels. Most of these methods are applicable to 3D images and require the use of the image's 3D gray level histogram. A representative iterative algorithm for automated threshold selection for 3D images, often referred to as the intermeans algorithm, is presented here [80]. An initial guess, typically the median intensity value, is used for the threshold at the start. This threshold is used to generate two categories, and the mean values of voxels in the two categories are calculated as [81]

$$\mu_1 = \frac{\sum_{k=0}^{T} kh_k}{\sum_{k=0}^{T} h_k} \quad \text{and} \quad \mu_2 = \frac{\sum_{k=T+1}^{N} kh_k}{\sum_{k=T+1}^{N} h_k} \tag{11.68}$$

where T is the median voxel value chosen such that $\sum_{k=0}^{T} h_k \geq \frac{n^2}{2} > \sum_{k=0}^{T-1} h_k$ and h_k specifies the number of voxels in the image with gray level value k, N is the maximum voxel value (typically 255), and n is the total number of voxels, i.e., $D_x \times D_y \times D_z$. Next, the threshold is computed to lie exactly halfway between the two means as

$$T = \left] \frac{\mu_1 + \mu_2}{2} \right[\tag{11.69}$$

where][indicates only integer values. The mean values are then calculated again, and the process is repeated to compute a new threshold, and so on, until the threshold stops changing value between consecutive computations.

Other iterative methods work similarly by iteratively splitting the image into regions by finding the peaks in the histogram due to each region, or by dividing the histogram based on specific criteria to pick the threshold. Another common segmentation approach is to determine multilevel thresholds, whereby different objects are segmented in different threshold bands. This is done by determining multiple threshold values by searching the global intensity histogram for peaks, or by using specific criteria to divide the histogram [82].

Thresholding is most suitable for images wherein the objects display homogeneous intensity values against a high-contrast uniform background. If the objects have large internal variation in brightness, thresholding will not produce the desired segmentation. In 3D images, intensity typically falls off deep within the specimen due to diffraction, scattering, and absorption of light, and a single threshold value is not applicable for the entire stack. Multilevel thresholding methods with localized threshold determination for individual 2D slices in the stack are typically more appropriate. Moreover, thresholding algorithms that are based solely on gray level and do not incorporate spatial information result in segmented images wherein objects are not necessarily connected. Most often the outcome of a thresholding operation is not used as the final result, but rather refinement procedures, such as morphological processing and region-based segmentation algorithms, are applied to further delineate regions.

11.12.2 Edge-Based Segmentation

In 2D edge-based segmentation, an edge filter is applied to the image, and pixels are classified as edge or nonedge pixels depending on the filter output (Chapter 7). Typical edge filters produce images in which pixels located near rapid intensity changes are highlighted. For 3D data, surfaces represent object edges or boundaries, and edge detection in 2D can be extended to 3D surface detection. For example, the Marr-Hildreth operator for 3D surface detection operates as follows [73]:

$$C(x, y, z) = \nabla^2(I(x, y, z) * G(x, y, z, \sigma)), \tag{11.70}$$

where

$$G(x, y, z, \sigma) = \frac{1}{\sqrt{(2\pi)^3} \cdot \sigma^3} \cdot e^{-(x^2+y^2+z^2)/2\sigma^2}$$

where $I(x, y, z)$ is the 3D image, $G(x, y, z, \sigma)$ is the Gaussian function, $C(x, y, z)$ is the resulting contour (surface) image, and ∇^2 is the Laplacian operator. Eq. (11.70) can be written as [73]

$$C(x, y, z) = I(x, y, z) * \nabla^2 G(x, y, z, \sigma) \tag{11.71}$$

where $\nabla^2 G$ is the Laplacian of a Gaussian operator. An implementation of the Difference of Gaussians (DOG) operator is shown in Eq. (11.72). The DOG operator is separable and can be implemented as [73]

$$\begin{aligned}\nabla^2 G(x, y, z, \sigma) &\approx G(x, y, z, \sigma_e) - G(x, y, z, \sigma_i) \\ &\approx G(x, \sigma_e) * G(y, \sigma_e) * G(z, \sigma_e) - G(x, \sigma_i) * G(y, \sigma_i) * G(z, \sigma_i)\end{aligned} \tag{11.72}$$

In order for the DOG operator to approximate the $\nabla^2 G$ operator, the values for σ_e and σ_i must be [83]

$$\sigma_e = \sigma \cdot \sqrt{\frac{1 - (1/r^2)}{2\ ln\, r}} \qquad \text{and} \qquad \sigma_i = r \cdot \sigma_e \tag{11.73}$$

where r is the ratio $\sigma_i : \sigma_e$.

The values of σ_e and σ_i are determined such that the bandwidth of the filter is small, and the sensitivity is high. As seen in Eq. (11.70), this edge detection filter uses a Gaussian filter to smooth the data and remove high-frequency components. The amount of smoothing can be controlled by varying the value of σ, which is the standard deviation (i.e., width) of the Gaussian filter. Moreover, the Laplacian component is rotationally invariant and allows the detection of edges at any orientation.

Overall, surface-based segmentation is most useful for images with "good boundaries," that is, the intensity value varies sharply across the surface, is homogeneous along the surface, and the objects that the surfaces separate are smooth and have a uniform surface. Generally, special surfaces, such as planar, spherical, or ellipsoidal, are detected as the boundaries between objects. The medical imaging field offers a number of promising algorithms that seek to optimize the surface of the object by balancing edge/region parameters measured from the image with a priori information about the shape of the object [84]. Despite their success in medical image analysis, boundary detection methods are seldom optimal for microscopy data, where the intensity changes are typically gradual and heterogeneous along the surface.

A major disadvantage of the edge-based algorithms is that they can result in noisy, discontinuous edges that require complex postprocessing to generate closed boundaries. Typically, discontinuous boundaries are subsequently joined using morphological matching or energy optimization techniques to find the surface that best matches the gradient map [85]. An advantage of edge detection is the relative simplicity of computational processing. This is due to the significant decrease in the number of voxels that must be classified and stored when considering only the voxels of the surface, as opposed to all the voxels in the volume of interest.

11.12.3 Region-Based Segmentation

Region-based methods are further classified into region growing, region splitting, and region merging methods. These algorithms operate iteratively by grouping together adjacent voxels that are connected as neighbors and have similar values, or by splitting groups of voxels that are dissimilar in value. Grouping of voxels is determined based on their connectivity.

11.12.3.1 Connectivity

The 2D connectivity concepts introduced in Chapter 7 are readily extended. For 3D images there are three widely used definitions of connectivity, namely, 6-neighborhood,

18-neighborhood, and 26-neighborhood connectivity. Given a voxel $\boldsymbol{v}=(x,y,z)$, the 6-neighborhood $N_6(\boldsymbol{v})$of $\boldsymbol{v}$ consists of the six voxels whose positions in the 3D space differ from $\boldsymbol{v}$ by ± 1 in only one coordinate. These voxels are called 6-neighbors of $\boldsymbol{v}$ and their coordinates (x',y',z') satisfy the following condition [73]:

$$N_6(\boldsymbol{v}) = \{\ (x',y',z'):\ |x-x'| + |y-y'| + |z-z'| = 1\ \} \tag{11.74}$$

Similarly, the 18-neighborhood, $N_{18}(\boldsymbol{v})$, and 26-neighborhood, $N_{26}(\boldsymbol{v})$, voxels satisfy the following conditions [73]:

$$\begin{aligned} N_{18}(\boldsymbol{v}) &= \{(x',y',z'): 1 < |x-x'| + |y-y'| + |z-z'| \leq 2 \text{ and} \\ &\qquad \max(|x-x'|,|y-y'|,|z-z'|) = 1\} \\ N_{26}(\boldsymbol{v}) &= \{(x',y',z'): \max < |x-x'| + |y-y'| + |z-z'| = 1\} \end{aligned} \tag{11.75}$$

Any voxel, p, in the 6-neighborhood of a voxel x shares one face, in the 18-neighborhood shares at least one edge, and in the 26-neighborhood shares at least one vertex with the voxel x. As in the 2D case, connectivity in 3D is defined in terms of the following adjacency pairs: (6, 26), (26, 6), (18, 6), and (6, 18) where the first component represents connectivity of the foreground and the second component represents connectivity of the background. Region-based methods typically utilize some characteristic, such as connectivity, intensity value, structural property, probabilistic correlation (e.g., Markov random field models), or correlation between voxels in a neighborhood, to attribute voxels to different objects or to the background.

11.12.3.2 Region Growing

Region growing segmentation techniques start with marked voxels, or small regions (called "seeds"), that have very similar properties based on chosen criteria. The seeds can be placed automatically or interactively. Then, moving along the boundary of each seed, neighboring (connected) unassigned voxels are examined sequentially to determine if each should be incorporated into the region. The procedure is repeated, and the region grows, until no further neighboring voxels qualify for incorporation. Notice that this procedure produces a segmentation in which no regions overlap, and many voxels may remain unassigned to any region.

11.12.3.3 Region Splitting And Merging

Region splitting methods start from a segmented region and recursively proceed toward smaller regions through a series of successive splits. Each region is checked for homogeneity based on some predefined criteria, and, if found to be nonhomogeneous, it is tentatively split into eight octant regions. These are examined, and octants with different properties remain as separate new regions, while similar octants are remerged. The splitting procedure is repeated for the newly created regions. The procedure will stop when

(1) all of the split regions remerge, (2) no more regions qualify for splitting, or (3) a minimum region size has been reached.

Region merging acts to merge touching regions that have similar properties based on predefined criteria. The process terminates when no further region merging is possible. This process can correct oversegmentation where objects have been broken up by the segmentation process.

Region splitting and merging can be performed together as a single algorithm known as the *split-and-merge* technique. This recursive method starts with the entire image as a single region and uses the splitting technique to divide nonhomogeneous regions based on predefined similarity criteria. Splitting continues until no further splitting is allowed. Then merging is used until it terminates. The resulting regions then constitute the segmentation results. This approach usually yields better segmentation results than the region splitting or merging methods alone. The region growing method frequently results in oversegmentation, and split-and-merge operations can be applied to improve the results.

11.12.4 Deformable Models

Deformable models, popularly known as active contours or snakes in 2D image processing (see Chapter 7), can be extended to 3D image processing where they are referred to as active surfaces or balloons [85]. These models represent geometric forms such as surfaces that are designed to simulate elastic sheets, so that when they are placed close to an object's boundary the models will deform under the influence of internal and external forces to conform to the shape of the object boundary. External forces, also known as *image forces*, are driven by image attributes and pull the active surface toward strong edges in the image. Internal forces typically enforce regularity based on specified properties of elasticity or rigidity that the surface must maintain.

There are two major classes of 3D deformable models, explicit [84] and implicit [86], and they differ in implementation. The surface of explicit deformable models is defined in a parametric form that uses global and local parameters. The surface S is defined in a parametric form. The two parameters s and r specify points that are on the surface as follows [73,84]:

$$u(s,r) = (u_1(s,r), u_2(s,r), u_3(s,r)) \quad s, r \in [0,1] \tag{11.76}$$

For every pair (s, r), u_1, u_2, and u_3 define the (x, y, z) coordinates of a point on the surface. The following energy functional, E, is associated with the surface [73]:

$$E(u) = \int_{\Omega} \left(w_{10} \left\| \frac{\partial u}{\partial s} \right\|^2 + w_{01} \left\| \frac{\partial u}{\partial r} \right\|^2 + 2w_{11} \left\| \frac{\partial^2 u}{\partial s \partial r} \right\|^2 + w_{20} \left\| \frac{\partial^2 u}{\partial s^2} \right\|^2 + w_{02} \left\| \frac{\partial^2 u}{\partial r^2} \right\|^2 + P(u(s,r)) \right) dsdr \tag{11.77}$$

where $\Omega = [0, 1] \times [0, 1]$, and w_{10}, w_{01} represent surface elasticity, w_{20}, w_{02} represent surface rigidity, and w_{11} represents its resistance to twist. The internal or image force, $P(u)$, which drives the surface toward the local gradient maxima, is given by [73]

$$P(u) = -\|\nabla I(u)\|^2 \tag{11.78}$$

The deformation process can be controlled by changing the values of the elasticity, rigidity, and resistance parameters or by using user-defined external forces, such as a pressure to drive the initial estimates until an edge is detected, or a weight that pulls the surface down until an object surface is detected. The Euler-LaGrange equation can be used to represent the local minimum of the energy functional E as [73]

$$-\frac{\partial}{\partial s}\left(w_{10}\frac{\partial u}{\partial s}\right) - \frac{\partial}{\partial r}\left(w_{01}\frac{\partial u}{\partial r}\right) + 2\frac{\partial^2}{\partial s \partial r}\left(w_{11}\frac{\partial^2 u}{\partial s \partial r}\right) + \frac{\partial^2}{\partial s^2}\left(w_{20}\frac{\partial^2 u}{\partial s^2}\right) + \frac{\partial^2}{\partial r^2}\left(w_{02}\frac{\partial^2 u}{\partial r^2}\right) = F(u) \tag{11.79}$$

where $F(u)$ is the sum of all forces acting on the surface. Typically, initial estimates are used to solve the following equation and compute the local minima [73] as

$$\frac{\partial u}{\partial t} - \frac{\partial}{\partial s}\left(w_{10}\frac{\partial u}{\partial s}\right) - \frac{\partial}{\partial r}\left(w_{01}\frac{\partial u}{\partial r}\right) + 2\frac{\partial^2}{\partial s \partial r}\left(w_{11}\frac{\partial^2 u}{\partial s \partial r}\right) + \frac{\partial^2}{\partial s^2}\left(w_{20}\frac{\partial^2 u}{\partial s^2}\right) + \frac{\partial^2}{\partial r^2}\left(w_{02}\frac{\partial^2 u}{\partial r^2}\right) = F(u) \tag{11.80}$$

The preceding equation is solved using the finite difference and finite element methods. This formulation of the explicit deformable model is also known as the energy-minimization formulation [87].

An alternative, known as the dynamic force formulation [85], provides the flexibility of applying different types of external forces to the deformable model. For example, the deformable model can start as a sphere inside the object and then use balloon force to push the model out to the boundary by adding a constant inflation pressure inside the sphere [85]. In general, explicit deformable models are easy to represent and have a fast implementation, but do not adapt easily to topology. Topological changes occurring during deformation are better handled by implicit deformable models [88] that represent the curves and surfaces implicitly as a level set of higher dimensional scalar function [89]. However, the implicit models' descriptions are not intuitive, and their implementations are computationally expensive.

Segmentation of cell and nuclei and their classification are two common tasks in microscope image analysis of biological specimens. Development of accurate and efficient algorithms for these tasks is a challenging problem because of the complexity and heterogeneity of biological samples, and several different approaches have been developed for 3D segmentation. The most popular approaches have combined region-based, edge-based, and morphological methods to achieve the desired results. The general consensus is that most applications require a customized set of specific segmentation algorithms, and most images require tuned parameters for optimal results, which limits providing an overview of published work in 3D segmentation. Thus we direct the interested reader to publications that provide a review of segmentation approaches for specific applications. For example, Magliaro et al. present a review of tools and procedures for segmenting single neurons [90], whereas methods for quantitative analyses of the 3D nuclear structures are reviewed by Schmid et al. [91] and Kraus et al. [92]. Reyer et al. [93] present an analysis package, Seg-3D, for the segmentation of bacterial cells in 3D images, based on local thresholding, shape analysis, concavity-based cluster splitting, and morphology-based 3D reconstruction. Software tools for 3D nuclei segmentation and quantitative analysis in multicellular aggregates were recently surveyed by Piccinini et al. [94]. Machine learning approaches for 3D segmentation are discussed in Chapter 15.

11.13 Comparing 3D Images

In some applications it is necessary to determine the similarity between two images, or to find the position of the best match between an image and a 3D template. The similarity between two three-dimensional images can be determined by computing the 3D correlation between them [73]:

$$R_{ab}(x', y', z') = \sum_{x=0}^{P_1-1} \sum_{y=0}^{P_2-1} \sum_{z=0}^{P_3-1} a(x, y, z)\, b(x + x', y + y', z + z') \tag{11.81}$$

where the input images a and b have regions of support $R_{P1,\ P2,\ P3}$, $R_{Q1,\ Q2,\ Q3}$, respectively. The correlation is defined in the region $(-Q_1, P_1) \times (-Q_2, P_2) \times (-Q_3, P_3)$. Correlation of two different images is called cross-correlation, whereas correlation of an image with itself is autocorrelation. The position of the peak in the cross-correlation function specifies the shift required for alignment, and the strength of that peak gives the degree of similarity.

11.14 Registration

In optical sectioning microscopy, where the sample is stationary during image acquisition (except in live cell imaging), all of the sections in the 3D optical stack will be well aligned. Thus, registration of the image slices requires no additional effort. For physically sectioned samples, however, the process of cutting the sections and fixing them to the glass slide creates distortion, and registration of the section images is required to form a 3D image. Cross-correlation, described earlier, can be used to determine the offset between adjacent section images. Another popular method to register images is based on the concept of *mutual information content* (or *relative entropy*) from the field of information theory. This has been used for multimodal images in medical imaging and is also effective for optical sections in a z-stack [95]. Mutual information content is a measure of the statistical dependence, or information redundancy, between the image intensities of corresponding voxels in adjacent optical slices. Maximal values are obtained if the images are geometrically aligned. The main advantage of this method is that feature calculation is straightforward since only gray level values are used, and the accuracy of the methods is thus not limited by segmentation errors.

11.15 Object Measurements in 3D

The ultimate objective of most microscopy investigations is to evaluate a sample in terms of features that convey essential information about the structure and function of the specimen. Image analysis algorithms are typically applied to measure object features such as topology, shape, contrast, and brightness. Intensity-based measurements for objects in 3D are similar to those described in Chapter 8 for 2D objects. Morphological and topological measurements for 3D objects are specialized since the third dimension is required, and some of the algorithms for performing these measurements are described here. Topological features convey information about the structure of objects, and they refer to those properties that are preserved through any deformation. Typically, definitions of adjacency and connectivity, described in Section 11.12.3.1, are the most common topological notions used in 3D image analysis. Another topological characteristic commonly measured for 3D objects is the Euler number, which is based on the number of cavities and tunnels in the object. Structural features include measures such as surface area, volume, curvature, and center of mass.

11.15.1 Euler Number

A *cavity* is the 3D analog of a hole in a 2D object and is defined as a background component that is totally enclosed within a foreground component [73]. The number of cavities is determined by applying the region growing algorithm (Section 11.12.3.2) to the background voxels in the image. This gives a count of the total number of background components, and the total number of cavities in the image is one less, since the background is also counted. Tunnels in 3D objects are more difficult to compute since a tunnel is not a separate background object. The first Betti number of the object is a value that is used to estimate the number of tunnels in an object. The first Betti number equals the number of nonseparating cuts one can make in an object without changing the number of connected components [73]. The Euler number of a 3D object is then defined as [73]

$$\chi(S) = (components) - (tunnels) + (cavities) \tag{11.82}$$

The Euler number corresponds to the topology of a closed surface rather than the topology of an object, and thus cannot differentiate between surfaces from distinct objects and surfaces originating from the same object, such as an object with a hole. The Euler number is used as a topology test in 3D skeletonization algorithms [96].

11.15.2 Bounding Box

The simplest method to measure the size of an object is to estimate the dimensions of its bounding box, that is, the smallest parallelepiped that contains the object. This can be done by scanning the 3D volume and finding the object voxels with the minimum and maximum coordinates along each dimension [73] as follows:

$$x = x_{max} - x_{min}, y = y_{max} - y_{min}, z = z_{max} - z_{min} \tag{11.83}$$

11.15.3 Center of Mass

The centroid of an object may be defined as the center of mass of an object of the same shape with constant mass per unit volume. The center of mass is, in turn, defined as that point where all the mass of the object could be concentrated without changing the first moment of the object about any axis [97]. In the 3D case the moments about the x, y and z axes are

$$\begin{aligned} X_c &= \iint_I x f(x, y, z) dx dy dz / M_o \\ Y_c &= \iint_I y f(x, y, z) dx dy dz / M_o \\ Z_c &= \iint_I z f(x, y, z) dx dy dz / M_o \\ M_o &= \iint_I f(x, y, z) dx dy dz \end{aligned} \tag{11.84}$$

where $f(x, y, z)$ is the value of the voxel at (x, y, z), M_o is the total mass, and (X_c, Y_c, Z_c) is the position of the center of mass. The integration is taken over the entire image. For discrete binary images the integrals become sums, and the center of mass for a 3D binary image can be computed as

$$X_c = \frac{\sum_i \sum_j \sum_k i f(i,j,k)}{\sum_i \sum_j \sum_k f(i,j,k)}, \quad Y_c = \frac{\sum_i \sum_j \sum_k j f(i,j,k)}{\sum_i \sum_j \sum_k f(i,j,k)}, \quad Z_c = \frac{\sum_i \sum_j \sum_k k f(i,j,k)}{\sum_i \sum_j \sum_k f(i,j,k)} \tag{11.85}$$

where $f(i, j, k)$ is the value of the 3D binary image (i.e., the intensity) at the point in the ith row, jth column and kth section of the 3D image (i.e., at voxel (i, j, k)). Intensities are assumed to be analogous to mass, so that zero intensity represents zero mass.

11.15.4 Surface Area Estimation

The surface area of 3D objects can be estimated by examining voxel connectivity in the 26-connected voxel neighborhoods of foreground voxels. A foreground voxel is considered to be a surface voxel if at least one of its 6-neighbors belongs to the background. The algorithm to find boundary voxels examines all object voxels to determine if there are any background voxels in their 6-neighborhood. The number of border voxels found is used as an estimate of the object's surface area. Other methods to estimate the surface have also been described [98].

11.15.4.1 Surface Estimation Using Superquadric Primitives

A parametric modeling method popular in the field of computer graphics can also be employed in digital image analysis for estimating the 3D surface area. This approach uses superquadric primitives with deformations to compute the 3D surface area. Superquadrics are a family of parametric shapes that are used as primitives for shape representation in computer graphics and computer vision. An advantage of using these geometric modeling primitives is that they allow complex solids and surfaces to be constructed and altered easily from a few interactive parameters. Superquadric solids are based on the parametric forms of quadric surfaces such as the superellipse and superhyperbola, in which each trigonometric function is raised to a power. The spherical product of pairs of such curves produces a uniform mathematical representation for the superquadric. This function is referred to as the inside-outside function of the superquadric, or the cost function. The cost function represents the surface of the superquadric that divides the 3D space into three distinct regions: inside, outside, and surface boundary. Model recovery may be implemented by using 3D data points as input. The cost function is defined such that its value depends on the distance of points from the model's surface and on the overall size of the model. A least-squares minimization method is used to recover model parameters, with initial estimates for minimization obtained from the rough position, orientation, and size of the object. During minimization, all the model parameters are adjusted iteratively to deform the model surface so that most of the input 3D data points are near the surface.

To summarize, a superquadric surface is defined by a single analytic function that can be used to model a large set of structures like spheres, cylinders, parallelepipeds, and shapes in between. Further, superquadric parameters can be adjusted to include such deformations as tapering, bending, and cavity deformation [99]. For example, superellipsoids may be used to estimate the 3D bounding surface of biological cells. The inside-outside cost function, $F(x, y, z)$, of a superellipsoid surface is defined by

$$F(x,y,z)=\left(\left(\left(\frac{x}{a_1}\right)^{2/\varepsilon_2}+\left(\frac{y}{a_2}\right)^{2/\varepsilon_2}\right)^{\varepsilon_2/\varepsilon_1}+\left(\frac{z}{a_3}\right)^{2/\varepsilon_1}\right)^{\varepsilon_1} \tag{11.86}$$

where x, y and z are the position coordinates in 3D space; a_1, a_2, a_3 define the superquadric size; and ε_1 and ε_2 are the shape parameters. The superquadric cost function is defined as [100]

$$F(x_W,y_W,z_W)=F(x_W,y_W,z_W:a_1,a_2,a_3,\varepsilon_1,\varepsilon_2,\varphi,\theta,\psi,c_1,c_2,c_3) \tag{11.87}$$

where a_1, a_2, a_3, ε_1 and ε_2 are described in the preceding text; φ, θ, ψ are orientation angles; and c_1, c_2, c_3 define the position in space of the center of mass. To recover a 3D surface it is necessary to vary the preceding 11 parameters to define a set of values such that most of the outermost 3D input data points lie on or near the surface. The orientation, size, and the shape parameters are varied, and the cost function is minimized using an algorithm such as the Levenberg-Marquardt method [47]. Typically, severe constraints are essential during minimization to obtain a unique solution, since different sets of parameter values can produce almost identical shapes. Fig. 11.7 presents an application of this approach for determining the separation distance in 3D between the fluorescent in situ hybridization (FISH) dots (i.e., signals) of a duplicated gene. Optical sections were obtained through FISH labeled cells at a z-interval of 0.1 μm. Object boundary points were determined for each of the two FISH dots, in each optical section of the z-stack, and used to estimate the 3D surface for each dot using Eq. (11.87). In the figure, the green symbols represent points on the estimated 3D surface for the two dots. The two dots are shown using blue and red symbols. In Fig. 11.7 the line segment PQ shows the separation distance between the two dots [40].

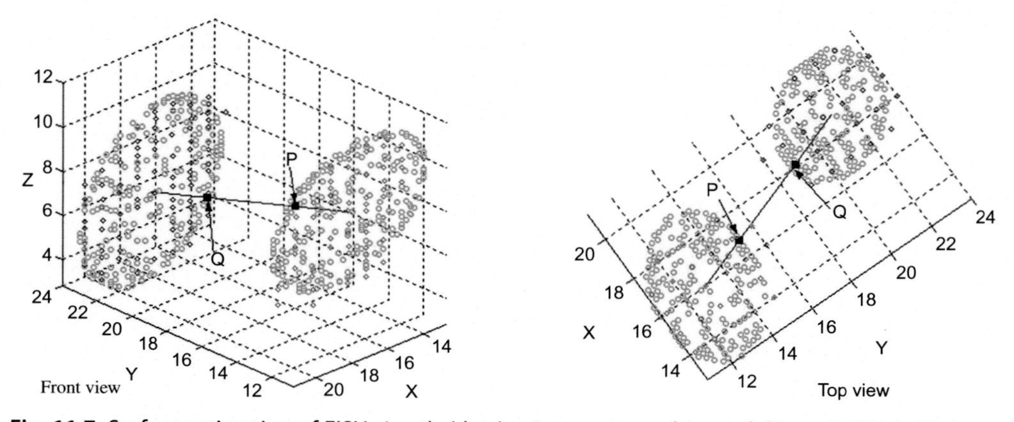

Fig. 11.7 Surface estimation of FISH signals (dots) using superquadric modeling primitives. Line segment PQ shows the separation distance between two dots. The distance between dots was computed after surface estimation.

11.15.4.2 Surface Estimation Using Spherical Harmonics

Surface estimation for cells and nuclei can also be performed using spherical harmonics (SPHARM), $Y_l^m(\theta, \phi)$, which is a frequency-space shape descriptor based on sphere coordinates at order, l, and degree, m:

$$Y_l^m(\theta, \phi) = \sqrt{\frac{2l+1}{4\pi} \frac{(l-m)!}{(l+m)}} P_l^m(\cos\theta) e^{im\phi} \tag{11.88}$$

where

$$P_l^m(x) = \frac{(-1)^m}{2^l l!} \left(1 - x^2\right)^{m/2} \frac{d^{l+m}}{dx^{l+m}} \left(x^2 - 1\right)^l \tag{11.89}$$

SPHARM, introduced by Brechbühler et al., [101] is a technique that combines three spherical functions with series of spherical harmonic coefficients to describe genus-zero surfaces. Yen et al. [102] implemented a modified SPHARM [103], which can define surface meshes from large data sets. SPHARM modeling was performed as follows. First a continuous mapping was conducted to transforms the object surface onto a unit sphere in form of spherical coordinates θ and ϕ. Next the surface was expanded into a set of spherical harmonic basis functions with coefficients for each spherical harmonic. The shape of the modeled objects is described by the sum of spherical harmonics in the different dimensions (degree) weighted by the coefficients

$$V(\theta, \phi) = (x(\theta, \phi), y(\theta, \phi), z(\theta, \phi))^T = \sum_{l=0}^{\infty} \sum_{m=-1}^{l} C_l^m Y_l^m(\theta, \phi). \tag{11.90}$$

C_l^m is the coefficient matrix corresponding to spherical harmonic, Y_l^m. Yen et al., [102] utilized SPHARM to model chromosome territories labeled via 3D fluorescent in situ hybridization (FISH) with whole chromosome panting (WCP). The surfaces of nuclei and the enclosed chromosome territories were reconstructed from SPHARM coefficients as shown in Fig. 11.8.

11.15.5 Length Estimation

Length measurement in 3D can be performed by two different approaches. The first is a straightforward extension of the chain-code based length estimator for 2D images described in Chapters 6 and 8. In this method, connected voxels along a straight line of N elements are segregated into three types: grid parallel, square diagonal, and cube diagonal. The numbers of elements in each class ($n_{grid}, n_{square}, n_{cube}$) are scaled using predefined weight coefficients (α, β, γ) to determine the 3D length as follows [104]:

$$L = \alpha n_{grid} + \beta n_{square} + \gamma n_{cube} \tag{11.91}$$

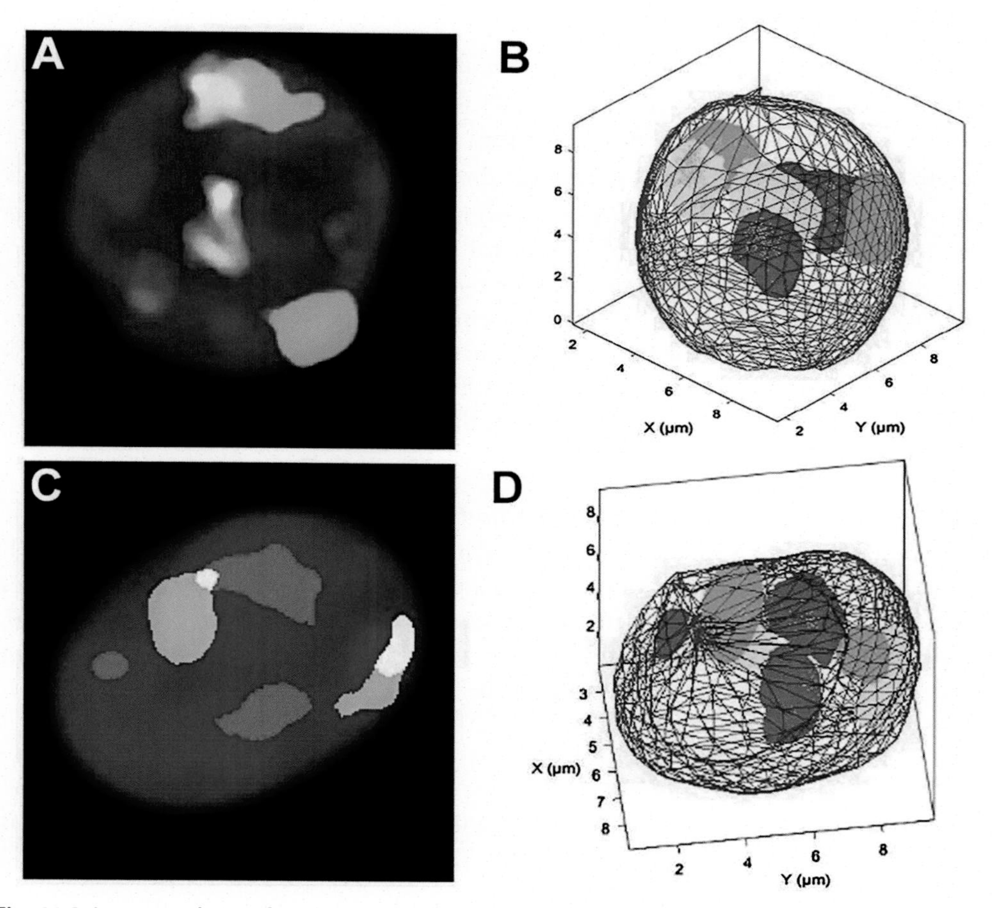

Fig. 11.8 Image analyses of nucleus and chromosome territories. (a) The maximum intensity projection of 22 optical sections of a diploid cell from a confocal microscope image. Overlay image showing the nucleus and enclosed chromosome territories for X *(green)*, 12 *(red)* and 8 *(yellow)*, (b) 3D surfaces of nucleus and chromosome territories of the cell shown in (a) estimated using SPHARM, (c) the maximum intensity projection of 29 optical sections for a trisomy 12 nuclei, and (d) 3D surfaces of nucleus and chromosome territories of the cell shown in (c) estimated using SPHARM. *(Reproduced with permission from [102])*

Several different values for the weight parameters, such as $\alpha = 0.9016$, $\beta = 1.289$, $\gamma = 1.615$, have been published [105–107].

The second method for measuring length in 3D is based on techniques from stereology. This approach works best for long curves or on sets of curves, yielding an estimate of the total length. The length is estimated by counting the number of intersections between the projection of the original curve onto a plane and a few equidistantly spaced straight lines in that plane [108].

11.15.6 Curvature Estimation

Curvature is increasingly used as a feature in object recognition and classification applications. The definitions of normal, Gaussian, and mean curvature are as follows. If $\vec{u}$ and $\vec{n}$ are the tangent and normal vectors, respectively, to any surface M at a point P, then the *normal curvature* of M at P, in the direction of $\vec{u}$, is the curvature of the space curve, C, formed by the intersection of M with the plane N spanned by $\vec{u}$ and $\vec{n}$. The minimum and maximum values of the normal curvature are known as the *principal curvatures*, p_1 and p_2, respectively. The *Gaussian curvature*, K, of a surface M at a point P is computed as the product of the two principal curvatures (i.e., $K = p_1 p_2$). Finally, the *mean curvature*, H, is the mean of the two principal curvatures, $H = 1/2[p_1 + p_2]$. Several techniques are available for curvature estimation [109]. Two approaches, surface triangulation and the cross patch method, are described in the following sections.

11.15.6.1 The Surface Triangulation Method

For a 3D surface mesh, a patch is approximated by a series of adjacent triangles. Since each triangle is flat, the Gaussian curvature is estimated at the common vertex of the triangles (the center of the patch). The Gaussian curvature, $K = \Delta\theta/A$, where $\Delta\theta = 2\pi - \sum_i \theta_i$ is a quantity called the angle deficit, θ_i is the vertex angle of the individual triangles in the series, and A is the sum of the areas of the triangles [109].

11.15.6.2 The Cross Patch Method

In this method, a discrete surface patch consisting of a cross (+) shaped set of points is modeled as a sampling of a continuous patch, $\vec{S}(u, v)$, such that

$$\vec{S}(u, v) = [x(u, v)\ y(u, v)\ z(u, v)]^T \tag{11.92}$$

where $x(u, v)$, $y(u, v)$ and $z(u, v)$ are polynomial functions whose coefficients are determined from the input data points using least squares fitting [109]. The parameters u and v represent traces along the surface in two orthogonal planes (the cross). These planes (1) must contain the point at which curvature is to be estimated (the inspection point), (2) must be parallel to the coordinate planes (the x-y, x-z, and y-z planes), and (3) must be the two such planes (out of three) that are most coincident with the surface normal at the inspection point, as indicated by the magnitude of the 2D vector resulting from projection of the normal onto the plane. The result of fitting coefficients for the $x(u, v)$, $y(u, v)$ and $z(u, v)$ polynomials is an estimate of the original continuous surface, and the parameters are used to determine the surface area and its curvature.

11.15.7 Volume Estimation

The simplest approach to obtain an estimate of the volume of a 3D object is an extension of the 2D procedure to measure area. The region of interest is initially outlined on each of the individual sections, and then the area of each segmented region on each section is multiplied by the section thickness (the z-interval). The total volume is then the sum of the volumes from each individual section. Alternatively, complex computer graphics models are estimated for the 3D surface of the region of interest, and the enclosed volume can be determined mathematically [98–100]. For example, polyhedral and curved surface representations allow piecewise integration to be used for volume measurement.

11.15.8 Texture

3D texture measures are used to describe the variations of image intensity value and contrast within the image. For example, a cell can contain directional texture indicative of the formation of fibers. Texture can be measured in 3D by simply extending work done in 2D analysis of textures to 3D images. A description of 2D texture measures is presented in Chapter 8.

11.16 3D Image Display

One of the greatest challenges in 3D imaging is the final step, that of visual representation and display of the images. Given the vast amount of information that exists in 3D datasets, it is difficult to extract useful information from the raw data, and well-designed visualization methods are essential to fully understand the 3D structures and their spatial relationships. Volumetric data may be displayed using techniques such as surface rendering, volume rendering, or maximum intensity projection. Traditional display devices, such as the computer monitor and paper prints, are inherently 2D, while the data is inherently 3D. A projection technique is commonly used to move data from higher dimensions to 2D for viewing. The use of various perceptual cues, such as shading or rotating a projection of the image volume, are needed to impart a sense of depth when displaying 3D data on a 2D device. Here we describe some of the popular techniques used to display microscopy data (normally a series of 2D slices) as a 2D projection image, and as a 3D rendering using stereo-pairs and anaglyphs.

11.16.1 Montage

A montage is often used to display a series of images. These images may be (1) the sequence of serial optical sections in a 3D image stack, (2) a series of projected images, each of which is created by rotating the 3D image to a specific angle, or (3) images from a time-lapse experiment. A montage display is a composite image created by combining several separate images. The images are tiled on the composite image, in a grid pattern, typically with the name of the image optionally appearing just below the individual tile. Each individual image in the montage is scaled to fit a predefined maximum tile size.

In 3D microscopy, because we are optically sectioning the specimen volume, a montage is the appropriate way to visualize all of the serial cross sections through the specimen simultaneously, rather than visualizing sequential presentations of single sections. The simultaneous presentation of images in a montage allows a visualization of continuity of image features in depth, and other spatial relationships.

Fig. 11.9 shows a montage of serial sections through a mouse 3T3 fibroblast obtained at an optical sectioning, z-interval of 1.8 µm. The labels 1–8 in Fig. 11.9 represent cell

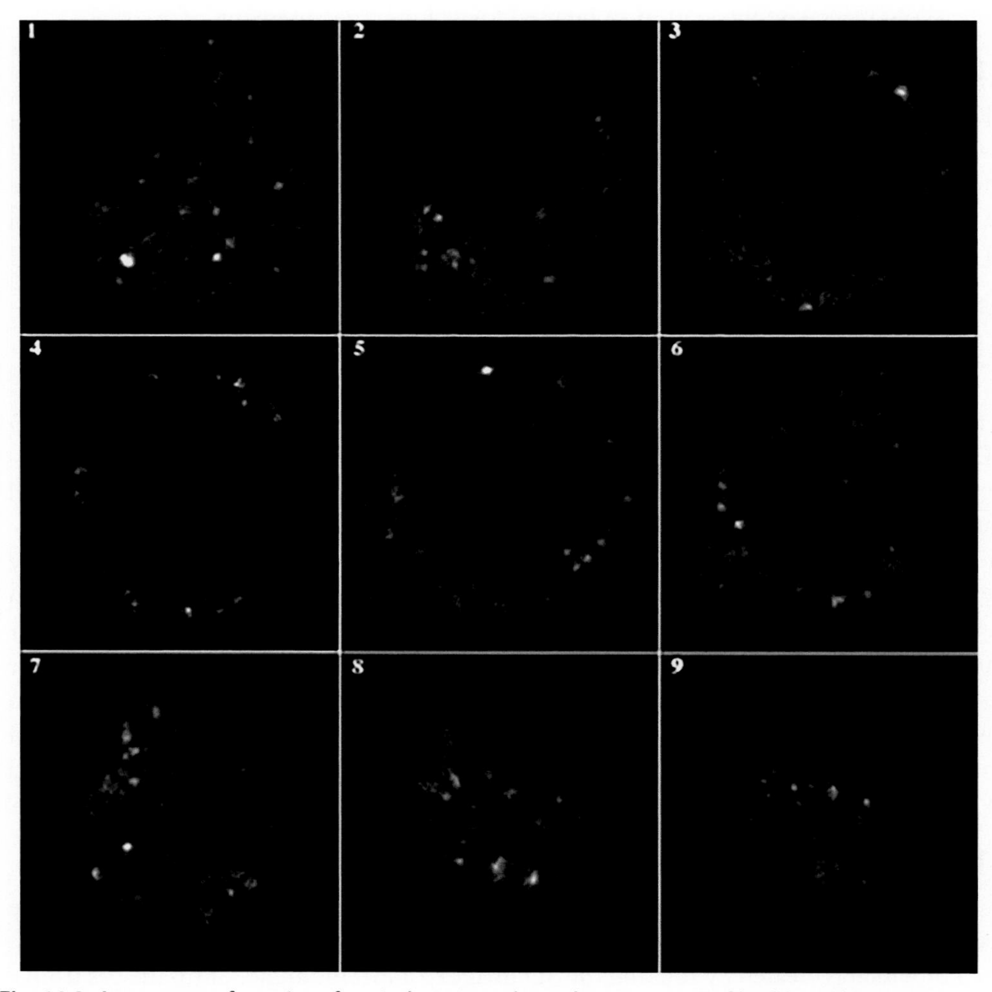

Fig. 11.9 A montage of a series of optical sections through a mouse 3T3 fibroblast. The images were obtained using a z-interval of 1.8 µm for optical sectioning. The labels 1–8 represent cell depths ranging from 0 to 14.4 µm. Panel 1 shows the top of the cell at 0 µm, panel 5 is the center of the cell at a depth of 7.2 µm, and panel 8 shows the bottom of the cell at a depth of 14.4 µm. The cell was treated with bacterial toxin H5 prior to immunofluorescence labeling.

depths ranging from 0 to 14.4 μm. Panel 1 shows the top of the cell at 0 μm, panel 5 is the center of the cell at a depth of 7.2 μm, and panel 8 shows the bottom of the cell at a depth of 14.4 μm. The cell was treated with bacterial toxin H5 prior to immunofluorescence labeling. The bacterial toxin interacts only with cell membranes. The montage display allows the 3D image data to be viewed simultaneously, thus permitting direct visualization of the peripheral localization of the toxin without cellular internalization.

11.16.2 Projected Images

A projected image offers a 2D view of a 3D image. It is created by projecting the 3D image data onto a plane. This is done by suitably combining the voxel intensity values along a set of rays extending through the image volume and onto a perpendicular plane. Different algorithms are available for generating such projections, two of which are discussed here: ray casting and voxel projection. Ray casting consists of tracing a line perpendicular to the viewing plane from every pixel in the serial sections into the 2D projection image. Voxel projection involves projecting voxels encountered along the projection line from the 3D image directly onto the 2D projection image. Of the two methods, voxel projection is computationally faster.

11.16.2.1 Voxel Projection

The most straightforward and computationally simple approach to image projection is to use the maximum intensity projection (MIP) algorithm. This algorithm uses a projection path that is orthogonal to the slices (2D image planes) in a 3D dataset. The MIP algorithm picks up the brightest voxel along a projection ray and displays it on the projected 2D image. The projected image is the same size as the slice images. MPI is most effective when the objects of interest in the 3D image have relatively simple structures with uniformly bright voxels. It produces images that have a particularly high contrast for small structures, but it does not represent fluorescence concentrations quantitatively and thus cannot be used prior to further analysis.

Alternatively, but at the risk of losing image sharpness, average (mean) intensity projection (AIP) can be used when quantification is required. AIP can be useful to measure relative fluorophore concentrations and their dynamic changes. The ranged intensity projection (RIP) method allows the user to select a range of intensity values that are of most interest in a specific application. It works like AIP except that only those voxel intensities that are within the specified range are accumulated during projection. A major advantage of these methods is that they do not require image segmentation. Their main drawback is their poor performance in noisy images. Fig. 11.10 presents a maximum intensity projection image of microvasculature at an islet transplant site in the rat kidney [110].

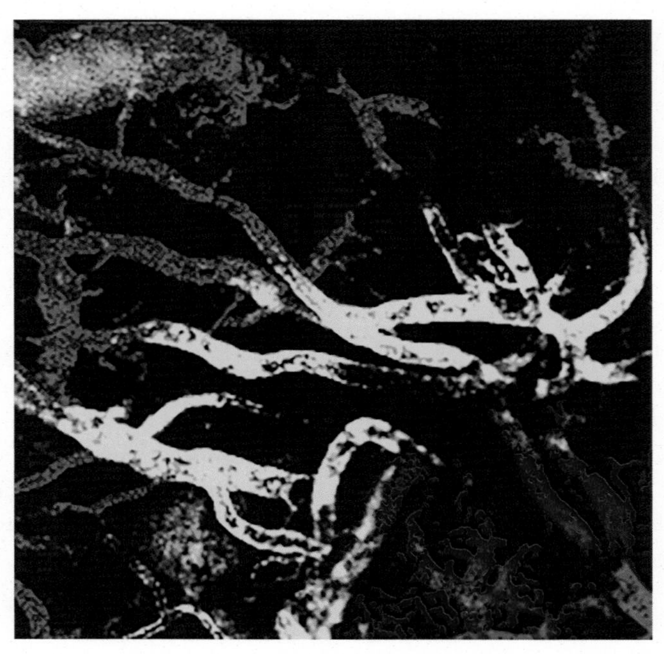

Fig. 11.10 Maximum intensity projection image of the microvasculature of an islet graft at the renal subcapsular site. The image is color coded to denote depth. The blood vessels appearing in the lower portion *(blue)* are at a depth of 30 µm, whereas those in the middle and upper portions of the image are at a depth of ~85 *(yellow)* and 135 µm *(violet)*, respectively.

11.16.2.2 Ray Casting

Ray casting requires segmented image data. It models the process of viewing the 3D object in space. Rays from a virtual light source are simulated as reflecting off the surface of the objects in the 3D image (see Fig. 11.11). The reflected rays are modeled to intersect a virtual screen between the dataset and the observer. The intensity and color of each reflected ray determines the intensity and color of the corresponding pixel on this screen. When this is done for all pixels in the output image, it becomes a 3D view of the object. A view from another direction can be calculated by moving the observer and screen to another point in the virtual space of the dataset. Views from several different viewing angles can be grouped together as sequential frames to form a movie of the object rotating in space.

11.16.3 Surface and Volume Rendering

Three-dimensional graphical rendering is popularly used in the field of computer graphics for displaying 3D objects. Volume rendering algorithms can be used to create 2D projection images showing stacks of 2D cross-sectional images from various points

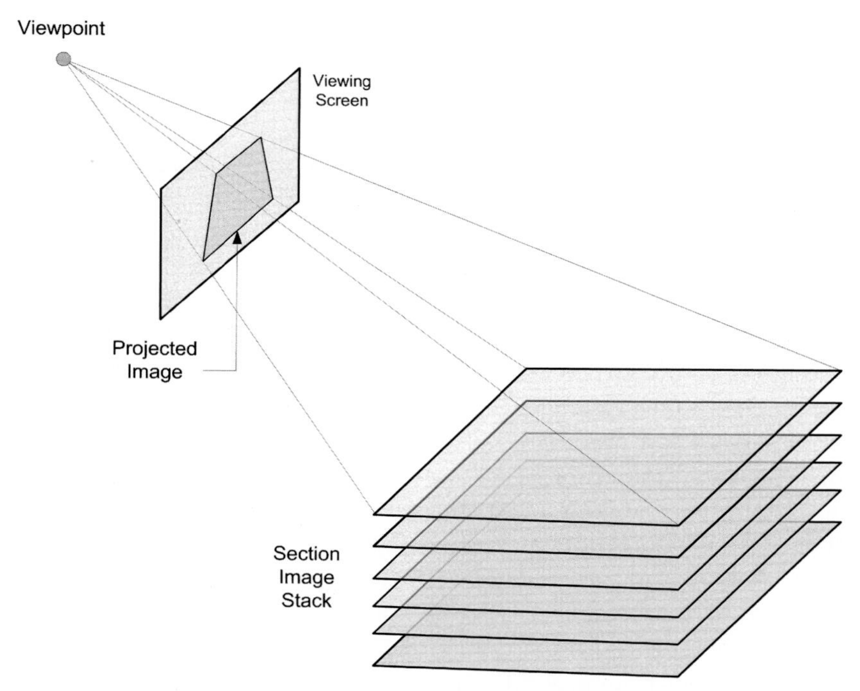

Fig. 11.11 Three-dimensional Image Projection. *(Courtesy of Kenneth R. Castleman, after [29], Fig. 22.20)*

of view. Two algorithms for rendering are briefly described here, and the reader is directed to published texts for an in-depth description [111,112].

11.16.3.1 Surface Rendering

Object surfaces in a segmented 3D image are portrayed by the surface rendering process. Standard 3D graphics techniques are used to render objects, wherein their surfaces are approximated by a list of polygons. Several algorithms are available to generate the polygons that represent an object's surface. The most widely used method to triangulate a 3D surface is known as the marching cube algorithm [113]. Following triangulation, the desired viewing angle is selected, and the object is displayed by projecting the polygons onto a plane perpendicular to the viewing direction. The object can be viewed from any desired angle either by applying a rotation matrix to the polygon list or by changing the viewing direction used in the projection process. Shading or changing the transparency of the rendered objects can be used to improve the aesthetics and enhance the 3D viewing effect. The quality of a display made using surface triangulation is highly dependent on the accuracy of the prior segmentation process, which, if poorly done, can result in a loss of detail in the rendered object.

11.16.3.2 Volume Rendering

An alternative approach that does not require defining the surface geometry with polygons is to render the entire volume by projecting colored, semitransparent voxels onto a chosen projection plane. There are three steps in volume rendering: classification, shading (also known as compositing), and projection. In the classification step, each voxel is assigned an opacity value ranging from zero (transparent) to one (opaque) based on its intensity value. In order to avoid uneven transitions between transparency and opacity, a hyperbolic tangent mapping of voxel intensities is often applied. Every voxel is then considered to reflect light, with the goal to determine the total amount of light it directs to the viewing screen. Shading is used to simulate the object's surface, including both the surface's position and orientation with respect to the light source and the observer. An illumination model is used to implement the shading function [114], and the color and opacity of each voxel is computed [115].

The final stage is the projection step. Here the colored semitransparent voxels are projected onto a plane perpendicular to the observer's viewing direction. Ray casting is used to project the data, as before. It involves tracing a line perpendicular to the viewing plane into the object volume. For each grid point (m, n) in the projection plane, a ray is cast into the volume, and the color and opacity are computed at evenly spaced sample locations. Compositing (shading) is applied in a back-to-front order to yield a single pixel color $C(m, n)$. A detailed description of this approach is given in [115].

Most of the major commercial 3D microscopy systems include a volume rendering module, and specialized software for 3D visualization [111]. Other commercial packages for 3D visualization include Volocity (Improvision, Perkin Elmer, Waltham, MA). Imaris (Bitplane, Oxford Instrument Company, Concord, MA, United States) [116], and Amira (Visage Imaging Inc.) [117]. There are also several free volume rendering programs available, including Voxx [118], and volume rendering plugins for the ImageJ software package such as VisBio, VolumeJ, and SurfaceJ [119–121]. Fig. 11.12 shows an example of volume rendering of the ureteric tree of a fetal mouse kidney. Optical sections of an E13 Hoxb7/GFP kidney were collected using a laser scanning confocal microscope, and volume rendered using the Volocity software.

Fig. 11.13 shows both a 3D rendering and 2D projection of a specimen where multiple fluorescent molecules are excited at different wavelengths. This example uses 2PE microscopy to prime multiple fluorophores that can be spectrally separated to map the different molecules [15].

11.16.4 Stereo Pairs

Stereoscopic techniques are used to retain the depth information when mapping 3D data into a 2D view for visualization. One method is to simultaneously present the observer with "left eye" and "right eye" views differing by a small parallax angle of about 6 degrees.

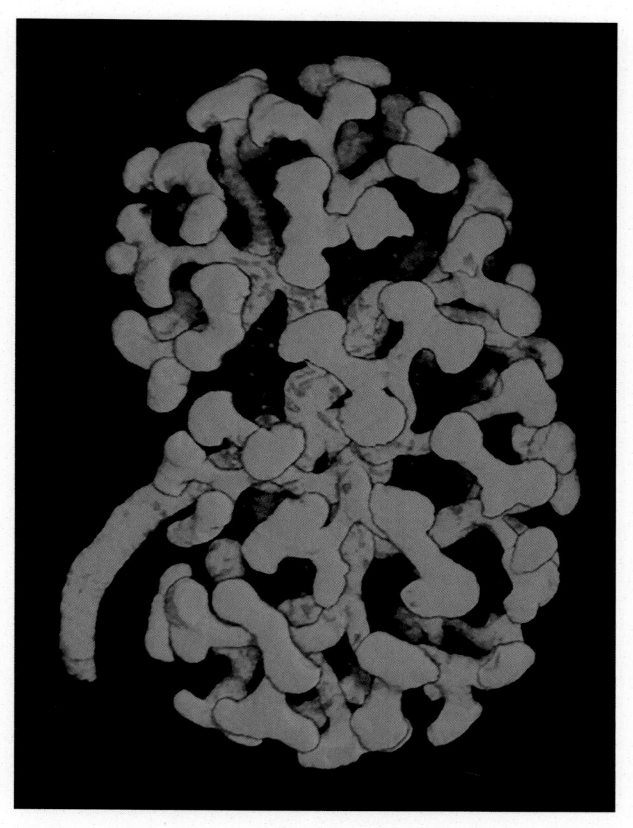

Fig. 11.12 Volume rendering of the ureteric tree of a fetal mouse kidney. *(Courtesy of Deborah Hyink)*

Stereo pairs can be generated from volumetric data in two ways. The first method is to rotate the full 3D dataset anywhere from ±3 to ±10 degrees and then project each of these rotated views onto a 2D plane [111]. Alternatively, instead of rotating the image data, the projection images can be created simply by summing the stack of serial images with an appropriate constant horizontal offset between the adjacent sections, to generate a top-down view. This can be described by the stereo "pixel shift" function [111] as

$$x_v = x_i + \delta x \times z_v, \qquad y_v = y_i, \qquad z_v = z_i \tag{11.93}$$

where ϕ_s is the parallax angle, $\delta x_{left} = \tan(0.5 \times \varphi_s)$ and $\delta x_{right} = -\delta x_{left}$. Alternatively, the optimal pixel shift can be computed using [122]

$$\delta_p = 2 \times nz_{calib} \times M \times \sin\left(\arctan\left(\delta x \times nz_i / nz_{calib}\right)\right) \tag{11.94}$$

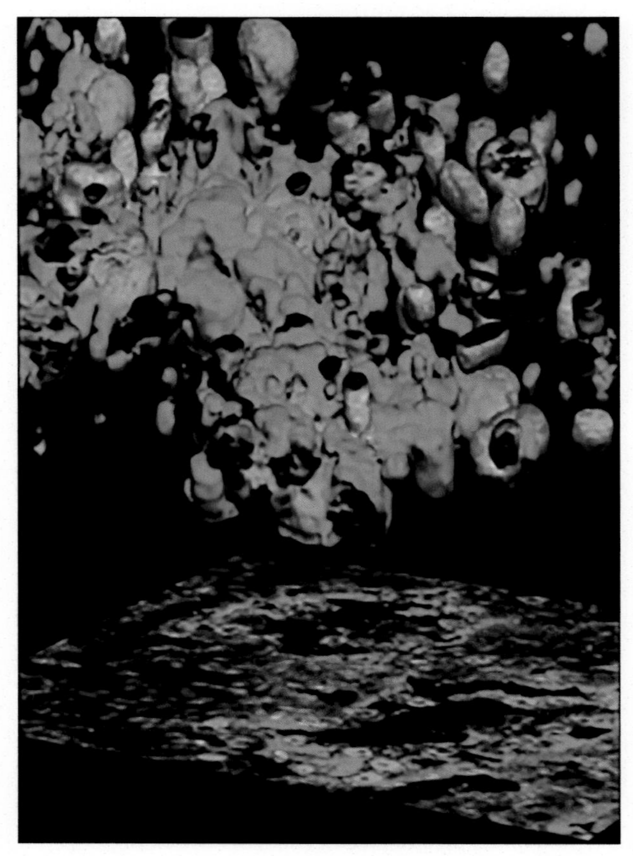

Fig. 11.13 3D and 2D fluorescence projections. Pictorial representation of the 3D and 2D projections of multiple fluorescence from a marine sponge, *Chondrilla nucula*. The specimen has been loaded with Alexa 488 fluorescent molecules specific for aminobutyric acid (GABA) emitting in green and DAPI *(blue)* for nuclear DNA. *Red* signals are due to the autofluorescence from symbiotic bacteria contamination. Imaging was acquired using a Chameleon XR ultrafast Ti-Sapphire laser (Coherent Inc., United States) coupled at LAMBS-MicroScoBio with a Spectral Confocal Laser Scanning Microscope, Leica SP2-AOBS. *(Courtesy Paolo Bianchini. Reproduced with permission from [15])*

where δ_p is the parallax shift between equivalent points in the two views, nz_{calib} is the calibrated z-size of the data (thickness of the specimen), and M is the magnification. The left view is made by sequentially shifting each image in the stack to the left, progressing from the farthest to the nearest section. The number of pixels to be shifted depends on the pixel size and the distance between sections.

The resulting left and right views are known as a "stereo pair." They must be visually "fused" into a 3D representation by positioning each view in front of the corresponding eye. The observer then perceives a single 3D view at the appropriate viewing distance.

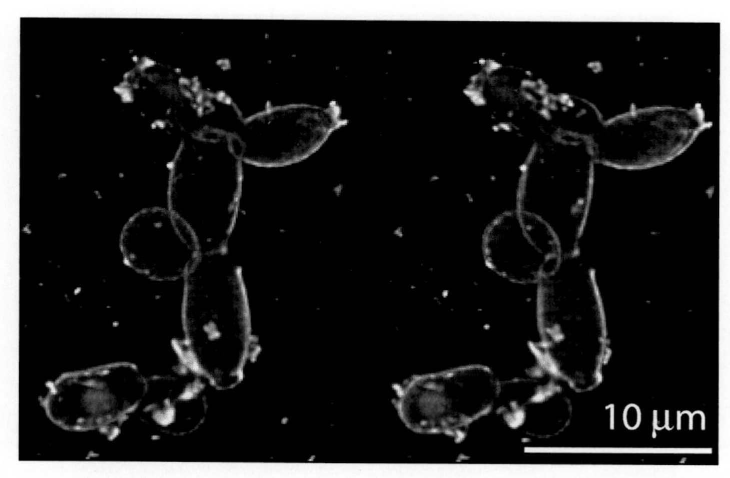

Fig. 11.14 Maximum intensity stereo pair image of the of hystoplasma capsulatum. *(Courtesy of José-Angel Conchello)*

Alternatively, side-by-side stereo pairs can also be viewed with a horizontal prismatic viewer that appropriately presents the images to each eye. Fig. 11.14 shows a maximum intensity stereo pair image of hystoplasma capsulatum.

11.16.5 Color Anaglyphs

Anaglyphing [123] is a method of presenting a stereo pair to the viewer's eyes. The anaglyph is a stereo pair in which the left eye image is printed in red and the right eye image in blue (or cyan). It is viewed through a special pair of glasses having a red lens over left eye and a blue lens over right eye. The two images combine to create the perception of a three-dimensional image. Fig. 11.15 shows a red-cyan anaglyph of a pancreatic islet.

11.16.6 Animations

Animations make use of the temporal display space for visualization. This technique involves the sequential display, in rapid succession, of rotated views of the object. The rotation angle is successively increased, and a rendered image is created at each angle. When the different views are presented as sequential frames, the result is a movie of the spinning object. The requirements for creating smooth animations are (1) small rotation steps, (2) a short persistence noninterlaced monitor, and (3) a display frame rate greater than 10 Hz [111]. Animations have proved to be a valuable display tool, especially for the assessment of the shape and relative position of 3D structures. A complex structure is much easier to visualize and understand in a rotating animation than in a series of cross-section images.

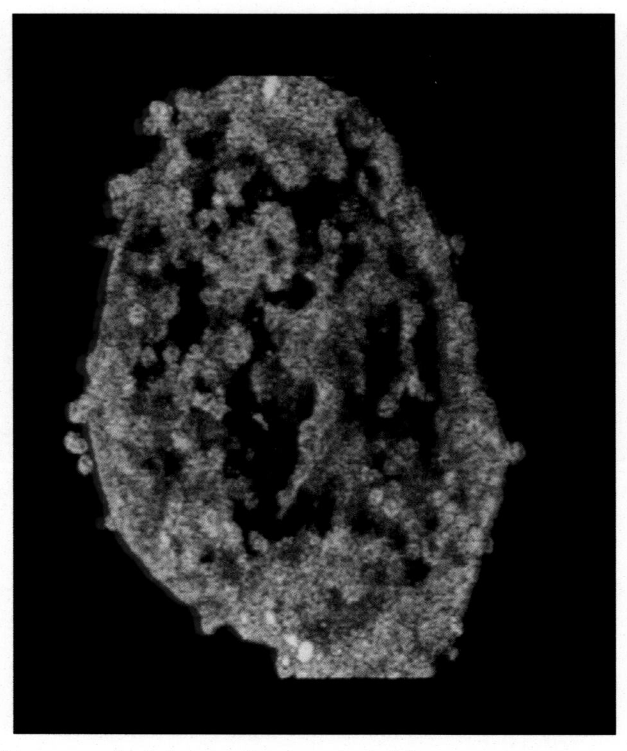

Fig. 11.15 *Red* and *cyan* anaglyph image of a pancreatic islet.

11.17 Summary of Important Points

1. Three-dimensional light microscopy offers a noninvasive, minimally destructive option for obtaining spatial and volumetric information about the structure and function of cells and tissues.
2. The most popular and widely commercialized approaches to 3D microscopy include the far-field techniques of wide-field, confocal, and multiphoton microscopy.
3. New methods for 3D microscopy that offer improved resolution include 4Pi-confocal, I^nM, HELM, OCT, OPT, and SPIM.
4. Three-dimensional data obtained from microscopes consist of a stack of two-dimensional images (optical sections), referred to as the "*z*-stack," "3D-stack" or "volumetric data."
5. In 3D imaging, light emitted from planes above and below the in-focus regions is also captured in each optical section, and it appears as out-of-focus blurring.

6. The out-of-focus light can be characterized by the point-spread function (psf), which is the 3D image of a point source of light.
7. Three methods are commonly used for estimating the psf of a microscope: experimental, theoretical, and analytical.
8. Image formation can be mathematically modeled as a convolution operation, whereby light emitted from each point in the specimen is convolved with the 3D psf and appears as a blurred region in the image.
9. The deconvolution process (deblurring or restoration) estimates the amount of out-of-focus light characterizing the particular microscope optics in use (i.e., psf), and then attempts to either subtract out this light or redistribute it back to its point of origin in the specimen volume.
10. Algorithms that are applied sequentially to each 2D plane of a 3D image stack, one at a time, are classified as deblurring procedures.
11. Image restoration algorithms operate simultaneously on every voxel in a 3D image stack and are implemented in 3D space to restore the actual distribution of light in the specimen.
12. There are six major categories of deconvolution algorithms: (1) no-neighbors methods, (2) neighboring plane methods, (3) linear methods, (4) nonlinear methods, (5) statistical methods, and (6) blind deconvolution.
13. The choice of the deconvolution procedure used is often dictated by the imaging system, experimental conditions, and the specific application.
14. The results obtained from deconvolution procedures depend heavily on image quality, specifically, the signal-to-noise ratio and the sampling density.
15. Deconvolution software is available both commercially and as freeware.
16. Image fusion is an approach frequently used in 3D microscopy to combine a set of optical section images into a single 2D image containing the detail from each optical section in the stack.
17. Three-dimensional image processing can be performed using three different approaches: (1) by performing 2D image processing operations on the individual optical sections, (2) by performing operations using the entire 3D dataset, and (3) by performing 2D operations on the fused image of a stack.
18. For most image processing techniques, the 2D counterpart can be easily extended to 3D.
19. Gray level mathematical operations such as subtraction and addition and binary operations such as OR, XOR, and AND can be expressed for 3D images, using the voxel as the brightness unit.
20. Image transformations, such as translation, reflection, rotation, and scaling can also be expressed in 3D.
21. Image-filtering operations are typically used either to reduce noise by smoothing or to emphasize edges.

22. Digital convolution filters are linear since the output pixel values are linear combinations of the input image pixels, but other types of filters are nonlinear.
23. Edge detection filters are commonly used for segmentation based on surface detection in 3D.
24. Mathematical morphological operators are a subclass of nonlinear filters that are used for shape and structure analysis, and filtering in binary and grayscale images.
25. Three-dimensional morphological operators are straightforward extensions of their 2D counterparts, with sets and functions defined in the 3D Euclidean grid $\mathbf{Z}^3$.
26. Three-dimensional image stacks are usually anisotropic, that is, the sampling interval along the axial dimension is larger than in the lateral dimension. For this reason the structuring element should be chosen as an anisotropic 3D kernel.
27. Segmenting regions of volumetric images is challenging, because, in 3D, regions are often touching each other or overlapping and irregularly arranged, with no definite shape, and intensity typically falls off deep within the specimen due to diffraction, scattering, and absorption of light.
28. There are three general approaches to segmentation: point-based, edge-based, and region-based methods.
29. For 3D segmentation the most popular approaches have combined region-based, edge-based, and morphological methods to achieve the desired results.
30. Two 3D images can be compared by computing the 3D correlation between them.
31. One popular method to register images in an optical stack is based on the concept of mutual information content (or relative entropy).
32. Intensity-based measurements for objects in 3D are similar to those for 2D objects.
33. Morphological and topological measurements, such as surface area, volume, curvature, length, Euler number, and center of mass for 3D objects are different from their 2D counterparts since the third dimension is required.
34. Traditional display methods are inherently 2D, so well-designed visualization methods are essential to display and understand 3D structures and their spatial relationships.
35. Typically perceptual cues, such as shading or rotating a projection of the image volume, are needed to impart a sense of depth when displaying 3D data.
36. Popular techniques used to display 3D microscopy data, include a montage of a series of 2D slices, 2D projection images, and 3D rendering with stereo-pairs and anaglyphs.

References

[1] J.A. Conchello, J.W. Lichtman, Optical sectioning microscopy, Nat. Methods 2 (12) (2005) 920–931.
[2] P.O. Bayguinov, et al., Modern laser scanning confocal microscopy, Curr. Protoc. Cytom. 85 (1) (2018), e39.
[3] H. Naora, Microspectrophotometry and cytochemical analysis of nucleic acids, Science 114 (2959) (1951) 279–280.

[4] M. Minsky, Memoir of inventing the confocal scanning microscope, Scanning 10 (1988) 128–138.
[5] M. Petran, et al., Tandem-scanning reflected-light microscope, J. Opt. Soc. Am. 58 (1968) 661–664.
[6] P.D. Davidovits, M.D. Egger, Scanning laser microscope, Nature 223 (1969) 831.
[7] T. Wilson, C. Sheppard, Theory and Practice of Scanning Optical Microscopy, Academic Press, London, 1984.
[8] G.J. Brakenhoff, et al., Three-dimensional imaging by confocal scanning fluorescence microscopy, Ann. N. Y. Acad. Sci. 483 (1986) 405–415.
[9] J.B. Pawley, Handbook of Confocal Microscopy, third ed., Springer, 2006.
[10] W. Becker, The bh TCSPC Handbook, ninth ed., 2021. Available on www.becker-hickl.com.
[11] R. Borlinghaus, The white confocal: continuous spectral tuning in excitation and emission, in: A. Diaspro (Ed.), Optical Fluorescence Microscopy, Springer, 2011.
[12] W. Denk, J.H. Strickler, W.W. Webb, Two-photon laser scanning fluorescence microscopy, Science 248 (1990) 73–76.
[13] A. Diaspro, G. Chirico, M. Collini, Two-photon fluorescence excitation and related techniques in biological microscopy, Q. Rev. Biophys. 38 (2) (2005) 97–166.
[14] W.R. Zipfel, R.M. Williams, W.W. Webb, Nonlinear magic: multiphoton microscopy in the biosciences, Nat. Biotechnol. 21 (2003) 1369–1377.
[15] A. Diaspro, et al., Multi-photon excitation microscopy, Biomed. Eng. Online 5 (2006) 36.
[16] R.M. Williams, D.W. Piston, W.W. Webb, Two-photon molecular excitation provides intrinsic 3-dimensional resolution for laser-based microscopy and microphotochemistry, FASEB J. 8 (1994) 804–813.
[17] C.R. Sheppard, M. Gu, Image formation in 2-photon fluorescence microscopy, Optik 86 (1990) 104–106.
[18] S. Hell, E.K. Stelzer, Fundamental improvement of resolution with a 4Pi-confocal fluorescence microscope using two-photon excitation, Opt. Commun. 93 (1992) 277–282.
[19] Z. Lavagnino, et al., Two-photon excitation selective plane illumination microscopy (2PE-SPIM) of highly scattering samples: characterization and application, Opt. Express 21 (2013) 5998–6008.
[20] L. Yu, M.K. Kim, Full-color 3-D microscopy by wide-field optical coherence tomography, Opt. Express 12 (26) (2004) 6632–6641.
[21] J. Sharpe, et al., Optical projection tomography as a tool for 3-D microscopy and gene expression studies, Science 296 (2002) 541–545.
[22] J. Swoger, J. Huisken, E.K. Stelzer, Multiple imaging axis microscopy improves resolution for thick-sample applications, Opt. Lett. 28 (18) (2003) 1654–1656.
[23] J. Langowski, Single plane illumination microscopy as a tool for studying nucleome dynamics, Methods 123 (2017) 3–10.
[24] E.H.K. Stelzer, Light-sheet fluorescence microscopy for quantitative biology, Nat. Methods 12 (1) (2014) 23–26.
[25] M.L. Gustafsson, D.A. Agard, J.W. Sedat, I^5M: 3-D widefield light microscopy with better than 100 nm axial resolution, J. Microsc. 195 (1999) 10–16.
[26] J.T. Frohn, H.F. Knapp, A. Stemmer, Three-dimensional resolution enhancement in fluorescence microscopy by harmonic excitation, Opt. Lett. 26 (2001) 828–830.
[27] D.A. Agard, J.W. Sedat, Three-dimensional architecture of a polythene nucleus, Nature 302 (1983) 676–681.
[28] K.R. Castleman, Three-dimensional image processing, in: Digital Image Processing, Prentice-Hall, 1979, pp. 351–360 (Chapter 17).
[29] K.R. Castleman, Digital Image Processing, second ed., Prentice-Hall, 1996.
[30] Y. Hiraoka, J.W. Sedat, D.A. Agard, Determination of the 3-D imaging properties of a light microscope system, Biophys. J. 57 (1990) 325–333.
[31] C. Preza, et al., Regularized linear method for reconstruction of 3-D microscopic objects from optical sections, J. Opt. Soc. Am. A 9 (1992) 219–228.
[32] P. Sarder, A. Nehorai, Deconvolution methods for 3-D fluorescence microscopy images, IEEE Signal Process. Mag. May (2006) 32–45.
[33] V.N. Mahajan, Zernike circle polynomials and optical aberrations of systems with circular pupils, Eng. Lab. Notes 17 (1994) S21–S24.

[34] P.J. Shaw, D.J. Rawlins, The point spread of a confocal microscope: its measurement and use in deconvolution of 3-D data, J. Microsc. 163 (1991) 151–165.
[35] W. Wallace, L.H. Schaefer, J.R. Swedlow, A workingperson's guide to deconvolution in light microscopy, Biotechniques 31 (2001) 1076–1097.
[36] H.H. Hopkins, The frequency response of a defocused optical system, Proc. R. Soc. A 231 (1955) 91–103.
[37] P.A. Stokseth, Properties of a defocused optical system, J. Opt. Soc. Am. A 59 (10) (1969) 1314–1321.
[38] J.R. Monck, et al., Thin-section ratiometric Ca^{2+} images obtained by optical sectioning of fura-2 loaded mast cells, J. Cell Biol. 116 (3) (1992) 745–759.
[39] M. Weinstein, K.R. Castleman, Reconstructing 3-D specimens from 2-D section images, Proc. Soc. Photo-Opt. Instr. Eng. 26 (1971) 131–138.
[40] F.A. Merchant, K.R. Castleman, Computerized microscopy, in: A.C. Bovik (Ed.), Handbook of Image and Video Processing, Academic Press, 2005.
[41] A. Chomik, et al., Quantification in optical sectioning microscopy: a comparison of some deconvolution algorithms in view of 3D image segmentation, J. Opt. 28 (1997) 225–233.
[42] G.P. van Kempen, Image restoration in fluorescence microscopy (Ph.D. thesis), Delft University Press, 1999.
[43] D.A. Agard, et al., Fluorescence microscopy in three-dimensions, Methods Cell Biol. 30 (1989) 353–377.
[44] W.A. Carrington, et al., Superresolution three-dimensional images of fluorescence in cells with minimal light exposure, Science 268 (1995) 1483–1487.
[45] P.J. Verveer, T.M. Jovin, Acceleration of the ICTM image restoration algorithm, J. Microsc. 188 (1997) 191–195.
[46] H.M. van der Voort, K.C. Strasters, Restoration of confocal images for quantitative image analysis, J. Microsc. 178 (1995) 165–181.
[47] W.H. Press, S.A. Teukolsky, W.T. Vetterling, Numerical Recipes in C, Cambridge University Press, 1992.
[48] J.A. Conchello, An overview of 3-D and 4-D microscopy by computational deconvolution, in: D. Stephens (Ed.), Cell Imaging: Methods Express, Scion Publishing Limited, 2006.
[49] N.M. Lashin, Restoration Methods for Biomedical Images in Confocal Microscopy (thesis), Technical University Berlin, 2005.
[50] L. Liang, Y. Xu, Adaptive Landweber method to deblur images, IEEE Signal Process Lett. 10 (5) (2003) 129–132.
[51] J.A. Conchello, E.W. Hansen, Enhanced 3D reconstruction from confocal scanning microscope images I: deterministic and maximum likelihood reconstructions, Appl. Optics 29 (1990) 3795–3804.
[52] T.J. Holmes, Y.H. Liu, Richardson-Lucy maximum likelihood image restoration algorithm for fluorescence microscopy: further testing, Appl. Optics 28 (1989) 4930–4938.
[53] J. Markham, J.A. Conchello, Tradeoffs in regulated maximum-likelihood image restoration, in: C.J. Cogswel, J.A. Conchello, T. Wilson (Eds.), 3D Microscopy: Image Acquisition and Processing IV, Proc. SPIE BIOS97, SPIE Press, 1997 (2984-18).
[54] S. Joshi, M.I. Miller, Maximum a posteriori estimation with Good's roughness for optical-sectioning microscopy, J. Opt. Soc. Am. A 10 (5) (1993) 1078–1085.
[55] L.I. Rudin, S. Osher, E. Fatemi, Nonlinear total variation based noise removal algorithm, Phys. D 60 (1992) 259–268.
[56] B.R. Frieden, Restoring with maximum likelihood and maximum entropy, J. Opt. Soc. Am. A 62 (1972) 511–518.
[57] T.J. Holmes, Blind deconvolution of quantum-limited incoherent imagery: maximum likelihood approach, J. Opt. Soc. Am. A 9 (1992) 1052–1061.
[58] E. Thiebaut, J.M. Conan, Strict a priori constraints for maximum-likelihood blind deconvolution, J. Opt. Soc. Am. A 12 (3) (1995) 485–492.
[59] L. Denis, et al., Fast approximations of shift-variant blur, Int. J. Comput. Vis. 115 (3) (2015) 253–278.
[60] Y. Chen, et al., Measure and model a 3-D space-variant PSF for fluorescence microscopy image deblurring, Opt. Express 26 (11) (2018) 14375–14391.

[61] E. Maalouf, B. Colicchio, A. Dieterlen, Fluorescence microscopy three-dimensional depth variant point spread function interpolation using zernike moments, J. Opt. Soc. Am. A 28 (9) (2011) 1864–1870.

[62] N. Patwary, C. Preza, Image restoration for three-dimensional fluorescence microscopy using an orthonormal basis for efficient representation of depth-variant point-spread functions, Biomed. Opt. Express 6 (10) (2015) 3826–3841.

[63] B. Kim, T. Naemura, Blind deconvolution of 3D fluorescence microscopy using depth-variant asymmetric PSF, Microsc. Res. Tech. 79 (6) (2016) 480–494.

[64] D. Sage, et al., DeconvolutionLab2: an open-source software for deconvolution microscopy, Methods 115 (2017) 28–41.

[65] C.A. Schneider, W.S. Rasband, K.W. Eliceiri, NIH image to ImageJ: 25 years of image analysis, Nat. Methods 9 (2012) 671–675.

[66] J. Schindelin, et al., Fiji: an open-source platform for biological-image analysis, Nat. Methods 9 (2012) 676–682.

[67] R. Dougherty, Extensions of DAMAS and benefits and limitations of deconvolution in beamforming, in: 11th AIAA/CEAS Aeroacoustics Conference, American Institute of Aeronautics and Astronautics, 2005.

[68] P. Wendykier, High Performance Java Software for Image Processing, Emory University, 2009.

[69] Z. Baster, Z. Rajfur, BatchDeconvolution: a Fiji plugin for increasing deconvolution workflow, Bio-Algoritms. Med-Syst. 16 (3) (2020) 20200027.

[70] B. Kim, DVDeconv: an open-source MATLAB toolbox for depth-variant asymmetric deconvolution of fluorescence micrographs, Cell 10 (2) (2021) 397.

[71] A. Griffa, N. Garin, D. Sage, Comparison of deconvolution software: a user point of view–part 1 and part 2, G.I.T. Imaging Microsc. 12 (2010) 41–45.

[72] Q. Wu, F.A. Merchant, K.R. Castleman, Microscope Image Processing, first ed., Academic Press, 2008 (Chapter 16).

[73] N. Nikolaidis, I. Pitas, 3-D Image Processing Algorithms, John Wiley and Sons, Inc, 2001.

[74] J. Serra, Image Analysis and Mathematical Morphology, Academic Press, 1998.

[75] F. Meyer, Mathematical morphology: from two-dimensions to three-dimension, J. Microsc. 165 (1992) 5–28.

[76] G. Lin, et al., A hybrid 3D watershed algorithm incorporating gradient cues and object models for automatic segmentation of nuclei in confocal image stacks, Cytometry A 56A (2003) 23–36.

[77] C.A. Glasbey, An analysis of histogram based thresholding operations, CVGIP: Graph. Models Image Process. 55 (1993) 532–537.

[78] P.K. Sahoo, S. Soltani, K.C. Wong, A survey of thresholding techniques, Comput. Vis. Graph. Image Process. 41 (1988) 233–260.

[79] E. Poletti, et al., A review of thresholding strategies applied to human chromosome segmentation, Comput. Methods Programs Biomed. 108 (2) (2012) 679–688.

[80] T.W. Ridler, S. Calvard, Picture thresholding using an iterative selection method, IEEE Trans. Syst. Man Cybern. 8 (8) (1978) 630–632.

[81] C.A. Glasbey, G.W. Horgan, Image Analysis for the Biological Sciences, John Wiley and Sons, Inc, 1995.

[82] F.J. Chang, J.C. Yen, S. Chang, A new criterion for automatic multilevel thresholding, IEEE Trans. Image Process. 4 (1995) 370–378.

[83] M. Bomans, et al., 3D segmentation of MR images of the head for 3D display, IEEE Trans. Med. Imaging 12 (1990) 153–166.

[84] T. McInerney, D. Terzopoulos, Medical image segmentation using topologically adaptable surfaces, in: Proc of the 1st Joint Conference, Comp Vis, Virtual Reality, and Robotics in Medicine and Med Robotics and Comp Assisted Surgery, 1997, pp. 23–32.

[85] L.D. Cohen, I. Cohen, Finite element methods for active contour models and balloons for 2D and 3D images, IEEE Trans. Pattern Anal. Mach. Intell. 15 (1993) 1131–1147.

[86] T. Chan, L. Vese, Active contours without edges, IEEE Trans. Image Process. 10 (2) (2001) 266–277.

[87] M. Kass, A. Witkin, D. Terzopoulos, Snakes: active contour models, Int. J. Comput. Vis. 1 (1987) 321–331.

[88] R. Malladi, J. Sethian, B.C. Vemuri, Shape modeling with front propagation: a Level set approach, IEEE Trans. Pattern Anal. Mach. Intell. 17 (2) (1995) 158–175.
[89] S. Osher, N. Paragios, Geometric Level Set Method in Imaging, Vision and Graphics, Springer, 2002.
[90] C. Magliaro, et al., Gotta Trace 'em all: a mini-review on tools and procedures for segmenting single neurons toward deciphering the structural connectome, Front. Bioeng. Biotechnol. 7 (2019) 202.
[91] V.J. Schmid, M. Cremer, T. Cremer, Quantitative analyses of the 3D nuclear landscape recorded with super-resolved fluorescence microscopy, Methods 123 (2017) 33–46.
[92] F. Kraus, et al., Quantitative 3D structured illumination microscopy of nuclear structures, Nat. Protoc. 12 (5) (2017) 1011–1028.
[93] M.A. Reyer, et al., An automated image analysis method for segmenting fluorescent bacteria in three dimensions, Biochemistry 57 (2) (2018) 209–215.
[94] F. Piccinini, et al., Software tools for 3D nuclei segmentation and quantitative analysis in multicellular aggregates, Comput. Struct. Biotechnol. J. 18 (2020) 1287–1300.
[95] F. Maes, et al., Multimodality image registration by maximization of mutual information, IEEE Trans. Med. Imaging 16 (2) (1997) 187–198.
[96] S. Lobregt, P.W. Verbeek, F.A. Groen, 3D skeletonization: principal and algorithm, IEEE Trans. Pattern Anal. Mach. Intell. **2** (1980) 75–77.
[97] F.A. Merchant, et al., Confocal microscopy, in: A.C. Bovik (Ed.), Handbook of Image and Video Processing, Academic Press, 2005.
[98] D. Gordon, J.K. Udupa, Fast surface tracking in 3D binary images, Comput. Vis. Graph. Image Process. 45 (2) (1989) 196–214.
[99] F. Solina, R. Bajcsy, Recovery of parametric models from range images: the case for superquadrics with global deformations, IEEE Trans. PAMI 12 (1990) 131–147.
[100] A. Shete, et al., Spatial quantitation of FISH signals in diploid versus aneuploid nuclei, Cytometry A 85 (4) (2014) 339–352.
[101] C. Brechbühler, G. Gerig, O. Kübler, Parametrization of closed surfaces for 3-D shape description, Comp. Vision Image Underst. (CVIU) 61 (2) (1995) 154–170.
[102] F.-Y. Yen, D. Pati, F. Merchant, 3D modeling of chromosomes territories in normal and aneuploid nuclei, in: Proc. SPIE 10578, Medical Imaging 2018: Biomedical Applications in Molecular, Structural, and Functional Imaging, 2018, p. 105782K.
[103] L. Shen, H. Farid, M.A. McPeek, Modeling three-dimensional morphological structures using spherical harmonics, Evolution (N. Y.) 63 (4) (2009) 1003–1016.
[104] L.J. Van Vliet, Gray Scale Measurements in Multidimensional Digitized Images (Ph.D. thesis), Delft University Press, 1993.
[105] A.D. Beckers, A.M. Smeulders, Optimization of length measurements for isotropic distance transformations in 3D, CVGIP Image Underst. 55 (3) (1992) 296–306.
[106] B.H. Verwer, Local distances for distance transformations in two and three dimensions, Pattern Recogn. Lett. 12 (11) (1991) 671–682.
[107] N. Kiryati, O. Kubler, On chain code probabilities and length estimators for digitized 3D curves, in: Pattern Recognition, Conference A: Computer Vision and Applications, vol. 1, 1992, pp. 259–262.
[108] L.M. Cruz-Orive, C.V. Howard, Estimating the length of a bounded curve in 3D using total vertical projections, J. Microsc. 163 (1) (1991) 101–113.
[109] M.H. Davis, A. Khotanzad, D.P. Flamig, S.E. Harms, Curvature measurement of 3D objects: evaluation and comparison of three methods, IEEE Proc. ICIP 95 (2) (1995) 627–630.
[110] F.A. Merchant, et al., Angiogenesis in cultured and cryopreserved pancreatic islet grafts, Transplantation 63 (11) (1997) 1652–1660.
[111] N.S. White, Visualization systems for multidimensional CLSM images, in: J.B. Pawley (Ed.), Handbook of Biological Confocal Microscopy, Plenum Press, 1995.
[112] R.A. Drebin, L. Carpenter, P. Hanrahan, Volume rendering, Computer Graphics 22 (4) (1988) 65–74.
[113] W.E. Lorensen, H.E. Cline, Marching cubes: a high resolution 3D surface reconstruction algorithm, in: Int Conf on Computer Graphics and Interactive Techniques, 1987, pp. 163–169.
[114] B.T. Phong, Illumination for computer generated pictures, Commun. ACM 18 (6) (1975) 311–317.

[115] H. Chen, et al., The collection, processing, and display of digital three-dimensional images of biological specimens, in: J.B. Pawley (Ed.), Handbook of Biological Confocal Microscopy, Plenum Press, 1995.
[116] J.E. Gilda, et al., A semiautomated measurement of muscle fiber size using the Imaris software, Am. J. Physiol. Cell Physiol. 321 (3) (2021) C615–C631.
[117] S. Handschuh, T. Schwaha, B.D. Metscher, Showing their true colors: a practical approach to volume rendering from serial sections, BMC Dev. Biol. 10 (2010) 41.
[118] J.L. Clendenon, et al., Voxx: a PC-based, near real-time volume rendering system for biological microscopy, Am. J. Physiol. Cell Physiol. 282 (1) (2002) C213–C218.
[119] C.T. Rueden, K.W. Eliceiri, J.G. White, VisBio: a computational tool for visualization of multidimensional biological image data, Traffic 5 (6) (2004) 411–417.
[120] B. Schmid, et al., A high-level 3D visualization API for Java and ImageJ, BMC Bioinf. 11 (2010) 274.
[121] C.T. Rueden, K.W. Eliceiri, Visualization approaches for multidimensional biological image data, Biotechniques 43(1 Suppl), 31 (2007) 33–36.
[122] P.C. Cheng, et al., 3D Image analysis and visualization in light microscopy and X-ray micro-tomography, in: A. Kriete (Ed.), Visualization in Biomedical Microscopies, VCH, 1992.
[123] R. Turnnidge, D. Pizzanelli, Methods of pre-visualizing temporal parallax suitable for making multiplex holograms. Part II: greyscale and color anaglyphs made in photoshop, Imaging Sci. J. 45 (1997) 43–44.

CHAPTER TWELVE

Superresolution Image Processing

David Mayerich and Ruijiao Sun

12.1 Introduction

Ever since the invention of the microscope, investigators have pushed to see more and more detail in smaller and smaller specimens. Optical diffraction places a strict limit on the resolution of the optical microscope (see Chapter 2). Superresolution microscopy is a collection of both classical and emerging techniques that are capable of imaging beyond the diffraction limit. This chapter discusses a number of methods that have been developed to bypass the diffraction limit and permit peering deeper into the secrets of microscopic specimens.

12.2 The Diffraction Limit

As discussed in Chapter 2, the diffraction of light places a rigid constraint on the spatial resolution of an image formed by an optical system. This constraint is controlled by two factors: (1) the wavelength of the light used for imaging and (2) the optical bandwidth of the imaging system. For general microscopy applications, this limitation is often expressed in terms of the Rayleigh criterion using the Airy distance (see Chapter 2, Section 2.6):

$$r_{Airy} = \frac{0.61\lambda}{NA}$$

where r_{Airy} is the spatial resolution of the image, λ is the wavelength, and NA is the numerical aperture of the imaging system [1–3]. This expression is an estimate of the minimum separation that two point sources imaged by an optical system can have and still be differentiated (Fig. 12.1).

The Rayleigh criterion is a convenient shorthand for estimating the resolving power of an imaging system, but it does not represent the best we can do. For example, one could resolve the position of point sources more precisely simply by fitting the point spread function (PSF) of the microscope to the images of the point sources. The ability to leverage additional information about the sample and the optics is used by superresolution techniques that seek to overcome the diffraction limit. These techniques can

Microscope Image Processing
https://doi.org/10.1016/B978-0-12-821049-9.00011-3
Copyright © 2023 Elsevier Inc.
All rights reserved.

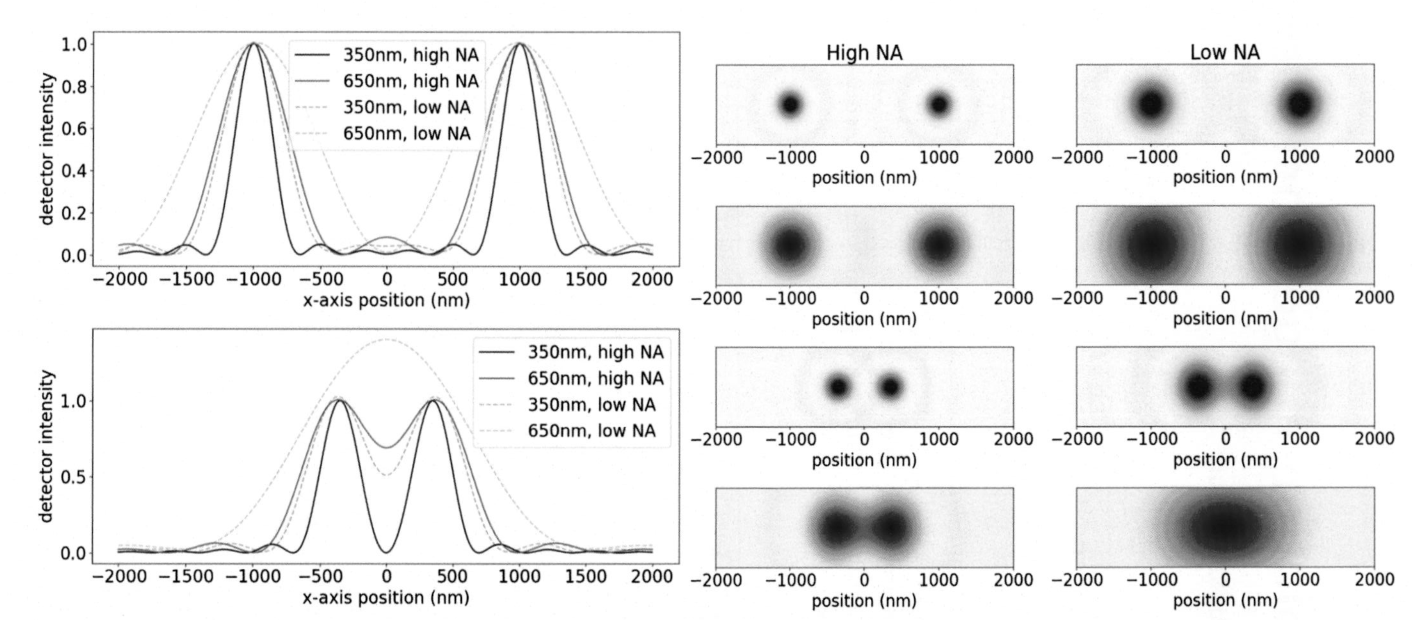

Fig. 12.1 This figure illustrates cases where points are emitting light at 350 and 650 nm and are separated by 2,000 and 700 nm. The points are resolvable as separate except in the case of the 650 nm wavelength and 700 nm separation, where the Rayleigh resolution criterion is not met.

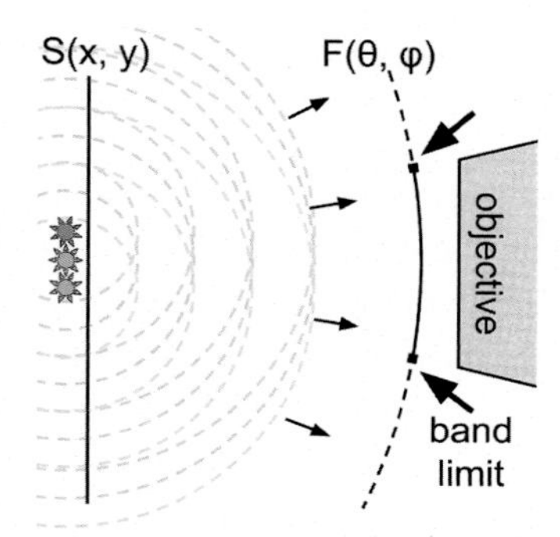

Fig. 12.2 Light emitted by points in a specimen emerge as expanding spherical wavefronts. These wavefronts are cropped as they enter the aperture of the objective lens. The blocked light contains the high-frequency detail in the image. The objective then forms an image using light that has had high-frequency information removed. This truncation of the wavefronts places a resolution limit on the optical system.

be implemented by digital image processing in an effort to improve the sharpness of the image.

As light expands outward from an emitting source, the objective aperture admits only a portion of the propagating wavefront (Fig. 12.2). Thus the most widely diffracted light waves do not enter the objective and thus do not contribute to the reconstructed image. The specimen detail that these waves represents is lost and cannot be recovered. This accounts for the diffraction limit, where only frequencies below the optical cutoff frequency $f_c = 2NA/\lambda$ are preserved. Any downstream reconstruction produced by light passing through the objective will be blurred due to this inherent low-pass filter (Fig. 12.3).

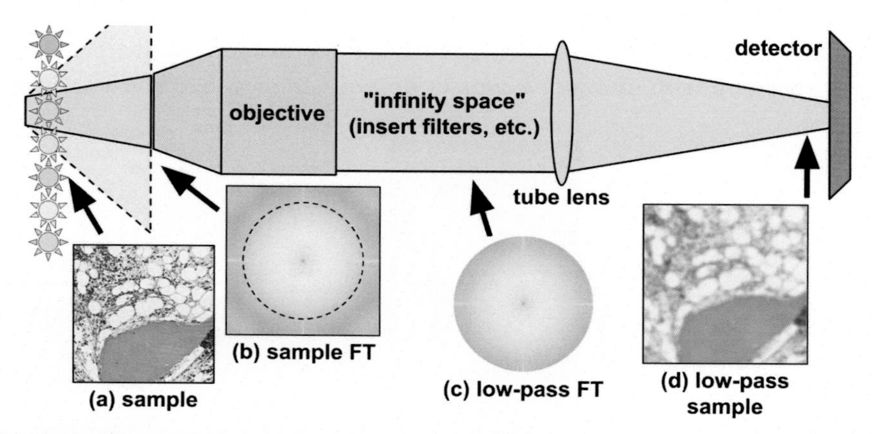

Fig. 12.3 A simple microscope model illustrating resolution limiting in an optical microscope. The expanding wavefront coming from the specimen is truncated by the entrance aperture of the objective. The remaining light waves pass on to the tube lens, which then forms an image of the specimen on the detector. The high-frequency information in the light waves that are blocked at the objective is lost, and this accounts for the resolution limit of the optical system.

12.3 Deconvolution

Given this description of an optical system (Fig. 12.3), it is natural to consider whether the filtering imposed by the microscope can be *reversed*, allowing us to calculate a true picture of the original sample. That is, can we recover the sample from the image more accurately by using information about the optical system?

Traditional methods of deconvolution are discussed in Chapter 5. Wiener deconvolution does an optimal job of both reversing the effects of the optical transfer function (OTF) (up to the diffraction limit) and suppressing noise, under a certain set of assumptions. The goal of superresolution, however, is to recover information that is not captured in the imaging process.

12.3.1 Signals and Noise

As discussed in Chapter 5, deconvolution algorithms generally consist of (1) estimating the PSF (or equivalently the OTF) of the imaging system and (2) using that to obtain a more accurate estimate of the specimen (or equivalently its Fourier spectrum). In its simplest form, deconvolution is simply an inverse filter, where the spectrum of the recorded image is multiplied by the inverse of the OTF. Thus one can, in theory, reconstruct the actual spectrum of an image all the way out to the diffraction limit. And since the OTF is the Fourier transform of the PSF, deconvolution can be done by a convolution operation implemented in the spatial domain, which may well be more computationally efficient.

Since imaging systems typically reduce the amplitude of the high-frequency components in the image spectrum (Fig. 12.4), the deconvolution process normally tends to amplify the high-frequency components and make the image look sharper. Unfortunately, the spectra of common noise sources do not fall off as quickly with increasing

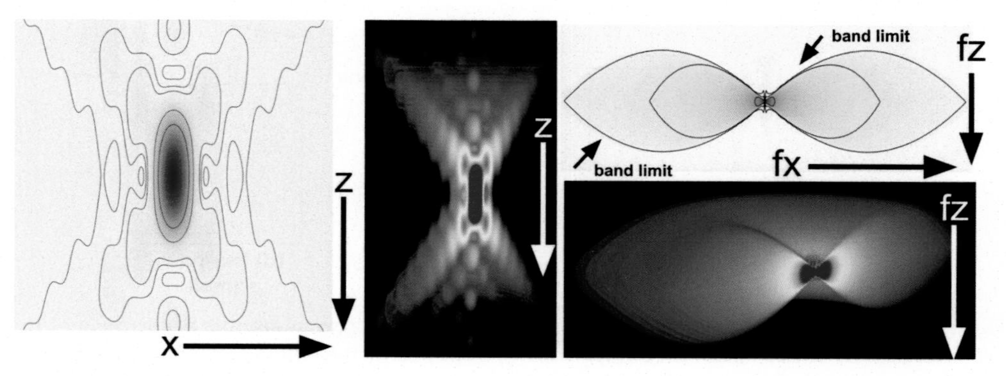

Fig. 12.4 Left and center: the 3D PSF of a microscope. Right: the 3D transfer function of a microscope.

frequency as signal spectra, if they decrease at all. Indeed, the most common noise sources can be assumed to be "white," that is, the amplitude of their spectrum is uniform with frequency. Simple inverse filters, then, significantly increase the amplitude of high-frequency components, and this can result in a very noisy image. Thus it is the signal-to-noise ratio (SNR) of an acquired image that places the primary limitation on the effectiveness of deconvolution.

While the diffraction limit places an upper limit on resolution, the SNR controls how close we can get to this limit with deconvolution. Most deconvolution approaches are therefore focused on improving image quality while suppressing high-frequency noise. Two of the most common approaches, Wiener filtering and regularization (discussed in Chapters 5 and 11), tightly integrate deconvolution with noise suppression.

In microscope imaging, the OTF tends to reduce or eliminate high-frequency components along all three frequency axes (Fig. 12.4), forming a *null space* where many different samples would produce the same image. Approaches for resolving this situation rely on finding the *most likely* sample among all those that would produce the observed image. The simplest and most common approach is to assume that all spatial frequencies above the diffraction limit have zero amplitude. This provides the "smoothest" or "simplest" reconstruction based on the available data.

In summary, the ability to reconstruct the amplitude of components at spatial frequencies below the diffraction limit is limited by the noise component. This is commonly considered to be additive noise with a Gaussian distribution in both the spatial and Fourier domains. Therefore, at spatial frequencies where the magnitude of the signal is small with respect to the noise, the true spectral amplitude will be ambiguous. Deconvolution techniques (Chapter 5) attempt to balance noise suppression with signal amplification, but trade-offs always exist. The general approach for deconvolution methods relies on (1) maximizing the signal, (2) minimizing the noise, and (3) providing the best possible approximation to the OTF. In practice, higher noise levels will amount to a reduced band limit (and a blurrier image), since high spatial frequencies usually correspond to small signal amplitude values.

12.3.2 Extrapolating Beyond the Diffraction Limit

At frequencies where the OTF is small or zero, the amplitude of the spatial frequency components is ambiguous. The simplest approach to address this is simply to set those values to zero. This provides a smooth reconstruction consisting of only image information that has been reliably collected by the imaging system. This type of processing is unlikely to introduce artifacts, but neither can it produce miracles of resolution improvement. The superresolution techniques discussed in this chapter seek to estimate those ambiguous values to form a sharper, more detailed image. They fall into two categories: statistical methods and machine learning methods.

12.3.2.1 Statistical Methods

The ideal superresolution reconstruction would fill in the ambiguous spectral components with their *most likely* values, to produce the *most likely image*. Computational techniques that employ statistical methods to extrapolate spatial frequencies beyond the band limit are the first type of superresolution methods we address.

Early superresolution techniques combined traditional deconvolution algorithms with maximum likelihood estimation [4,5]. These methods use a simple statistical model of the unknown data to select optimal values for spatial frequencies where the spectral amplitude is ambiguous. More recent sparse optimization methods used in signal processing provide better approximations [6] at the cost of increased computational complexity. Microscopy images are generally sparse signals, and these methods take advantage of that to provide tighter optimization constraints using an L1 or L0 norm. Such optimizations are significantly more challenging computationally than the L2 norm used in traditional applications.

12.3.2.2 Machine Learning Methods

Superresolution techniques that employ machine learning methods expose an algorithm to a large training set made up of high-resolution and low-resolution image pairs. The algorithm learns common relationships between the available spatial frequencies and those that are missing, and this enables the computer to "guess" the amplitude of the missing components. These methods require a significant amount of computational effort for training. But once the model is optimized it can be applied to new images with minimal computational effort. Recent research using convolutional neural networks (see Chapter 15) has shown significant success learning these relationships. However, such methods require some form of training data collected using physically based superresolution imaging techniques. This approach is addressed in more detail later in this chapter.

It is important to note that these probabilistic approaches are, at best, an "artist's rendition" of the high-resolution image. Barring confirmation with a better imaging system, we cannot know for sure that the resulting image is correct—just that it is the most likely portrayal of the specimen.

12.4 Superresolution Imaging Techniques

A wide range of methods have been proposed to improve resolution beyond the diffraction limit. These superresolution methods side-step basic limitations in optical bandwidth by modifying the imaging process to increase the amount of information passed through the imaging system—usually by buffering information over time. Superresolution methods differ in the amount of computational complexity required. Some methods are purely computational while others, such as stimulated emission and depletion (STED) microscopy, are performed solely through sample preparation and instrumentation. In this section, we discuss some of these methods in terms of their computational complexity.

12.4.1 Analytic Continuation

We begin with a purely computational technique for extending the band limit. As discussed previously, the OTF of an optical system is necessarily bandlimited: that is, it goes to zero for all frequencies above the optical cutoff frequency $f_c = 2NA/\lambda$. Further, the image itself is spatially bounded: that is, it goes to zero for all locations outside its borders. It is a property of the Fourier transform that if a function is zero everywhere outside some finite interval, then its Fourier spectrum is an analytic function, which means that all of its derivatives exist everywhere. Reversing the domains, since the spectrum is bandlimited, the image will also be an analytic function. Thus all images and their spectra are analytic, and this is a very limited and well-behaved class of functions.

A well-known property of analytic functions is that if one is known over a finite interval, then it is known everywhere [7]. This means that, given a curve defined over a particular interval, no more than one analytic function (of infinite extent) can be fitted exactly to the given curve over that interval. Thus it is theoretically possible to reconstruct the analytic spectrum of an image at all frequencies, given its spectrum below the diffraction limit. All we have to do is create an analytic function that exactly matches the spectrum of the image within the passband, and then we have (after an inverse Fourier transform) the image with unlimited resolution. The process of reconstructing an analytic function in its entirety, given its values over a specified interval, is called the extrapolation of bandlimited functions, or *analytic continuation*.

The superresolution technique advanced by Harris [8] and restated by Goodman [1] involves applying the sampling theorem, with domains reversed, to obtain a system of linear equations that can be solved for values of the signal spectrum that lie outside the diffraction limited passband. This generates an estimate of the spectrum that is bandlimited at a frequency higher than the diffraction limit of the imaging system. An inverse Fourier transform then yields an image with improved resolution. This technique does not require any additional information beyond the image itself.

There are other analytic continuation approaches that work by alternately forcing the analyticity property upon the image and its spectrum [9]. While simulations have produced interesting results, these methods are all very sensitive to the unavoidable noise that contaminates digitized images, and useful examples of this type of superresolution are rare. However, adding the constraint of analyticity on the image and its spectrum into other superresolution techniques could perhaps improve performance.

12.4.2 Stimulated Emission Depletion Microscopy

One approach, first proposed in 1994, achieves resolution beyond the diffraction limit by spatially deactivating fluorophores outside of a central, illuminated focal point by a technique termed stimulated emission depletion (STED) microscopy [10]. Stimulated emission is the process of selectively depleting fluorescence by shifting excited-state fluorophores to a higher vibration state than the fluorescence transition it would normally enter [11]. This shift to a higher vibrational state lowers the energy and increases the wavelength of the emitted photon so that it is released in the far-red spectrum and can be avoided during detection [11,12].

A two pulsed laser design is used to achieve simultaneous excitation and depletion. The depletion laser forms a doughnut-shaped deexcitation pattern around the circular excitation pattern of the other laser, resulting in a small focal spot actively emitting fluorescence. The doughnut-shaped STED pattern comes from a phase mask placed in the light path to modulate its phase-spatial distribution in the x, y plane, providing an x, y resolution down to ~30 nm [13]. Due to the use of two high-intensity lasers, STED imaging requires the use of highly photostable fluorescent dyes; those from the Atto dye family are most often used. It also requires a high-quality antifade mountant to inhibit photobleaching, ProLong mountants are recommended for this application.

12.4.3 Expansion Microscopy

Expansion microscopy (ExM) is a recent development in the field of high-resolution imaging techniques [14]. ExM stands apart from the methods described earlier in that it focuses on physically enlarging biological samples. It uses a polymer system to make nanometer resolution possible on conventional compound and confocal microscopes [15]. This innovation broke the barrier to accessible nanoscale imaging by circumventing the need for expensive, high-precision microscopes, such as electron microscopes [16]. Adding more to its cost effectiveness, the materials required for expansion cost relatively little and are compatible with traditional immunofluorescence probes, thus avoiding the need for costly, assay-specific reagents. This technique allows users to obtain a 4X to 8X enlargement of a tissue sample with little to no loss of protein structure [17].

To achieve this, the tissue is first fixed and then labeled for proteins of interest with fluorescent probes. An acryloyl-X compound is introduced that binds to the lysine residues of proteins and antibodies that anchor the proteins into a framework. The sample is then equilibrated in a gel matrix followed by a proteinase-K digestion that breaks the proteins into fragments, so they can expand. Water washes complete the procedure by washing out the remaining salts and expanding the hydrogel to four times its original size. The isotropic swelling of the polymer network allows fine structures in the sample to be imaged with high accuracy using a less powerful microscope than would be needed without sample expansion [18].

An added benefit of ExM is the ability to perform expansion on nonlabeled tissue by incubating the sample with antibody postexpansion. This is valuable in cases where the tissue is too thick or too dense for antibodies to penetrate. Limitations to this technique come into play when one or more of the four processing steps (labeling, protein anchoring, sample digestion, and water expansion) are incompletely carried out. Incomplete processing can cause uneven cell digestion leading to broken cells and disproportionate expansion, leaving the sample unusable. Moreover, some fluorophore markers may experience significant bleaching during the polymerization process, and preliminary studies are required to optimize the signal yield after expansion. Major commercialized dyes such as Alexa or Dylight fluorophores are robust enough to survive ExM [17,19].

12.4.4 Single Molecule Localization Microscopy

A group of microscopy techniques termed single-molecule localization microscopy (SMLM) [20] has achieved resolution on a nanoscale for point source emitters (see Chapter 13). SMLM can pinpoint an emitter's position by stimulating only a sparse number of emitters at a time. Then that subset is deactivated, and another subset of emitters is activated until a full image is formed. This permits precise localization of each emitter, even in densely labeled samples that would otherwise be difficult to resolve if all emitters were active at the same time.

One of the first and most popular wide-field SMLM methods is photoactivated localization microscopy (PALM) [21]. This technique overcomes the diffraction limit by limiting the number of spatial components passed through the imaging system at any time. PALM is a probe-based superresolution technique that can reach resolutions as small as 30 nm. This technique uses either (1) photoactivatable dyes that exist in an "off" state of no fluorescence until they become activated by irradiation with light, or (2) photoconvertible dyes that can reside in and can convert between two different fluorescent states by irradiation with light.

Using photoactivatable dyes, an experiment first starts with a region of interest being stimulated with a low frequency of an activator laser. This low-frequency stimulation allows for only a few fluorophores in the region to become active, and then a readout laser is used to capture the image and photo bleach the active emitters (or in the case of convertible probes, turn them back to the "off" state). The objective of this method is to capture single emitting fluorophores one at a time to then determine their individual position by fitting a Gaussian function. The process of activation, imaging, and photobleaching is then repeated until all the emitters in the field have been exhausted or enough data has been collected for the desired purpose. The coordinates of each single-molecule emitter are then artificially reconstructed into an image with subdiffraction-limit resolution. The process of repeated activation and

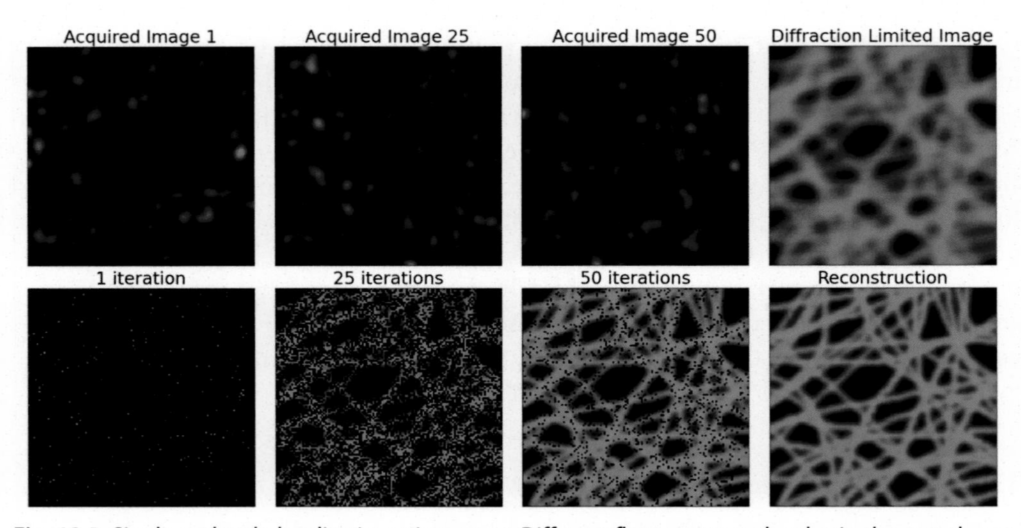

Fig. 12.5 Single molecule localization microscopy. Different fluorescent molecules in the sample are made to flash randomly, producing a series of sparsely populated images. The top row shows three frames of the image sequence and the diffraction-limited image. Then a narrow PSF (e.g., Gaussian) is fitted to each individual molecule image to determine its precise location. The bottom row shows the blank (initially *black*) reconstruction image being filled in as PSFs are inserted at the location of each detected molecule, producing the high-resolution reconstructed image at the bottom right.

imaging at one region of interest severely limits PALM acquisition speed, but this superresolution technique has become very popular in fluorescence imaging (Fig. 12.5).

Variations of this imaging concept create a diverse group that includes (1) stochastic optical reconstruction microscopy (STORM) [22], (2) fluorescence PALM (FPALM) [23], (3) direct STORM (dSTORM), (4) spectral precision distance microscopy (SPDM), and (5) point accumulation for imaging in nanoscale topography (PAINT). However, these approaches are limited by their inability to perform live-cell imaging quickly and effectively.

12.4.5 Structured Illumination Microscopy

Structured illumination microscopy (SIM) has its roots in enabling three-dimensional optical sectioning in traditional wide-field microscopes [24]. While these initial techniques were diffraction limited, they provided the analytical framework for super-resolution SIM (SR-SIM) [25], which is used today for both two- and three-dimensional techniques to overcome the diffraction limit in wide-field microscopes without modification of sample preparation. Like SMLM, SIM introduces sparsity in excitation, sending multiple pieces of the sample image through the optical system separately over time. The early work in both SIM optical sectioning and SR-SIM make this one of the earliest SR imaging methods [26]; however, unlike STED, SR-SIM leverages image postprocessing as a critical step.

SR-SIM requires that the emitted light is incoherent from the excitation, and it is most commonly used in fluorescence microscopy. SR-SIM works by controlling the excitation in the sample plane, often by generating interference patterns that can be readily combined as a postprocessing step [27,28]. We first demonstrate the intuition behind these principles in the spatial domain, and then extend the theory into the Fourier domain to describe reconstruction methods.

In the spatial domain—at the sample and image planes—SR-SIM relies on the fact that interference between coherent light sources produces fringes. The input beams (Fig. 12.6a) and the output fringes projected onto the sample (Fig. 12.6c) are related by the Fourier transform (Fig. 12.6b)

$$P(x, y) = \mathcal{F}\{\widehat{P}(u, v)\}$$

where $P(x, y)$ is the complex illumination pattern projected onto the sample and $\widehat{P}(u, v)$ is the image of the interfering beams at the back aperture of the objective. Placing a two-dimensional sample $S(x, y)$ into this formulation, the *convolution theorem* describes the relationship between the collected image and its Fourier transform:

$$P(x, y)\, S(x, y) = \mathcal{F}\{\widehat{P}(u, v) * \widehat{S}(u, v)\}$$

where the $*$ represents the convolution operation. This relationship is key to the intuition behind SIM. By placing points in the Fourier domain near the diffraction limit, spatial components outside the diffraction limit (Fig. 12.6h) are encoded near those points inside the diffraction limit (Fig. 12.6i). However, these high-frequency components overlap low-frequency components within the band limit that are collected by the central beam—a phenomenon known as *aliasing* (Chapter 3). These high-frequency components can be computationally separated from their low-frequency counterparts.

Algorithms for separating the aliased high-frequency components rely on collecting multiple images while varying the fringing patterns (Fig. 12.6d–f). The overlapping signals can be linearly separated by making a sufficient number of independent measurements [29]. But far more robust algorithms use statistical methods [30], least-squares optimization [31], or direct spatial reconstruction [32] to extract the high-frequency information.

12.4.6 Synthetic Superresolution with Machine Learning

Superresolution microscopy is a series of emerging techniques capable of imaging below the diffraction limit. Traditional superresolution microscopy techniques are divided into

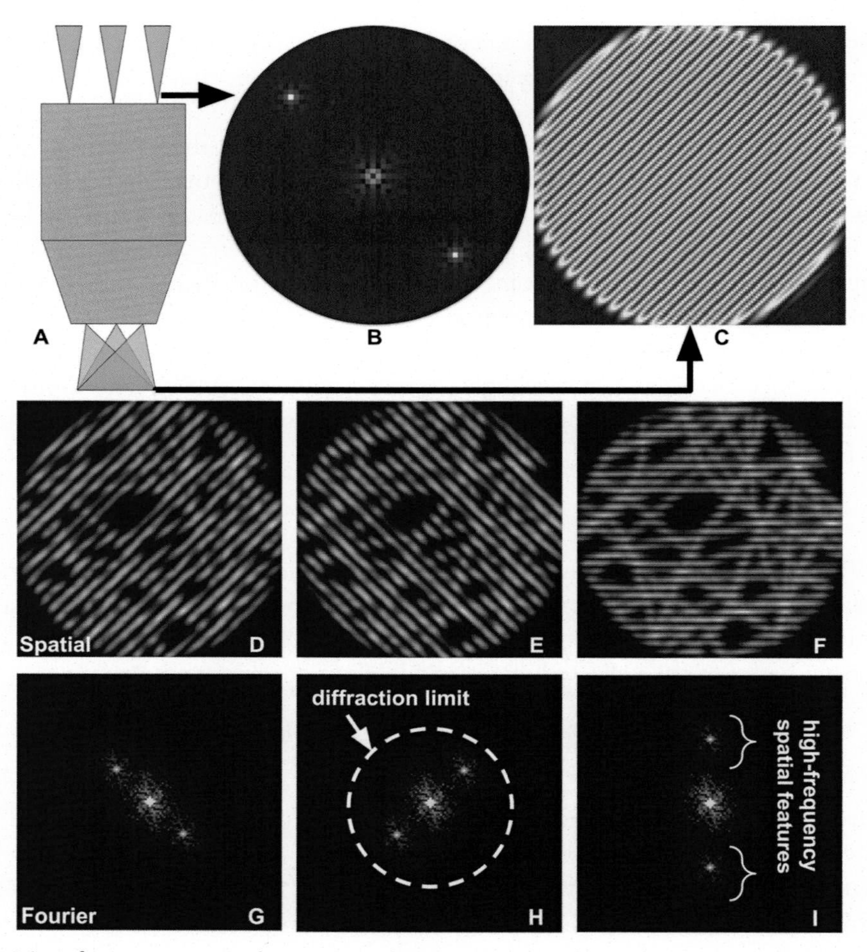

Fig. 12.6 Interference among coherent input beams (a) creates fringes in the illumination projected onto the sample (c) and produces spikes in the Fourier transform of the image (b). Image components lying outside the diffraction limit (h) are encoded inside the diffraction limit near those spikes (i). Multiple images are collected while varying the orientation of the fringing patterns (d–f). Algorithms separate the aliased high-frequency components from the remainder of the Fourier transform.

two categories: deterministic super-resolution [10,33–35] and stochastic superresolution [21,22]. But these methods are time-consuming and sometimes cost prohibitive, and they are often limited to small samples. Methods based on deep learning (see Chapter 15) circumvent these problems by enabling feature mapping between superresolution images and low-resolution images.

The superresolution convolutional neural network (SRCNN) [36] exploited by Dong et al. was the first deep learning method used to achieve superresolution.

The success of SRCNN gave rise to many deep learning superresolution algorithms for microscope image analysis. Heinrich et al. [37] have adapted Fast SRCNN [38] to the anisotropic 3D case for superresolution. They have also introduced a 3D U-Net model for comparative studies, which was integrated with an additional fractionally strided convolution layer to make the output of this layer isotropic. By implementing two networks on a distortion-free scanning electron microscopy dataset with fine-tuned hyperparameters, 3D Fast SRCNN has been verified to produce visually better results in most cases. However, 3D U-Net has been shown to provide better performance given bigger image size, larger field-of-view, and a multiscale representation. Both networks have outperformed the sparse coding approach [39].

Deep learning (Chapter 15) may also be embedded into traditional methods to obtain more precise and faster performance, such as Deep-STORM [40], and deep-STORM3D [41]. Nehme et al. [40] have introduced the parameter-free method termed Deep-STORM, based on a fully convolutional encoder-decoder network for obtaining super-resolution images from blinking emitters, such as fluorescent molecules. Compared to a fast multiemitter fitting algorithm (FALCON) [42], Deep-STORM obtained higher resolution in both simulated microtubule and experimental datasets obtained from Sage et al. [43]. Nehme et al. [40] recently established another method called deepSTORM3D [41] by incorporating a physical simulation layer in the CNN model with an adjustable phase modulation to realize precise 3D localization for dense U2OS cells.

One classical network architecture for microscopy superresolution is the generative adversarial network (GAN). Wang et al. [44] have trained a GAN model to improve the resolution of wide-field microscopy images acquired with low numerical aperture objectives without any prior knowledge about the sample and the image formation process. That approach in bovine pulmonary artery endothelial cell structures has been demonstrated to produce reliable superresolution images in line with corresponding ground truth images. In addition, GAN was also widely applied in other microscopy superresolution tasks [45,46].

"Digital staining" or "semantic mapping" is another emerging task that GAN is commonly used for. Digital staining refers to the process of generating a virtual histology appearance from intact tissue images. This method offers a less expensive alternative to existing histology methods that preserves precious tissue for downstream analysis. Rivenson et al. [47] have proposed a digital staining technique called PhaseStain, in which they trained a GAN model. This framework has demonstrated the effectiveness of digital staining in different tissue sections such as human skin, kidney, and liver tissue. Comparison with brightfield images of histologically stained tissue has suggested that deep learning-based digital staining is less sensitive to common tissue preparation imperfections.

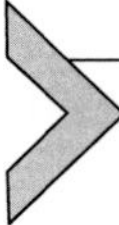

12.5 Summary of Important Points

1. Optical imaging systems, such as microscopes, are limited by physical constraints in the information that they can acquire.
2. The wave nature of light, optical aberrations, and noise reduce the quality of the images acquired by microscopes.
3. Image processing algorithms, such as deconvolution, can compensate for many optical aberrations but can only carry us up to the diffraction limit.
4. Extrapolation algorithms allow us to select the most likely candidates from the infinite set of samples that could produce that image.
5. Machine-learning approaches examine thousands of images and learn the relationships between a high-resolution object and its corresponding low-resolution image.
6. The most practical methods for acquiring data beyond the diffraction limit rely on a combination of sample preparation, optical system development, and image processing.
7. Selecting an appropriate superresolution method is largely based on the application and the user's capabilities.
8. Expansion microscopy is simple, but it requires a significant amount of expertise in sample preparation and immunohistochemistry.
9. Structured illumination microscopy (SIM) is relatively simple, but exceeding a factor of 2X above the diffraction limit becomes time-consuming and subject to noise limitations.
10. Stimulated emission depletion (STED) imaging provides the opportunity for extremely high-resolution imaging, but the instrumentation is complex and expensive.
11. Single molecule localization microscopy (SMLM, discussed in Chapter 13) provides molecular-level resolution, but it requires specialized instrumentation and expertise in sample preparation.

References

[1] J.W. Goodman, Introduction to Fourier optics, Roberts and Company Publishers, 2005.
[2] J. Peatross, M. Ware, Physics of Light and Optics, Brigham Young University, Department of Physics, 2011.
[3] B.E. Saleh, M.C. Teich, Fundamentals of Photonics, John Wiley & sons, 2019.
[4] T.J. Holmes, Maximum-likelihood image restoration adapted for noncoherent optical imaging, J. Opt. Soc. Am. A 5 (5) (1988) 666–673.
[5] T.J. Holmes, Y.-H. Liu, Richardson-Lucy/maximum likelihood image restoration algorithm for fluorescence microscopy: further testing, Appl. Opt. 28 (22) (1989) 4930–4938.
[6] S. Hugelier, et al., Sparse deconvolution of high-density super-resolution images, Sci. Rep. 6 (1) (2016) 21413, https://doi.org/10.1038/srep21413.

[7] H. Cartan, Elementary Theory of Analytic Functions of One or Several Complex Variables, Dover Publications, 1995.
[8] J.L. Harris, Diffraction and resolving power, J. Opt. Soc. Am. 54 (7) (1964) 931–936.
[9] K.R. Castleman, Digital Image Processing, Prentice Hall, 1996.
[10] S.W. Hell, J. Wichmann, Breaking the diffraction resolution limit by stimulated emission: stimulated-emission-depletion fluorescence microscopy, Opt. Lett. 19 (11) (Jun. 1994) 780–782, https://doi.org/10.1364/OL.19.000780.
[11] Cell 143 (7) (2010) 1047–1058, https://doi.org/10.1016/j.cell.2010.12.002.
[12] ChemPhysChem 13 (8) (2012) 1986–2000, https://doi.org/10.1002/cphc.201100986.
[13] V. Westphal, S.W. Hell, Nanoscale resolution in the focal plane of an optical microscope, Phys. Rev. Lett. 94 (14) (2005) 143903.
[14] F. Chen, P.W. Tillberg, E.S. Boyden, Expansion microscopy, Science 347 (6221) (2015) 543–548.
[15] F. Chen, et al., Nanoscale imaging of RNA with expansion microscopy, Nat. Methods 13 (8) (2016), https://doi.org/10.1038/nmeth.3899, 8.
[16] A.T. Wassie, Y. Zhao, E.S. Boyden, Expansion microscopy: principles and uses in biological research, Nat. Methods 16 (1) (2019), https://doi.org/10.1038/s41592-018-0219-4, 1.
[17] P.W. Tillberg, et al., Protein-retention expansion microscopy of cells and tissues labeled using standard fluorescent proteins and antibodies, Nat. Biotechnol. 34 (9) (2016), https://doi.org/10.1038/nbt.3625, 9.
[18] Y. Zhao, et al., Nanoscale imaging of clinical specimens using pathology-optimized expansion microscopy, Nat. Biotechnol. 35 (8) (2017) 757–764, https://doi.org/10.1038/nbt.3892.
[19] T.J. Chozinski, et al., Expansion microscopy with conventional antibodies and fluorescent proteins, Nat. Methods 13 (6) (2016), https://doi.org/10.1038/nmeth.3833, 6.
[20] I.M. Khater, I.R. Nabi, G. Hamarneh, A review of super-resolution single-molecule localization microscopy cluster analysis and quantification methods, Patterns 1 (3) (2020) 100038, https://doi.org/10.1016/j.patter.2020.100038.
[21] E. Betzig, et al., Imaging intracellular fluorescent proteins at nanometer resolution, Science 313 (5793) (2006) 1642–1645, https://doi.org/10.1126/science.1127344.
[22] M.J. Rust, M. Bates, X. Zhuang, Sub-diffraction-limit imaging by stochastic optical reconstruction microscopy (STORM), Nat. Methods 3 (10) (2006), https://doi.org/10.1038/nmeth929, 10.
[23] S.T. Hess, T.P. Girirajan, M.D. Mason, Ultra-high resolution imaging by fluorescence photoactivation localization microscopy, Biophys. J. 91 (11) (2006) 4258–4272.
[24] M.A. Neil, R. Juškaitis, T. Wilson, Method of obtaining optical sectioning by using structured light in a conventional microscope, Opt. Lett. 22 (24) (1997) 1905–1907.
[25] R. Heintzmann, T. Huser, Super-resolution structured illumination microscopy, Chem. Rev. 117 (23) (2017) 13890–13908, https://doi.org/10.1021/acs.chemrev.7b00218.
[26] B. Bailey, D.L. Farkas, D.L. Taylor, F. Lanni, Enhancement of axial resolution in fluorescence microscopy by standing-wave excitation, Nature 366 (6450) (1993) 44–48.
[27] R. Heintzmann, T.M. Jovin, C. Cremer, Saturated patterned excitation microscopy—a concept for optical resolution improvement, J. Opt. Soc. Am. A 19 (8) (2002) 1599–1609, https://doi.org/10.1364/JOSAA.19.001599.
[28] R. Heintzmann, Saturated patterned excitation microscopy with two-dimensional excitation patterns, Micron 34 (6) (2003) 283–291, https://doi.org/10.1016/S0968-4328(03)00053-2.
[29] M.G.L. Gustafsson, et al., Three-dimensional resolution doubling in wide-field fluorescence microscopy by structured illumination, Biophys. J. 94 (12) (2008) 4957–4970, https://doi.org/10.1529/biophysj.107.120345.
[30] F. Orieux, E. Sepulveda, V. Loriette, B. Dubertret, J.-C. Olivo-Marin, Bayesian estimation for optimized structured illumination microscopy, IEEE Trans. Image Process. 21 (2) (2012) 601–614, https://doi.org/10.1109/TIP.2011.2162741.
[31] J. Luo, et al., Super-resolution structured illumination microscopy reconstruction using a least-squares solver, Front. Phys. (2020), https://doi.org/10.3389/fphy.2020.00118.
[32] S. Tu, et al., Fast reconstruction algorithm for structured illumination microscopy, Opt. Lett. 45 (6) (2020) 1567–1570, https://doi.org/10.1364/OL.387888.

[33] S.W. Hell, M. Kroug, Ground-state-depletion fluorscence microscopy: a concept for breaking the diffraction resolution limit, Appl. Phys. B 60 (5) (1995) 495–497, https://doi.org/10.1007/BF01081333.
[34] M. Hofmann, C. Eggeling, S. Jakobs, S.W. Hell, Breaking the diffraction barrier in fluorescence microscopy at low light intensities by using reversibly photoswitchable proteins, PNAS 102 (49) (2005) 17565–17569, https://doi.org/10.1073/pnas.0506010102.
[35] U. Böhm, S.W. Hell, R. Schmidt, 4Pi-RESOLFT nanoscopy, Nat. Commun. 7 (1) (2016), https://doi.org/10.1038/ncomms10504, 1.
[36] C. Dong, C.C. Loy, K. He, X. Tang, Learning a deep convolutional network for image super-resolution, in: Computer Vision—ECCV 2014, 2014, pp. 184–199, https://doi.org/10.1007/978-3-319-10593-2_13. Cham.
[37] L. Heinrich, J.A. Bogovic, S. Saalfeld, Deep Learning for Isotropic Super-Resolution from Non-Isotropic 3D Electron Microscopy, Jun. 2017, arXiv:1706.03142 [cs]. [Online]. Available: http://arxiv.org/abs/1706.03142. (Accessed 16 August 2020).
[38] C. Dong, C.C. Loy, X. Tang, Accelerating the super-resolution convolutional neural network, in: *Computer Vision—ECCV* 2016, 2016, pp. 391–407, https://doi.org/10.1007/978-3-319-46475-6_25. Cham.
[39] A. Veeraraghavan, et al., Increasing depth resolution of electron microscopy of neural circuits using sparse tomographic reconstruction, in: 2010 IEEE Computer Society Conference on Computer Vision and Pattern Recognition, Jun. 2010, pp. 1767–1774, https://doi.org/10.1109/CVPR.2010.5539846.
[40] E. Nehme, L.E. Weiss, T. Michaeli, Y. Shechtman, Deep-STORM: super-resolution single-molecule microscopy by deep learning, Optica 5 (4) (2018) 458–464, https://doi.org/10.1364/OPTICA.5.000458.
[41] E. Nehme, et al., DeepSTORM3D: dense 3D localization microscopy and PSF design by deep learning, Nat. Methods 17 (7) (2020) 734–740, https://doi.org/10.1038/s41592-020-0853-5.
[42] J. Min, et al., FALCON: fast and unbiased reconstruction of high-density super-resolution microscopy data, Sci. Rep. 4 (1) (2014), https://doi.org/10.1038/srep04577, 1.
[43] D. Sage, et al., Quantitative evaluation of software packages for single-molecule localization microscopy, Nat. Methods 12 (8) (2015), https://doi.org/10.1038/nmeth.3442, 8.
[44] H. Wang, et al., Deep learning enables cross-modality super-resolution in fluorescence microscopy, Nat. Methods 16 (1) (2019), https://doi.org/10.1038/s41592-018-0239-0, 1.
[45] K. de Haan, Z.S. Ballard, Y. Rivenson, Y. Wu, A. Ozcan, Resolution enhancement in scanning electron microscopy using deep learning, Sci. Rep. 9 (1) (2019), https://doi.org/10.1038/s41598-019-48444-2, 1.
[46] H. Zhang, X. Xie, C. Fang, Y. Yang, D. Jin, P. Fei, High-throughput, High-resolution Registration-free Generated Adversarial Network Microscopy, Oct. 2018, *arXiv:1801.07330 [physics, q-bio]*. [Online]. Available: http://arxiv.org/abs/1801.07330. (Accessed 13 August 2020).
[47] Y. Rivenson, T. Liu, Z. Wei, Y. Zhang, K. de Haan, A. Ozcan, PhaseStain: the digital staining of label-free quantitative phase microscopy images using deep learning, Light Sci. Appl. 8 (1) (2019), https://doi.org/10.1038/s41377-019-0129-y, 1.

CHAPTER THIRTEEN

Localization Microscopy

Christian Franke

13.1 Introduction

Single-molecule localization microscopy (SMLM) is a collection of techniques that can produce ultra-high resolution images of biological specimens that have been labeled with a fluorescent dye. As shown in Fig. 13.1, SMLM works by (1) disabling the fluorescence of all but small subsets of the label molecules, (2) capturing images of the sparsely located emitters, (3) fitting a 2D surface that models the PSF (typically a Gaussian) to the image of each of the isolated spots, (4) using the parameters of the surface fit to establish the precise position of each emitter, (5) iterating that procedure with different subsets of the emitters activated, and (6) composing a high-resolution image of the specimen using the known locations of the label molecules. A number of different implementations of this technique are addressed in this chapter with a focus on the image processing aspects. Quality control and precision estimation are also addressed.

Over the past 15 years the field of SMLM has grown into a vibrant and productive conglomerate of method developers, early and advanced practitioners, software and hardware development engineers, as well as all-around laboratories. This has resulted in a diverse and growing single-molecule super-resolution community. Every year, several thousand scientific articles are published related to the various aspects of SMLM. Today, no one can oversee every detail in the SMLM landscape, and any attempt at reviewing the entirety of SMLM will fail. This chapter gives an overview of the most important SMLM concepts that should be mastered in order to understand and apply SMLM. Although not a complete list of achievements in this field, the referenced works are recommended for further reading. Today we see specimens through the nanoscopic filter of "single molecules." It can take a while, even for microscopists with expertise in macroscopic optics, to adjust to the nanometer scale. But it is very satisfying to see "single blinking molecules" on the screen, particularly when they appear in a clean super-resolved image of a structure of interest.

13.1.1 A Brief History of Localization Microscopy

The ability to see with one's own eyes is limited to objects roughly the size of a thin human hair ($\sim$ 50–200 μm), depending on what is considered *resolution* (see also Chapter 2, Section 2.1. and Chapter 5, Section 5.2) [2]. Unfortunately, the spatial

Microscope Image Processing
https://doi.org/10.1016/B978-0-12-821049-9.00016-2
Copyright © 2023 Elsevier Inc.
All rights reserved.

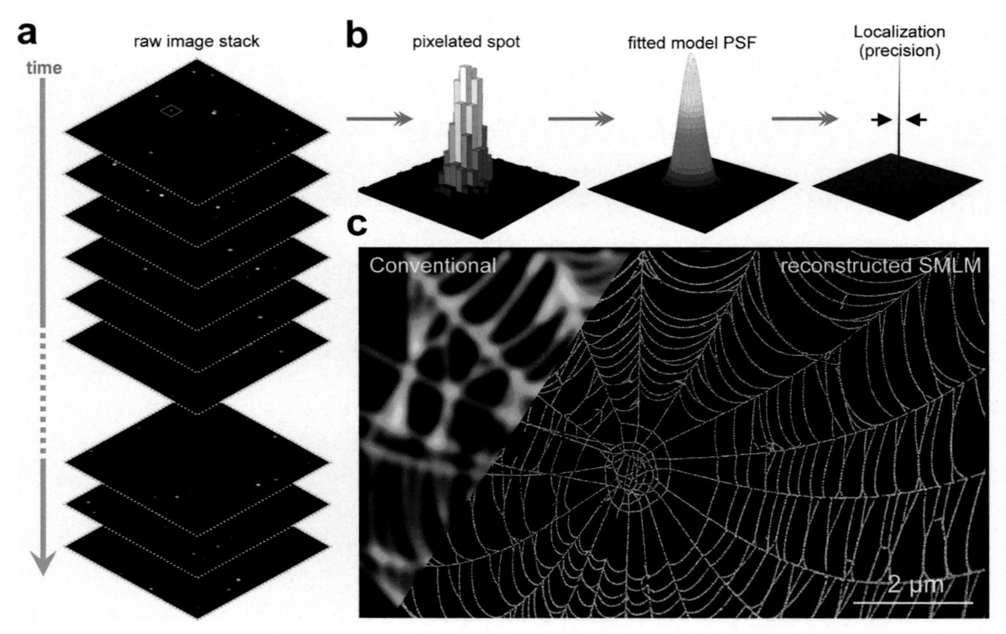

Fig. 13.1 An Overview of SMLM. (a) The fluorophore ensemble is controlled at the single molecule level so that, in each image taken, only a sparse subset of emitters is activated. Thousands of frames may be captured while the fluorophores switch on and off randomly. (b) Each spot image in each frame is fitted with a model PSF function, such as a Gaussian. The fitting process produces a nanometer-precise determination of the molecular position, i.e., the corresponding molecule's localization. (c) Each of the localizations is assigned to a digital image having significantly smaller pixel spacing (e.g., 10 nm) than the raw images. This results in an output image with resolution well beyond the diffraction limit. *(After Ref. [1])*

resolution of any optical system is also limited (Chapter 2). The diffraction of light causes a blurring in the image plane, resulting in a limited resolution that is approximately scaled by the wavelength of the illumination. In other words, the "bluer" the light used, the better the resolution (see Chapter 2, Section 2.1, and Chapter 5, Section 5.2) [3,4].

The connection between illumination wavelength and spatial resolution led to the discovery of fluorescence microscopy itself. August Koehler attempted to resolve subcellular structures in a transmitted-light bright-field microscope by using ultraviolet light of much shorter wavelength than usual. To his surprise he found it virtually impossible to image his specimens due to autofluorescence of cellular components (e.g., some amino acids capable of UV fluorescence) [5].

Fluorescence microscopy subsequently became an entire universe of techniques and the method of choice to visualize subcellular architecture, driven by the idea that structure dictates function and vice versa. However, due to the diffraction limit when working in the visible spectrum, other avenues to improve spatial resolution were needed to visualize cellular organelles and machineries that exist at the subdiffraction (nanometer) scale.

Over the past 20 years, a plethora of ideas and concepts to bypass the far-field diffraction barrier has been developed, creating the field of super-resolution fluorescence microscopy. This led to the 2014 Nobel Prize in Chemistry, awarded to Stefan W. Hell for the realization of STED (stimulated emission depletion) [6,7] and Eric Betzig and W.E. Moerner, for their work on single-molecule localization microscopy (SMLM) [8–12]. STED is based on classic confocal microscopy, and SMLM uses a wide-field illumination and detection system.

In this chapter we focus on SMLM, although the approach of achieving super-resolution by spatiotemporally decoupling the emission of single emitters from the bulk emission is also used by STED (Chapter 12).

The paradigm shift toward subdiffraction imaging originated from the idea of distinguishing individual emitters from the bunch, thus circumventing the overall blurring effect that occurs when detecting all emitters in a sample simultaneously. It introduced using fluorescence lifetime, absorption, polarization, and emission shift to differentiate and localize individual molecules [8,13,14].

The idea to isolate single emitters and thereby enable the precise localization of their three-dimensional position was based on pioneering discoveries made in the late 1980s in the field of single-molecule spectroscopy. Single molecules were first optically detected in 1989 by sampling their absorption in solid matter [10] and by sensing their fluorescence emission in crystals and in liquid solution in 1990 [15,16]. However, in these experiments, the single-molecule environment was created artificially by the strong dilution of the fluorescent probe in the medium—a tactic not suited for any structural imaging approach.

There are a limited number of distinct spectroscopic features for fluorescent molecules (on the order of 10). There is also a need for a biological specimen to be densely labeled (10^3–10^5 per diffraction limited area) in order to achieve a sufficient sampling of the target structure. The simultaneous detection of all fluorescent molecules in a biological sample would lead to an irresolvable overlapping of their individual emission patterns, no matter which spectroscopic feature is used. The key to subdiffraction imaging is the spatiotemporal decoupling of the bulk fluorescence emission: in other words, allowing only one emitter per diffraction limited area to be active at any given time. This enables the computational postprocessing steps to complete the imaging process.

In contrast to the previously published deterministic concepts of STED [6,17] and structured illumination (SIM) [18–20], SMLM was always based on a stochastic modulation of the temporally dynamic fluorescence emission of individual fluorophores. Lidke et al. proposed in 2005 to utilize the intrinsic fluorescence fluctuation of semiconductor quantum dots and called their approach *pointilism* [21]. While sparsely seeded quantum dots were localized with nanometer precision in this way, the uncontrollable photoblinking of quantum dots cannot be used to resolve densely labeled structures (see Section 13.3.5). Shortly after Lidke, alternate approaches utilizing directed forms of photoactivation, photoconversion, and photoswitching, or a combination thereof, were deployed. The number of SMLM methods and acronyms quickly grew, based on various implementations of photoactivation, photoconversion, and photoswitching. These were used in photoactivated localization microscopy (PALM), utilizing photoconvertible fluorescent proteins such as PA-GFP or mEOS [9,22,23], in fluorescence photoactivation localization microscopy (FPALM) [24,25], in stochastic optical reconstruction microscopy (STORM) [26–29], *direct* STORM (*d*STORM), [30–33], and related localization microscopy approaches based on the photochemical reduction of small organic fluorophores [34–41]. Today, there are some slightly novel approaches, and dozens more acronyms, all sharing the same basic underlying concept of stochastically switching the overall majority of emitters into a semistable dark state and localizing the remaining few very precisely.

In this chapter, we focus on the technical and computational aspects of SMLM, which are substantial beyond the localization step itself. Of course, even the most sophisticated analysis software is pointless if the preparation of the biological sample or the imaging process is deficient.

13.2 Overcoming the Diffraction Limit

In this section we review two concepts that are fundamental to SMLM: the diffraction limit and photoswitching of fluorescent molecules.

13.2.1 Diffraction-Limited Resolution

Since an understanding of the diffraction limit of far-field optical microscopy has significant implications for super-resolution, photoswitching artifacts, and directed localization quantification (e.g., molecular counting), we now take a brief look at the *Rayleigh resolution criterion* for self-luminous point emitters, which is discussed in more detail in Chapter 2.

Here we assume fluorescence microscopy and focus on active point emitters. One can imagine their emission as a spherical wave emanating from the object plane. Only a limited solid angle of this emission is passed by the aperture of the objective lens, thus causing a truncation of the expanding spherical wave. Consequently, less than the entire spherical wave is used in the reconstruction of the image, and light converges not to a point, but to a small spot in the image plane. The shape of that spot is called the point spread function (PSF). Thus every optical system works as a low-pass filter in Fourier frequency space.

The emission pattern of an infinitesimally small fluorophore therefore has a distinct shape and is approximated in an ideal, i.e., aberration-free and in-focus, case with the scaled Airy distribution:

$$I_{transverse}(0, v) = I_0 \left(\frac{2J_1(v)}{v} \right)^2 \qquad v = \frac{2\pi x}{\lambda NA} \qquad D_{Airy} = \frac{1.22\lambda}{NA}$$
$$I_{axial}(0, v) = I_0 \left(\frac{\sin(u/4)}{u/4} \right)^2 \qquad u = \frac{2\pi z n}{\lambda NA^2} \qquad R_u = \frac{n\lambda}{NA^2} \tag{13.1}$$

where J_1 is the Bessel function of the first kind and order one, x and z are the lateral and axial coordinates, v and u are the normalized lateral and axial coordinates, respectively, and n is the refractive index. D_{Airy} is the Airy diameter and R_u the Rayleigh distance.

So far, we have only considered one isolated emitter. In order to define a spatial resolution, we consider two independent emitters at a distance d apart and pose the question: what is the minimum distance d_{min} that still allows one to distinguish (i.e., resolve) the two point sources from each other? There is no objective answer to that question, as several distance criteria can be used. The most common is the *Rayleigh criterion* [4] (Chapter 2), which states that two images are resolvable if the maximum of one emission pattern coincides with the first side-minimum of the other. This consideration is primarily for the lateral dimensions, and the axial part of the PSF scales differently, resulting in a significantly worse axial resolution. This lateral resolution limit d_{min} can be written as a function of the observed wavelength λ and the numerical aperture NA of the optical system:

$$d_{\min 2D\ Rayleigh} = \frac{0.61\ \lambda}{NA} = \frac{0.61\ \lambda}{n\ sin(\alpha)} \tag{13.2}$$

The Rayleigh resolution criterion is compared to total separation in Fig. 13.2. See also Figs. 2.7 and 12.1.

13.2.2 Photoswitching Mechanisms

Every SMLM approach relies on the spatiotemporal separation of a large ensemble of fluorescent emitters within a diffraction limited area. In principle, a state is required where only a single emitter is active within the area of a PSF, since spots will otherwise overlap and create a blurred image of the emission pattern. There are many fundamental papers and reviews on the subject, but since this is one of the pivotal requirements for SMLM, we include here a brief overview of the different mechanisms that can be used to achieve single-molecule spatial density. In the remainder of the chapter we assume that the sparsity requirement is met, unless otherwise specified.

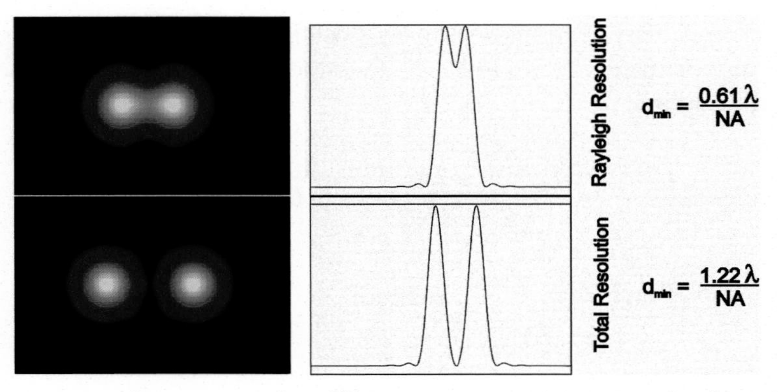

Fig. 13.2 Two-point resolution criteria for self-luminous point emitters. *Top*: The classic Rayleigh criterion is widely accepted as a reasonable definition of two-point resolution. *Bottom*: Total resolution requires the complete separation of the two emitters, resulting in a twofold larger resolution value compared to the Rayleigh criterion.

In principle, *photoswitches* are molecules that can be toggled between two or more distinct states in response to their excitation light (Fig. 13.3). Generally any fluorophore that exhibits a reversible or irreversible transition between a nonfluorescent and a fluorescent state can be used. Photon yield, switching ratio, and state duration, however,

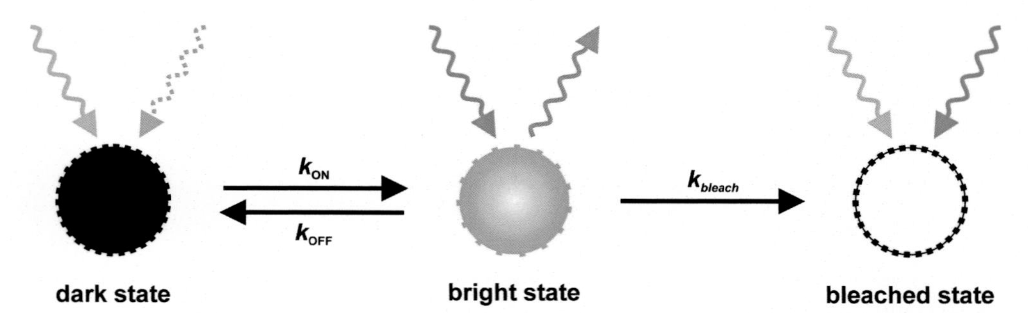

Fig. 13.3 General scheme of SMLM photoswitching mechanisms. In PALM imaging approaches, fluorophores are photoconverted from a nonfluorescent *dark state (left)* to a fluorescent *bright state (middle)*, whose extinction is usually redshifted relative to the dark state. Then the fluorescence is read out until the fluorophore is photobleached *(right)* into a permanent inability to fluoresce, regardless of extinction wavelength. In *d*STORM imaging approaches, most of the fluorescent emitters *(middle)* undergo a repetitive cycle of photoswitching to a chemically reduced *dark state (left)* (whose extinction is usually blueshifted relative to the *bright state*) and back until photobleaching occurs *(right)*. In PAINT imaging modalities, emitters that are not bound to the receptor are far away from the detection plane and therefore effectively *dark*. When binding to their receptor, dyes effectively turn to the bright state, which is detectable until they are *bleached* or the transient binding state is terminated. *(Adapted from Ref. [1])*

play a vital role in their applicability to SMLM. One can distinguish fluorophores/ photoswitches that are [42]:

(1) irreversibly photoactivated from a nonfluorescent (*dark*, D) state to a fluorescent state (F), termed *photoactivatable fluorophores*: D → F
(2) irreversibly photoconverted from a fluorescent state (F) to a red-shifted fluorescent state (F′), termed *photoconvertible fluorophores*: F → F′, or
(3) reversibly switched between a fluorescent *on*- and nonfluorescent *off*-state, termed *photoswitchable fluorophores*: D ⇌ F.

Depending on the photoswitching mechanism used to achieve the condition of emitter sparsity, we can distinguish a couple of SMLM classes: PAINT [43] (points accumulation for imaging in nanoscale topography) relies upon the photogenic properties of some dyes, which was initially demonstrated with the twisted intermolecular charge transfer turn-on behavior of Nile Red [44]. An extension to the original method is DNA-PAINT [45], where the transient binding of short dye-labeled oligonucleotides to a complementary target within the sample is used to create a blinking effect, where the emitter density is directly proportional to the DNA concentration in the medium.

In the *d*STORM (*direct* stochastic optical reconstruction microscopy) approach [30–33], small organic molecules (e.g., Cy5, Alexa647) are reduced into a dark state by photoinduced reactions with a reducing agent, e.g., a thiol. The dark state is created because the reduced form of the dye has a blue-shifted absorbance and is no longer excited by the original wavelength. Only a small subset of emitters remains in their original fluorescent state. The emitter density can be controlled by the excitation laser power density, the thiol concentration, the removal of oxygen from the sample medium, the pH value of the solution, and by optional excitation with a near UV wavelength (e.g., a 405 nm laser) to induce a back-transition from the dark state to the ground state.

In the original (f)PALM (fluorescence photoactivated localization microscopy) approach [9,22–25], a photoactivatable/photoconvertible fluorescent protein is detected, and congruent to the (re-)activation in dSTORM, a distinct activation laser is used to switch a small subset of emitters in the emission state. This process will be repeatable if the fluorescent protein can cycle between the two states, or it will be irreversible if the active state is terminated by photobleaching.

In general, organic dyes yield much higher photon counts than fluorescent proteins, which is a very desirable feature for SMLM. However, fluorescent proteins can be genetically encoded, and they allow SMLM with much less interference into the native state of the biological target. Today, many extensions of these three classical principles are available, but the basic concept of switching between an on state and an off state while only leaving a small minority of the emitter ensemble active remains the same in all of them.

13.3 Localizing Molecular Position

The ultimate computational task in any form of SMLM is to calculate the position of a fluorophore (i.e., the localization) with subdiffraction level accuracy and precision, based on the analysis of its single-molecule emission pattern (spot image) on a two-dimensional detector. There are a few general approaches for achieving localization from the single-molecule emission pattern:

- Comparing the individual spot to a predetermined set of high-resolution calibration data (accurately simulated or experimentally gathered)
- Deploying statistical properties like a center of mass estimation on isolated spots
- Fitting a model PSF to the experimental data with subpixel accuracy

The latter option quickly emerged as the primary method starting from the early days of SMLM, and it is still going strong today.

As previously established, the image of a point source emitter corresponds to the PSF of the optical system. But the PSF is dependent on the dipole orientation of the single-molecule emitter. The PSF for a dipole point emitter with a static orientation has a parallel (||) and an orthogonal component ($\perp$):

$$PSF(x', y') \propto sin^2\beta I_{\|}\left(p_{\|} + \cos(2\phi' - 2\alpha)\Delta p_{\|}\right) + sin\beta\, cos\beta \cos(\phi' - \alpha) I_{\perp} p_{\perp} + cos^2\beta I_{\perp} p_{\perp} \text{ with } (x', y') = \rho'(\cos\phi', sin\,\phi') \tag{13.3}$$

With the azimuthal angle α, the polar angle β, $p_{\perp}$ and $p_{\|}$ are the orthogonal and parallel PSF components relative to the image plane [46]. For this formulation of the PSF, a closed expression does not exist, so fitting an accurate analytical approximation is computationally very costly.

Nowadays there are elaborate and computationally costly fitting approaches that account for the dipole orientation and approximate the experimental emission pattern accurately. But the vast majority of dedicated single-molecule localization programs in use today rely on a simplified PSF-fitting model, the two-dimensional Gaussian function with a constant background offset. There are two main reasons for the emergence of that solution to the modeling problem: (1) the historical connection of SMLM to classic astronomical peak-fitting approaches and (2) the mathematical and computational simplicity of the Gaussian model. The latter was especially important in the early 2000s when real-time computing was unavailable in most microscopy research labs. Another justification for this simple model, compared to the more accurate but lengthy form in Eq. (13.3), is that, in most cases, the dipole orientation of common SMLM dyes is not fixed over the timespan of a single on-state. Thus dipole effects on experimental spots are negligible in most cases [46]. The Gaussian fitting of single-molecule spots on a noisy background is still the most common way of generating super-resolved images from the raw data, and it is the backbone of the majority of SMLM procedures. Since virtually all subsequent steps of analysis and postprocessing, as well as all further considerations on how to improve SMLM, heavily rely on the quality of the model-fit, the following sections give an overview of its intricacies.

13.3.1 Spot Candidate Selection

Spot fitting is computationally costly. The brute force approach of simply testing every pixel in the source image as a potential candidate for a spot is out of the question. A reasonable preselection of spot candidates for the spot-fitting algorithm can be estimated by applying simple local maximum finders, nonmaximum suppression approaches, or image segmentation techniques (Chapter 7). Any real spot should occupy an area of several square pixels. In the case of segmented objects, the center of mass (centroid) can be used as the starting position. A variety of low-pass and band-pass filters can be applied to improve the accuracy of the spot-finding algorithms [47–49].

The selection of reasonable candidates and starting parameters for the subsequent fitting attempts are crucial for a practical approach to SMLM. As always, a priori knowledge of the sample, especially of the fluorescent probe, helps to choose reasonable preselection parameters. Desired peak brightness, the axial extent of the sample (and therefore the possible range of defocus), and the expected width of the emitter when in focus are useful. The latter can be derived from knowledge of the refractive indices of the sample and immersion medium, the NA of the objective, and the emission wavelength of the dye in use. But in principle, the preselection should also allow a semiblind approach. Some of the most common and straightforward preprocessing considerations are listed in the following sections.

13.3.1.1 Local Intensity Maxima

One simple method is based on sliding a kernel through each image pixel location and determining if the pixel intensity $I_{P_xP_y}$ meets a basic threshold condition. At the same time it is checked whether $I_{P_xP_y}$ is greater than or equal to all values within the (typically a 4- or 8-connected) kernel neighborhood. If this condition holds, the point is further processed as a candidate pixel for a valid emitter.

13.3.1.2 Nonmaximum Suppression

There are many implementations of nonmaximum suppression filtering. One example based on *morphological grayscale dilation* (see Chapter 6) is proposed by Sonka et al. [50] and can be written as

$$(I \oplus S)(x, y) = \max_{(x', y') \in D_S} [I(x - x', y - y') + S(x', y')] \tag{13.4}$$

With the image I, S is a structuring element of size $l \times l$ and $l = 2r + 1$, with r being the dilation radius, and D_S the domain of S [49]. This form of grayscale dilation is more broadly applicable than local maximum detection because the structuring element can have an arbitrary form. In this way a simple binary image can be constructed where all pixels but the local maxima are set to 0 and the pixels set to 1 are candidates for the spot-fitting algorithm.

13.3.1.3 Centroid Estimation

If we consider a single-molecule emission pattern to be an extended and consistent bright object on the camera detector, we can approach the candidate approximation from a classical feature segmentation point of view. Image segmentation is a field of its own and is discussed extensively in Chapter 7, but one straightforward approach to spot-finding is to combine a simple intensity threshold with a subsequent watershed-based segmentation algorithm [48,50,51] (see Chapter 7, Section 7.2.2). Then, potential starting positions $\overline{x_p}$ and $\overline{y_p}$ for the localization algorithm can be obtained by:

$$\overline{x_p} = \frac{1}{c_p} \sum_{i=1}^{c_p} x_{i,p}, \quad \overline{y_p} = \frac{1}{c_p} \sum_{i=1}^{c_p} y_{i,p} \tag{13.5}$$

with p as an object index; c_p is the total number of pixel elements in the objects, and $x_{i,p}$ and $y_{i,p}$ are the plain pixel coordinates. This centroid estimation approach (also described in Chapter 8) can also be used to calculate the precise localization itself, as described later.

13.3.1.4 The Intensity Threshold

Setting a reasonable intensity threshold before even starting with any substantial image analysis is crucial for any SMLM localizer. It removes background signal and other potential artifacts and thus reduces computation time. But it also has an impact on the number of false negative and false positive detections. There are many possible ways of defining intensity thresholds (discussed in Chapter 7). Most commonly used are either user-defined global total intensity, or local signal-to-background and signal-to-noise ratio-based thresholds. There are more complex ways to semiobjectively define thresholds. Izeddin et al. suggested applying wavelet filtering [52] as a preprocessing step on the raw image, then using the second wavelet level of the image as a probe and choosing the threshold value between 0.5 and 2 times the standard deviation of the intensity values from the first wavelet level of the raw image [48].

13.3.2 Gaussian Model Fitting

The 2D Gaussian plus a background constant PSF model has been shown to approximate the experimentally detected single-molecule spots very well, assuming no distinct dipole orientation and an axial position near the focal plane [53–55]. This is particularly true for spatial coordinate determination but less so for other spot parameters like total intensity. This is shown in more recent studies deploying photometric methods for intensity determination. They demonstrated a systematic underestimation of the actual intensity, and this can be used to determine the axial position of the emitter [56–58]. Its most basic form $PSF_{basic2D}$ for a two-dimensional approach can be written with only six free parameters as:

$$PSF_{basic\,2D}(x_0, y_0, \sigma_x, \sigma_y, I, b) = bg + \frac{I}{2\pi\sigma_x\sigma_y} \exp\left[-\frac{1}{2}\left(\frac{(x-x_0)^2}{\sigma_x^2} + \frac{(y-y_0)^2}{\sigma_y^2}\right)\right] \quad (13.6)$$

There are extensions of the basic Gaussian PSF model that use more free parameters, but the simplified Gaussian model has proven to be a good compromise between complexity, speed, and accuracy of the localization process. Importantly, effects of noise from different sources lead to imperfect localization and limit its precision along all parameter scales. Later we discuss the theoretical limits of SMLM in terms of localization precision and its effect on structural resolution. For now, we focus on the fitting process, in which the noise deprives us of an exact location. Instead, we obtain the most likely answer to the question, "Where is the fluorophore located?" There are several ways to estimate the correct answer.

13.3.2.1 Least Squares Fitting

In general, one can define a metric to assess the quality of a model-based estimator. We use the so-called residue—the difference between any given theoretical model $M(x_k, y_k)$ and the experimental data S_k itself. The most likely solution is then defined as the solution for which the squared sum of all residues $\chi^2_{\text{LeastSquared}}$ is minimized. This can be written as:

$$\chi^2_{LeastSquared} = \frac{1}{\sigma^2}\sum_k (M(x_k, y_k) - S_k)^2 \quad (13.7)$$

This likelihood condition is strictly fulfilled when the noise, contributing to S_k, is normally distributed. Although this need is adequately met for most cases of SMLM raw data, we will later see that the usage of modern EMCCD detectors introduces additional factors into the equation, especially when considering nanometric localization precision. For now, we consider only the pure minimization problem in Eq. (13.7) because the spot is usually formed by more than 10^3–10^5 photons, and the low intensity side-lobes of the emission pattern have little or no impact on the localization accuracy [53].

The minimization problem is formalized by comprising all parameters of the PSF into a parameter vector p and considering $\chi^2_{\text{LeastSquared}}$ to be dependent on p. For nonlinear functions like the Gaussian PSF, there is no closed-form solution for the minima of $\chi^2_{\text{LeastSquared}}(p)$. Therefore the solution is calculated iteratively by assuming an initial value frame p_0 and iteratively minimizing $\chi^2_{\text{LeastSquared}}$ by updating according to $p_{i+1} = p_i + \delta_i$. The following sections discuss several commonly used iterative algorithms that provide reasonable values for δ_i [59].

13.3.2.2 The Method of Steepest Descent

From the pool of possible approaches to find suitable values for δ_i, the saddle-point, or steepest descent method (Eq. 13.8), is the simplest solution. For every iteration, the gradient of $\chi^2_{\text{LeastSquared}}$ at the current position is calculated and used to obtain the direction for the next position. The length of the step is set as a constant c, and its choice is nontrivial. High values risk the possibility of missing or overshooting the minimum, while low values can result in many iterations, or even worse, getting stuck at a local minimum. Thus, a smart compromise must be found for c, and it may vary from one dataset to another [60].

$$\delta_i = -c\nabla\chi^2_{LeastSquared}(p_i) \tag{13.8}$$

13.3.2.3 Newton's Method

Newton's method omits the tricky c constant by determining the minimum of the locally approximated paraboloid, which can be expressed as a linear equation system of first and second derivatives. Besides not relying on c, Newton's method is also parameter-free, and it normally converges relatively quickly. The major downside to Newton's method is that, under certain unfavorable conditions, the quickly convergent solution can shift into divergent oscillations [60]. Further, the matrix of second-order derivatives sometimes can be singular, and thus not always solvable. This problem can be circumvented by defining the second-order terms in the Hessian matrix $H(p)$, which are weighted by the residues, as neglectable. Here the (most often valid) assumption is that residues are small and random when the approximation is close to a minimum. We can assume this because the residues are caused not by systematic errors, but by noise effects not having any significant spatial correlation. The assumption is that all noise and background contributions are "well behaved" and show no spatial correlation on the scale of a typical PSF width. Any deviation from this assumption can complicate the fitting process significantly. With this, the problem is always solvable since this reduced Hessian matrix, $H^*(p)$, is positive definite [59].

$$\begin{aligned} H\delta_i &= -\nabla\chi^2_{LeastSquared}(p_i) \qquad H_{jk} = \left(\nabla^2\chi^2_{LeastSquared}\right)_{jk} \\ &= \sum_i \frac{1}{\sigma_i}\frac{\partial G}{\partial p_j}\frac{\partial G}{\partial p_k} - \sum_i \frac{(S_i - G)}{\sigma_i}\frac{\partial^2 G}{\partial p_j \partial p_k} \cong \sum_i \frac{1}{\sigma_i}\frac{\partial G}{\partial p_j}\frac{\partial G}{\partial p_k} = H^*_{jk} \end{aligned} \tag{13.9}$$

13.3.2.4 The Levenberg-Marquardt Method

Finally, the *Levenberg-Marquardt* method cleverly combines the saddle-point and Newton's method approaches. It does so by multiplying the diagonal elements of the reduced Hessian by an adaptive scaling parameter $D^* = D+1$ [60]. For low values of D, the resulting expression is equal to Newton's method with H^*. For high values of D, the scaled diagonal values of the matrix dominate, and the equation switches to steepest descent with $c = D^{-1}$:

$$\left[H^*(p_i) + D\, diag\left(H^*(p_i)\right)\right]\delta_i = -\nabla\chi^2_{LeastSquared}(p_i) \tag{13.10}$$

The key point with this approach is the way the scaling parameter D is handled dynamically. Initially, D is set to a semirandom or educated guess value and then iteratively adjusted. It is decreased whenever a step also decreased the value of $\chi^2_{LeastSquared}$, and increased whenever the computed step would enhance the residues. A feedback loop is thereby implemented, avoiding overshooting or getting stuck on a local minimum in $\chi^2_{LeastSquared}$. Effectively, the Levenberg-Marquardt method boils down to Newton's method when the function is "well-behaved" and to the saddle-point method for tricky functions where a lot of small iterative steps take over.

13.3.2.5 Maximum Likelihood Fitting

The maximum likelihood estimator (MLE) is a statistical estimator that optimizes a set of model parameters in such a way that there should be no other choice of the model parameters that yields a higher probability of producing the observed data [59]. The previously described least-squares fitting can be seen as an MLE when we assume the noise originating from different sources to be normally distributed. In practice, this condition is almost perfectly fulfilled for spot center pixels of a bright and in-focus fluorophore. However, this approximation becomes less valid for low-intensity cases, strongly defocused emission spots, and other nonlinear effects like the previously mentioned dipole effect on the three-dimensional PSF [46].

But also for every *normal* single-molecule spot that is sufficiently well sampled (i.e., for pixel sizes well below the emission wavelength), the peripheral pixels of a spot receive very few photons and thus show a Poissonian rather than Gaussian noise distribution. These effects are not well accounted for by least-squares algorithms but can be more reliably treated with an MLE method. For example, Poissonian statistics can be implemented for MLE with the reliable Levenberg-Marquardt method by weighting the differences between model function and measured data [59,61] according to:

$$\chi^2_{MLE} = 2\sum_i G(x_i) - S_i\left(1 + ln\,\frac{G_i}{S_i}\right) \tag{13.11}$$

In their 2010 SMLM cornerstone paper, Mortensen et al. [46] describe an extensive effort comparing different MLE and non-MLE estimators for single-molecule PSF localization. This includes MLE with a theoretical PSF (MLEwT), a Gaussian mask estimator (GME), and a weighted least-squares (WLS) Gaussian fit. They found that the best and most robust classic estimator, even for SMLM borderline cases, was an MLE with a 2D Gaussian plus a constant background, at least for reasonably large photon numbers.

13.3.3 Localization Methods

Today, there are as many dedicated localization software packages, stand-alone programs, and plug-ins as there are acronyms for SMLM-related techniques, and even that is probably a huge understatement. Many of the current state-of-the-art programs are open-source available and ready to use to a certain degree. There is a wide variety of complexity and user friendliness among them. Selecting the software can be a matter of taste and previous experience of the user, and it can make or break the success of the localizer in specific hands. However, there have been two comprehensive evaluation competitions for SMLM software packages. They assessed detection rate, accuracy, quality of image reconstruction, resolution, software usability and computational resources for 2D and 3D data [62,63]. The 2015 and 2019 "super-resolution fight club" assessments not only provide a broad assessment of 3D SMLM software in realistic conditions, but also give potential first-time users a resource to identify optimal analytical software for their experiments using simulated and experimentally aquired raw data [62,63]. These two resources are excellent, as different applications warrant specific or dedicated localization software, and there are many things that can go wrong during an SMLM experiment.

13.3.3.1 Spot Centroid Calculation

The centroid method can be used not only to select candidates for spot fitting as described earlier, but also to calculate the subpixel precise molecular position from the spot candidate itself [64]. The determination of the centroid in a local image neighborhood is a very fast method for subpixel molecule localization, but this approach does not estimate the intensity of emitters, an important parameter for quantitative SMLM, and it is sensitive to noise. In order to use centroid calculation as a spot localizer, one must weigh the pixel coordinates (x,y) with their respective intensity I_{xy}:

$$\overline{x} = \frac{\sum_{x,y\in D} x\, I_{xy}}{\sum_{x,y\in D} I_{xy}}, \quad \overline{y} = \frac{\sum_{x,y\in D} y\, I_{xy}}{\sum_{x,y\in D} I_{xy}} \tag{13.12}$$

13.3.3.2 The Radial Symmetry Method

If we assume a well-behaved microscope PSF, we can also assume that the resulting emission pattern exhibits significant radial symmetry around the molecular position.

Therefore, one can estimate this coordinate by determining the point with maximum radial symmetry. This is defined as the point with minimum distance from gradient-oriented lines passing through all data points [65]. Like the centroid method, this approach is very fast, and it has the additional advantage of being robust against noise. But this algorithm does not estimate the intensity or the imaged size of a molecule. One possible implementation of this localizer was described by Ovesny et al. [49], and it starts by determining the intensity cogradient $\nabla\widetilde{I}_{x,y}$:

$$\nabla\widetilde{I}_{x,y} = \left[\frac{\partial\widetilde{I}(x,y)}{\partial u}, \frac{\partial\widetilde{I}(x,y)}{\partial v}\right]^T \tag{13.13}$$

for every pixel coordinate (x,y) from the initial set D. The u,v coordinates are rotated by 45 degrees from the x,y coordinate system of the image, because the partial derivatives are determined by the Roberts cross operator [50] as:

$$\begin{aligned}\frac{\partial\widetilde{I}(x,y)}{\partial u} &= \widetilde{I}(x+1,y+1) - \widetilde{I}(x,y)\\ \frac{\partial\widetilde{I}(x,y)}{\partial v} &= \widetilde{I}(x,y+1) - \widetilde{I}(x+1,y)\end{aligned} \tag{13.14}$$

Then, the calculated cogradient $\nabla\widetilde{I}_{x,y}$ corresponds to the midpoint $(x+0.5,\ y+0.5)$ and the slope $s_{x,y}$ of a line passing through this point can be written as:

$$s_{x,y} = \left(\frac{\partial I(x,y)}{\partial u} + \frac{\partial I(x,y)}{\partial v}\right)\left(\frac{\partial I(x,y)}{\partial u} - \frac{\partial I(x,y)}{\partial v}\right)^{-1} \tag{13.15}$$

Parthasarathy et al. showed that the analytical solution to determine the origin of radial symmetry is the point $(x_0^*,\ y_0^*)$ that minimizes the sum of weighed distances of all considered lines to that point [49,65]. It can be written as.

$$\begin{aligned}x_0^* &= a^{-1}\left[\left(\sum_{j\in D} s_j w_j b_j\right)\left(\sum_{j\in D} w_j\right) - \left(\sum_{j\in D} w_j b_j\right)\left(\sum_{j\in D} s_j w_j\right)\right]\\ y_0^* &= a^{-1}\left[\left(\sum_{j\in D} s_j w_j b_j\right)\left(\sum_{j\in D} s_j w_j\right) - \left(\sum_{j\in D} w_j b_j\right)\left(\sum_{j\in D} s_j^2 w_j\right)\right]\\ a &= \left(\sum_{j\in D} s_j w_j\right)^2 - \left(\sum_{j\in D} w_j\right)\left(\sum_{j\in D} s_j^2 w_j\right),\quad b = y - s_j x,\quad w_j = \frac{\widetilde{w}_j}{s_j^2+1}\\ j &= x(2r+1) + y,\ \ s_j = s_{x,y},\ \ w_j = w_{x,y},\ \ b_j = b_{x,y},\ \ x,y \in D\end{aligned} \tag{13.16}$$

The point-to-line distances are then weighted according to

$$\widetilde{w_j} = \widetilde{w_{x,y}} = \frac{|\nabla I_{x,y}|^2}{d(A_{x,y,C})} \tag{13.17}$$

with $d(A_{x,y,C})$ as the Euclidean distance of point to the centroid C. The centroid is then computed from the gradient magnitudes [49,65], analogously to Eq. (13.12).

13.3.3.3 Spline and Complex Model Fitting

The original justification to use a Gaussian model to approximate the microscope PSF was largely based on its mathematical simplicity and the resulting low computational cost of the fitting process. However, today's abundance of computational speed allows for more elaborate procedures and models to approximate the three-dimensional geometry of single-molecule emission patterns with much higher detail. For example, spline functions (i.e., piecewise polynomials for which high-order derivatives are continuous at the knots) recently emerged as a new tool for nanometer-precise single-molecule localization analysis [66–69]. Most commonly used are cubic splines, also widespread in modern computer graphics. For instance, Li et al. implemented a cspline-based algorithm to evaluate the three-dimensional PSF of a standard microscope so precisely that they could generate a high-quality model to then later derive the 3D nanometer coordinates of experimental emission patterns [66]. For this, they first measure a bead sample that is moved through the focal plane of the microscope and then calculate average PSFs from that. Their software is based on csplines with 64 coefficients in each voxel of the 3D PSF stack. The relation between the 3D PSF and the 3D cspline for voxel (i, j, k) can be written as.

$$f_{i,j,k} = \sum_{m=0}^{3}\sum_{n=0}^{3}\sum_{p=0}^{3} a_{i,j,k,m,n,p} \left(\frac{x - x_i}{\Delta x}\right)^m \left(\frac{y - y_j}{\Delta y}\right)^n \left(\frac{z - z_k}{\Delta z}\right)^p \tag{13.18}$$

where Δx and Δy are the pixel size of the PSF in the object space in the x and y directions, respectively; Δz is the step size in the objective space in the z direction; and x_i, y_j, and z_k are the starting positions of voxel (i, j, k) in the x, y, and z directions [66]. The required 64 cspline coefficients are calculated by upsampling (using cspline interpolation) each voxel three times in the x, y, and z directions, respectively. The 64 upsampled coordinates (including the boundary of neighboring voxels) are then used to calculate the 64 cspline coefficients [66]. Li et al. achieved unprecedented three-dimensional SMLM resolutions without any additional optics. This approach is of course also applicable to two-dimensional data and is extendable to other splines, polynomial functions, and complex theoretical PSF models, which were too computationaly intensive in the early days of SMLM.

13.3.4 Visualization of Localization Data

The key data structure of SMLM is the multidimensional dataset that is calculated from the raw image stack by the localization algorithm. From this, a multitude of further processing steps is possible. This can be much more complex than the initial localization step itself, depending on the basic intent of the experiment. However, in almost every case the first task is to render the localization set into a super-resolved image. As we have already established, observing the structure of interest with significantly improved spatial resolution is, by far, the biggest benefit in most SMLM experiments. Since seeing is believing, the choice of visualization can improve or even negate the resolution enhancement achieved by SMLM and thus influence the conclusions to be drawn from the data. There are a variety of ways to visualize localization data, and there is no gold standard for choosing one (Fig. 13.4). In most localization software, there are multiple options to render the dataset with different parameters, and SMLM users can experiment with them to get a feel for the differences they make. In the end, the choice of visualization is also a matter of individual preference.

In the interests of reproducibility, it is useful to record and report the exact parameters used, both for the localization step and for the final image rendering. Changing a single parameter, such as the intensity threshold in the localizer, can lead to huge changes in the resulting output data and image. The most commonly used localization software offers the option of generating a log file where all relevant algorithmic parameters are stored. SMLM users can store these log-files together with the raw data and the final images in a repository for the sake of reproducibility.

13.3.4.1 Scatterplots

A quick and easy way to visualize localization data is to use scatterplots. They give a useful overview but do not always provide high-quality results. There are two common ways to render localization data in a scatterplot representation:

(1) A large binary image is rendered, where the pixel size is chosen so that no more than one localization can occupy each pixel. Then, the intensity of all localization-positive pixels is set to one and all other pixels are set to zero.

(2) A smaller grayscale image is rendered, where the pixel size is chosen so that many of the pixels will contain more than one localization. The intensity of each pixel is set to the number of localizations that fall within its region.

Although scatter plots are often seen as inferior to other representations, they might be the only option that can avoid rendering induced alterations of the raw data.

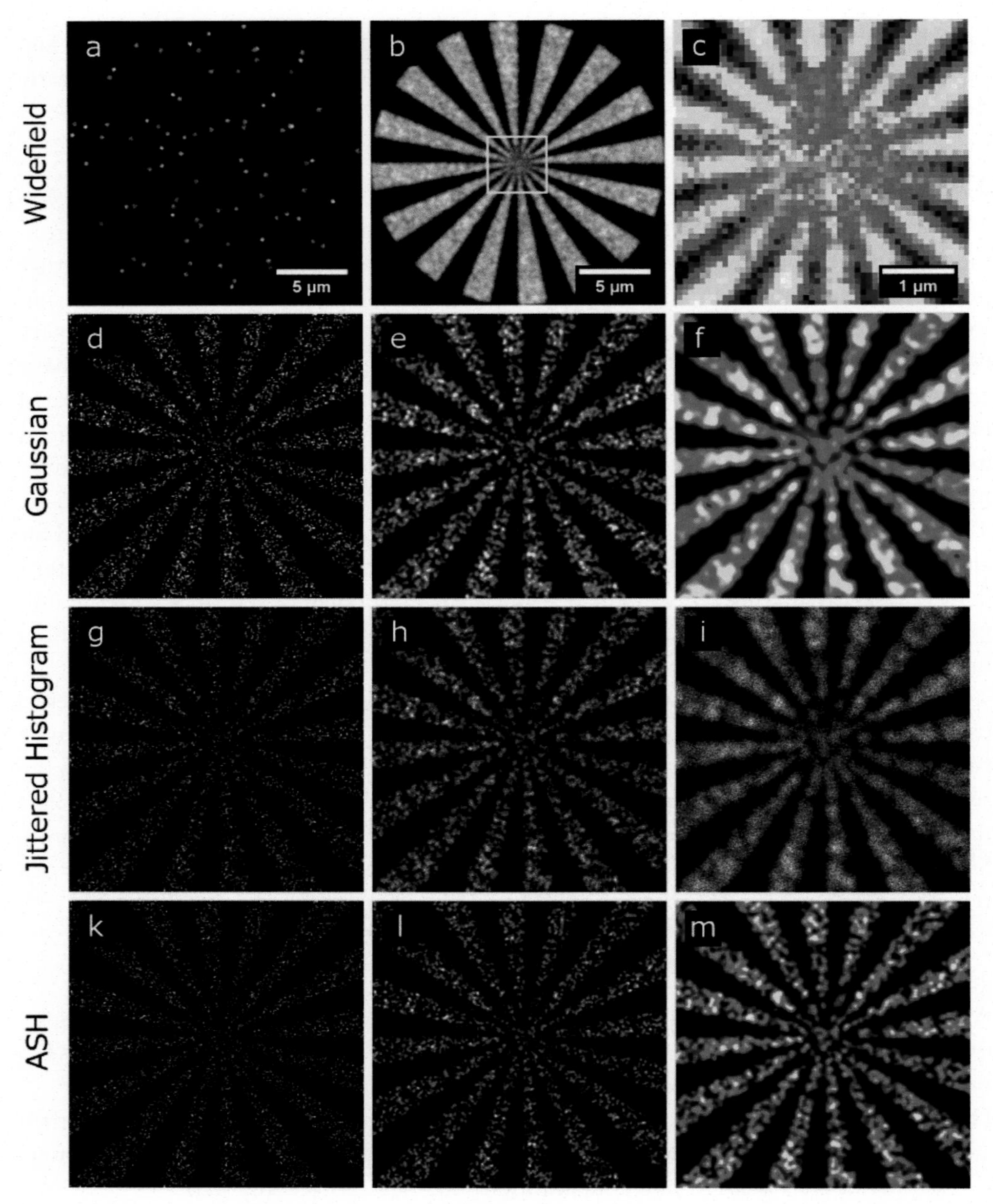

Fig. 13.4 Overview of different SMLM visualization methods. Widefield images: (a) a single frame of raw data, (b) average intensity time projection, (c) zoomed region indicated in (b). (d–m) Super-resolution SMLM images of the region indicated in (b) magnified 10 ×. Gaussian rendering with lateral uncertainty set to: d) computed localization uncertainty, (e) 20 nm, (f) 50 m. Jittered histogram with number of averages set to 100 and with lateral uncertainty set to: (g) computed localization uncertainty, (h) 20 nm, (i) 50 nm. Average shifted histogram with the number of lateral shifts set to: (k) 2, (l) 4, (m) 8. The *Thunderstorm* plugin [48] provides the displayed rendering options, while additional alternatives exist in other localization software. *(From Ref. [48] with permission)*

13.3.4.2 Two-dimensional Histograms

Two-dimensional histograms are the most commonly used rendering method for SMLM data, because they are the direct, resolution-enhanced, equivalent of the raw image. They represent signal (i.e., localization) density by counting the number of instances that fall into each of the histogram bins. The bins represent the pixel size of the final super-resolution image [70], and bin size should be chosen with respect to the desired resolution of the enhanced image. The optical pixel size of the camera should be in the range of 100 nm so, dependent on the SMLM experiment, a corresponding histogram bin size should be set. Assuming a tenfold resolution increase, a bin size in the range of 10 nm is generally acceptable, although some experiments might require a different value.

In the simplest case, the final pixel intensity value simply represents the number of localizations that fall within the current bin. There are many ways to perform an additional weighting step—most commonly the integrated localization intensity is used, as it is inversely related to the theoretical localization precision. This way more precise localizations contribute more toward the final image resolution. A 2D localization histogram generated in this way can be modified additionally to improve perception of the enhanced resolution. This can involve smoothing or shading the structure.

Jittering

An easy way to "smooth" the image is to introduce an artificial uncertainty to every localization and therefore create a more filled-out image [47]. For example, a random number drawn from a normal distribution can be added to the actual coordinates of *every* localization prior to histogramming. The standard deviation can be defined by the user or set as the localization precision theoretically determined based on the single-molecule brightness. If this procedure is performed multiple times and all jittered histograms are averaged together, the final image can appear more populated and thus more complete. With increasing numbers of jitters, the resulting image converges to the image obtained by Gaussian rendering, which is described later. Jittering is computationally inexpensive and fast for small numbers of iterations, but the resulting images can appear noisy [47].

13.3.4.3 Intensity Interpolation to Neighboring Pixels

Another way of achieving a more filled-out appearance of the image is to distribute the initial pixel intensity among neighboring pixels. If the molecular position of the emitter is calculated to localize not to the exact center of the final pixel, the intensity is interpolated into the neighboring pixels. This can be performed with, for example, bilinear or quadratic interpolation.

Averaged Shifted Histograms

A related approach to achieve a pleasantly filled final image is to create a range of histograms with the same bin size, but with the center coordinate of each histogram shifted incrementally from the previous histogram. In order to calculate a final image, $2n$ histograms are averaged for n shifts in each of the two dimensions. The number of shifts n in the lateral and axial directions are independent and should be kept within reasonable bounds, as exessive shifting will lead to a washed-out and degraded structural resolution.

13.3.4.4 Gaussian Rendering

Apart from the classical histogram approach, there is the aforementioned notion that the inherent precision of every individual localization should be taken into account to render a final image that is as true to the experimental conditions as possible. To this end, Gaussian rendering can be created by placing a normalized symmetric 2D Gaussian function over the voxel volume for every localized molecule, with the standard deviation equal to the assumed localization uncertainty. As an example, to calculate the contribution of one molecule to the voxel intensity, v, at the three-dimensional integer position, Ovesny et al. compute

$$v(x, y, z \mid \boldsymbol{\theta}_p) = E_x E_y E_z \tag{13.19}$$

where p indexes the individual molecules and the parameters $\boldsymbol{\theta}_p = [x^*_p, y^*_p, z^*_p, \sigma^*_{xyp}, \sigma^*_{zp}]$. Here (x^*_p, y^*_p, z^*_p) defines the position of the molecule and $(\sigma^*_{xyp}, \sigma^*_{zp})$ are the lateral and axial localization uncertainties, respectively. Also.

$$\begin{aligned}
E_x &= \frac{1}{2} erf\left(\frac{x - x^* + \frac{1}{2}}{\sqrt{2}\sigma^*_{xy}}\right) - \frac{1}{2} erf\left(\frac{x - x^* - \frac{1}{2}}{\sqrt{2}\sigma^*_{xy}}\right) \\
E_y &= \frac{1}{2} erf\left(\frac{y - y^* + \frac{1}{2}}{\sqrt{2}\sigma^*_{xy}}\right) - \frac{1}{2} erf\left(\frac{y - y^* - \frac{1}{2}}{\sqrt{2}\sigma^*_{xy}}\right) \\
E_z &= \frac{1}{2} erf\left(\frac{z - z^* + \frac{\Delta z}{2}}{\sqrt{2}\sigma^*_{z}}\right) - \frac{1}{2} erf\left(\frac{z - z^* - \frac{\Delta z}{2}}{\sqrt{2}\sigma^*_{z}}\right)
\end{aligned} \tag{13.20}$$

where Δ_z is the size of a voxel in the axial direction. In the original implementation the user is also given the option to limit the contribution distance to molecules within a circle of radius $\mathbf{k}\sigma^*_{xyp}$ around the molecule position in the lateral dimension and by $\mathbf{k}\sigma^*_{zp}$ in the axial direction, with $\mathbf{k}$ as a vectorial scaling factor [49]. Fig. 13.5 summarizes the principle of SMLM described in the previous section.

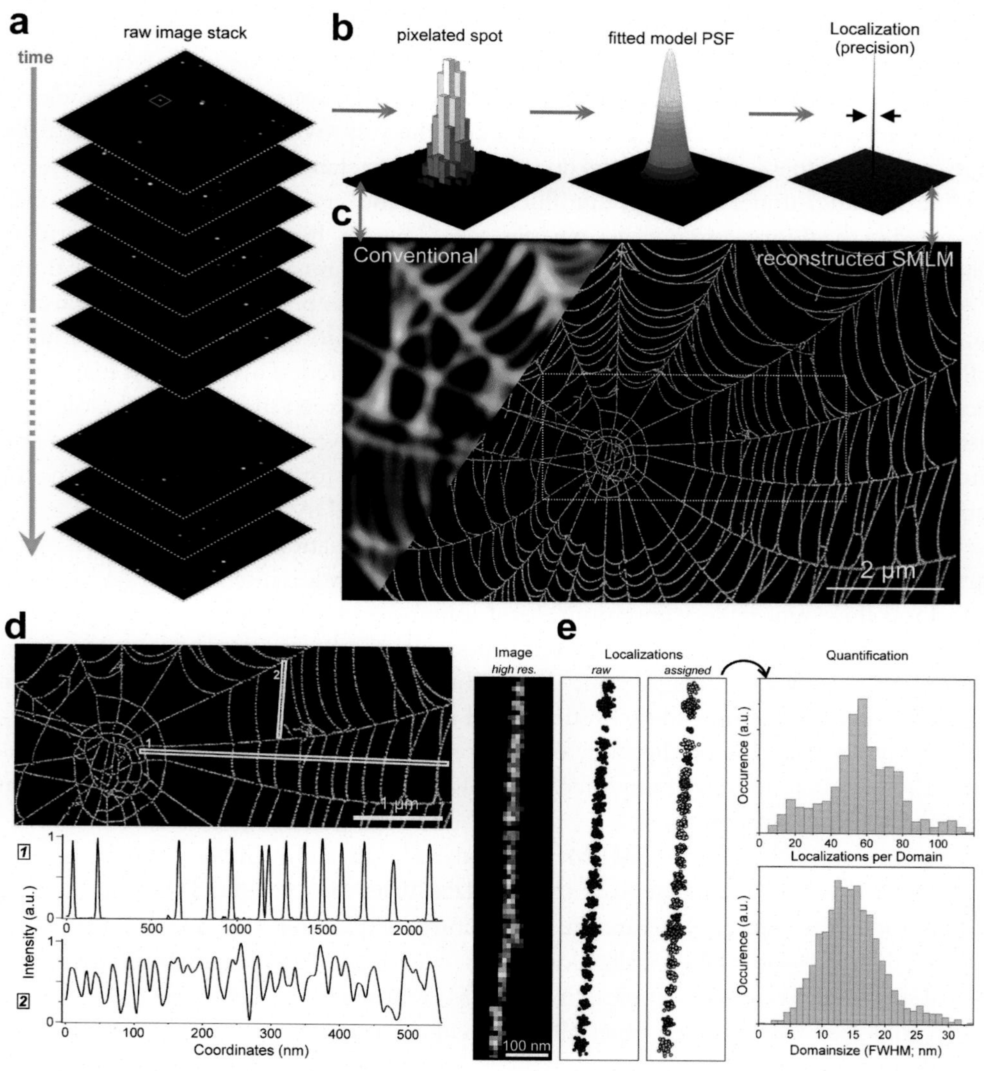

Fig. 13.5 General schematic overview of SMLM. (a) The fluorophore ensemble is tuned to the single molecule level, so that in every individual camera frame only a sparse subset is allowed to be active. To collect all emitters, usually tens of thousands of frames are gathered while fluorophores switch between the On and Off state stochastically. (b) The pixelated emission pattern is fitted with a model PSF function, allowing nanometer-precise determination of the molecular position, i.e., the localization. (c) All collected localizations are assigned to a pixel array with significantly smaller pixel spacing (e.g., 10 nm) than the raw images. (d) SMLM results in a molecular information map, enabling sophisticated image-based (d) and localization-based (e) postprocessing approaches that can lead to a structural information gain beyond the super-resolution limit. *(Adapted from Ref. [1])*

13.3.5 Localization and Image Artifacts in SMLM

Like all other fluorescence microscopy techniques, SMLM is prone to artifacts. Indeed, SMLM is a standard widefield microscopy technique that suffers from the same drawbacks as its diffraction-limited predecessors and the improved resolution can actually worsen some of these effects. But we do not discuss here standard problems such as nonspecific labeling, undersampling, autofluorescence, and crosstalk, since these are well described in the fluorescent microscopy literature. There is, however, one type of artifact that is unique to SMLM: the effect of high emitter density introducing false localizations.

Until now, we have assumed the sparsity requirement to be met whenever discussing nanometric localization. If, however, more than one fluorophore is active within a diffraction-limited area, then overlapping PSFs are generated. This results in inaccurate localizations since the center of gravity of the spot lies between the individual molecular positions [42,71,72]. For standard SMLM localization software, a rough estimate of one emitter per μm^2 can be taken as a guideline [73], except for specialized algorithms for fitting multiple-emitter PSFs, like compressed sensing or DAO-Storm [74,75].

The main causes of high emitter densities are low irradiation intensity densities, suboptimal buffer compositions (*d*STORM), high photoactivation intensities (PALM), very high local fluorophore concentrations, local inhomogeneities in photoswitching rates, slightly defocused signals, and background fluorescence from other image planes that can seriously deteriorate image quality [76]. The study of these artifacts is an entire field of its own, and it takes time to theoretically and experimentally understand them. Therefore we highly recommend the study of the existing literature on SMLM artifacts, especially the many renderings of imperfect SMLM datasets [71,73,76–78], before taking on serious SMLM experiments.

A common problem with SMLM and modern localization software is that the user will almost always get an image from the raw data. Whereas a veteran SML microscopist would notice the typical ghost and fuzzy structures, exploding stars, and webbings that indicate density-related false localizations, a less-experienced user might mistake the artifactual structures as a new biological finding. SMLM artifacts are quite common, even in many published works today, especially with experimental dyes or uncooperative biological samples.

13.4 Three-Dimensional Localization Microscopy

Most of the earlier discussion of nanometer-precise localization was directed to two-dimensional structures and PSFs. Chapter 12 gives an excellent overview of diffraction limited fluorescence microscopy techniques, and all discussions there of three-dimensional far-field fluorescence microscopy apply fully to SMLM. However, the nanoscopic approach warrants some further considerations of the three-dimensional nature of the PSF. As described previously, the axial dimensions of the diffraction limited, aberration-free PSF are significantly larger than the lateral counterparts. Moreover, the shape of this standard emission pattern (e.g., its measurable FWHM) changes relatively slowly near the focal plane, where SMLM requires us to work due to the limited photon budget of typical dyes. The PSF is symmetric about the focal plane, which makes it impossible to assign a unique axial position without additional workarounds. One is to set the focal plane exactly beneath the first detectable signal plane, thus restricting any axial extent to only one direction. This simple approach has disadvantages, such as the further limitation of the overall axial detection range and the effective reduction of lateral localization precision, since all detectable emitters will be defocused. Using extensions to the standard microscopy setup can result in isometric three-dimensional resolutions in the range of 10–20 nm for certain applications.

Today we can group the different approaches to 3D SMLM into three groups, based on the technological concept they employ:

- Multiplane Imaging
- PSF Engineering
- Intensity Sensing Imaging

Nowadays, there is a plethora of dedicated three-dimensional SMLM methods, varying in complexity and in their quality metrics. In principle, one has to define in advance the desired resolution and axial range, taking into account the sample type and the applicabilty of each method. However, the three general approaches, mentioned earlier and described in the following sections, offer a starting point to decide which three-dimensional SMLM approach is the most promising.

13.4.1 Calibration Measurements

Independent of the method chosen to achieve a unique axial localization, one key concept is the same for every approach: namely, aligning the experimentally derived parameter to a well-defined calibration curve. This curve can be theoretically calculated based on a proper PSF model, but the most common way is to perform an actual measurement in which a monolayer of fluorescent molecules is moved through the microscope's focal plane in a defined manner (Fig. 13.6B–D). By relating the sets' axial position relative to the unique axial parameter, one can interpolate a precise axial look-up table to correct any potential experimental value.

Although these experiments can be performed in many different ways, it is recommended to use as close an approximation to the biological sample as possible for the calibration sample, i.e., use the same antibodies and/or dyes to create a single-molecule surface and label the biological target structure [56,58,79]. If, however, the use of fluorescent beads cannot be avoided, their physical size should be small enough ($\leq$40 nm) not to affect the PSF.

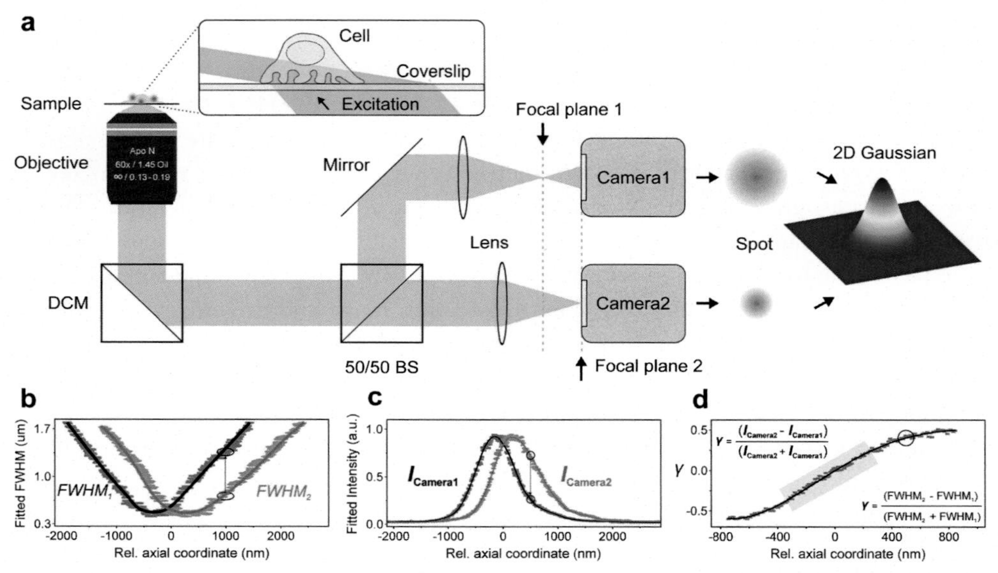

Fig. 13.6 The principle of multiplane SMLM imaging. (a) Overview of the experimental setup used to perform multiplane SMLM imaging. HILO illumination of the sample: *blue* beam (shown only in enlarged box) triggers fluorescence emission *(orange)*, which is split by a 50/50 nonpolarizing beamsplitter (50/50 BS) to acquire biplane images on two separate (EM-CCD) cameras. The respective imaging lenses are shifted along the optical axis to induce a relative defocus of the image detection on synchronized cameras. Spots, apparent in both detection planes, are fitted, e.g., by a Gaussian PSF model. (b–d) Using a piezo-driven stage, the focal plane is linearly moved through the sample plane while imaging a single-molecule surface with synchronized cameras. Fitting the raw emission pattern with independent Gaussians with variable FWHM yields the classical, axially dependent FWHM curves (b). Fitting with constant FWHM Gaussians yields single-molecule intensity curves (c), resulting in dTRABI data. The relative change of position of the imaging lens in the reflection path is mirrored by the relative shift of the respective curve (indicated by circles). An axially precise and unique calibration function γ is derived either from the FWHM or the molecular raw intensities, as indicated in (d). *(Adapted from Ref. [79])*

13.4.2 Multiplane Imaging

Axial slicing is a common approach in confocal microscopy to build up the three-dimensional image iteratively. In widefield microscopy, we can adapt this concept and construct an optical detection path that allows us to image the same lateral region with multiple different focal planes. The simplest and most common implementation of this approach is also known as biplane imaging. Here, we divide the fluorescence image into two parts, e.g., with a 50/50 beamsplitter, and subsequently alter one of the detection paths, resulting in a slightly shifted focal plane (Fig. 13.6). This can be achieved by imaging the two channels with two independent detectors, using either two separate cameras [80] or a single split camera chip [81,82]. One is shifted along the optical axis, either physically or by optical means. The optimal distance between the two planes is a compromise between the desired axial range and the requirement to associate one emitter with its counterpart in the other channel. In general, the ideal spacing can be calculated by CRLB considerations [83], but in practice a plane-to-plane distance between 300 and 500 nm has proved to be best for axial precision, while still achieving an axial range of ~1.5 μm.

Practically speaking, we break the axial symmetry of the PSF by calculating the dynamic relation between the two channels. The most straightforward way is simply to use the fitted width of the individual spots, since biplane data is usually fitted following the standard Gaussian approach. There are different ways of combining the channel information, but a common formulation to achieve a unique axial parameter P_z can be written as

$$P_z = \frac{FWHM_1 - FWHM_2}{FWHM_1 + FWHM_2} \tag{13.21}$$

Biplane-based 3D SMLM imaging has been applied to a wide range of biological questions and structures because of its technically simple implementation [56,58,79,82,84–86]. In principle, it is possible to extend the axial range of this method by adding additional planes [54,87,88]. Howe*v*er, due to the limited photon budget of SMLM dyes, excessive extension of multiplane methods can result in too much signal spread and therefore a loss of localization quality. Another interesting way to further improve the quality and applicability of multiplane imaging is to combine it with other 3D approaches, such as astigmatism [89,90], other PSF engineering techniques, and photometric methods [56,58,79,86].

13.4.3 Point Spread Function Engineering

Whereas multiplane imaging accepts the limitations of the diffraction-limited PSF and finds a high-quality work-around, there is an alternative means to improving microscope resolution. This approach is to encode the emitter's axial position into the shape of the emission pattern. The main concept relies on the addition of phase aberrations by inserting a phase-modulating element into the Fourier plane of the microscope. There are two general approaches to achieve this, and they differ in terms of technical demands and in the variety of applications:

- Using basic optical elements like astigmatic lenses [27,28] or glass wedges [91] can be used to introduce simple, static phase patterns.
- Using complex phase modulators like lithographically etched dielectric phase plates [92], deformable mirrors (DM) [93], or liquid crystal spatial light modulator (SLM) [94–97], offering a fine-tuned control over the phase pattern.

SLMs and DMs can be programmed to generate a wide variety of phase patterns, but also have their disadvantages in practical applications. For instance, SLMs are optimized for a specific linear polarization of light, thus limiting significantly the number of collected photons, which degrades not only localization precision but also the effective axial range of the microscope. DMs suffer from a limited number of assessable pixels and are therefore limited in recording fine features [98].

The general idea of PSF engineering is to alter the lateral shape of the PSF in an axially dependent fashion to generate a uniquely measurable shape parameter along the optical axis. There is an extensive mathematical framework for this task, stemming from Fourier optics [99], and it can be used to calculate the phase pattern $P(x',y')$ necessary to generate the electromagnetic field $E(x',y')$ in the backfocal plane of the objective for the desired intensity distribution I in the sample

$$I(u, v; x, y, z) \propto |\mathrm{FT}\{E(x', y'; x, y, z)P(x', y')\}|^2 \tag{13.22}$$

This general equation implies that an axial displacement of the fluorophore introduces a curvature in the phase of $E(x',y')$, and the introduced phase pattern $P(x',y')$ changes how this manifests in the image $I(u,v)$ [98]. This z-dependent curvature of the electromagnetic field at the back focal plane can be approximated from the scalar approximation in the Gibson-Lanni model [100] as

$$E(x', y') \propto \mathrm{e}^{-ikn_1 z_1\sqrt{1-\left(\frac{\mathrm{NA}}{n_1\rho}\right)^2} + ikn_2 z_2\sqrt{1-\left(\frac{\mathrm{NA}}{n_2\rho}\right)^2}} \tag{13.23}$$

assuming a typical two-layer system with a sample refractive index n_2 (e.g., water) and the intermediate (glass/immersion oil) refractive index of n_1. Then, z_1 denotes the distance between the focal plane and the sample interface, and z_2 is the distance between the emitter and this interface, with k being the wavenumber $2\pi/\lambda$. PSF engineering for SMLM has become an entire field on its own. There are several designs that have been used to increase either the axial precision, the range, or the detection robustness. This includes the astigmatic PSF [28,101,102], the rotating double-helix PSF (DH-PSF) [94,103], the corkscrew PSF [104], the phase ramp PSF [91], the self-bending PSF [95], and the Tetrapod PSF [96,97] (Fig. 13.7). Here, the axially unique parameters range from simple geometric width, to angular displacement or other asymmetry descriptors that can either be assessed by the standard Gaussian or a more catered fitting approach, based on the a priori known model function.

For a more in-depth overview of the various PSF engineering approaches, including an extensive CRLB evaluation of the techniques and other 3D SMLM methods, see the review article by Diezman, Shechtman and Moerner [98]. One of the most impressive applications of PSF engineering in recent years was the simultaneous encoding of not only the axial coordinate of the emitter, but also its emission wavelength [105,106].

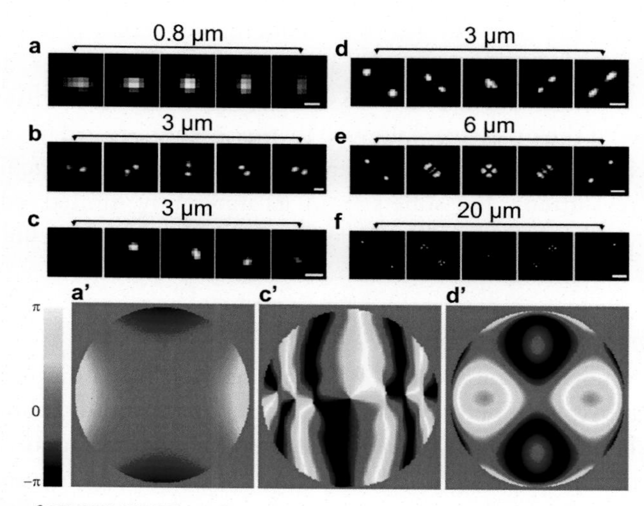

Fig. 13.7 Overview of 3D SMLM PSF engineering approaches. *Top:* PSF shape as a linear function of the axial emitter position. The applicable z-range is indicated above each panel. (a) Astigmatism. Scale bar = 500 nm. (b) Double helix. Scale bar = 2 μm. (c) Corkscrew. Scale bar = 1 μm. (d–f) various Tetrapod PSFs. Scale bars = 1, 2, and 5 μm, respectively. *Bottom:* Corresponding phase patterns for astigmatic PSF (a,a′), doublehelix PSF (c,c′) and lower range Tetrapod PSF (d,d′). *(Adapted from Ref. [98] with permission)*

13.4.4 Intensity-Based Approaches

So far we have mainly looked at 3D SMLM approaches that utilize the measured shape of the emission pattern to derive the axial position of each molecule. However, another important parameter is the single molecule intensity, i.e., the integrated amplitude over the entire emission pattern. We omit discussion of the 3D SMLM methods relying on interferometry [107,108] and the sensing of fluorescence life-time changes [109,110], which pose significant technical challenges to implement and apply. We now consider two commonly used approaches, supercritical angle intensity detection and photometry-based microscopy (TRABI).

13.4.4.1 Supercritical Angle Localization

Supercritical angle localization microscopy (SALM) uses the effects of the emitter's angular signal to extract the relative axial position. A common implementation relies on the inherent refractive index mismatch between the sample and the coverslip. Molecules that are further away from the glass coverslip than $\sim\lambda$ will have that portion of their emission that has incident angles relative to the substrate surface greater than $\theta_{\mathrm{critical}} \approx 61$ degrees totally internally reflected back into the sample. For emitters closer to the coverslip, supercritical light above this angle will be coupled into the coverslip and detectable by a high-NA objective. The intensity of this supercritical emission fraction exhibits an axial dependence comparable to that of total internal reflection (TIR) excitation with highly inclined illumination. To extract the axial position of the emitter, the supercritical and normal fractions of the fluorescence are separated by a circular aperture inserted into the backfocal plane of the objective, followed by the straightforward fraction calculation and scoring against a calibration measurement. SALM can be used to achieve very good axial and lateral resolutions, but suffers from its inherently limited axial range. The method was originally used with 3D origami nanostructures and cellular targets near the coverslip, e.g., cellular membrane proteins [111,112].

13.4.4.2 Photometric Localization

Another way of achieving nanometer-scale axial precision based on measuring emitter intensity is to employ photometric methods. Rooted in astronomy, photometry is easy to implement and computationally cheap, measuring the emitter intensity within a circular area with a subsequent background correction. This method was introduced as temporal, radial-aperture-based intensity estimation (TRABI, Fig. 13.8) [56,58,113].

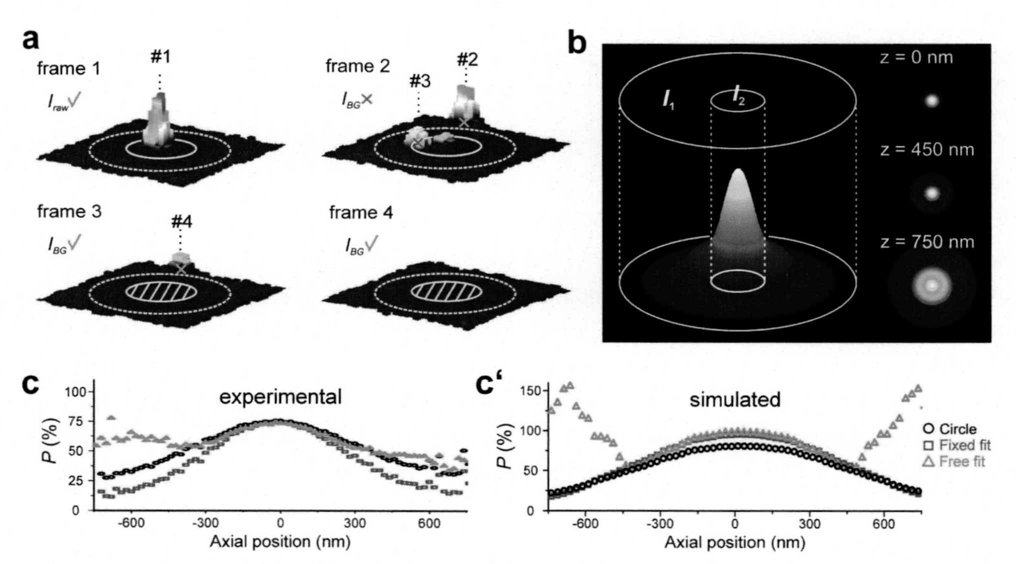

Fig. 13.8 Photometric 3D SMLM enabled by TRABI. (a) TRABI background correction uses a circular aperture with radius r_1 (solid circle) and the EZ defined by $2r_1$ (dashed circle). I_1 is determined by subtracting background (I_{BG}) from raw intensity (Iraw,1). $I_{raw,1}$ and I_{BG} are not estimated in frames in which localizations from neighboring spots interfere. (b–d) 3D single-plane TRABI imaging. (b) The *z*-dependent photometric parameter P is determined by the intensities in two apertures of the molecule's emission pattern according to $P = I_2 I_1^{-1}$. (c, d) P as a function of the axial position calculated from experimental pixel size (c) and simulated data (d). Dots represent the median of the distribution. I_1 was determined by TRABI ($r_{1,exp} = 865$ nm, $r_{1,sim} = 750$ nm), and I_2 was determined either by TRABI ($r_{2,exp} = 333$ nm, $r_{1,sim} = 250$ nm, black circles) or through free (green) or fixed fitting (magenta, *FW HM* = 300 nm). *(Adapted from Refs. [56] and [1])*

TRABI uses the fluorescent on-states and nonfluorescent off-states of photoswitchable or photoactivatable fluorophores. A circular aperture with radius r_1 and area A_1 encircles the center of the spot. The integrated signal in this aperture gives the raw intensity, $I_{raw,1}$. The same aperture is then used to determine the background intensity, I_{BG}, in a subsequent frame, when the fluorophore resides in its off-state or bleached state. In this way, the local background can be accurately determined and the corrected intensity calculated, $I_1 = I_{raw,1} - I_{BG}$, without the need to expand the aperture. The axial position of the emitter can then be derived by a straightforward model mismatch intensity calculation, based on the fact that Gaussian fit models tend to systematically underestimate the total emitter intensity in an axially dependent fashion.

Alternatively, TRABI can be combined with a biplane detection scheme, resulting in a significantly improved axial localization precision compared to the classical methods relying on the FWHM evaluation of emitters. TRABI is open source and easy to use. It has been applied to numerous biological structures to perform high-quality 3D SMLM [1,56,58,79,86], to assess emitter intensities [1,114–116], and to evaluate novel SMLM dyes [1,86]. It gives an accurate estimate of the single-molecule brightness in an experimental setting, which is a very important parameter because it is directly connected to the localization precision.

13.5 Quantitative Localization Microscopy

The key data structure of SMLM is the multidimensional localization set. It enables a variety of data analysis approaches, unique within super-resolution microscopy techniques. In the following section the quantitative evaluation of SMLM data to measure and ensure the quality of the dataset is discussed. Several approaches can be employed to unravel quantitative molecular and structural information from the localization dataset, often inaccessible by image data alone, even in the super-resolved case.

13.5.1 Quality Control of Localization Data

13.5.1.1 Temporal Drift Correction

SMLM raw image stacks typically comprise from 10^3 to 10^5 single camera frames in order to cover the structure of interest adequately by maximizing the probability of detecting every fluorophore at least once. As in every diffraction-limited microscopy application, long time series often suffer from sample drift. But it is even more of a concern in the SMLM case, since drifts even in the nanometer range can alter the result. There are a couple of ways to detect and correct even a slight sample drift. The basic idea is to compile a precise trace of the apparent drift and subtract it from the localization dataset. The principle and effect of SMLM drift and its correction are illustrated in Fig. 13.9.

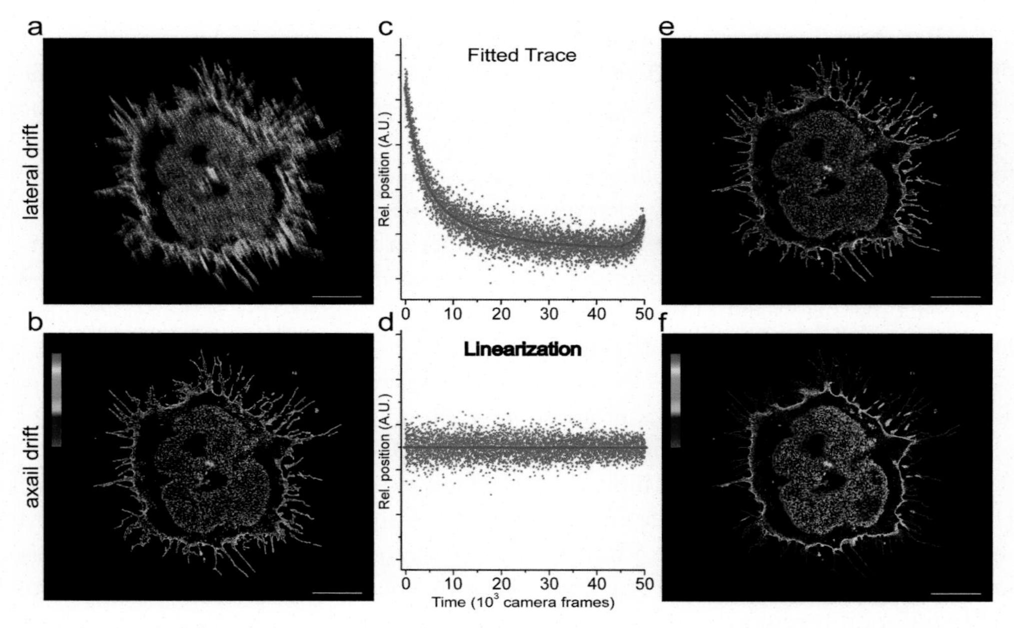

Fig. 13.9 SMLM drift correction. SMLM drift correction exemplified with a *d*STORM (dTRABI) dataset of the fluorescently labeled T-cell membrane receptor CD4 in a dying cell [79]. The effect of lateral (a) and axial (b) sample drift during the imaging process. (a) Lateral sample drift often results in a fuzzy image and the resolution enhancement is negated. (b) Axial sample drift can be noted when color-coding the z-coordinate, resulting in a dull image with a lot of white pixels. By following the drift and building a trace (c) with a subsequent spatiotemporal linearization (d), the three-dimensional drift can be back-calculated, resulting in clean localization datasets and images (e, f). Scalebars = 5 μm. *(Adapted from Ref. [79])*

Fiducial Markers

A common approach for correcting drift is performed by tracking fiducial markers under the assumption that, without any drift, the spatial mean coordinate is stationary. The main difference from diffraction limited fiducial-based methods is that the marker of choice needs to be localizable with at least the same precision as the fluorescent label. It is clear that the usefulness of this method is limited by the localization precision of the markers. Fiducial markers are usually artificial structures that are incubated with the biological sample, i.e., lay on top or beside them, and have to be uniquely detectable in at least one feature. One way of realizing this is to choose a marker that can be separated spectrally. Although this seems to be the easiest way, it increases the experimental complexity, since chromatic aberrations between the sample and fiducial channel must be considered, as well as the need for a second detection path altogether.

Alternatively, fiducial markers might be recognized by a significantly increased brightness compared to the fluorescent labels, or by the fact that they don't exhibit any blinking. One can also manually define certain areas or spots that are then treated as likely fiducials. Typically, the spatiotemporal sample drift is then calculated by averaging the trajectories of all markers. The time-stamped apparent sample drift $D_{\text{sample}}(\text{frame})$ can be defined as

$$D_{sample}(frame) = \frac{1}{F}\sum_{i=1}^{F}\left(C_{i,frame} - \theta_i\right) \tag{13.24}$$

where F is the total number of fiducial markers, $C_{i,t}$ is the absolute position of the ith marker at the respective frame and $i \in [1, F]$. θ_i can be a user-defined or unknown parameter, which in the latter case can be estimated by least-squares minimization of the sum of squared differences between the relative marker positions and the relative sample drift, summed over all markers and frames, [49] according to

$$\boldsymbol{\theta} = \min_{\boldsymbol{\theta}=[\theta_1,\ldots,\theta_F]} \sum_{frame=1}^{Frame}\sum_{i=1}^{F}\left(\left(C_{i,frame} - \theta_i\right) - D_{sample}(frame)\right)^2 \tag{13.25}$$

with $\boldsymbol{\theta} = [\theta_1, \ldots, \theta_F]$ as the estimated offsets for each marker. Finally, the calculated drift trajectory can be smoothed by, e.g., robust locally weighted regression, and it can be used to correct the localization dataset and the resulting high-resolution image [49,117].

When utilizing fiducial markers this way, it is important to keep in mind that constantly active, very bright spots are apparent in the source images. If these fiducials are located within or very close to the structure of interest, either physically or by proxy of projection, the previously described typical photoblinking/high-density artifacts are very likely to appear in the final images as well. This is, unfortunately, almost unavoidable and can result in structural holes or ambiguities in certain areas of the image.

Alternatively, one can assign distinct features of the structure itself, such as crossing points of filaments or the centers of circular organelles, as intrinsic fiducial markers [114]. Here the advantage lies in the avoidance of artificial markers and connected disturbances of the sample. The main drawback lies in the limited applicability of this method, since a lot of a priori knowledge is needed to robustly define intrinsic markers for drift correction.

Self-Alignment

Following this notion of sample self-alignment, one can define not only distinct small features as intrinsic fiducial markers but, in certain cases, the entire structure itself. For example, when imaging a membrane marker, one can track the spatiotemporal axial footprint of the entire super-resolved membrane over time, resulting in a spatiotemporal drift trace. This trace is then fitted with a high-order polynomial that serves as the correction term for the raw localizations by linearization. Here the valid assumption is that, in a thin sample layer like the plasma membrane, the spatiotemporal distribution of active photo-switches that reside in their on-state is constant over time. The selection of appropriate regions in axially more extended samples can include structures where the local z-dimension is restricted, e.g., the outermost areas of adherent cells.

By simply subtracting the frame-precise, fitted drift case from the raw localization dataset, a straightforward, fiducial-free, and three-dimensional drift correction can be achieved [79]. This technique is somewhat superior to the next described correlation-based approach, because it does not rely on the discrete splitting of the localization set. However, cross-correlation is applicable to all kind of structures, whereas virtually flat structures are required here.

Cross-Correlation Analysis

An alternative to fiducial-based drift correction lies in the usage of a spatiotemporal cross-correlation analysis of the localization dataset, as introduced by Mlodzianoski et al. [84]. Here, the localization set is sorted into $n+1$ temporal batches, which is possible due to the fact that every localization has its own timestamp, i.e., the camera frame from which it originated. From those, $n+1$ individual super-resolved images are rendered and analyzed under the assumption that the photophysical properties of the sample are reasonably constant over the span of the entire recording. Thus the same structures will appear in all reconstructed images. This is the same notion used in the previously described method, and it is valid in the vast majority of SMLM cases.

Then cross-correlation analysis, e.g., based on the Fourier transform of batch images, is applied to determine the spatiotemporal shift between all images, relative to the first image. This results in n cross-correlation images where the positional difference is assumed to be caused by the spatial drift. By interpolating the drift between batch centers, one can create a frame-precise drift trace and therefore correct the original molecular coordinates by simply subtracting these estimated values.

Obviously, there is a sweet spot for this kind of analysis that falls between maximal theoretical precision and applicability. The batch images still need to be reasonably saturated by localizations, and the structures of interest should be resolved. On the other hand, the more batches, the better the estimation of the actual drift will be, especially for nonlinear behavior.

13.5.2 Localization Precision and Image Resolution

13.5.2.1 Theoretical Localization Precision

The precision (or variance) of the position determination by Gaussian approximation can be obtained by several methods. A theoretical limit of precision can be calculated for each localization, depending on the size and intensity of the single-molecule spot. An often-used analytical expression was developed for the localization precision of

Gaussian fitting by Thompson et al., with a later extension for pixelation and noise effects by Mortensen et al. [46,118–120]:

$$Variance(\mu_x) = \sigma_{loc}^2 = \frac{\sigma_{PSF}^2}{N}\left(1 + \int_0^1 \frac{\ln t}{1 + Na^2t/2\pi\sigma_{PSF}^2 b^2}\,dt\right)^{-1}$$
$$Variance(\mu_x) = \sigma_{loc}^2 = \frac{\sigma_{PSF}^2}{N}\left(\frac{16}{9} + \frac{8\pi\sigma_{PSF}^2 b^2}{Na^2}\right) \tag{13.26}$$

with the pixel area a^2, the background b, and the wavelength-dependent width of the PSF σ_{PSF}^2. Notably, with typical emitter intensities and reasonable background, this equation boils down in such a way that the effective localization precision can be estimated by the inverse square root of the integrated spot intensity, i.e., the number of photons, N,

$$\sigma_{loc} \propto \frac{1}{\sqrt{N}} \tag{13.27}$$

which has been widely accepted as a reasonable estimation for single-molecule localization precision. However, this approximation only holds for the ideal case of an aberration-free, in-focus, and strictly two-dimensional PSF, which is not the case in any experimental imaging setup.

A more general approach to estimating the theoretical limit, i.e., the best possible precision achievable for an arbitrary three-dimensional PSF, is rooted in general information theory and is based on the calculation of Fisher information and its inverse, i.e., the Cramér-Rao lower bound (CRLB). This is the lower bound on the variance with which the a parameter of Eq. (13.26) can be estimated using any unbiased estimator. This makes sense, since the quantification of localization precision is basically a parameter estimation problem [98].

As Raimund Ober's group pioneered and adapted it from general information theory for SMLM, Fisher information is a mathematical measure of the sensitivity of any observable quantity (e.g., the emission spot pattern) to changes in its underlying parameters (i.e., the molecular coordinate and emitter brightness) [54,120,121]. Mathematically, Fisher information, as the information content of an image or image fragment that is dependent on the parameter set, is given by:

$$I(\theta) = E\left[\left(\frac{\partial \ln f(s;\theta)}{\partial\theta}\right)^T \left(\frac{\partial \ln f(s;\theta)}{\partial\theta}\right)\right] \tag{13.28}$$

where $f(s;\theta)$ is the probability density, i.e., the probability of measuring a signal s given the underlying parameter vector θ; T denotes the transpose operation; and E denotes expectation over all possible values of s. In the special case where Poisson noise is dominant and there is spatially uniform background, this equation becomes:

$$I(\theta) = \sum_{k=1}^{N} \frac{1}{\mu(k) + \beta} \left(\frac{\partial \mu(k)}{\partial \theta} \right)^T \left(\frac{\partial \mu(k)}{\partial \theta} \right) \tag{13.29}$$

where μ(k) is the model of the PSF in pixel k and β is the number of background photons per pixel. More sophisticated models that account for the noise characteristics of EMCCD [122] and sCMOS detectors [123] can be used. Calibration of the pixel-to-pixel variation of sCMOS detectors is necessary for accurate localization.

As previously mentioned, the inverse of the Fisher information is the CRLB, which is the lower bound on the variance with which the set of parameters θ can be estimated using any unbiased estimator [124]. It is therefore an ideal metric to estimate localization precision in SMLM. Specifically, given the position coordinates (x,y,z), this translates to the optimal precision with which the fluorescent emitter can be localized for a given imaging modality, signal, background, and noise model. Interestingly, the CRLB is an important theoretical metric for the assessment of novel imaging modalities, but it has also been shown that the CRLB can be achieved by experimental means [46,125,126]. This has a special implication, since localization precision is an often-stated claim that must be reasonable in the context of its theoretical limit for any given circumstance.

As for the computational implications of modern CRLB calculations, it can be shown that the precision of MLE is always superior to LS methods and will generally achieve the CRLB, assuming a reasonably good match between the mathematical model and the experiment. However, LS estimators do not require a sophisticated noise model, are simpler to implement, and can achieve precisions close to the CRLB for high background levels, where noise statistics are similar between pixels [125,127]. CRLB calculations allow us to systematically evaluate different SMLM methods and imaging modalities, and they give the theoretically best-case scenario for any given experimental SMLM setup in two and three spatial dimensions [98]. However, the experimental conditions often will not allow us to achieve this theoretical limit. Emitter brightness, background and noise levels, and photoswitching kinetics often degrade the quality of the localization. If an experiment predicts precision close to the CRLB, one should check the experimental details to determine if the results are reasonable.

13.5.2.2 Experimental Precision and Resolution

Analyzing Isolated Emitter Spots

The seemingly easiest and most straightforward way to get an experimental measure of localization precision is to analyze the pattern of isolated single-molecule emitters that were detected in several frames. By either fitting a 2D Gaussian to the localization cloud or image spot, or just simply calculating the standard deviation from the comprised localization set, one can derive a somewhat reasonable metric to assess the variance of the localization process. However, this method is not very robust nor objective because it strongly relies on the definition of what constitutes an *isolated emitter*. Of course, if the a priori knowledge of the biological sample suggests a purely monomeric underlying structure, this approach might be usable, but the vast majority of relevant SMLM experiments will not meet this assumption.

Tracing and Tracking

SMLM data offers a unique way to determine the precision of the system experimentally. It inherently produces spatiotemporal single-molecule tracks that are active over n consecutive frames and therefore become localized n times. Under the reasonable assumption that the actual molecular position is stationary, one can then derive a localization precision from these persistent emitters by tracking the spatial coordinate over time and calculating the standard deviation along the track, assuming zero for a perfect system. A possible approach is to utilize established single-molecule tracking algorithms, for which an extensive literature exists. As a practical example used for SMLM data, one can apply the Kalman filter [128] to assess these traces. The most common assumption for all tracking approaches is a single, closed track, i.e., a continuous track without interruption for even one frame. Then, for every frame, the tracking model computes the expected position and the position uncertainty for all known traces. Every localization that is within a defined distance confidence interval of an active trace is appended to the active trace. Calculating the mean standard deviation along each trace then gives a good estimate for the experimentally achieved localization precision. This value should scale according to the accumulated photon number. If not, this can be an indication of irregularities in the imaging process. Instead of applying a classic probability tracing, one can also set a defined fixed distance threshold that neither consecutive localizations, nor the entire track, are allowed to exceed. This only defines another break condition for an active trace, while the precision determination itself stays the same.

There are many alternative extensions of this concept, e.g., by calculating the probability for a positional displacement between spatiotemporal nearest neighboring localizations [129] or by pair correlation approaches. Interestingly, experimentally determined localization precisions often clash with the theoretically determined values that are based on the photon numbers. One reason might be that Gaussian PSF models often significantly underestimate the actual photon count, thus also underestimating localization precision in three-dimensionally extended samples [56–58].

Localization Precision, Resolution, and Sampling

The notion of using the localization precision as a good estimate for the general quality of the SMLM dataset, and the implied direct connection to image resolution, has one big problem—the Nyquist sampling theorem. The Nyquist criterion states that in order to preserve the spatial resolution within a rendered image from the raw data, a sampling interval equal to twice the highest specimen spatial frequency is required. Equivalent to Nyquist's considerations is Shannon's sampling theorem, which essentially states that the sampling interval of the measuring device, i.e., the super-resolved image, must be no more than half as large as the desired resolution. From there, two considerations must be taken into account:

(1) The pixel size of the super-resolved image needs to be half the desired or theoretically achievable resolution, based on the localization precision or CRLB.

(2) The structure of interest needs to be fluorescently labeled in a saturated way, such as that labels are twice as dense than the desired or theoretically achievable resolution.

The first point has a somewhat trivial solution, i.e., the choice of a suitable pixel size. This is usually driven by the notion of a tenfold resolution enhancement and is often set in the range of 5–15 nm. However, the second point poses a more serious issue, since one can calculate or measure localization precisions close to or equal to the CRLB from isolated emitters, even without underlying structures. Consequently, a poorly labeled sample can give the same overall localization precision, and therefore theoretical resolution, as a saturated one. Nevertheless, there is a crucial distinction between precision and actual structural resolution: a singular emitter can be localized precisely, but not resolved. Here, the same considerations made earlier to define diffraction limited resolution are valid, where, by definition, at least two distinct features have to be taken into account. There exists an unpleasant practice all over super-resolution microscopy to simply state the smallest observable feature, or cross-sections thereof, as a measure of image resolution. This is not correct in the slightest, neither by general resolution definitions, nor by the Nyquist theorem or Shannon's theorem. To be more tangible, let us consider a circle with radius R where only two isolated, opposing points are fluorescently labeled and subsequently localized infinitely precise. We have not resolved the circle as a structure, but only the two points. The best resolution is then defined by R, whereas we cannot define any resolution when only labeling a single point on the circle.

Fourier-Ring Correlation

So far, we have mostly covered approaches that are essentially limited to determining the theoretically best-case localization precision, or the experimental localization precision, and thereby producing some estimate of the resolution. Unfortunately, the exact knowledge of the statistical localization precision does not directly relate to an accurate estimate of actual structural resolution of a sample of interest. A lot of additional factors like labeling density, localization density (Nyquist), false-positive and false-negative localizations, other localization artifacts, biological heterogeneities, and even the choice of visualization method can have huge effects on the relationship between localization precision and image resolution. Also, sometimes a global value might not be reasonable if the structure (i.e., the underlying biology) in question can cause significant local variations in these parameters. Obviously, the more a priori knowledge one has about the structure of interest, the closer the theoretical estimation will approximate the experimental reality.

However, knowing everything in advance about a sample raises the question, "Why do the experiment at all?" This becomes even clearer when thinking about a virtually blind approach to an unknown biological sample, where the only thing assumed is that the protein of interest is fluorescently labeled to some degree. Nieuwenhuizen et al. proposed a resolution quantification approach not depending on any a priori information. It is based on using Fourier-ring correlation (FRC) to directly measure the inherent structural resolution from the image data [130]. This procedure is similar to the spectral signal-to-noise ratio (SSNR) commonly used in electron microscopy image assessment [131–133]. The original algorithm and metric was named *FIRE* (Fourier image resolution) and is based on the division of the original single-molecule localization dataset into two equal and statistically independent subsets. From those, two corresponding high-resolution images $f_1(\boldsymbol{r})$ and $f_2(\boldsymbol{r})$ can be rendered, Fourier transformed ($\widehat{f}_1(\boldsymbol{q})$ *and* $\widehat{f}_2(\boldsymbol{q})$) and finally correlated over the pixels on the perimeter of circles of constant spatial frequency magnitude $q = |\boldsymbol{q}|$ [130] according to

$$FRC(|\boldsymbol{q}|) = \frac{\sum_{\boldsymbol{q} \in circle} \widehat{f}_1(\boldsymbol{q}) \widehat{f}_2(\boldsymbol{q})^*}{\sqrt{\sum_{\boldsymbol{q} \in circle} \widehat{f}_1(\boldsymbol{q})^2 \sum_{\boldsymbol{q} \in circle} \widehat{f}_2(\boldsymbol{q})^2}} \quad (13.30)$$

From a typical FRC curve (Fig. 13.10), the objective image resolution can then be extracted by setting an absolute threshold. The point where the FRC value drops below this threshold is then defined as its inverse spatial frequency. In the original work, several threshold criteria were tested, and nowadays the semiempirical FRC threshold of $1/7 \cong 0.14$ is used for most SMLM applications. It is assumed that, without a priori knowledge of the sample and its inherent structure, smaller details in the image cannot be resolved. This is true for 2D as well as for 3D SMLM data. One of the main advantages of this metric, besides its blind approach to the resolution evaluation, is the possibility of creating a pixel-precise map of local image resolution. This also gives an indication of regions in the sample that might be affected by density-induced artifacts (too low or too high emitter density). This feature is discussed in more detail at the end of this section.

13.5.3 Localization-Based Cluster Analysis

Life is three-dimensional, and it begins with molecules [1,134]. Due to the inherent pinpointing nature of SMLM, i.e., localizing single fluorescent molecules, one could assume that it would be straightforward to translate the total localization count in a given structure to a quantitative number of target proteins. However, this approach is almost always unsuccessful for a number of reasons. First, photoactivatable fluorescent proteins

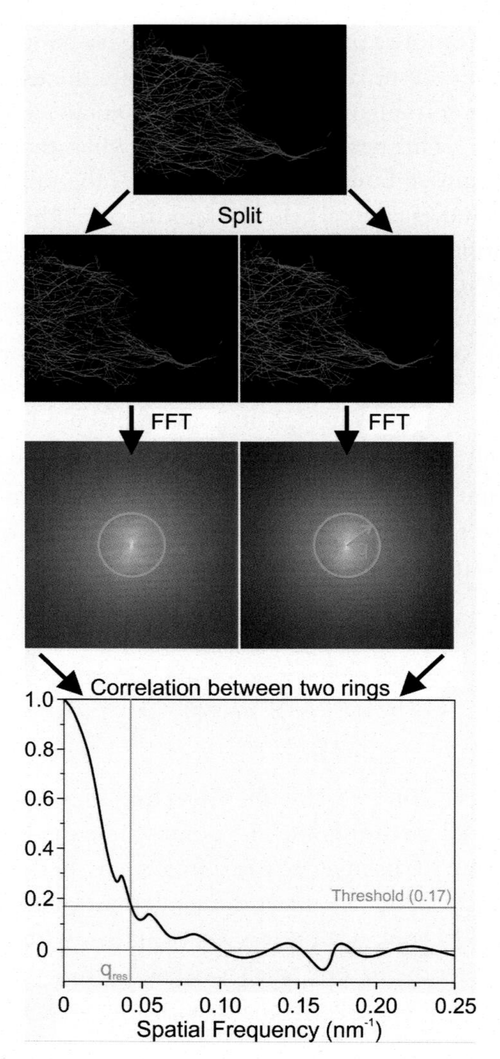

Fig. 13.10 The FRC Principle. All localizations from the original image are split randomly into two equal halves, and the correlation between their Fourier transforms (FFTs) is calculated iteratively for a range of circles with radius q in Fourier frequency space. The desired image resolution is the inverse of the FRC value at the threshold frquency of $1/7 \approx 0.143$, so a threshold value at $q = 0.0378\,\text{nm}^{-1}$ is equivalent to a 26.5 nm resolution in this image. *(Adapted based on Ref. [130])*

(PAFPs), and, even more so, organic fluorophores used for SMLM, exhibit transient repetitive On/Off switching (blinking), possibly leading to a recounting of the same fluorophore [135,136]. Additional overcounting can arise from fluorophores being active over several frames. Although the latter can be adjusted by temporal tracking approaches, this has to be taken into account, especially when dealing with densely labeled samples that make reasonable tracking approaches impossible. Furthermore, dense samples are prone to a variety of fitting artifacts due to overlapping emission patterns, which make any attempt of quantifying localization densities and counts obsolete [76].

Since a direct deduction of protein amount based on localization count is highly inadvisable, one could suggest using calibration experiments conducted with isolated fluorophores. This would entail measuring the mean number of localizations one could expect from a single fluorophore, taking photoblinking and photobleaching into account. Then the number of fluorophores in a given sample would be the total number of localizations divided by that amount. Unfortunately, it has been shown that those approaches are also largely invalid because the photophysical behavior, like the switching performance, strongly depends on the nanoenvironment of the fluorophores. This leads to different localization counts per fluorophore in calibration samples and target samples [137]. Still, isolated emitters can be used as benchmarks to calibrate the total number of localizations to a certain target protein count, as long as they are in the same experimental environment [138], or are imaged simultaneously.

In principle, one can extract single emitters out of a bulky dataset if strict control mechanisms are in place, e.g., algorithmically as published in [139]. Another experimentally challenging option to assure quantitative dependability is to perform measurements on the same sample type but with different label concentrations. This kind of titration experiment, usually done by varying the ratio of primary to secondary antibodies, reduces the impact of switching ratio artifacts and ensures quantitative labeling [140].

Assuming these issues can be accounted for, one can further operate in the native data type of SMLM, namely the set of localizations, treating them as the main source of information rather than the prelude to a high-resolution image [141]. Nevertheless, the superior spatial resolution of rendered images allows for some degree of quantitative analysis of organization and pattern that eludes diffraction-limited microscopy. There is an extensive toolbox of classical image analysis methods, almost all applicable to super-resolved images, with varying degrees of complexity and sophistication. But the necessary image-rendering step (from localization data to a pixelated image) can conceal or even lose important information about the underlying structure, depending on the rendering method and parameters [70,139]. SMLM data is, by itself, a multidimensional array of molecular parameters, which is the starting point for any quantitative analysis in modern-day SMLM.

One of the driving forces in super-resolution based quantitative biology is the question of whether an imaged set of target proteins is organized in clusters. This is of special significance in membrane biology, since the clustering of membrane proteins is thought to regulate cell physiology, function, and stability as well as intercellular recognition [79,142–144]. In recent years, a number of different approaches to localization clustering have been deployed with varying benefits and drawbacks. In general there are two classes of clustering analysis:

(1) Global, statistical ensemble approaches, that can elucidate the general state of spatial clustering in a localization dataset. These methods include several approaches of nearest neighbor analysis, Ripley's K, L and H function, and pair correlation.

(2) Local segmentation approaches, including density-based methods like DBSCAN and k-means clustering, as well as mesh representation-based techniques like Voronoi diagrams and Delaunay triangulation.

Some of the widely used concepts are schematically shown in Fig. 13.11, selected from the perspective article by Nicovich, Owen and Gaus [141].

13.5.3.1 Statistical SMLM Cluster Analysis

A number of statistical methods have been used for cluster analysis in SMLM. Two important ones are discussed in this section.

Ripley's Functions

Ripley's K, H, and L functions are an easily implemented and informative entry point to quantify the spatial pattern formed by SMLM localization data. In general, Ripley's functions can reveal local density fluctuations in multidimensional scattered point data within a global context [145]. Specifically, Ripley's K function gives what is known as a second moment property, i.e., the relationships between two or more point patterns, by finding the expected number of points N within a distance r of another point. It is often used to analyze two-dimensional localization data, and has been extended to three-dimensional SMLM data [146–149]. Formally, the K function can be written as

$$K(r) = \frac{1}{n}\sum_{i=1}^{n}\frac{N_{p_i}(r)}{\lambda} \tag{13.31}$$

where $\lambda = N/A$ is the number, N, of points per area A, and p_i is the ith point in the dataset. For a random, homogeneous distribution, this equation gives the expected curve simply as the number of points in a circle of radius r:

$$K(r) = \pi r^2 \tag{13.32}$$

Consequently, any nonrandom distribution of scattered points, either clustered or dispersed, must show a deviation from this value. Therefore Ripley's K function of the experimental data always has to be compared to either theoretical norms, or to other experimental datasets.

Another important constraint for its applicability is the edge effect that occurs when analyzing points very close to the border. Any edge point will have only half the number of neighboring localizations and will thus distort the overall distribution. To account for

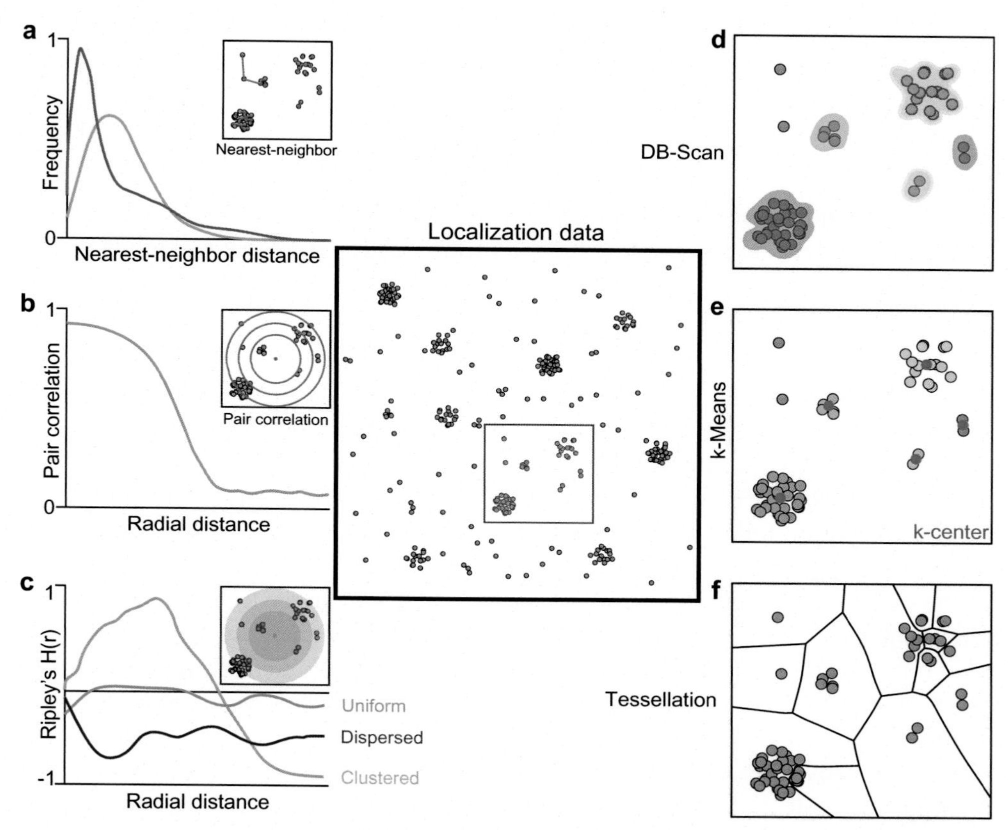

Fig. 13.11 Overview of quantitative SMLM and cluster analysis methods. This figure shows global (left) and segmentation-based (right) cluster analysis approaches on simulated data (middle). (a) Nearest neighbor analysis measures the closest pairwise distance between neighboring localizations. The ensemble histogram (orange) shows a significant deviation from a randomized dataset (purple). (b) Pair correlation is conceptually closely related to the Ripley functions (c), but utilizes circular segments instead of full circles to determine an area density metric. (c) Ripley's functions measure the area-dependent localization density around every localization (inset). Averaging over all points in the dataset leads to a peaked curve (orange), indicating clustering, whereas a randomized dataset would show a flat line (green) or a dispersed one a negative curve (purple). (d) DBSCAN groups localizations into clusters and background by assuming a minimum number of points to constitute a cluster and a confidence interval. Colored areas indicate the convex hulls of individual clusters. (e) K-means clustering relies on the appropriate choice of possible cluster candidates (red dots) and groups localizations to the most likely cluster center. (f) Tessellation approaches, like the displayed Voronoi diagram, facilitate segmentation by another form of density metric, the inverse relationship between local density and area tile size. *(Inspired by Ref. [141])*

that problem, many different correction approaches based on a weighting function, w_{ij}, have been proposed [150–153]. The corrected K-function K^* can be written generally as

$$K^*(r) = \frac{A}{N} \sum_i \sum_{i \neq j} \frac{\delta_{ij}(r)}{w_{ij}} \tag{13.33}$$

In order to provide a more intuitive grasp of the underlying structural information, we introduce two often-used and straightforward normalization steps that yield the L and H functions as

$$L(r) = \sqrt{\pi^{-1} K(r)} \text{ and } H(r) = L(r) - r \tag{13.34}$$

This way, the H function can yield three principal results for any value of r:

- $H(r) > 0$, indicating clustering of points over the spatial scale of r
- $H(r) < 0$, indicating dispersion of points over the spatial scale of r
- $H(r) \cong 0$, indicating a random distribution of points, i.e., Poissonian noise

The third condition follows intuitively from $L(r) = r$ for all values of r, for a completely random point distribution.

This characteristic is illustrated in Fig. 13.12, where we can see that a somewhat clustered localization dataset gives us a pronounced peak in the H function. The r value of this peak is not a precise metric for the presumed cluster diameter, but it is more of an estimate thereof. It is generally accepted that the H-peak value provides a starting point for more in-depth cluster analysis. However, if the goal of the localization quantification is a binary or semiquantitative answer to the question of whether a specific dataset shows signs of clustering, or if this dataset might cluster more than another, Ripley's functions might be the easiest approach. A comprehensive overview of the combination of Ripley's functions with a subsequent, more in-depth analysis of SMLM data can be found in Khater et al.'s 2020 review of quantitative SMLM, [146] upon which this section draws.

Correlation-Based Clustering

As described previously, correlation-based approaches can be used to correct sample drift and other related technical problems that can distort image and data quality. One of the key aims for quantitative SMLM cluster analysis is the notion of molecular counting. Overcounting caused by repeated blinking events of a single fluorophore can cause severe interpretation problems. To overcome this, several approaches to temporal grouping have been proposed, some of which are correlation based. For example, Malkusch et al. pioneered a coordinate-based rather than image-based colocalization approach to assign a colocalization likelihood between −1 to 1 for every detected localization. Here −1 signifies completely segregated, 0 means uncorrelated (randomly distributed), and +1 implies perfect colocalization between two localizations [154]. Consequently, this idea has also been used for classical cluster analysis [155–157] and many other

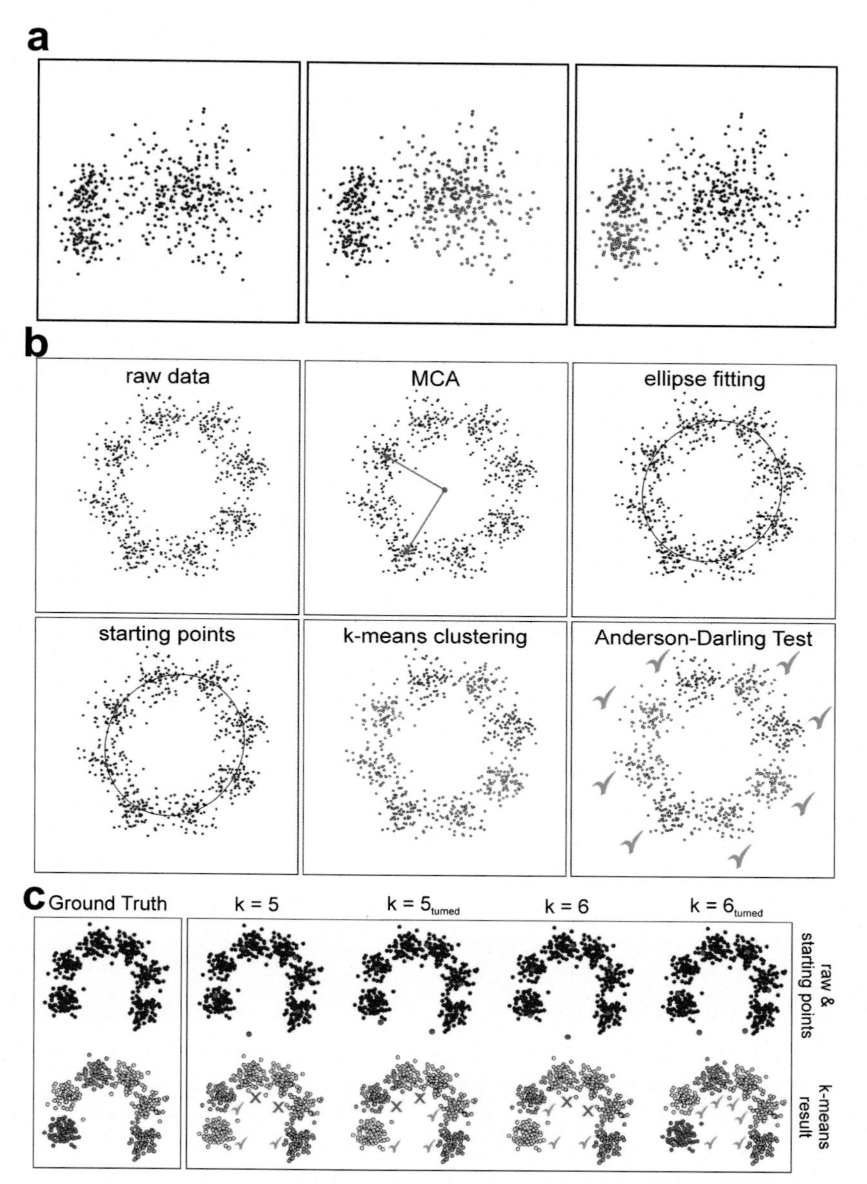

Fig. 13.12 Directed *k*-means clustering. (a) When *k*-means clustering is applied to localization data *(left)* without a priori knowledge, it tends to split the input data into even partitions *(middle)*. If provided with meaningful starting points, it performs well to separate the simulated three domains from each other *(right)*. (b) The general scheme of an optimal multilevel *k*-means algorithm [139] is applied to separate individual gp210 domains from the whole pore raw data. Following a Canny edge filtering operation, a mean component analysis of the localizations is performed to test the ellipticity of the preseparated data. Sufficiently nonelliptical MCA parameters are used to fit an ellipse to the data and set starting points equidistant along that ellipse. The result of a *k*-means analysis is tested for "Gaussianess" via an Anderson-Darling test. (c) Iterative *k*-means separation of gp210 domains. Left: Simulated ground truth localization set, representing a NPC with only 6 gp210 domains *(top)*, color-coded according to their actual affiliation *(bottom)*. Right: Iterative *k*-means process, based on starting points calculated from the *MCA* (indicated as red circles in the upper panel). Beginning with $k = 5$, the configuration is iteratively changed depending on the *ADT* of each domain candidate. Domains passing the test are indicated with a green check mark, and failing candidates with red crosses *(bottom panel)*. Configuration changes may include the simultaneous turning of starting points or increasing *k*. Once all candidates pass the test, sufficient domain separation can be assumed. *(Adapted from Ref. [1])*

correlation functions, such as pair correlation, autocorrelation, cross-correlation, and colocalization. These have been used to derive quantitative information on either the general existence of biologically relevant protein clusters or their geometric properties [154–161].

13.5.3.2 Density-Based Clustering (DBSCAN)

Many SMLM concepts are derived from other disciplines, and popular density-based clustering methods are no exception, as they stem from data mining and spatial data clustering concepts. An often-used form introduced by Ester et al. [162] is called density-based spatial clustering of applications with noise (DBSCAN). Its two key advantages for SMLM data are its ability to detect clustered localizations with an arbitrary underlying shape while filtering out noise, i.e., unspecific or nonclustered localizations, and the fact that it depends on only two parameters, a neighborhood radius ε and the minimum number of localizations N_{min} within ε required to qualify as a cluster. The algorithm iteratively checks every localization and propagates through all other localizations until the breaking condition is reached, namely, when the connectivity of the molecules of the in-process clusters should maintain N_{min} in ε within the same cluster. All other localizations are then considered outliers.

DBSCAN has certain limitations in analyzing super-resolution SMLM data. First of all, common implementations of the algorithm are slow scaling with the number of localizations ($>O(n \log(n))$). Second, the selection of the starting parameters can influence the outcome of the clustering process significantly, and the algorithm is susceptible to forming pseudoclusters in response to typical SMLM artifacts. Different implementations of DBSCAN for the analysis of SMLM data are applied to different biological targets [155,157,159,163–169], in an effort to overcome these limitations. DBSCAN methods have been made faster and less subjective by reducing the number of required inputs, but there is still no gold standard for SMLM cluster analysis, and high-density issues that are often inherent in biologically relevant SMLM data still exist.

13.5.3.3 K-means Clustering

The k-means clustering method is an alternative method of cluster segmentation that is related to the previously described approaches [170]. As discussed earlier, there are a number of different classical clustering algorithms for more or less sparse datasets that are adaptable to SMLM; k-means clustering iteratively minimizes the variation, V, of allocated localizations within a cluster by optimizing the function

$$V = \sum_{i=1}^{k} \sum_{x_j \in S_i} \|x_j - \mu_i\|^2 \tag{13.35}$$

13.5.3.4 Voronoi Tessellation

Like DBSCAN and *k*-means, the Voronoi diagram, or tessellation of point clouds, originated outside the SMLM field and was previously applied in computational geometry, computational physics, astrophysics, computational chemistry, and biology. In the context of SMLM, this method is used to create polygonal regions (Voronoi cells) where each localization sits in the center of one cell. In principle, one can understand the Voronoi approach to be the creation of a density map with arbitrary pixel shapes, where the area of each cell is inversely proportional to the nearest neighbor distances of each localization, i.e., the local density environment. Basically, the Voronoi edges are created by finding the nearest neighbor for any given localization, projecting a perpendicular equidistant line between them that breaks when touching another edge.

Voronoi tessellation alone won't provide cluster quantification easily, but it can be used in conjunction with other methods to segment candidates and analyze their shape discriptors [171–173]. Similarly to the approach using Ripley's functions, the distribution of Voronoi areas and shapes can then be compared to reference distributions, such as uniform or dispersed localization patterns [173]. Alternatively, it can be used in combination with complementary 2D or 3D cluster analysis approaches [174–176]. Voronoi tessellation and such variations as Delaunay triangulation have been applied to a variety of biological structures of different shapes, including noncanonical cluster shapes like tubular structures [172,173].

13.5.3.5 Bayesian Cluster Analysis

Many of the approaches for quantitative SMLM cluster analysis rely on input parameters provided by the user. This can introduce a high degree of subjectivity, and it reduces the reproducibility and solidity of the results. The main goal of any Bayesian approach is to design and implement a method that omits the need for arbitrary or subjective parameters. The Bayesian model is used to evaluate the assignment of every molecule to clusters by its marginal posterior probability [146,177]. Generally, cluster candidates are generated beforehand by statistical methods such as Ripley's K function, and they are then scored against the Bayesian model in an attempt to find the best-fitting number of clusters [56,177–180]. For example, a model could assume that the dataset consists of clustered and nonclustered localizations and that clustered localizations follow a spherical Gaussian distribution where the assumed radius can be derived from Ripley's K/L/H functions [177–180].

13.5.4 Particle Averaging

As discussed earlier, one of the main limitations of SMLM, even assuming unlimited localization precision of single emitters, is the effective underlabeling of the structure due to either steric hindrance between fluorescent probes or ineffective labeling approaches. The labeling molecule or protein is often much larger than the target protein, introducing an inherent steric mismatch. On top of that, SMLM inherits from classic, diffraction-limited fluorescence microscopy all of the common problems regarding non-specific or incomplete labeling—effects that can be increased due to the improved spatial resolution.

One way to deal with this is to combine many super-resolved structures into an avaraged metaparticle of saturated signal and shape with improved signal-to-noise and structural resolution. This approach, derived from cryo-EM, can usually be applied only when the sample consists of many identical copies of the structure of interest [181]. To date, most of the particle-averaging approaches used in SMLM are either template-based methods or single particle analysis (SPA) algorithms adapted from cryo-electron microscopy. Template-based methods are computationally efficient, but are susceptible to template bias artifacts [139,182,183]. Methods derived from SPA for cryo-EM [165,184] have been employed to generate 3D reconstructions from 10^4 to 10^6 2D projections of random viewing angles of a structure [181]. Because of the different data and image types and 2D to 3D transferability issues, there are several caveats in the adaptation of established SPA approaches from EM to SMLM. For a more in-depth discussion of this problem, see the recent article by Heydarian et al. [181].

Due to its highly symmetrical metastructure, the nuclear pore complex has emerged as the prime paradigm to perform SMLM-based particle analysis. In the first ever successful attempt to super-resolve this well-known structure from EM, the authors applied a template-based averaging approach, generating a super-precise and saturated representation of the outer and inner ring [183]. Based on that groundbreaking work, the nuclear pore emerged as one of the prime control structures for high-end 2D and 3D SMLM and as a playground for particle averaging studies. In a follow-up paper, Loeschberger et al. employed a *k*-means based algorithm to unravel structural information well beyond the super-resolution limit [139]. From there the particle averaging, especially of the nuclear pore, evolved and spawned manifold iterations. Here Bernd Rieger's group, among others, pioneered an entire range of SMLM-based averaging approaches, ranging from the original work on nuclear pores and their offspring [183,185], to template-free [186] and 3D symmetry sensing of macromolecules [181], to name just a few.

In principle, every structure that is self-consistent enough to benefit from an averaging approach, and that can be imaged in a high-throughput fashion, can be assessed by following the examples from modern cryo-EM studies. Asymmetric protein complexes can also be evaluated in that fashion, given that a sufficient alignment pipeline can be implemented [165,182]. Additionally, the molecular localization set of SMLM allows for complex meta-analyses of multiprotein complexes. This follows the same ideas as classical averaging approaches while omitting the need for strict symmetry and likeness between individual samples. It can reveal the same level of detail and insight as with the synaptonemal complex [187].

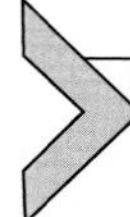

13.6 Implementation and Applications of SMLM

13.6.1 Machine and Deep Learning for SMLM

As in all things image processing, machine learning methods based on multilayer artificial neural networks (see Chapter 15) have arrived in SMLM [188] (see Chapter 12). Deep learning has been applied to take over the localization process and has proved especially useful under high-density conditions [189]. Deep learning was also used to fill in super-resolved structures that were incomplete due to either underlabeling artifacts or too-short imaging times. This effectively speeds up the imaging process in some applications [190,191]. Apart from the localizer, the use of deep learning will have a big impact on quantitative SMLM in the near future. Until now, there have been no practical applications in classical SMLM cluster analysis. A recent work by Tobin et al., however, applied machine learning to separate background localizations from localizations that belong to the structure of interest and are effectively clustered [160]. Further, Sieben et al. used machine learning to analyze SMLM data in order to classify the imaged biological structures [165].

13.6.2 MINFLUX

MINFLUX [192] is a concept that combines the inherent photoblinking kinetics used in SMLM with a patterned illumination scheme. In the original implementation of MINFLUX the molecular position of the emitter is probed with an illumination intensity minimum, and a triangulation approach to analyzing the fluorescence emission can reveal the position of the molecule with precision superior to SMLM techniques. While well-defined test structures like DNA-origami and the nuclear pore complex can be imaged with unprecedented results, labeling strategies and imaging protocols for unknown, dense, or complex structures are still in development. In addition, the reliability of structural imaging by MINFLUX has to be proven in a variety of real biological structures before broad applicability can be claimed [193].

13.6.3 Applications of SMLM

The field of SMLM has a large number of applications. The molecular nature of SMLM allows for not only the construction of stunning images, but also for the execution of complex biological studies and even high-throughput screenings. This chapter lists a number of references to excellent works concerning advanced applications of SMLM to answer biological questions. The review by Schermelleh and colleagues [194] gives an excellent overview of selected applications of SMLM in cell biology and provides context in the wider super-resolution field. The review by Werner et al. [195] is excellent for further reading regarding the applications of SMLM in neuroscience. Whereas this chapter focuses more on the computational challenges of SMLM, the more practical guide to SMLM by Lelek et al. [196] serves as a useful supplement. And, finally, an excellent and practical introduction to the intricacies of SMLM-related sample preparation is found in the article by Jimenez and coworkers [197].

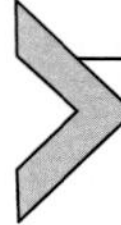

13.7 Summary of Important Points

(1) The diffraction of light causes an infinitisemal point emitter to be imaged as a blurred spot by an optical system.

(2) The size of this spot depends on the imaging system's parameters, such as its numerical aperture, the refractive indices of the media involved, and the wavelength of the emitted light.

(3) The smaller the wavelength of the observed light, the smaller is the size of the spot.

(4) Spatial and structural resolution must be measured with at least two independent features and cannot be defined by a single value.

(5) The spatial resolution of an optical system can be specified with different criteria, such as the Rayleigh criterion for two self-luminous point emitters.

(6) Single-molecule localization microscopy (SMLM) circumvents the diffraction limit by a spatiotemporal separation of the emission of individual fluorophores. At any given time, only a sparse subset of emitters is allowed to reside in the fluorescent ON state, while the vast majority are switched to an inactive OFF state.

(7) There is a multitude of different photophysical and photochemical switching mechanisms, but the key paradigm of one active emitter per diffraction limited resolution cell must always be met.

(8) Isolated single-molecule spots, detected by a sensitive camera, can be analyzed to determine the nanometer-precise coordinates of each active fluorescent molecule.

(9) To sample the underlying structure sufficiently, many thousands to many hundreds of thousands of camera frames are acquired, each imaging a different set of single molecule spots due to the temporally stochastic cycling of emitters between the ON and OFF state.

(10) The accumulation of emitter locations can be used to render a super-resolved image, e.g., by sorting the molecular coordinates into a fine pixel grid with, e.g., tenfold decreased pixel size.

(11) There are many different ways to render a super-resolved image from the localization dataset, with significant variations in resolution and structural perception.

(12) SMLM is susceptible to high emitter density artifacts caused by the overlapping of spots, and they result in the typical washed-out appearance of the resulting image.

(13) So-called switching artifacts can be reduced by optimizing the buffer conditions (pH value, thiol concentration, etc.), the excitation power, and the detection rate to match the fluorophore and sample of choice.

(14) SMLM artifacts can cause serious misinterpretation of the underlying structure of a specimen and therefore its function or mechanism.

(15) Three-dimensional SMLM can be achieved by a variety of approaches, like multiplane imaging, PSF engineering, or intensity-based methods, with a broad range of axial resolution and range, applicability in various biological samples, and technical complexity.

(16) The key data structure in SMLM is the multidimensional localization dataset, consisting of the molecular coordinates and other spot parameters like brightness, timestamp, and other descriptors.

(17) The inherent nature of SMLM is to reveal molecular information, allowing for a broad range of postprocessing approaches that often reveal structural information that is inaccessible even from just the super-resolved image.

(18) Quantitative SMLM has become a subgenre in its own right, and approaches like complex cluster analysis workflows and machine-learning based approaches allow an even deeper look into the molecular organization of cellular organelles.

(19) Although SMLM already offers a very diverse and powerful toolbox for the study of cell-biological questions at the nanoscale, there is still room for improvement on both the microscopy and computational sides.

References

[1] C. Franke, Advancing Single-Molecule Localization Microscopy: Quantitative Analyses and Photometric Three-Dimensional Imaging, PhD Thesis, Bayerische Julius-Maximilians-Universitaet Wuerzburg, 2019.

[2] A. Ul-Hamid, Introduction, in: A Beginners' Guide to Scanning Electron Microscopy, Springer, Cham, 2018.

[3] E. Abbe, Beitrage zur theorie des mikroskops und der mikroskopischen wahrnehmung, Archiv für mikroskopische Anatomie 9 (1) (1873) 413. 418.

[4] F.R.S. Lord Rayleigh, Investigations in optics, with special reference to the spectroscope, Philos. Mag. 8 (49) (1879) 261–274. 5.

[5] A. Koehler, Mikrophotographische Untersuchungen mit ultraviolettem Licht, Zeitschrift für wissenschaftliche Mikroskopie und für mikroskopische Technik 21 (1904).

[6] S.W. Hell, J. Wichmann, Breaking the diffraction resolution limit by stimulated emission: stimulated-emission-depletion fluorescence microscopy, Opt. Lett. 19 (11) (1994) 780–782.

[7] S.W. Hell, Nobel Lecture, NobelPrize.org, 28 Jul 2021. Nobel Prize Outreach AB 2021. Wed.

[8] E. Betzig, Proposed method for molecular optical imaging, Opt. Lett. 20 (3) (1995) 237–239.

[9] E. Betzig, G.H. Patterson, R. Sougrat, O.W. Lindwasser, S. Olenych, J.S. Bonifacino, M.W. Davidson, J. Lippincott-Schwartz, H.F. Hess, Imaging intracellular fluorescent proteins at nanometer resolution, Science 313 (2006) 1642–1645.

[10] W.E. Moerner, L. Kador, Optical detection and spectroscopy of single molecules in a solid, Phys. Rev. Lett. 62 (21) (1989) 2535.

[11] E. Betzig, Facts, NobelPrize.org, 2021. Nobel Prize Outreach AB 2021. Wed.

[12] W.E. Moerner, Facts, NobelPrize.org, 2021. Nobel Prize Outreach AB.

[13] A.M. Van Oijen, J. Köhler, J. Schmidt, M. Müller, G.J. Brakenhoff, 3- dimensional super-resolution by spectrally selective imaging, Chem. Phys. Lett. 292 (1) (1998) 183–187.
[14] M. Heilemann, D.P. Herten, R. Heintzmann, C. Cremer, C. Müller, P. Tinnefeld, K.D. Weston, J.-u. Wolfrum, M. Sauer, High-resolution colocalization of single dye molecules by fluorescence lifetime imaging microscopy, Anal. Chem. 74 (14) (2002) 3511–3517.
[15] M. Orrit, J. Bernard, Single pentacene molecules detected by fluorescence excitation in a p-terphenyl crystal, Phys. Rev. Lett. 65 (21) (1990) 2716.
[16] E. Brooks Shera, N.K. Seitzinger, L.M. Davis, R.A. Keller, S.A. Soper, Detection of single fluorescent molecules, Chem. Phys. Lett. 174 (6) (1990) 553–557.
[17] K.I. Willig, S.O. Rizzoli, V. Westphal, R. Jahn, S.W. Hell, STED microscopy reveals that synaptotagmin remains clustered after synaptic vesicle exocytosis, Nature 440 (2006) 935–939.
[18] R. Heintzmann, C.G. Cremer, Laterally modulated excitation microscopy: improvement of resolution by using a diffraction grating, Proc. SPIE 3568 (1999) 185–196.
[19] M.G. Gustafsson, Surpassing the lateral resolution limit by a factor of two using structured illumination microscopy, J. Microsc. 198 (2000) 82–87.
[20] M.G. Gustafsson, Nonlinear structured-illumination microscopy: wide-field fluorescence imaging with theoretically unlimited resolution, Proc. Natl. Acad. Sci. U. S. A. 102 (2005) 13081–13086.
[21] K.A. Lidke, B. Rieger, T.M. Jovin, R. Heintzmann, Superresolution by localization of quantum dots using blinking statistics, Opt. Express 13 (18) (2005) 7052–7062.
[22] S. Manley, J.M. Gillette, G.H. Patterson, H. Shroff, H.F. Hess, E. Betzig, J. Lippincott-Schwartz, High-density mapping of single-molecule trajectories with photoactivated localization microscopy, Nat. Methods 5 (2008) 155–157.
[23] H. Shroff, C.G. Galbraith, J.A. Galbraith, H. White, J. Gillette, S. Olenych, M.W. Davidson, E. Betzig, Dual-color superresolution imaging of genetically expressed probes within individual adhesion complexes, Proc. Natl. Acad. Sci. U. S. A. 104 (2007) 20308–20313.
[24] S.T. Hess, T.P. Girirajan, M.D. Mason, Ultra-high resolution imaging by fluorescence photoactivation localization microscopy, Biophys. J. 91 (2006) 4258–4272.
[25] S.T. Hess, T.J. Gould, M.V. Gudheti, S.A. Maas, K.D. Mills, J. Zimmerberg, Dynamic clustered distribution of hemagglutinin resolved at 40 nm in living cell membranes discriminates between raft theories, Proc. Natl. Acad. Sci. U. S. A. 104 (2007) 17370–17375.
[26] M. Bates, B. Huang, G.T. Dempsey, X. Zhuang, Multicolor super-resolution imaging with photoswitchable fluorescent probes, Science 317 (2007) 1749–1753.
[27] B. Huang, S.A. Jones, B. Brandenburg, X. Zhuang, Whole-cell 3D STORM reveals interactions between cellular structures with nanometer-scale resolution, Nat. Methods 5 (2008) 1047–1052.
[28] B. Huang, W.Q. Wang, M. Bates, X.W. Zhuang, Three-dimensional super-resolution imaging by stochastic optical reconstruction microscopy, Science 319 (2008) 810–813.
[29] M.J. Rust, M. Bates, X.W. Zhuang, Sub-diffraction-limit imaging by stochastic optical reconstruction microscopy (STORM), Nat. Methods 3 (2006) 793–795.
[30] M. Heilemann, S. van de Linde, M. Schuttpelz, R. Kasper, B. Seefeldt, A. Mukherjee, P. Tinnefeld, M. Sauer, Subdiffraction-resolution fluorescence imaging with conventional fluorescent probes, Angew. Chem. Int. Ed. Engl. 47 (2008) 6172–6176.
[31] M. Heilemann, S. van de Linde, A. Mukherjee, M. Sauer, Super-resolution imaging with small organic fluorophores, Angew. Chem. Int. Ed. Engl. 48 (2009) 6903–6908.
[32] S. van de Linde, A. Löschberger, T. Klein, M. Heidbreder, S. Wolter, M. Heilemann, M. Sauer, Direct stochastic optical reconstruction microscopy with standard fluorescent probes, Nat. Protoc. 6 (2011) 991–1009.
[33] S. van de Linde, S. Aufmkolk, C. Franke, T. Holm, T. Klein, A. Löschberger, S. Proppert, S. Wolter, M. Sauer, Investigating cellular structures at the nanoscale with organic fluorophores, Chem. Biol. 20 (2013) 8–18.
[34] D. Baddeley, I.D. Jayasinghe, C. Cremer, M.B. Cannell, C. Soeller, Light-induced dark states of organic fluochromes enable 30 nm resolution imaging in standard media, Biophys. J. 96 (2009) L22–L24.

[35] H. Bock, C. Geisler, C.A. Wurm, C. von Middendorff, S. Jakobs, A. Schönle, A. Egner, S.W. Hell, C. Eggeling, Two-color far-field fluorescence nanoscopy based on photoswitchable emitters, Appl. Phys. B Lasers Opt. 88 (2007) 161–165.

[36] J. Fölling, M. Bossi, H. Bock, R. Medda, C.A. Wurm, B. Hein, S. Jakobs, C. Eggeling, S.W. Hell, Fluorescence nanoscopy by ground-state depletion and single-molecule return, Nat. Methods 5 (2008) 943–945.

[37] C. Geisler, A. Schönle, C. von Middendorff, H. Bock, C. Eggeling, A. Egner, S.W. Hell, Resolution of $\lambda/10$ in fluorescence microscopy using fast single molecule photo-switching, Appl. Phys. A Mater. Sci. Process. 88 (2007) 223–226.

[38] P. Lemmer, M. Gunkel, D. Baddeley, R. Kaufmann, A. Urich, Y. Weiland, J. Reymann, P. Muller, M. Hausmann, C. Cremer, SPDM: light microscopy with single-molecule resolution at the nanoscale, Appl. Phys. B Lasers Opt. 93 (2008) 1–12.

[39] C. Steinhauer, C. Forthmann, J. Vogelsang, P. Tinnefeld, Superresolution microscopy on the basis of engineered dark states, J. Am. Chem. Soc. 130 (2008) 16840–16841.

[40] J. Vogelsang, T. Cordes, C. Forthmann, C. Steinhauer, P. Tinnefeld, Controlling the fluorescence of ordinary oxazine dyes for single-molecule switching and superresolution microscopy, Proc. Natl. Acad. Sci. U. S. A. 106 (2009) 8107–8112.

[41] T. Klein, S. Proppert, M. Sauer, Eight years of single-molecule localization microscopy, Histochem. Cell Biol. 141 (2014) 561–575.

[42] S. van de Linde, M. Sauer, How to switch a fluorophore: from undesired blinking to controlled photoswitching, Chem. Soc. Rev. 2014 (43) (2014) 1076–1087.

[43] A. Sharonov, R.M. Hochstrasser, Wide-field subdiffraction imaging by accumulated binding of diffusing probes, Proc. Natl. Acad. Sci. U. S. A. 103 (2006) 18911–18916.

[44] E. Mei, F. Gao, R.M. Hochstrasser, Controlled biomolecular collisions allow sub-diffraction limited microscopy of lipid vesicles, Phys. Chem. Chem. Phys. 8 (2006) 2077–2082.

[45] J. Schnitzbauer, M. Strauss, T. Schlichthaerle, et al., Super-resolution microscopy with DNA-PAINT, Nat. Protoc. 12 (2017) 1198–1228.

[46] K.I. Mortensen, L.S. Churchman, J.A. Spudich, H. Flyvbjerg, Optimized localization analysis for single-molecule tracking and super-resolution microscopy, Nat. Methods 7 (2010) 377–381.

[47] P. Křížek, et al., Minimizing detection errors in single molecule localization microscopy, Opt. Express 19 (2011) 3226–3235.

[48] I. Izeddin, et al., Wavelet analysis for single molecule localization microscopy, Opt. Express 20 (2012) 2081–2095.

[49] M. Ovesný, P. Křížek, J. Borkovec, Z. Švindrych, G.M. Hagen, ThunderSTORM: a comprehensive ImageJ plugin for PALM and STORM data analysis and super-resolution imaging, Bioinformatics 30 (16) (2014) 2389–2390.

[50] M. Sonka, V. Hlavac, R. Boyle, Image Processing, Analysis, and Machine Vision, third ed., Cengage Learning, 2007.

[51] D.E. Knuth, The Art of Computer Programming, third ed., vol. 1, Addison-Wesley, Boston, 1997.

[52] J.-L. Starck, F. Murtagh, Astronomical Image and Data Analysis, SpringerVerlag, 2002.

[53] S. Stallinga, B. Rieger, Accuracy of the gaussian point spread function model in 2D localization microscopy, Opt. Express 18 (24) (2010) 24461–24476.

[54] S. Ram, P. Prabhat, J. Chao, E.S. Ward, R.J. Ober, High accuracy 3D quantum dot tracking with multifocal plane microscopy for the study of fast intracellular dynamics in live cells, Biophys. J. 95 (2008) 6025–6043.

[55] B. Zhang, J. Zerubia, J. Olivo-Marin, Gaussian approximations of fluorescence microscope point-spread function models, Appl. Opt. 46 (2007) 1819–1829.

[56] C. Franke, M. Sauer, S. van de Linde, Photometry unlocks 3D information from 2D localization microscopy data, Nat. Methods 14 (2017) 41–44.

[57] R.Ø. Thorsen, C.N. Hulleman, M. Hammer, et al., Impact of optical aberrations on axial position determination by photometry, Nat. Methods 15 (2018) 989–990.

[58] C. Franke, S. van de Linde, Reply to 'Impact of optical aberrations on axial position determination by photometry', Nat. Methods 15 (2018) 990–992.

[59] S. Wolter, Single-Molecule Localization Algorithms in Super-Resolution Microscopy, PhD Thesis, Bayerische Julius-Maximilians-Universitaet Wuerzburg, 2015.
[60] W.H. Press, B.P. Flannery, S.A. Teukolsky, W.T. Vetterling, Numerical Recipes in C: The Art of Scientific Computing, second ed., Cambridge University Press, October 1992.
[61] T.A. Laurence, B.A. Chromy, Efficient maximum likelihood estimator fitting of histograms, Nat. Methods 7 (5) (2010) 338–339.
[62] D. Sage, H. Kirshner, T. Pengo, N. Stuurman, J. Min, S. Manley, M. Unser, Quantitative evaluation of software packages for single-molecule localization microscopy, Nat. Methods 12 (2015) 717–724.
[63] D. Sage, T.A. Pham, H. Babcock, et al., Super-resolution fight club: assessment of 2D and 3D single-molecule localization microscopy software, Nat. Methods 16 (2019) 387–395.
[64] R. Henriques, M. Lelek, E.F. Fornasiero, F. Valtorta, C. Zimmer, M.M. Mhlanga, QuickPALM: 3D real-time photoactivation nanoscopy image processing in ImageJ, Nat. Methods 7 (5) (2010) 339–340.
[65] R. Parthasarathy, Rapid, accurate particle tracking by calculation of radial symmetry centers, Nat. Methods 9 (7) (2012) 724–726.
[66] Y. Li, M. Mund, P. Hoess, et al., Real-time 3D single-molecule localization using experimental point spread functions, Nat. Methods 15 (2018) 367–369.
[67] A. Tahmasbi, E.S. Ward, R.J. Ober, Determination of localization accuracy based on experimentally acquired image sets: applications to single molecule microscopy, Opt. Express 23 (2015) 7630–7652.
[68] H. Kirshner, C. Vonesch, M. Unser, 2013 IEEE 10th International Symposium on Biomedical Imaging, IEEE, 2013, pp. 588–591.
[69] H.P. Babcock, X. Zhuang, Sci. Rep. 7 (2017) 552.
[70] D. Baddeley, M.B. Cannell, C. Soeller, Visualization of localization microscopy data, Microsc. Microanal. 16 (1) (2010) 64–72.
[71] S. van de Linde, S. Wolter, M. Heilemann, M. Sauer, The effect of photoswitching kinetics and labeling densities on super-resolution fluorescence imaging, J. Biotechnol. 149 (2010) 260–266.
[72] M. Sauer, Localization microscopy coming of age: from concepts to biological impact, J. Cell Sci. 126 (2013) 3505–3513.
[73] S. Wolter, U. Endersfelder, S. van de Linde, M. Heilemann, M. Sauer, Measuring localization performance of super-resolution algorithms on very active samples, Opt. Express 19 (2011) 7020–7033.
[74] L. Zhu, W. Zhang, D. Elnatan, B. Huang, Faster STORM using compressed sensing, Nat. Methods 9 (2012) 721–723.
[75] S.J. Holden, S. Uphoff, A.N. Kapanidis, DAOSTORM: an algorithm for high-density super-resolution microscopy, Nat. Methods 8 (2011) 279–280.
[76] A. Burgert, S. Letschert, S. Doose, et al., Artifacts in single-molecule localization microscopy, Histochem. Cell Biol. 144 (2015) 123–131.
[77] D. Whelan, T. Bell, Image artifacts in single molecule localization microscopy: why optimization of sample preparation protocols matters, Sci. Rep. 5 (2015) 7924.
[78] S. Culley, D. Albrecht, C. Jacobs, et al., Quantitative mapping and minimization of super-resolution optical imaging artifacts, Nat. Methods 15 (2018) 263–266.
[79] C. Franke, T. Chum, Z. Kvíčalová, D. Glatzová, A. Rodriguez, D.A. Helmerich, O. Frank, T. Brdička, S. van de Linde, M. Cebecauer, Unraveling nanotopography of cell surface receptors, bioRxiv (2019). 2019.12.23.884460.
[80] P. Prabhat, S. Ram, E.S. Ward, R.J. Ober, Simultaneous imaging of different focal planes in fluorescence microscopy for the study of cellular dynamics in three dimensions, IEEE Trans. Nanobiosci. 3 (2004) 237–242.
[81] E. Toprak, H. Balci, B.H. Blehm, P.R. Selvin, Three-dimensional particle tracking via bifocal imaging, Nano Lett. 7 (2007) 2043–2045.
[82] M.F. Juette, T.J. Gould, M.D. Lessard, M.J. Mlodzianoski, B.S. Nagpure, B.T. Bennett, S.T. Hess, J. Bewersdorf, Three-dimensional sub-100 nm resolution fluorescence microscopy of thick samples, Nat. Methods 5 (2008) 527–529.
[83] A. Tahmasbi, S. Ram, J. Chao, A.V. Abraham, F.W. Tang, E. Sally Ward, R.J. Ober, Designing the focal plane spacing for multifocal plane microscopy, Opt. Express 22 (2014) 16706–16721.

[84] M.J. Mlodzianoski, J.M. Schreiner, S.P. Callahan, K. Smolková, A. Dlasková, J. Šantorová, P. Ježek, J. Bewersdorf, Sample drift correction in 3D fluorescence photoactivation localization microscopy, Opt. Express 19 (2011) 15009–15019.
[85] J. Ries, C. Kaplan, E. Platonova, H. Eghlidi, H. Ewers, A simple, versatile method for GFP-based super-resolution microscopy via nanobodies, Nat. Methods 9 (2012) 582–584.
[86] M.S. Michie, R. Götz, C. Franke, M. Bowler, N. Kumari, V. Magidson, M. Levitus, J. Loncarek, M. Sauer, M.J. Schnermann, Cyanine conformational restraint in the far-red range, J. Am. Chem. Soc. 139 (36) (2017) 12406–12409.
[87] S. Ram, D. Kim, R. Ober, E.S. Ward, 3D single molecule tracking with multifocal plane microscopy reveals rapid intercellular transferrin transport at epithelial cell barriers, Biophys. J. 103 (2012) 1594–1603.
[88] P.A. Dalgarno, H.I.C. Dalgarno, A. Putoud, R. Lambert, L. Paterson, D.C. Logan, D.P. Towers, R.J. Warburton, A.H. Greenaway, Multiplane imaging and three dimensional nanoscale particle tracking in biological microscopy, Opt. Express 18 (2010) 877–884.
[89] B. Hajj, M. El Beheiry, M. Dahan, PSF engineering in multifocus microscopy for increased depth volumetric imaging, Biomed. Opt. Express 7 (2016) 726–731.
[90] L. Oudjedi, J. Fiche, S. Abrahamsson, L. Mazenq, A. Lecestre, P. Calmon, A. Cerf, M. Nöllmann, Astigmatic multifocus microscopy enables deep 3D super-resolved imaging, Biomed. Opt. Express 7 (2016) 2163–2173.
[91] D. Baddeley, M.B. Cannell, C. Soeller, Three-dimensional sub-100 nm super-resolution imaging of biological samples using a phase ramp in the objective pupil, Nano Res. 4 (2011) 589–598.
[92] A. Gahlmann, J.L. Ptacin, G. Grover, S. Quirin, A.R.S. von Diezmann, M.K. Lee, M.P. Backlund, L. Shapiro, R. Piestun, W.E. Moerner, Quantitative multicolor subdiffraction imaging of bacterial protein ultrastructures in 3D, Nano Lett. 13 (2013) 987–993.
[93] I. Izeddin, M. El Beheiry, J. Andilla, D. Ciepielewski, X. Darzacq, M. Dahan, PSF shaping using adaptive optics for three-dimensional single-molecule super-resolution imaging and tracking, Opt. Express 20 (2012) 4957–4967.
[94] S.R.P. Pavani, M.A. Thompson, J.S. Biteen, S.J. Lord, N. Liu, R.J. Twieg, R. Piestun, W.E. Moerner, Three-dimensional, single-molecule fluorescence imaging beyond the diffraction limit by using a double-helix point spread function, Proc. Natl. Acad. Sci. U. S. A. 106 (2009) 2995–2999.
[95] S. Jia, J.C. Vaughan, X. Zhuang, Isotropic three-dimensional super-resolution imaging with a self-bending point spread function, Nat. Photonics 8 (2014) 302–306.
[96] Y. Shechtman, S.J. Sahl, A.S. Backer, W.E. Moerner, Optimal point spread function design for 3D imaging, Phys. Rev. Lett. 113 (2014).
[97] Y. Shechtman, L.E. Weiss, A.S. Backer, S.J. Sahl, W.E. Moerner, Precise 3D scan-free multiple-particle tracking over large axial ranges with tetrapod point spread functions, Nano Lett. 15 (2015) 4194–4199.
[98] L. von Diezmann, Y. Shechtman, W.E. Moerner, Three-dimensional localization of single molecules for super-resolution imaging and single-particle tracking, Chem. Rev. 117 (11) (2017) 7244–7275.
[99] J.W. Goodman, Introduction to Fourier Optics, Roberts and Co. Publishers, Greenwood Village, CO, 2005.
[100] S.F. Gibson, F. Lanni, Experimental test of an analytical model of aberration in an oil-immersion objective lens used in three-dimensional light microscopy, J. Opt. Soc. Am. A 8 (1991) 1601–1613.
[101] H.P. Kao, A.S. Verkman, Tracking of single fluorescent particles in three dimensions: use of cylindrical optics to encode particle position, Biophys. J. 67 (1994) 1291–1300.
[102] L. Holtzer, T. Meckel, T. Schmidt, Nanometric three-dimensional tracking of individual quantum dots in cells, Appl. Phys. Lett. 90 (2007), 053902.
[103] R. Piestun, Y.Y. Schechner, J. Shamir, Propagation-invariant wave fields with finite energy, J. Opt. Soc. Am. A 17 (2000) 294–303.
[104] M.D. Lew, S.F. Lee, M. Badieirostami, W.E. Moerner, Corkscrew point spread function for far-field three-dimensional nanoscale localization of pointlike objects, Opt. Lett. 36 (2011) 202–204.
[105] C. Smith, M. Huisman, M. Siemons, D. Grünwald, S. Stallinga, Simultaneous measurement of emission color and 3D position of single molecules, Opt. Express 24 (2016) 4996–5013.

[106] Y. Shechtman, L.E. Weiss, A.S. Backer, M.L. Lee, W.E. Moerner, Multicolor localization microscopy by point-spread-function engineering, Nat. Photonics 10 (2016) 590–594.
[107] G. Shtengel, J.A. Galbraith, C.G. Galbraith, J. Lippincott-Schwartz, J.M. Gillette, S. Manley, R. Sougrat, C.M. Waterman, P. Kanchanawong, M.W. Davidson, Interferometric fluorescent super-resolution microscopy resolves 3D cellular ultrastructure, Proc. Natl. Acad. Sci. U. S. A. 106 (2009) 3125–3130.
[108] C.V. Middendorff, A. Egner, C. Geisler, S.W. Hell, A. Schönle, Isotropic 3D nanoscopy based on single emitter switching, Opt. Express 16 (2008) 20774–20788, https://doi.org/10.1364/OE.16.020774.
[109] N. Karedla, A.I. Chizhik, I. Gregor, A.M. Chizhik, O. Schulz, J. Enderlein, Single-molecule metal-induced energy transfer (smMIET): resolving nanometer distances at the single-molecule level, ChemPhysChem 15 (2014) 705–711.
[110] R.R. Chance, A. Prock, R.J. Silbey, Molecular fluorescence and energy transfer near interfaces, Adv. Chem. Phys. 37 (1978) 1–65.
[111] J. Deschamps, M. Mund, J. Ries, 3D superresolution microscopy by supercritical angle detection, Opt. Express 22 (2014) 29081–29091.
[112] N. Bourg, C. Mayet, G. Dupuis, T. Barroca, P. Bon, S. Lécart, E. Fort, S. Lévêque-Fort, Direct optical nanoscopy with axially localized detection, Nat. Photonics 9 (2015) 587–593.
[113] C. Franke, M. Sauer, S. van de Linde, Method and Microscope for Determining a Fluorescence Intensity, US Patent App. 16/461,908, 2019.
[114] C. Franke, U. Repnik, S. Segeletz, et al., Correlative single-molecule localization microscopy and electron tomography reveals endosome nanoscale domains, Traffic 20 (2019) 601–617.
[115] L. Belicova, U. Repnik, J. Delpierre, E. Gralinska, S. Seifert, J.I. Valenzuela, H.A. Morales-Navarrete, C. Franke, H. Räägel, E. Shcherbinina, T. Prikazchikova, V. Koteliansky, M. Vingron, Y.L. Kalaidzidis, T. Zatsepin, M. Zerial, Anisotropic expansion of hepatocyte lumina enforced by apical bulkheads, J. Cell Biol. 220 (10) (2021), e202103003.
[116] P. Paramasivam, C. Franke, M. Stöter, A. Höijer, S. Bartesaghi, A. Sabirsh, L. Lindfors, M.Y. Arteta, A. Dahlén, A. Bak, S. Andersson, Y. Kalaidzidis, M. Bickle, M. Zerial, Endosomal escape of delivered mRNA from endosomal recycling tubules visualized at the nanoscale, biorxiv (2021), https://doi.org/10.1101/2020.12.18.423541.
[117] W. Cleveland, Robust locally weighted regression and smoothing scatterplots, J. Am. Stat. Assoc. 74 (368) (1979) 829–836, https://doi.org/10.2307/2286407.
[118] N. Bobroff, Position measurement with a resolution and noise-limited instrument, Rev. Sci. Instrum. 57 (1986) 1152–1157.
[119] R.E. Thompson, D.R. Larson, W.W. Webb, Precise nanometer localization analysis or individual fluorescent probes, Biophys. J. 82 (5) (2002) 2775–2783.
[120] R.J. Ober, S. Ram, E.S. Ward, Localization accuracy in single-molecule microscopy, Biophys. J. 86 (2) (2004) P1185–P1200.
[121] E. Jerry Chao, S. Ward, R.J. Ober, Fisher information theory for parameter estimation in single molecule microscopy: tutorial, J. Opt. Soc. Am. A 33 (2016) B36–B57.
[122] J. Chao, E.S. Ward, R.J. Ober, Localization accuracy in single molecule microscopy using electron-multiplying charge-coupled device cameras, in: Proc. SPIE 8227, Three-Dimensional and Multidimensional Microscopy: Image Acquisition and Processing XIX, 2012, p. 82271P.
[123] F. Huang, T. Hartwich, F. Rivera-Molina, et al., Video-rate nanoscopy using sCMOS camera-specific single-molecule localization algorithms, Nat. Methods 10 (2013) 653–658.
[124] S.M. Kay, Fundamentals of Statistical Signal Processing: Estimation Theory, Englewood Cliffs, NJ, Prentice-Hall PTR, 1993.
[125] A.V. Abraham, S. Ram, J. Chao, E.S. Ward, R.J. Ober, Quantitative study of single molecule location estimation techniques, Opt. Express 17 (2009) 23352–23373.
[126] C.S. Smith, N. Joseph, B. Rieger, K.A. Lidke, Fast, single-molecule localization that achieves theoretically minimum uncertainty, Nat. Methods 7 (2010) 373–375.
[127] B. Rieger, S. Stallinga, The lateral and axial localization uncertainty in super-resolution light microscopy, ChemPhysChem 15 (2014) 664–670.

[128] R.E. Kalman, A new approach to linear filtering and prediction problems, ASME. J. Basic Eng. 82 (1) (1960) 35–45.
[129] U. Endesfelder, S. Malkusch, F. Fricke, M. Heilemann, A simple method to estimate the average localization precision of a single-molecule localization microscopy experiment, Histochem. Cell Biol. 141 (6) (2014) 629–638.
[130] R. Nieuwenhuizen, K. Lidke, M. Bates, et al., Measuring image resolution in optical nanoscopy, Nat. Methods 10 (2013) 557–562.
[131] W.O. Saxton, W. Baumeister, The correlation averaging of a regularly arranged bacterial cell envelope protein, J. Microsc. 127 (1982) 127–138.
[132] M. van Heel, Similarity measures between images, Ultramicroscopy 21 (1987) 95–100.
[133] M. Unser, B.L. Trus, A.C. Steven, A new resolution criterion based on spectral signal-to-noise ratio, Ultramicroscopy 23 (1987) 39–52.
[134] P.E. Bourne, Life is three-dimensional, and it begins with molecules, PLoS Biol. 15 (2017), e2002041.
[135] P. Annibale, S. Vanni, M. Scarselli, U. Rothlisberger, A. Radenovic, Uantitative photo activated localization microscopy: unraveling the effects of photoblinking, PLoS One 6 (7) (2011), e22678.
[136] D. Lando, U. Endesfelder, H. Berger, L. Subramanian, P.D. Dunne, J. McColl, D. Klenerman, A.M. Carr, M. Sauer, R.C. Allshire, et al., Quantitative single-molecule microscopy reveals that cenp-acnp1 deposition occurs during g2 in fission yeast, Open Biol. 2 (7) (2012), 120078.
[137] U. Endesfelder, S. Malkusch, B. Flottmann, J. Mondry, P. Liguzinski, P.J. Verveer, M. Heilemann, Chemically induced photoswitching of fluorescent probes|a general concept for super-resolution microscopy, Molecules 16 (4) (2011) 3106–3118.
[138] R. Jungmann, M.S. Avenda~no, M. Dai, J.B. Woehrstein, S.S. Agasti, Z. Feiger, A. Rodal, P. Yin, Quantitative super-resolution imaging with qpaint, Nat. Methods 13 (5) (2016) 439–442.
[139] A. Loschberger, C. Franke, G. Krohne, S. van de Linde, M. Sauer, Correlative super-resolution fluorescence and electron microscopy of the nuclear pore complex with molecular resolution, J. Cell. Sci. 127 (20) (2014) 4351–4355.
[140] N. Ehmann, S. Van De Linde, A. Alon, D. Ljaschenko, X.Z. Keung, T. Holm, A. Rings, A. DiAntonio, S. Hallermann, U. Ashery, et al., Quantitative super-resolution imaging of bruchpilot distinguishes active zone states, Nat. Commun. 5 (2014).
[141] P.R. Nicovich, D.M. Owen, K. Gaus, Turning single-molecule localization microscopy into a quantitative bioanalytical tool, Nat. Protoc. 12 (3) (2017) 453–460.
[142] H.H. Freeze, Genetic defects in the human glycome, Nat. Rev. Genet. 7 (7) (2006) 537–551.
[143] G.W. Hart, M.P. Housley, C. Slawson, Cycling of o-linked β-n-acetylglucosamine on nucleocytoplasmic proteins, Nature 446 (7139) (2007) 1017–1022.
[144] S. Letschert, A. Ğohler, C. Franke, N. Bertleff-Zieschang, E. Memmel, S.o. Doose, J.u. Seibel, M. Sauer, Superresolution imaging of plasma membrane glycans, Angew. Chem. Int. Ed. 53 (41) (2014) 10921–10924.
[145] B.D. Ripley, The second-order analysis of stationary point processes, J. Appl. Probab. 13 (2) (1976) 255–266.
[146] I.M. Khater, I.R. Nabi, G. Hamarneh, A review of super-resolution single-molecule localization microscopy cluster analysis and quantification methods, Patterns 1 (3) (2020) 100038 (ISSN 2666-3899).
[147] P.M. Dixon, Ripley's K Function, in: Wiley StatsRef: Statistics Reference Online, 2014.
[148] K. Hansson, M. Jafari-Mamaghani, P. Krieger, RipleyGUI: software for analyzing spatial patterns in 3D cell distributions, Front. Euroinf. 7 (2013) 5.
[149] M.A. Kiskowski, J.F. Hancock, A.K. Kenworthy, On the use of Ripley's K-function and its derivatives to analyze domain size, Biophys. J. 97 (2009) 1095–1103.
[150] P. Haase, Spatial pattern analysis in ecology based on Ripley's Kfunction: introduction and methods of edge correction, J. Veg. Sci. 6 (1995) 575–582.
[151] E. Marcon, F. Puech, Generalizing Ripley's K Function to Inhomogeneous Populations, 2009. https://halshs.archives-ouvertes.fr/halshs-00372631/document.
[152] A.J. Baddeley, R.A. Moyeed, C.V. Howard, A. Boyde, Analysis of a three-dimensional point pattern with replication, J. R. Stat. Soc.: Ser. C: Appl. Stat. 42 (1993) 641–668.

[153] F. Goreaud, R. Pelissier, On explicit formulas of edge effect correction for Ripley's K-function, J. Veg. Sci. 10 (1999) 433–438.
[154] S. Malkusch, U. Endesfelder, J. Mondry, M. Gelléri, P.J. Verveer, M. Heilemann, Coordinate-based colocalization analysis of single-molecule localization microscopy data, Histochem. Cell Biol. 137 (2012) 1–10.
[155] S.V. Pageon, P.R. Nicovich, M. Mollazade, T. Tabarin, K. Gaus, Clus-DoC: a combined cluster detection and colocalization analysis for single-molecule localization microscopy data, Mol. Biol. Cell 27 (2016) 3627–3636.
[156] F.B. Lopes, Š. Bálint, S. Valvo, J.H. Felce, E.M. Hessel, M.L. Dustin, D.M. Davis, Membrane nanoclusters of FcγRI segregate from inhibitory SIRPα upon activation of human macrophages, J. Cell Biol. 216 (2017) 1123–1141.
[157] S. Malkusch, M. Heilemann, Extracting quantitative information from single-molecule super-resolution imaging data with LAMA—LocAlization microscopy analyzer, Sci. Rep. 6 (2016) 34486.
[158] J. Rossy, E. Cohen, K. Gaus, D.M. Owen, Method for co-cluster analysis in multichannel single-molecule localisation data, Histochem. Cell Biol. 141 (2014) 605–612.
[159] J. Schnitzbauer, Y. Wang, S. Zhao, M. Bakalar, T. Nuwal, B. Chen, B. Huang, Correlation analysis framework for localization-based superresolution microscopy, Proc. Natl. Acad. Sci. U. S. A. 115 (2018) 3219–3224.
[160] S.J. Tobin, D.L. Wakefield, V. Jones, X. Liu, D. Schmolze, T. Jovanović-Talisman, Single molecule localization microscopy coupled with touch preparation for the quantification of trastuzumab-bound HER2, Sci. Rep. 8 (2018) 1515.
[161] M.B. Stone, S.L. Veatch, Steady-state cross-correlations for live two-colour super-resolution localization datasets, Nat. Commun. 6 (2015) 7347.
[162] M. Ester, H.P. Kriegel, J. Sander, X. Xu, A density-based algorithm for discovering clusters in large spatial databases with noise, in: KDD-96 Proceedings, 1996, pp. 226–231.
[163] A. Mazouchi, J.N. Milstein, Fast optimized cluster algorithm for localizations (FOCAL): a spatial cluster analysis for super-resolved microscopy, Bioinformatics 32 (2015) 747–754.
[164] T. Pengo, S.J. Holden, S. Manley, PALMsiever: a tool to turn raw data into results for single-molecule localization microscopy, Bioinformatics 31 (2014) 797–798.
[165] C. Sieben, N. Banterle, K.M. Douglass, P. Gönczy, S. Manley, Multicolor single-particle reconstruction of protein complexes, Nat. Methods 15 (2018) 777.
[166] L. Barna, B. Dudok, V. Miczán, A. Horváth, Z.I. László, I. Katona, Correlated confocal and super-resolution imaging by VividSTORM, Nat. Protoc. 11 (2016) 163.
[167] T. Lagache, A. Grassart, S. Dallongeville, O. Faklaris, N. Sauvonnet, A. Dufour, L. Danglot, J.C. Olivo-Marin, Mapping molecular assemblies with fluorescence microscopy and object-based spatial statistics, Nat. Commun. 9 (2018) 698.
[168] M. Mollazade, T. Tabarin, P.R. Nicovich, A. Soeriyadi, D.J. Nieves, J.J. Gooding, K. Gaus, Can single molecule localization microscopy be used to map closely spaced RGD nanodomains? PLoS One 12 (2017), e0180871.
[169] Y. Zhang, M. Lara-Tejero, J. Bewersdorf, J.E. Galán, Visualization and characterization of individual type III protein secretion machines in live bacteria, Proc. Natl. Acad. Sci. U. S. A. 114 (2017) 6098–6103.
[170] S. Lloyd, Least squares quantization in PCM, IEEE Trans. Inf. Theory 28 (2) (1982) 129–137.
[171] A. Okabe, B. Boots, K. Sugihara, S.N. Chiu, Spatial Tessellations: Concepts and Applications of Voronoi Diagrams, John Wiley, 2009.
[172] F. Levet, E. Hosy, A. Kechkar, C. Butler, A. Beghin, D. Choquet, J.B. Sibarita, SR-Tesseler: a method to segment and quantify localization-based super-resolution microscopy data, Nat. Methods 12 (2015) 1065.
[173] L. Andronov, I. Orlov, Y. Lutz, J.L. Vonesch, B.P. Klaholz, ClusterViSu, a method for clustering of protein complexes by Voronoi tessellation in super-resolution microscopy, Sci. Rep. 6 (2016) 24084.
[174] R. Peters, J. Griffié, G.L. Burn, D.J. Williamson, D.M. Owen, Quantitative fibre analysis of single-molecule localization microscopy data, Sci. Rep. 8 (2018) 10418.
[175] K.T. Haas, M. Lee, A. Esposito, A.R. Venkitaraman, Single-molecule localization microscopy reveals molecular transactions during RAD51 filament assembly at cellular DNA damage sites, Nucleic Acids Res. 46 (2018) 2398–2416.

[176] L. Andronov, Y. Lutz, J.L. Vonesch, B.P. Klaholz, SharpViSu: integrated analysis and segmentation of super-resolution microscopy data, Bioinformatics 32 (2016) 2239–2241.
[177] P. Rubin-Delanchy, G.L. Burn, J. Griffié, D.J. Williamson, N.A. Heard, A.P. Cope, D.M. Owen, Bayesian cluster identification in single-molecule localization microscopy data, Nat. Methods 12 (2015) 1072.
[178] J. Griffié, G.L. Burn, D.J. Williamson, R. Peters, P. Rubin-Delanchy, D.M. Owen, Dynamic Bayesian cluster analysis of live-cell single molecule localization microscopy datasets, Small Methods 2 (2018).
[179] J. Griffié, L. Shlomovich, D.J. Williamson, M. Shannon, J. Aaron, S. Khuon, G. Burn, L. Boelen, R. Peters, A.P. Cope, et al., 3D Bayesian cluster analysis of super-resolution data reveals LAT recruitment to the T cell synapse, Sci. Rep. 7 (2017) 4077.
[180] J. Griffié, M. Shannon, C.L. Bromley, L. Boelen, G.L. Burn, D.J. Williamson, N.A. Heard, A.P. Cope, D.M. Owen, P. Rubin-Delanchy, A Bayesian cluster analysis method for single-molecule localization microscopy data, Nat. Protoc. 11 (2016) 2499.
[181] H. Heydarian, M. Joosten, A. Przybylski, et al., 3D particle averaging and detection of macromolecular symmetry in localization microscopy, Nat. Commun. 12 (2021) 2847.
[182] J. Broeken, et al., Resolution improvement by 3D particle averaging in localization microscopy, Methods Appl. Fluoresc. 3 (2015), 014003.
[183] A. Löschberger, S. van de Linde, M.-C. Dabauvalle, B. Rieger, M. Heilemann, G. Krohne, M. Sauer, Super-resolution imaging visualizes the eightfold symmetry of gp210 proteins around the nuclear pore complex and resolves the central channel with nanometer resolution, J. Cell Sci. 125 (3) (2012) 570–575.
[184] D. Salas, et al., Angular reconstitution-based 3D reconstructions of nanomolecular structures from superresolution light-microscopy images, Proc. Natl. Acad. Sci. U. S. A. 114 (2017) 9273–9278.
[185] A. Szymborska, A. de Marco, N. Daigle, V.C. Cordes, J.A.G. Briggs, J. Ellenberg, Nuclear pore scaffold structure analyzed by super-resolution microscopy and particle averaging, Science (2013) 655–658.
[186] H. Heydarian, et al., Template-free 2D particle fusion in localization microscopy, Nat. Methods 15 (2018) 781–784.
[187] K. Schücker, T. Holm, C. Franke, M. Sauer, R. Benavente, Elucidation of synaptonemal complex organization by super-resolution imaging with isotropic resolution, Proc. Natl. Acad. Sci. 112 (7) (2015).
[188] L. Mockl, A.R. Roy, W.E. Moerner, Deep learning in single-molecule microscopy: fundamentals, caveats, and recent developments invited, Biomed. Opt. Express 11 (2020) 1633–1661.
[189] E. Nehme, et al., DeepSTORM3D: dense 3D localization microscopy and PSF design by deep learning, Nat. Methods 17 (2020) 734–740.
[190] W. Ouyang, A. Aristov, M. Lelek, X. Hao, C. Zimmer, Deep learning massively accelerates super-resolution localization microscopy, Nat. Biotechnol. 36 (2018) 460–468.
[191] L. von Chamier, R.F. Laine, J. Jukkala, et al., Democratising deep learning for microscopy with ZeroCostDL4Mic, Nat. Commun. 12 (2021) 2276.
[192] F. Balzarotti, Y. Eilers, K.C. Gwosch, A.H. Gynnå, V. Westphal, F.D. Stefani, J. Elf, S.W. Hell, Nanometer resolution imaging and tracking of fluorescent molecules with minimal photon fluxes, Science (2017) 606–612.
[193] K. Prakash, A. Curd, Assessment of 3D MINFLUX data for quantitative structural biology in cells, bioRxiv (2021), https://doi.org/10.1101/2021.08.10.455294. 2021.08.10.455294.
[194] L. Schermelleh, A. Ferrand, T. Huser, et al., Super-resolution microscopy demystified, Nat. Cell Biol. 21 (2019) 72–84.
[195] C. Werner, M. Sauer, C. Geis, Super-resolving microscopy in neuroscience, Chem. Rev. (2021), https://doi.org/10.1021/acs.chemrev.0c01174.
[196] M. Lelek, M.T. Gyparaki, G. Beliu, et al., Single-molecule localization microscopy, Nat. Rev. Methods Primers 1 (2021) 39.
[197] A. Jimenez, K. Friedl, C. Leterrier, About samples, giving examples: optimized single molecule localization microscopy, Methods 174 (2020) 100–114.

CHAPTER FOURTEEN

Motion Tracking and Analysis

Erik Meijering, Ihor Smal, Oleh Dzyubachyk, and Jean-Christophe Olivo-Marin

14.1 Introduction

By their very nature, biological systems are dynamic, and a proper understanding of the cellular and molecular processes underlying living organisms and how to manipulate them is a prerequisite to combatting diseases and improving human health care. One of the major challenges of current biomedical research, therefore, is to unravel not just the spatial organization of these complex systems, but their *spatiotemporal* relationships as well [1]. Catalyzed by substantial improvements in optics hardware, electronic imaging sensors, and a wealth of fluorescent probes and labeling methods, light microscopy has, over the past decades, matured to the point that it enables sensitive time-lapse imaging of cells and even single molecules [2–5]. These developments have had a profound impact on how research is conducted in the life sciences.

A consequence of these developments is that the size and complexity of image data is ever increasing. Datasets generated in time-lapse experiments commonly contain hundreds to thousands of images, each containing hundreds to thousands of objects to be analyzed. Fig. 14.1 shows example frames from image sequences acquired for specific time-lapse imaging studies containing large numbers of cells or subcellular particles to be tracked over time. These examples illustrate the complexity of typical time-lapse imaging data and the need for automated image analysis, but they also show the difficulty of the problem.

Such huge amounts of data cannot be analyzed by visual inspection or manual processing within any reasonable amount of time. It is now generally recognized that automated methods are necessary, not only to handle the growing rate at which images are acquired, but also to provide a level of sensitivity and objectivity that human observers cannot match [6].

In general, time-lapse imaging studies consist of four successive steps: (1) planning the experiment and acquiring the image data, (2) preprocessing the data to correct for systemic and random errors and to enhance relevant features, (3) analyzing the data by detecting and tracking the objects relevant to the underlying biological questions of the study, and (4) analyzing the resulting trajectories to test predefined hypotheses or detect new phenomena. Fig. 14.2 gives a topical overview of the process. The circular structure of the diagram reflects the iterative nature of the imaging

Microscope Image Processing
https://doi.org/10.1016/B978-0-12-821049-9.00013-7
Copyright © 2023 Elsevier Inc.
All rights reserved.

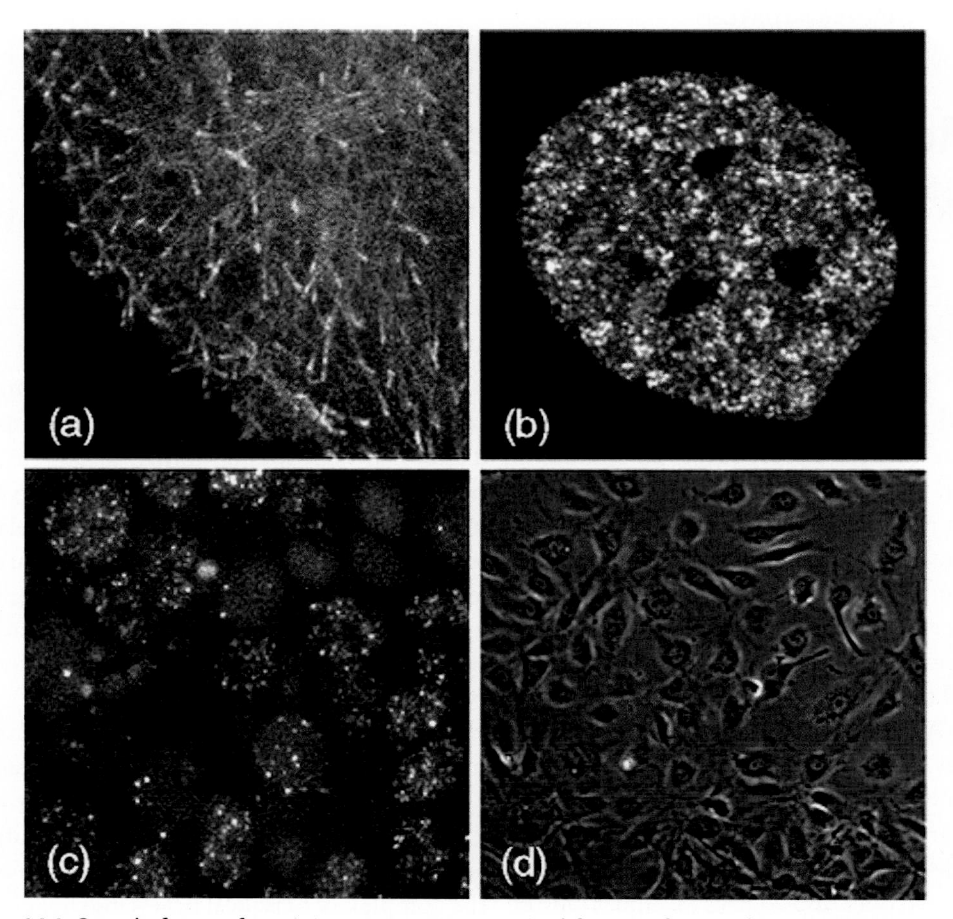

Fig. 14.1 Sample frames from image sequences acquired for specific time-lapse imaging studies. The sequences contain large numbers of cells or subcellular particles to be tracked over time. (a) Single frame ($36 \times 36\,\mu m$) from a fluorescence microscopy image sequence (1 s between frames) showing labeled microtubule plus-ends moving in the cytoplasm of a single COS-7 cell (only partly visible). (b) Single frame ($30 \times 30\,\mu m$) from a fluorescence microscopy image sequence (about 12 s between frames) showing labeled androgen receptors moving in the nucleus of a Hep3B cell. (c) Single frame ($73 \times 73\,\mu m$) from a fluorescence microscopy image sequence (about 16 min between frames) showing labeled Rad54 proteins in the nuclei of mouse embryonic stem cells. (d) Single frame (about $500 \times 500\,\mu m$) from a phase contrast microscopy image sequence (12 min between frames) showing migrating human umbilical vein endothelial cells in a wound healing assay. *(Images (a)–(d) courtesy of N. Galjart, A. Houtsmuller, J. Essers, and T. Ten Hagen, respectively)*

process, in that the results of previous experiments usually trigger the planning of new studies. In this chapter we sketch the current state of the art in time-lapse microscopy imaging, in particular cell and particle tracking, from the perspective of automated image processing and subsequent motion analysis. While it is clear that biological research is increasingly relying on computerized methods, it also follows

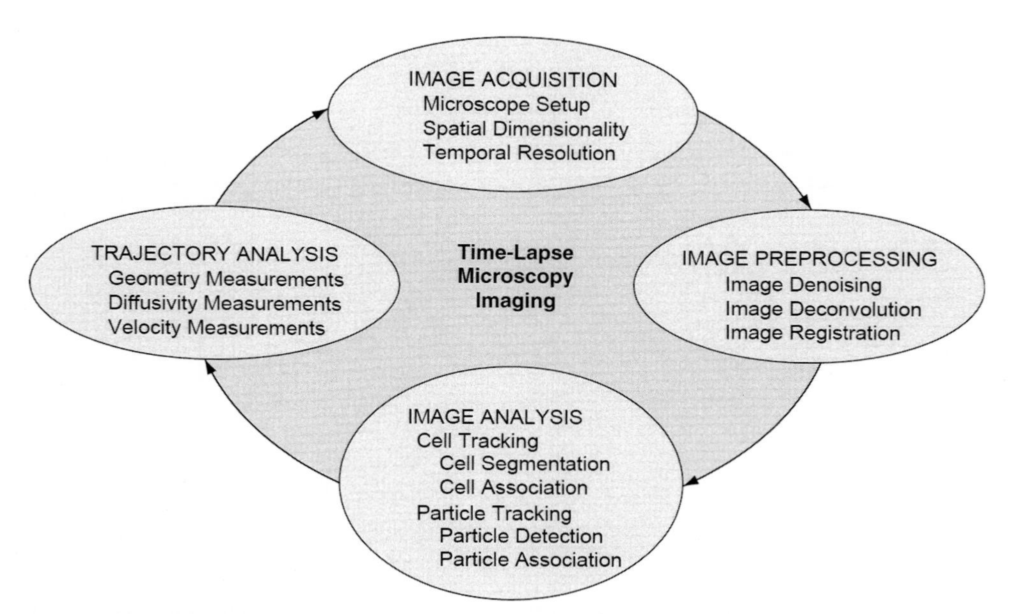

Fig. 14.2 The circle of life in time-lapse microscopy imaging. This diagram depicts the successive steps in the imaging process and gives an overview of the topics addressed in this chapter. Following image acquisition, preprocessing is often required to increase the success of subsequent automated analysis of the images and, eventually, of the resulting trajectories. The circular structure of the diagram reflects the iterative nature of the imaging process.

that such methods are still developing and there is much room for improvement. Although there is growing consensus on the strengths and weaknesses of specific approaches, there is currently no single best solution to the problem in general [7,8]. The main conclusion emerging from the literature is that, with currently available image processing and analysis methodologies, any given tracking application requires its own dedicated algorithm to achieve the best results.

This chapter addresses each of these issues from a computational perspective. It focuses on methodological rather than hardware or software aspects and gives examples of image processing and analysis methods that have been used successfully for specific applications. The ultimate goal of this chapter is to prepare the reader to select methods intelligently.

14.2 Image Acquisition

Time-lapse imaging experiments involve the acquisition of not only spatial information, but also temporal information, and often spectral information as well, resulting in up to five-dimensional (x, y, z, t, c) image datasets. Fig. 14.3 shows different configurations and dimensionalities in time-lapse microscopy imaging. Notice that dimensionality, as used here, does not necessarily describe the imaging configuration unambiguously. For example,

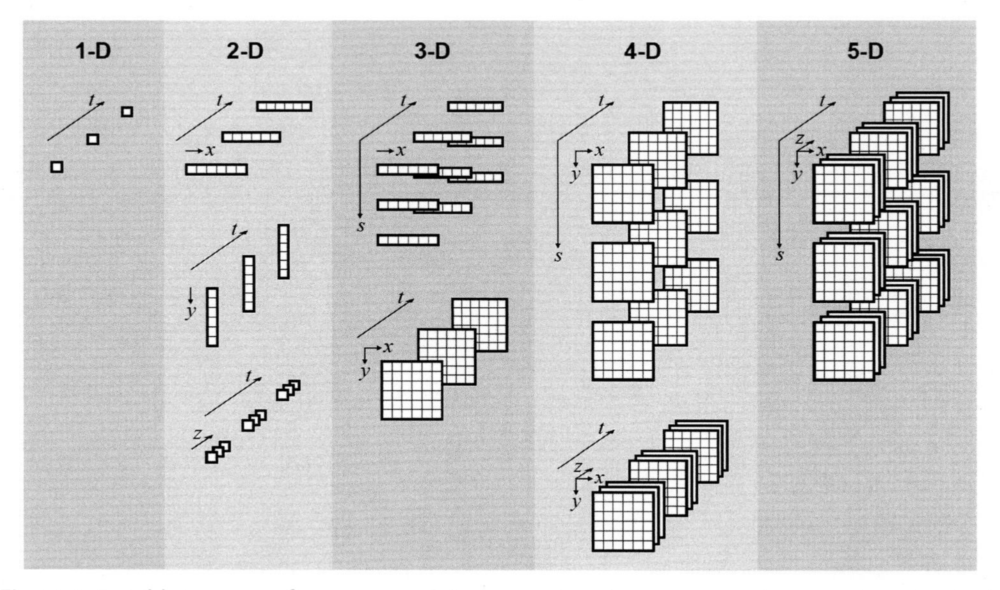

Fig. 14.3 Possible image configurations and dimensionalities in time-lapse microscopy imaging. Each dimension corresponds to an independent physical parameter or coordinate: *x* and *y* commonly denote the in-plane spatial coordinates, *z* the depth or axial coordinate, *t* the time coordinate, and *s* denotes any spectral parameter, such as wavelength.

4D imaging may refer to either spatially 2D multispectral time-lapse imaging, or to spatially 3D time-lapse imaging. To avoid confusion, it is better to use the abbreviations 2D and 3D for referring to spatial dimensionality only, and to indicate explicitly whether the data also involves a temporal or spectral coordinate. Therefore, in this chapter, we indicate spatially 2D and 3D time-lapse imaging by 2D + t and 3D + t, respectively, rather than by 3D and 4D.

Regardless of the imaging technique being used, a careful design of the microscope setup is imperative, as shortcomings may require additional pre- or postprocessing of the resulting image data or may even lead to artifacts that cannot be removed and that hamper data analysis. We include here a few general remarks concerning the choice of microscopy, spatial dimensionality, and temporal resolution from the perspective of subsequent data analysis.

14.2.1 Microscope Setup

Time-lapse imaging experiments generally involve living cells and organisms. A fundamental concern is keeping the specimen alive during the acquisition of hundreds or thousands of images over a period of time that may range from minutes to hours or even days. This not only calls for a suitable environment with controlled temperature, humidity, and a stably buffered culture medium [9], but it also requires economizing light exposure, since living cells are sensitive to photodamage [10]. In fluorescence microscopy, excessive illumination bleaches fluorophores, and this limits their emission time span and generates free radicals that are toxic for living cells.

Two very important factors determine whether automated methods can be applied successfully, and they strongly affect accuracy. They are *signal contrast* (the intensity difference between objects and background) and *noise*, which, in light microscopy, is signal dependent. These two factors are usually combined into a single measure, the signal-to-noise ratio (SNR), calculated as the difference in mean intensity between the object, I_o, and background, I_b, divided by a representative noise level, σ, that is, $\mathrm{SNR}=(I_o - I_b)/\sigma$. Ideally, experiments should be designed so as to maximize SNR to allow robust and accurate automated image analysis, and the only way to accomplish this is with high light exposure levels.

Decreasing light exposure to minimize photodamage or fluorophore bleaching, and increasing light exposure to maximize SNR, are conflicting aims in need of a compromise. The optimal solution depends on the type of objects to be imaged, their dimensions, motility, and viability, which determine the best type of microscopy to be used. For example, living cells in a culture medium produce poor contrast with standard bright-field illumination, and they often require contrast-enhancing imaging techniques, such as phase contrast or differential interference contrast microscopy. Intracellular particles are hardly visible without contrast enhancement and are better studied using fluorescence microscopy (Tables 14.1 and 14.2). In all cases, the system should make the best use of the available light, implying the use of high numerical aperture objectives in conjunction with highly sensitive detectors. In many cases this may also mean that widefield microscopy is preferable over confocal microscopy [5].

In practice, for any biological application, there is often no single best microscope setup. A compromise must be found between sufficient (but not destructive) illumination, and resolution (spatial and temporal), so that the maximum number of acceptable images (optical slices and frames) can be acquired before the specimen is completely photobleached or damaged [9]. Here *acceptable* means having the minimum SNR required by automated image analysis techniques (discussed later in this chapter). To this end, a good understanding of different microscope systems is needed, for which we refer to excellent introductory texts [5].

14.2.2 Spatial Dimensionality

One of the fundamental questions to be addressed when setting up an experiment is whether imaging needs to be performed in two or in three spatial dimensions over time (denoted as 2D + t and 3D + t, respectively, and the latter is also referred to as 4D). Despite much 4D imaging reported in the literature these days, many experiments are still performed in 2D + t (Tables 14.1 and 14.2). Often this is due to limitations imposed by photodamage and photobleaching, discussed previously. In other studies, in particular those concerning intracellular dynamic processes, acquiring multiple optical slices would simply take too much time relative to the motions of interest, resulting in intrascan motion artifacts. And sometimes 3D + t imaging simply does not provide much additional relevant

Table 14.1 Selected cell tracking methods and some of their features and applications.

Ref.	Year	Dim.	Segmentation	Association	Microscopy	Application	Auto.
[11]	1985	2D + t	Edge detection	Matching motion parameters	PC	Sperm cells	FA
[12]	1988	2D + t	Size + intensity thresholding	Nearest unique cell	PC	Sperm cells	FA
[13]	1993	2D + t	Manual outlining	Active contour evolution	BF	Fibroblasts	MI
[14]	1994	3D + t	Multiple thresholding	Template matching	F	*Dictyostelium discoideum*	FA
[15]	1997	2D + t	Space-time filtering + thresholding	Space-time trace angle + distance	PC	Leukocytes	FA
[16]	1999	2D + t	Manual outlining + watersheds	Unspecified matching algorithm	PC	PC-3 cells	MI
[17]	2002	2D + t	Manual indication	Template matching	PC	Leukocytes	MI
[18]	2002	3D + t	Thresholding	Nearest cell	HMC	Cancer cells	FA
[19]	2002	2D + t	Active contours	Active contour evolution	F	Dictyostelium cells	FA
[20]	2002	2D + t	Manual outlining	Active contour evolution	PC	*Entamoeba histolytica*	MI
[21]	2003	2D + t	Manual indication	Template matching	PC	HSB-2 T-cells	MI
[22]	2004	2D + t	Manual indication + active contours	Active contour evolution	PC	Endothelial cells	MI
[23]	2004	2D + t	Level sets	Level set evolution	PC	Leukocytes	FA
[24]	2005	2D + t	Manual indication	Coupled mean shift processes	PC	Cancer and endothelial cells	MI
[25]	2005	3D + t	Multiple level sets	Level set evolution	F	*Entamoeba histolytica* and epithelial cells	FA
[26]	2006	2D + t	Otsu thresholding + watersheds	Distance + area + overlap	F	HeLa cells	FA

[27]	2006	2D + t	Template matching	Probabilistic association	PC	Hematopoietic stem cells	FA
[28]	2006	2D + t	Active contours	Monte Carlo tracking	F	HeLa cells	FA
[29]	2006	2D + t	Manual outlining + active contour	Fourier-based template matching	PC	Keratocytes	MI
[30]	2006	2D + t	Marker controlled watersheds	Mean shift + Kalman filtering	F	HeLa cells	FA
[31]	2010	3D + t	Multiple level sets	Level set evolution	F	HeLa cells	FA
[32]	2013	3D + t	Chan-Vese model + graph cuts	Model evolution	F	ADMSC and H157 cells	FA
[33]	2016	2D + t	Thresholding + watershed	Greedy minimum cost	PC	Various bacterial cells	FA
[34]	2018	2D + t	Convolutional neural network	Multiple hypothesis Kalman filtering	PC	Endothelial cells	FA
[35]	2020	2D + t	Convolutional neural network	Convolutional neural network	PC	*Escherichia coli*	FA

From left to right the columns indicate the reference number describing the method and the year of publication, the dimensionality of the data for which the method was designed, the main features of spatial segmentation and temporal association used by the method, the type of microscopy used and the applications considered in the described experiments, and the level of automation of the method. *BF*, bright-field; *CNN*, convolutional neural network; *F*, fluorescence; *FA*, fully automatic, meaning that in principle no user interaction is required, other than parameter tuning; *HMC*, Hoffman modulation contrast; *MI*, manual initialization, meaning that more user interaction is required than parameter tuning; *PC*, phase contrast.

Table 14.2 Selected particle tracking methods and some of their features and applications.

Ref.	Year	Dim.	Detection	Association	Microscopy	Application	Auto.
[36]	1992	2D + t	Gaussian fitting	Distance + intensity	F	Lipoproteins, influenza viruses	FA
[37]	1992	2D + t	Thresholding + centroid	Area + major/minor axes + distance	F	Actin filaments	FA
[38]	1994	2D + t	Thresholding + centroid	Nearest particle	F	Low density lipoprotein receptors	FA
[39]	1996	2D + t	Thresholding + centroid	Nearest particle	F	Actin filaments	FA
[40]	1999	3D + t	Thresholding + centroid	Template matching	F	Subchromosomal foci	FA
[41]	1999	2D + t	Gradient magnitude + tracing	Fuzzy logic	F	Secretory vesicles, speckles	FA
[42]	2000	2D + t	Thresholding + centroid	Nearest particle	F	Microspheres, actin filaments	FA
[43]	2000	2D + t	Local maxima	Nearest particle	F	R-phycoerythrin	FA
[44]	2001	3D + t	Thresholding	Fuzzy logic	F	Chromosomes, centrosomes	FA
[45]	2001	2D + t	Gaussian fitting	Nearest particle	F	P4K proteins	FA
[46]	2002	3D + t	Gaussian mixture fitting	Global weighted distance minimization	F	Chromosomes, spindle pole body	FA
[47]	2003	2D + t	Local maxima selection	Multilayered graphs	F	Actin and tubulin fluorescent speckles	FA
[48]	2004	3D + t	Pyramid linking	Intensity + velocity + acceleration	F	Microspheres, vimentin	FA
[49]	2004	3D + t	Thresholding + centroid	Nearest particle	F	Secretory granules	FA
[50]	2005	2D + t	Template matching	Minimum cost paths	F	Quantum dots, glycine receptors	FA

[51]	2005	3D + t	Gaussian mixture fitting	Template matching	F	Kinetochore microtubules	FA
[52]	2005	2D + t	Artificial neural networks	Velocity and virtual flow	F	Microspheres	FA
[53]	2005	3D + t	Laplacian of Gaussian	Dynamic programming	F	Telomeres	FA
[54]	2005	2D + t	Local maxima	Distance + intensity moments	TIRF	Lipoproteins, adenovirus-2, quantum dots	FA
[55]	2006	3D + t	Multiscale wavelet products	Interacting multiple models	F	Quantum dots, endocytic vesicles	FA
[56]	2008	2D + t	Multiscale wavelet products	Two-step linear assignment	TIRF	Endocytic CCPs and CD36 receptors	FA
[57]	2008	2D + t	Curvature-based probabilistic sampling	Multimodality particle filtering	F	Microtubule tips	FA
[58]	2015	2D + t	Multiscale Laplacian of Gaussian	Two-step multiframe association	F	Avian leukosis virus particles	FA
[59]	2016	2-D + t	Mathematical morphology	Probability hypothesis density filtering	F	Microtubule tips + stem cells	FA
[60]	2018	3-D + t	Convolutional neural network	Unspecified linking method	F	Carboxylated polystyrene nanoparticles	FA

From left to right the columns indicate the reference number describing the method and the year of publication, the dimensionality of the data for which the method was designed, the main features of spatial detection and temporal association used by the method, the type of microscopy used and the applications considered in the described experiments, and the level of automation of the method. *F*, fluorescence; *FA*, fully automatic, meaning that in principle no user interaction is required, other than parameter tuning; *TIRF*, total internal reflection fluorescence.

information for analysis. For example, when studying cell migration in monolayers or microtubule dynamics in neurons, the structures of interest are sufficiently flat to allow 2D + t imaging by widefield microscopy to provide a good understanding of a process. The improved light collection and the lower number of optical slices in such cases yields a better SNR and allows for a higher temporal resolution.

However, most cellular and intracellular processes do not occur in a single plane but in three dimensions over time and require 3D + t imaging to fully characterize cell morphodynamics [61]. For example, it is known that tumor cells treated with drugs that block migration on 2D substrates can still move inside an artificial 3D collagen matrix by means of a very different type of motility [62]. Regarding intracellular processes, studies into kinetochore microtubule dynamics have revealed that trajectories obtained from 2D + t imaging may differ significantly from those obtained from 3D + t imaging and may lead to severe misinterpretation of the underlying processes [51]. These findings suggest that the validity of 2D + t imaging should always be confirmed by additional experiments. This could be similarly important as making sure that fluorescent probes, in fluorescence microscopy imaging, do not alter physiology.

14.2.3 Temporal Resolution

Another issue of great importance in time-lapse experiments is the rate at which images should be acquired over time, also referred to as the *temporal sampling rate*, or the *frame rate*, or *temporal resolution*. Ideally, this rate should be sufficiently high to capture the relevant details of the object motion. However, similar to spatial dimensionality and resolution, the temporal resolution is constrained by the total light exposure allowed without impacting the biological processes under study.

From sampling theory (discussed in Chapter 3) it is known that in order to be able to reconstruct a continuous signal from a set of discrete samples, the sampling rate must be at least twice the highest frequency component in the signal. This minimum sampling rate, often called the *Nyquist rate*, also applies to sampling in time. Object position as a function of time is a continuous signal, and high-fidelity reconstruction of this signal and any motion parameters derived from it, such as velocity and acceleration, is possible only if temporal sampling is done at a rate that complies with the theory.

In practice, however, establishing the optimal temporal sampling rate is a chicken-and-egg problem. Before sampling, one must have knowledge of the expected velocities to be estimated, which can be obtained only by sampling at or above the proper rate in the first place. A series of experiments at different sampling rates is often necessary to establish the optimal rate. Several studies can be found in the literature [11,12,51] discussing temporal resolution for specific applications. They clearly demonstrate how undersampling may have a significant effect on velocity estimation. From the point of view of image analysis, it should also be realized that many automated cell tracking algorithms

(Table 14.1) fail if the displacement between time points is larger than the cell diameter [62], especially in cases involving cell contact. Similar limitations exist for particle tracking algorithms (Table 14.2), particularly in cases of high particle densities or when trying to characterize Brownian motion.

14.3 Image Preprocessing

During image acquisition there are many factors that may cause image quality degradation. Since illumination levels must be kept to a minimum to avoid photobleaching and photodamage in fluorescence microscopy, the SNR in the acquired images is often rather low. Furthermore, any optical imaging device has limited resolution due to diffraction, which manifests itself as blurring. In addition, out-of-focus light causes a loss of contrast for in-focus objects, especially in widefield microscopy imaging. Finally, even if the microscope setup is perfectly stable, unwanted motion may occur in the specimen. This is the case, for example, when studying intracellular dynamic processes while the cells themselves are migrating. In this section we discuss methods developed specifically for reducing such artifacts. A more in-depth discussion of these methods can be found elsewhere in this volume and in other works [5].

14.3.1 Image Denoising

Any image acquired with a physical device will be contaminated with noise. In this chapter we refer to *noise* as any random fluctuation in image intensity, as distinct from systematic distortions, such as hot or cold pixels in charge-coupled devices (CCDs), or background shading, which can be compensated for by other methods. In optical microscopy imaging, noise originates from sources that fall into four categories: (1) the quantum nature of light, which gives rise to *photon shot noise* (Poisson distributed), (2) random electron generation due to thermal vibrations, called *thermal noise* (also Poisson distributed), (3) random fluctuations in the analog electric signals in the imaging sensors before digitization, referred to as *readout noise* (Gaussian distributed), and (4) round-off error introduced by converting the analog signal into digital form, known as *quantization noise* (uniformly distributed). Whereas thermal, readout, and quantization noise can be controlled by careful electronic design and proper operating conditions, photon shot noise is inherent to optical imaging and can be reduced during acquisition only by increasing the light exposure level or time [63].

Although noise cannot be avoided during acquisition, it can be reduced afterwards, to some degree, by image processing. Many denoising techniques are available for this purpose, which can be roughly divided into linear and nonlinear filtering methods (discussed in Chapter 5). Linear methods are typically implemented by convolution filtering. Examples include uniform local averaging [43] and Gaussian smoothing [64]. While effective in reducing noise, these methods also blur relevant image structures. This can be avoided by using nonlinear methods. The most common of these is median filtering [40,65].

More sophisticated methods that are sometimes used in time-lapse imaging [41,66] are based on the principle of anisotropic diffusion filtering [67]. By avoiding blurring near object edges, these methods usually yield superior results. Methods based on gray-level morphological operations (discussed in Chapter 6) have also been used to remove not only noise, but small-scale image structures as well [17]. Other state-of-the-art denoising methods include total-variation minimization, nonlocal filtering, Wiener filtering, wavelet filtering, and sparsity-based methods, all of which have their specific strengths and weaknesses [68].

14.3.2 Image Deconvolution

Conventional widefield microscopes are designed to image specimens at the focal plane of the objective lens, but they also collect light emanating from out-of-focus planes. This reduces the contrast of in-focus image structures. In confocal microscopes (discussed in Chapter 11), this out-of-focus light is largely rejected by the use of pinholes, resulting in clearer images and increased resolution, both laterally and axially [5]. In either case, however, diffraction occurs as the light passes through the finite-aperture optics of the microscope, introducing a blurring effect. For a well-designed imaging system, this blurring can be modeled mathematically as a convolution of the true incident light distribution with the point spread function (PSF) of the system. If the PSF is known, it is possible, in principle, to reverse this operation at least partially by image processing. This process is called *deconvolution* [69].

Various methods are available for deconvolution, which vary greatly in computational demand, the requirement for accurate knowledge of the PSF, and their ability to reduce blur, improve contrast, increase resolution, and suppress noise [69–72]. Similar to denoising, they can be roughly divided into linear and nonlinear methods. The former category includes the *nearest-neighbor* and Fourier-based *inverse filtering* algorithms, which are conceptually simple and computationally fast, but have the tendency to amplify noise and can even introduce artifacts. Generally, they are not successful when studying small, intracellular structures and dynamic processes. More sophisticated, nonlinear methods involve *iterative constrained* algorithms, which allow enforcing specific constraints. The nonlinear category also includes *blind deconvolution* algorithms [72], which do not require knowledge of the PSF, as it is estimated from the data during the process. Time-lapse imaging of thick samples may require even more sophisticated, space-variant deconvolution methods.

While some authors have advocated always to deconvolve image data if possible [69], the question whether deconvolution, as a separate preprocessing step, is really necessary or beneficial depends on the application. Particularly in studies requiring tracking of sub-resolution particles, explicit deconvolution seems less relevant, except perhaps when widefield microscopes are used [9]. This is because the localization of such particles, which appear in the images as diffraction-limited spots, can be done with much higher accuracy and precision than the resolution of the imaging system [73–75]. Indeed, when the detection and localization process involves fitting a model of the PSF, this implicitly solves the deconvolution problem.

14.3.3 Image Registration

One of the difficulties frequently encountered in quantitative motion analysis is the presence of unwanted movements confounding the movements of interest. In time-lapse imaging of living specimens, the observed movements are often a combination of global displacements and deformations of the specimen as a whole, superposed upon the local movements of the structures of interest [65]. For example, in intravital microscopy studies, which involve living animals, the image sequences may show cardiac, respiratory, or other types of global motion artifacts [15,52,76]. But even when imaging live cell cultures, the dynamics of intracellular structures may be obscured by cell migration, deformation, or division [48,53,77,78]. In these situations, prior motion correction is necessary. This can be achieved by global or local image alignment, also referred to as *image registration*.

Many image registration methods have been developed for a wide variety of applications, notably in clinical medical imaging [79–81]. The main aspects of a registration method that determine its suitability for a specific registration problem include: (1) the type of geometrical transformation allowed (which can be divided into rigid, being translation and rotation, versus nonrigid deformation, which also includes scaling, affine, and elastic deformations), (2) the measure (such as cross-correlation or mutual information) used to quantify the similarity of images, and (3) the optimization strategy used (including implementational issues such as interpolation and discretization). The conclusion from the large body of literature on the subject seems to be that there is no universal best method for all possible image registration problems. In practice, the optimal choice of transformation model, similarity measure, and optimization strategy, will depend on the given registration problem. Also, specific implementational choices may have a large impact on the results and require careful consideration.

In biological imaging, image registration methods are less common than in clinical medical imaging, but a variety of techniques are increasingly used in both 3D and time-lapse microscopy. For tracking leukocytes in phase-contrast images, for example, normalized cross correlation of edge information has been used to achieve translational background registration [15,17,76]. By contrast, tracking of intracellular particles in fluorescence microscopy (discussed in Chapter 10) usually requires correction for translation and rotation, or even for more complex deformations. This can be done by intensity-based cross correlation [40] or by using the labeled proteins as landmarks in an iterative point-based registration scheme [44]. In fluorescence microscopy, large parts of the images often contain no relevant information, and thus the use of landmarks can improve the robustness and accuracy of registration [82]. At present, similar to medical image registration, no clear consensus has emerged about which method generally works best for biological image registration, and the choices to be made are application dependent.

14.4 Image Analysis

The ultimate goal of time-lapse imaging experiments is to gain insight into cellular and intracellular dynamic processes. Inevitably this requires quantitative analysis of motion patterns. Approaches to accomplish this fall into three categories. The first consists of real-time, single-target tracking techniques. These usually involve a microscope setup containing an image-based feedback loop controlling the positioning and focusing of the system, to keep the object of interest locked in the center of the field-of-view [83–85]. Only a small portion of the specimen is illuminated this way, which reduces photodamage and allows imaging to be done faster or over a longer period of time. The second category consists of ensemble tracking approaches, such as fluorescence recovery after photobleaching (FRAP), or fluorescence loss in photobleaching (FLIP) [10]. While useful for assessing specific dynamic parameters (such as diffusion coefficients and association/dissociation rates of labeled proteins) they are limited to yielding averages over larger populations. The third category, which we focus on here, consists of approaches that aim to track all individual objects of interest in the data. These are usually performed off-line.

Possible levels of computerization of object tracking range from simply facilitating image browsing and manual analysis, to manual initialization followed by automated tracking, to full automation (Tables 14.1 and 14.2). From the perspective of efficiency, objectivity, and reproducibility, full automation is to be preferred. However, given the large variety of imaging and labeling techniques, the objects to be tracked may have widely differing and even time-varying appearances. As a consequence, full automation in any specific application usually can be achieved only by developing very dedicated and well-tuned algorithms. This explains why commercial tracking software tools, which are typically developed using very general assumptions, often fail to yield satisfactory results for specific tracking tasks.

In this section we discuss published approaches to automated object tracking in time-lapse microscopy images. A distinction is made between *cell tracking* and *particle tracking*. Two different strategies exist for both problems. The first consists of the identification of the objects of interest in the entire image sequence, separately for each frame, followed by temporal association, which tries to relate identified objects either globally over the entire sequence or from frame to frame. In the second strategy, objects of interest are identified only in a first frame and are subsequently followed in time by image matching or by model evolution. In either case, the algorithms usually include a detection or segmentation stage, and a temporal association stage. Both are essential to performing motion analysis of individual objects. Alternative approaches based on optic flow have also been studied [65,66,86–86], but these are limited to computing collective cell motion and intracellular particle flows, unless additional detection algorithms are applied.

14.4.1 Cell Tracking

Cell motility and migration are of fundamental importance to many biological processes [1,61,62,89–91]. In embryonic development, for example, cells migrate and differentiate into specific cell types to form different organs. Failures in this process may result in severe congenital defects and diseases. In adult organisms, too, cell movement plays a crucial role. In wound healing, for example, several interrelated cell migration processes are essential in regenerating damaged tissue. The immune system consists of many different proteins and cells interacting in a dynamic network to identify and destroy infectious agents. Many disease processes, most notably cancer metastasis, depend heavily on the ability of cells to migrate through tissue and reach the bloodstream. Because of its importance for basic cell biology and its medical implications, cell migration is a very active field of research. Automated methods for segmenting cells and following them over time (Table 14.1) are essential in quantifying cell movement and interaction under normal and perturbed conditions.

14.4.1.1 Cell Segmentation

The simplest approach for separating cells from the background is intensity thresholding (discussed in Chapter 7). This involves a single threshold parameter that can be set manually or derived automatically from the data, based on the intensity histogram. Although used in many past cell-tracking algorithms [12,18,92], this approach is successful only if cells are well separated and their intensity levels differ markedly and consistently from the background. In practice, however, this condition is rarely met. In phase-contrast microscopy, for example, cells may appear as dark regions surrounded by a bright halo, or vice versa, depending on their position relative to the focal plane. In the case of fluorescence microscopy, image intensity may fall off over time due to photobleaching. While the situation may be improved by using adaptive thresholding or some sort of texture filtering [14,93], thresholding based on image intensity alone is generally inadequate.

A fundamentally different approach to cell detection and segmentation that is particularly relevant to phase-contrast and differential interference contrast microscopy is to use a predefined cell-intensity profile, also referred to as a *template*, to be matched to the image data. This works well for objects that do not change shape significantly, such as certain blood cells, algal cells, or cell nuclei [27,94,95]. However, most cell types are highly plastic and move by actively changing shape. Keeping track of such morphodynamic changes would require the use of a very large number of different templates, which is impractical from both algorithm design and computational considerations.

Another well-known approach to image segmentation is to apply the watershed transform [90]. By considering the image as a topographic relief map and by flooding it from its local minima, this transform completely subdivides the image into regions (*catchment basins*) with delimiting contours (*watersheds*). Fast implementations exist for this intuitively sensible method, which is easily parallelized. The basic algorithm has several drawbacks, however, such as sensitivity to noise and a tendency toward over-segmentation [96]. Carefully designed pre- and postprocessing strategies are required for acceptable results. By using marking, gradient-weighted distance transformation, and model-based merging methods, several authors have successfully applied the watershed transform to cell segmentation in microscopy [16,30,33,97,98].

More recently, cell segmentation has been performed using model-based approaches [19,20,22,23,25,28,31,32,99]. These include classical deformable models such as parametric *active contours* or "snakes" [100] in 2D and implicit *active surfaces* or level sets [101] in 3D. They start with an approximate boundary and iteratively evolve in the image domain to optimize a predefined energy functional. Typically, this functional consists of both image-dependent and boundary-related terms. The former may contain statistical measures of intensity and texture in the region enclosed by the developing boundary, or gradient magnitude information along the boundary. Image-independent terms concern properties of the boundary shape itself represented by the front, such as boundary length, surface area, and curvature, and the similarity to reference shapes. This mixture of terms, which enables incorporation of both image information and prior knowledge, makes deformable models easily adaptable to specific applications, but relies on the user to define these terms. Avoiding this nontrivial step, data-driven models obtained by deep learning of artificial neural networks are increasingly replacing such classical model-based approaches [35,102–104].

14.4.1.2 Cell Association

Several strategies exist for performing interframe cell association. The simplest is to associate each segmented cell in one frame with the nearest cell in a subsequent frame. Here, *nearest* may refer not only to spatial distance between boundary points or centroid positions [11,18,93], but also to similarity in terms of average intensity, area or volume, perimeter or surface area, major and minor axis orientation, boundary curvature, angle or velocity smoothness, and other features [26,52]. Generally, the more features involved, the lower the risk of ambiguity. However, matching a large number of features may be as restrictive as template matching, since cell shape changes between frames are less easily accommodated [14,17,21,29]. Some applications may not require keeping track of cell shape features, and following only the cell center position is sufficient, which may be achieved by mean-shift processes [24,30].

In addition to performing cell segmentation, classical deformable model approaches are also naturally suited for capturing cell migration and cell shape changes over time [62]. At any point in an image sequence, the contours or surfaces obtained in the current frame can be used as initialization for the segmentation process in the next frame [13,20,23,25]. With standard algorithms, however, this usually works well only if cell displacements are limited to no more than one cell diameter from frame to frame [25,99]. Otherwise, more sophisticated approaches are required, such as the use of gradient-vector flow [20,99] or incorporating known or estimated dynamics [22,28].

Rather than using explicitly defined models (parametric active contours), some cell tracking methods have been based on implicitly defined models (through level sets), as they can easily handle topological changes such as cell division and can be extended readily to deal with higher-dimensional image data [25]. In either case, however, several adaptations of the standard algorithms are usually needed to be able to track multiple cells simultaneously and to handle cell appearances, disappearances, and touches. While this is certainly feasible [20,23,25], it usually introduces a number of additional parameters that must be tuned empirically for each specific application. This increases the risk of making errors and reducing reproducibility, requiring further postprocessing to validate tracking results [105]. Alternative cell association approaches involve cell motion estimation by Bayesian filtering or deep learning [34,106,107].

14.4.2 Particle Tracking

The ability of cells to migrate, perform a variety of specialized functions, and reproduce is the result of a large number of intracellular processes involving thousands of differently sized and shaped biomolecular complexes, collectively termed *particles* in this chapter. Since many diseases originate from a disturbance or failure of one or more of these processes, they have become the subject of intense research by academic institutes and pharmaceutical companies. Fluorescent probes allow selective visualization of intracellular particles [108,109]. Combined with time-lapse optical microscopy, they enable studying the dynamics of virtually any protein in living cells. Automated image analysis methods for detecting and following fluorescently labeled particles over time (Table 14.2) are indispensable in extracting motion information from the image data acquired in such studies [1].

14.4.2.1 Particle Detection

In fluorescence microscopy imaging, the particles of interest are never observed directly, but their position is revealed indirectly by the fluorescent molecules attached to them. Typically, these fluorophores are cylindrically shaped molecules having a length and diameter on the order of a few nanometers only. In most experiments it is unknown how many fluorescent molecules are actually attached to the particles of interest. Commonly, however, a fluorescently labeled particle will be much smaller than the optical resolution of the imaging system. Even though recent advances in light microscopy have led to significantly improved resolution [110–112], in practice many studies of intracellular dynamic processes are done using typical widefield or confocal microscopes, which can resolve up to about 200 nm laterally and 600 nm axially. Therefore, fluorescently labeled particles effectively act as point sources of light, and they appear in the images as diffraction-limited spots, also called *foci*.

Several studies [73–75] have shown that the accuracy of localizing single particles and the resolvability of multiple particles depend on a number of factors. If magnification and spatial sampling are properly matched to satisfy the Nyquist sampling criterion, the limiting factor is the SNR, or effectively the photon count, with higher photon counts yielding higher accuracy and resolvability. The consensus emerging from these studies is that a localization accuracy for single particles of around 10 nm is achievable in practice. Estimation of the distance between two particles is possible with reasonable levels of accuracy for distances of about 50 nm and larger. Smaller distances can be resolved, but with rapidly decreasing accuracy. In order to improve accuracy in such cases, the number of detected photons would have to be increased substantially, which typically is not possible in time-lapse imaging experiments without causing excessive photobleaching.

A number of approaches to particle detection and localization exist. Similar to cell segmentation, the simplest approach to discriminate between objects and background is to apply intensity thresholding. The localization of a particle is often accomplished by computing the local centroid, or center of intensity, of image elements with intensity values above a certain threshold [37,38,42]. Clearly such threshold-based detection and localization will be successful only in cases of very limited photobleaching unless some form of time-adaptive thresholding is used. More robustness can be expected from using the intensity profile of an imaged particle in one frame in a template matching process to detect the same particle in subsequent frames [113,114]. This approach can be taken one step further by using a fixed template representing the theoretical profile of a particle. In the case of diffraction-limited particles, this profile is simply the PSF of the microscope, which in practice is often approximated by the Gaussian function [36,45,115,116]. Extensions of this approach involving Gaussian mixture model fitting have been used for detecting multiple closely spaced particles simultaneously [46,51]. For larger particles with varying shapes and sizes, detection schemes using wavelet-based multiscale products have been used successfully [55,56,117]. More recently, deep learning methods have been adopted for particle detection with promising results [60,118].

Multiple studies have been carried out to quantitatively evaluate common algorithms for particle detection and localization. One of the earliest [119] assessed performance as a function of SNR and object diameter in terms of both accuracy (determinate errors or bias) and precision (indeterminate errors). The algorithms included two threshold-based centroid detection schemes, Gaussian fitting, and template matching using normalized cross correlation or the sum of absolute differences as similarity measures. It was concluded that for particles with diameter less than the illumination wavelength, Gaussian fitting is the best approach by several criteria. For particles having much larger diameter, cross correlation-based template matching appears to be the best choice. It was also concluded that the SNR constitutes the limiting factor of algorithm performance. As a rule of thumb, the SNR should be at least 5 in order to achieve satisfactory results using these algorithms. A subsequent evaluation study [120] even mentioned SNR values of 10 and higher.

Later studies [121–124] considered more sophisticated methods, including wavelet-based multiscale products, top-hat filtering, h-dome filtering, Laplacian-of-Gaussian fitting, Hessian-based methods, various supervised machine-learning methods, and many others. Especially h-dome based detectors performed favorably in all studies, although other methods may be superior for specific applications. Also, the differences in performance between methods vanish as the SNR increases to above 5. Since such levels are quite optimistic in practice, especially in time-lapse imaging experiments, the quest for more robust detection schemes is likely to continue for some time to come.

14.4.2.2 Particle Association

Similar to cell association, the simplest approach to particle association is to use a nearest-neighbor criterion, based on spatial distance [42,43,45,49]. While this may work well in specimens containing very limited numbers of well-spaced particles, it will fail to yield unambiguous results in cases of higher particle densities. In order to establish the identity of particles from frame to frame in such cases, additional cues are needed. For example, when tracking subresolution particles, the identification may be improved by taking into account intensity and spatiotemporal features (such as velocity and acceleration) estimated in previous frames [36,48,54]. Larger particles may also be distinguished by using spatial features such as size, shape, or orientation [37]. Ultimately, matching a large number of spatial features is similar to performing template matching [40,51].

Rather than finding the optimal match for each particle on a frame-by-frame basis, the temporal association problem can also be solved in a more global manner. This is especially favorable in more complex situations of incomplete or ambiguous data. For example, particles may temporarily disappear, either because they move out of focus for some time, or (as in the case of quantum dots) the fluorescence of the probe is intermittent. For single or well-spaced particles, this problem has been solved by translating the tracking task into a spatiotemporal segmentation task and finding optimal paths through the entire data [50,53].

The problem becomes more complicated in the case of high particle densities and the possibility of particle interaction. For example, two or more subresolution particles may approach each other so closely at some point in time that they appear as a single spot that cannot be resolved by any detector. Then they may separate at some later time to form multiple spots again. Keeping track of all particles in such cases requires some form of simultaneous association and optimization. Several authors have proposed to solve the problem using graph-theoretic approaches, in which the detected particles and all possible correspondences and their likelihoods from frame to frame together constitute a weighted graph [46,47,54,114]. The subgraph representing the best overall solution is obtained by applying a global optimization algorithm. More sophistication can be achieved by multistage approaches, which first link detected particles to form track segments that are then further linked into complete trajectories [56,125,126]. Generally, methods that use information from multiple frames in solving the data association problem from frame to frame perform better than methods that consider less temporal context [58,127,128].

Most published particle tracking algorithms are deterministic, in that they make hard decisions about the presence or absence of particles in each image frame and the correspondence of particles between frames. An alternative is to use probabilistic approaches to reflect the uncertainty in the image data and any measurements from the data [55,57,59,129]. Typically, these approaches consist of a Bayesian filtering framework, and involve models of object dynamics, to be matched to the data. It has been argued that incorporating assumptions about the kinematics of object motion is risky in biological tracking since little is known about the laws governing that motion, and the purpose of tracking is to deduce this [14]. However, biological investigation is an iterative endeavor, leading to more and more refined models of cellular and molecular structure and function, so it makes sense at each iteration to take advantage of knowledge previously acquired. The most recent trend is to avoid having to explicitly define such models and to let deep-learning methods extract them from the data [130].

14.5 Trajectory Analysis

The final stage in any time-lapse microscopy imaging experiment is the analysis of the obtained cell or particle trajectories, to confirm or reject predefined hypotheses about dynamic processes or to discover new phenomena. Qualitative analysis by visual inspection of computed trajectories may already give hints about trends in the data, but usually does not provide much more information than can be obtained by directly looking at the image data itself, or projections thereof in the form of kymographs [131,132]. Quantitative analyses of the trajectories are required in order to achieve higher sensitivity in data interpretation and to be able to perform statistical tests. Of course, which parameters to measure and analyze depends very much on the underlying research questions of the experiment. Here we briefly discuss examples of measurements frequently reported in the literature.

14.5.1 Geometry Measurements

The most straightforward measurements concern the geometry of the trajectories and the objects themselves. An example is the maximum distance of the object over time relative to its own initial position[16,22,24]. Other examples are the length of the trajectory (the total distance traveled by the object) and the distance between the start and the end point (the net distance traveled) [133]. The ratio between the latter and the former is known as the confinement ratio, or the meandering index, or the straightness index [134]. A related measure is the so-called McCutcheon index [93], which is often used in chemotaxis studies to quantify the efficiency of cell movement. It is defined as the ratio between the net distance traveled in the direction of increasing chemoattractant concentration and the total distance traveled. Derived parameters, such as the directional change per time interval and its autocorrelation [89], are indicative of the directional persistence and memory of a translocating cell. Information about the cell contour or surface at each time point allows the computation of a variety of shape features, such as diameter, perimeter or surface area, area or volume, circularity or sphericity, convexity or concavity [89], elongation or dispersion [19], and how they change over time.

14.5.2 Diffusivity Measurements

A frequently studied parameter, especially in particle tracking experiments, is the mean square displacement (MSD) [36,40,42,43,49,51,53,135]. It is a convenient measure to study the diffusion characteristics of individual particles [136–141] and it allows assessment of the viscoelastic properties of the media in which they move [138,142,143]. By definition, the MSD is a function of time lag, and the shape of the MSD-time curve for a given trajectory is indicative of the mode of motion of the corresponding particle (Fig. 14.4). For example, in the case of normal diffusion by pure thermally driven Brownian motion, the MSD will increase linearly as a function of time, where the diffusion constant determines the slope of the line. In the case of flow or active transport, on the other hand, the MSD will increase more rapidly, and in a nonlinear (quadratic) fashion. The contrary case of anomalous subdiffusion, which is characterized by a lagging MSD-time curve compared to normal diffusion, occurs if the motion is impeded by obstacles. Confined motion, caused by corrals or tethering or other restrictions, manifests itself by a converging curve, where the limiting MSD value is proportional to the size of

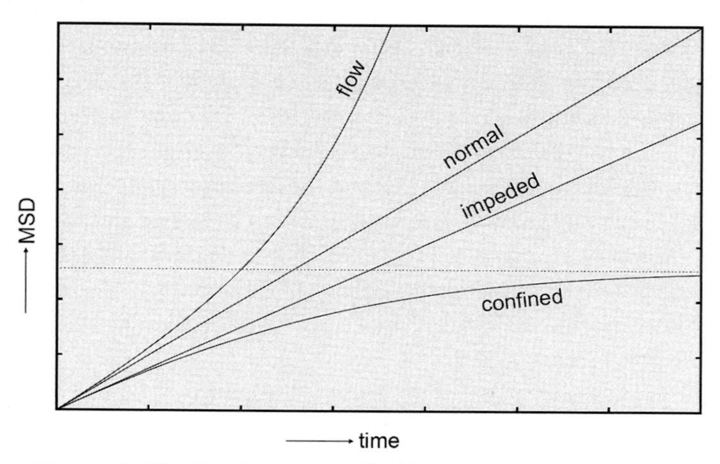

Fig. 14.4 Different types of diffusivity characterized by the MSD as a function of time lag. The idealized curves apply to the case of noise-free measurements and consistent object motion. In the case of localization errors and inconsistent motion, the curves will show offsets and irregularities.

the region accessible for diffusion. Mathematically, the MSD is the second-order moment of displacement. A more complete characterization of a diffusion process may be obtained by computing higher-order moments of displacement [54,144].

Some prudence is called for in diffusivity measurements. In isotropic media, where the displacements in each of the three spatial dimensions may be assumed to be uncorrelated, the 2D diffusion coefficient is equal to the 3D diffusion coefficient [138]. In practical situations, however, it may be unknown in advance whether the medium is isotropic. In this context we stress again the importance of experimental verification of one's assumptions [51,62,90]. Furthermore, the diffusivity of a particle may depend on its diameter relative to the microstructure of the biological fluid in which it moves. Here, a distinction must be made between microscopic, mesoscopic, and macroscopic diffusion [138]. Also, in the case of normal diffusion, the relation between the slope of the MSD-time line and the diffusion constant strictly holds only for infinite trajectories [136]. The shorter the trajectories, the larger the statistical fluctuations in the diffusivity measurements, and the higher the relevance of studying distributions of diffusion constants rather than single values. But even for very long trajectories, apparent subdiffusion patterns may arise at short time scales, caused by uncertainty in particle localization in noisy images [135]. Finally, care must be taken in computing the MSD over an entire trajectory, as it may obscure transitions between diffusive and nondiffusive parts [137].

14.5.3 Velocity Measurements

Another commonly studied parameter in time-lapse imaging experiments is velocity [16,18,93,133,134]. It is computed simply as distance over time. Instantaneous object velocity can be estimated as the distance traveled from one frame to the next divided by the time interval. Average velocity, also referred to as curvilinear velocity, is then computed as the sum of the frame-to-frame distances traveled, divided by the total time elapsed. If the temporal sampling rate is constant, this is the same as averaging the instantaneous velocities. The so-called straight-line velocity, another type of average velocity, is computed as the distance between the first and the last trajectory position divided by the total time elapsed between the two. The ratio between the latter and the former, known as the linearity of forward progression [11,12,92,133], is reminiscent of the McCutcheon index mentioned earlier. Histograms of velocity [39,47,66,93,136] are often helpful in gaining insight into motion statistics. Object acceleration can also be estimated from velocity [89], but it is rarely studied.

Several cautions are in order regarding velocity estimation. In the case of cell tracking, motion analysis is tricky, due to the possibility of morphological changes over time. Often, to circumvent the problem, a center position is tracked [11,12,18,92]. However, with highly plastic cells, centroid-based velocity measurements can be very deceptive [89]. For example, an anchored cell may extend and retract pseudopods, thereby continuously changing its centroid position and generating significant centroid velocity, while the cell is not actually translocating. In the contrary case, a cell may spread in all directions at high velocity in response to some stimulant, while the cell centroid position remains essentially unchanged. Another caution concerns the accuracy of velocity estimation in relation to the temporal sampling rate [89]. Theoretically, the higher this rate, the more detailed the movements captured, and the closer the velocity estimates will be to the true values. In practice, however, the situation is more complicated, as uncertainties in the object localization step cause errors in velocity estimations, which increase rather than decrease when increasing the frame rate [145].

14.6 Sample Algorithms

In the previous section we have discussed many methodologies for cell and particle tracking. To illustrate the intricacies of the tracking problem and some of the solutions that have been proposed, we now describe two specific algorithms in more detail, one for cell tracking and one for particle tracking applications. Both are based on the use of models, and they represent some of the more involved approaches, variants of which have performed favorably in international open challenges on cell tracking [7] and particle tracking [8].

14.6.1 Cell Tracking

One of the more extensively studied classical model-based approaches to image segmentation in many areas is the use of level-set methods [101]. These methods have also been explored for cell segmentation and tracking [23,25,31,32,146,147], with promising results. Other model-based approaches involve cumbersome explicit representations of objects by parametric contours or surfaces controlled by point markers [100,148]. By contrast, level-set methods conveniently define object boundaries in an implicit way as the zero-level set of a scalar function, denoted by $\varphi(.)$ here. This level-set function is defined such that $\varphi(\mathbf{x}) > 0$ when $\mathbf{x}$ lies inside the object, $\varphi(\mathbf{x}) < 0$ when $\mathbf{x}$ is outside the object, and $\varphi(\mathbf{x}) = 0$ at the object boundary, where $\mathbf{x}$ denotes spatial position within the image domain. The important advantages of this representation over explicit representations are its ability to handle objects of any topology and data of any dimensionality without needing conceptual modifications.

The idea of level-set based image segmentation is to evolve the level-set function $\varphi(.)$ iteratively so as to minimize a predefined energy functional. In principle, it is possible to define $\varphi(.)$ in such a way that its zero-level includes the boundaries of all objects of interest in the image, and to evolve these boundaries concurrently by evolving this single function. However, in order to have better control over the interaction between object boundaries when segmenting multiple cells, and to conveniently keep track of individual cells, it is advisable to define a separate level-set function, $\varphi_i(.)$, for each object, $i=1 \ldots N$. Using this approach, we can define the energy functional as.

$$E(\varphi_1, \ldots, \varphi_N) = \int_\Omega \sum_{i=1}^{N} [\alpha\delta(\varphi_i(\mathbf{x}))|\nabla\varphi_i(\mathbf{x})| + H(\varphi_i(\mathbf{x}))e_i(\mathbf{x}) + e_0(\mathbf{x})\frac{1}{N}\prod_{j=1}^{N}\left(1 - H\left(\varphi_j(\mathbf{x})\right)\right) + \gamma\sum_{i<j} H(\varphi_i(\mathbf{x}))H\left(\varphi_j(\mathbf{x})\right)\Bigg] d\mathbf{x} \quad [(14.1)]$$

where $\delta(.)$ is the Dirac delta function, $H(.)$ denotes the Heaviside step function, α and γ are positive parameters, the integral is over the entire image domain, denoted by $\boldsymbol{\Omega}$, and $e_i(.)$ are object energy functions, with $e_0(.)$ denoting the background energy function. The model-based aspect of the level-set approach lies primarily in the latter functions.

The core of this equation consists of four terms, each with an intuitive meaning. The first, with weight α, accounts for the size of the object boundary (contour length in 2D and surface area in 3D) and its edge strength (gradient magnitude). The second term adds energy values for positions inside the boundary, the third is the total background energy, and the fourth, with weight γ, is a penalty term for overlapping boundaries. The formula for iterative evolution of the level-set functions corresponding to the N objects follows from the Euler-Lagrange equations associated with the minimization of the functional in Eq. (14.1) as

$$\partial\varphi_i(\mathbf{x}) = \delta(\varphi_i(\mathbf{x}))\left[\alpha\nabla \cdot \frac{\nabla\varphi_i(\mathbf{x})}{|\nabla\varphi_i(\mathbf{x})|} - e_i(\mathbf{x}) + e_0(\mathbf{x})\prod_{j\neq i}\left(1 - H\left(\varphi_j(\mathbf{x})\right)\right) - \gamma\sum_{j\neq i} H\left(\varphi_j(\mathbf{x})\right)\right]\partial\tau \quad [(14.2)]$$

where $\partial\tau$ denotes the step size in artificial (evolution) time, that is, for segmentation carried out within a single image frame, not to be confused with the real time interval between image frames. Once the energy functional is minimized, and thus a segmentation (boundary) has been obtained for a given image frame, the resulting level-set functions can be used to compute any morphological feature of interest and can also serve as initialization for the minimization procedure for the next image frame.

In summary, the main steps of a level-set based tracking algorithm, and the associated points of attention concerning its application to multiple cell tracking in time-lapse microscopy, are:

1. Define the object and background energy functions, $e_i(.)$ and $e_0(.)$, respectively. These functions mathematically describe the deviation of object and background features from their desired values. This allows one to incorporate prior knowledge about cell and background appearance. In practice, it often suffices to model appearance in terms of simple image statistics, such as the deviation from the mean intensity within the cell or background, and intensity variance.
2. Specify the parameters α and γ. These determine the influence of the boundary magnitude and overlap penalty terms, respectively, relative to the object and background energy terms in the total energy functional (Eq. 14.1) and are necessarily application dependent. Optimal values for these parameters will have to be obtained by experimentation.
3. Segment the first image of the sequence. This is done by defining a single level-set function $\varphi(.)$ and evolving it according to the single-object version of Eq. (14.2) until it converges. Since proper initialization is crucial to achieving fast convergence and arriving at the global optimum, the initial level-set function chosen must be as close as possible to the true boundaries. For example, one could apply a simple segmentation scheme and initialize $\varphi(.)$ based on that outcome.
4. Initialize the level-set functions in the first image of the sequence. Cell objects are obtained by finding connected components in the segmentation resulting from step 3. For each detected object O_i, a level-set function $\varphi_i(.)$ is computed from the signed distance function applied to the boundaries of O_i, with positive values inside and negative values outside O_i.
5. Evolve the level-set functions $\varphi_i(.)$ concurrently according to Eq. (14.2) until convergence. The time step $\partial\tau > 0$ in the discretized version of the evolution equation should be chosen with care. Values too small may cause unnecessarily slow convergence. Conversely, values too large may cause object boundaries to be missed. In practice, values between 0.01 and 0.1 give satisfactory results. To speed up the computations, one could choose to update the level-set functions only for positions $\mathbf{x}$ in a narrow band around the current zero-level sets, for which $\varphi_i(\mathbf{x}) = 0$.
6. Detect incoming and dividing cells. An additional level-set function could be used to detect cells that enter the field of view from the boundaries of the image. Cell division could be detected by monitoring cell shape over time. Drastic morphological changes are indicative of approaching division. If, just after such an event, a level-set function contains two disconnected components, one could decide to replace the function with two new level-set functions.
7. Initialize the level-set functions for the next frame of the sequence. This can be done simply by taking the functions from the previous frame. Notice that in order for this

approach to work, in practice, cells should not move more than their diameter from one frame to the next. To prevent the level-set functions from becoming too flat, it may be advantageous to reinitialize them to the signed distance to their zero-level after a fixed number of iterations.

8. Repeat steps 5–7 until all frames of the image sequence have been processed. The resulting level-set functions $\varphi_i(.)$ as a function of real time enable estimation of the position and morphology of the corresponding cells for each frame in the sequence.

Sample results with specific implementations of this cell tracking algorithm [25,31] applied to the tracking of the nuclei of proliferating HeLa and Madin-Darby canine kidney (MDCK) cells are shown in Fig. 14.5. The examples illustrate the ability of the algorithm to yield plausible contours even in the presence of considerable object noise and strongly varying intensities, as caused, for example, by photobleaching in the case of FRAP experiments (Fig. 14.5a–c). The renderings also demonstrate the ability of the algorithm to keep track of cell division (Fig. 14.5d).

14.6.2 Particle Tracking

An interesting approach to particle tracking is to cast the temporal association problem into a Bayesian estimation problem [55,57,129]. In general, Bayesian tracking deals with the problem of inferring knowledge about the true state of a dynamic system based on a sequence of noisy measurements, or observations. The state vector, denoted by $\mathbf{x}_t$, contains all relevant information about the system at any time t, such as position, velocity, acceleration, intensity, and shape features. Bayesian filtering consists of recursive estimation of the time-evolving posterior probability distribution $p(\mathbf{x}_t|\mathbf{z}_{1:t})$ of the state $\mathbf{x}_t$, given all measurements up to time t, denoted as $\mathbf{z}_{1:t}$. Starting with an initial prior distribution, $p(\mathbf{x}_0|\mathbf{z}_0)$, with $\mathbf{z}_0=\mathbf{z}_{1:0}$ being the set of no measurements, the filtering first predicts the distribution at the next time step:

$$p(\mathbf{x}_t|\mathbf{z}_{1:t-1}) = \int D(\mathbf{x}_t|\mathbf{x}_{t-1})p(\mathbf{x}_{t-1}|\mathbf{z}_{1:t-1})d\,\mathbf{x}_{t-1} \qquad [(14.3)]$$

based on a Markovian model, $D(\mathbf{x}_t|\mathbf{x}_{t-1})$, for the evolution of the state from time $t-1$ to time t. Next, it updates the posterior distribution of the current state by applying Bayes' rule:

$$p(\mathbf{x}_t|\mathbf{z}_{1:t}) \propto L(\mathbf{z}_t|\mathbf{x}_t)p(\mathbf{x}_{t-1}|\mathbf{z}_{1:t-1}) \qquad [(14.4)]$$

using a likelihood, $L(\mathbf{z}_t|\mathbf{x}_t)$, which models the probability of observing $\mathbf{z}_t$ given state $\mathbf{x}_t$. The power of this approach lies not only in the use of dynamics and observation models, but also in the fact that at any time t, all available information up to that time is exploited.

These preceding recurrence relations are analytically tractable only in a restricted set of cases, such as when dealing with linear dynamic systems and Gaussian noise, for which

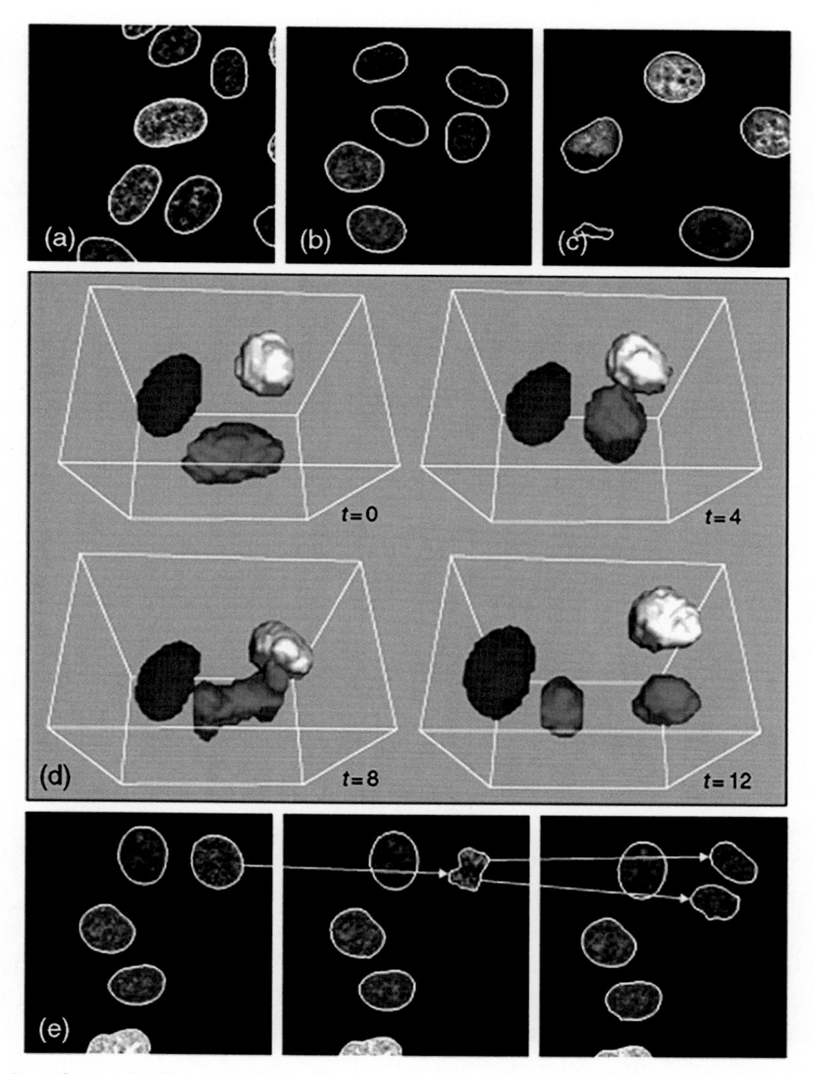

Fig. 14.5 Sample results from applying the described level-set based cell tracking algorithm. (a–c): Segmentations (*white* contours) for arbitrary frames taken from three different 2D+t image sequences. (d) Visualization of segmented surfaces of cell nuclei in four frames of a 3D+t image sequence. (e) Illustration of tracking cell division in a 2D+t image sequence. *(Images a, b, c, e courtesy of G. van Cappellen)*

an optimal solution is provided by the so-called *Kalman filter* [149]. In most biological imaging applications, where the dynamics and noise can be expected to be nonlinear and non-Gaussian, efficient numerical approximations are provided by sequential Monte Carlo (SMC) methods, such as the *condensation algorithm* [150]. In that case, the posterior distribution is represented by N_s random state samples and associated weights [151]:

$$p(\mathbf{x}_t|\mathbf{z}_{1:t}) \approx \sum_{i=1}^{N_s} w_t^{(i)} \delta\left(\mathbf{x}_t - \mathbf{x}_t^{(i)}\right) \quad [(14.5)]$$

where $\delta(.)$ is the Dirac delta function, and the weights $w_t^{(i)}$ of the state samples $\mathbf{x}_t^{(i)}$ sum to 1. The problem of tracking large numbers of objects using this framework is conveniently solved by representing the filtering distribution by an M-component mixture model:

$$p(\mathbf{x}_t|\mathbf{z}_{1:t}) = \sum_{m=1}^{M} \omega_{m,t} p_m(\mathbf{x}_t|\mathbf{z}_{1:t}) \quad [(14.6)]$$

where the weights $\omega_{m,t}$ of the components p_m sum to 1, the total number of samples is $N = M \cdot N_s$, and each sample is augmented with a component label denoted by $c_t^{(i)}$, with $c_t^{(i)} = m$ if state sample i belongs to mixture component m. This representation is updated in the same fashion as the single-object case. At each time step, statistical inferences about the state, such as the expected value, or the maximum a posteriori (MAP) or minimum mean square error (MMSE) values, can easily be approximated from the weighted state samples for each object.

In summary, the main steps of an SMC-based tracking algorithm, and the associated points of attention concerning its application to multiple particle tracking in time-lapse microscopy, are:

1. Define the state vector $\mathbf{x}_t$. In most experimental situations it will be sufficient to include position (x_t, y_t, z_t), velocity ($v_{x,t}$, $v_{y,t}$, $v_{z,t}$), acceleration ($a_{x,t}$, $a_{y,t}$, $a_{z,t}$), and intensity (I_t).
2. Define the state evolution model $D(\mathbf{x}_t|\mathbf{x}_{t-1})$. This model mathematically describes the probability of a particle to jump from the previous state $\mathbf{x}_{t-1}$ to state $\mathbf{x}_t$. It allows one to incorporate prior knowledge about the dynamics of the objects to be tracked and is therefore necessarily application dependent. An example of such a model is the Gaussian weighted deviation of $\mathbf{x}_t$ from the expected state at time t based on $\mathbf{x}_{t-1}$ (for instance, expected position at time t based on velocity at time $t-1$, and expected velocity at t based on acceleration at $t-1$). This framework also permits the use of multiple interacting models to deal with different types of motion concurrently (such as random walk, or directed movement, with constant or changing velocity). Notice that having I_t as part of the state vector allows one to also model the evolution of particle intensity (photobleaching).
3. Define the observation model $L(\mathbf{z}_t|\mathbf{x}_t)$. This model mathematically describes the likelihood or probability of measuring state $\mathbf{z}_t$ from the data given the true state $\mathbf{x}_t$. It allows one to incorporate knowledge about the imaging system, in particular the PSF, as well as additional static object information, such as morphology. Such a model could, for example, be defined as the Gaussian weighted deviation of the total measured intensity in a neighborhood around $\mathbf{x}_t$ from the expected total intensity, based

on a shape model of the imaged objects, and from the background intensity, taking into account the noise levels in the object and the background.

4. Specify the prior state distribution $p(\mathbf{x}_0|\mathbf{z}_0)$. This can be based on information available in the first frame. For example, one could apply a suitable detection scheme to localize the most prominent particles, and for each detected particle add a Gaussian shaped mixture component p_m with weight $\omega_{m,0}$ proportional to size or intensity. Each component is then sampled N_s times to obtain state samples $\mathbf{x}_t^{(i)}$ and associated weights $w_t^{(i)}$ for that component. The detection step does not need to be very accurate because false positives will be filtered out rapidly as the system evolves.
5. Use the posterior state distribution $p(\mathbf{x}_{t-1}|\mathbf{z}_{1:t-1})$ at time $t-1$ to compute the predicted state distribution $p(\mathbf{x}_t|\mathbf{z}_{1:t-1})$ at time t according to Eq. (14.3), and subsequently use this prediction to compute the updated posterior state distribution $p(\mathbf{x}_t|\mathbf{z}_{1:t})$ according to Eq. (14.4). Effectively this is done by inserting each sample $\mathbf{x}_{t-1}^{(i)}$, $i=1, \ldots, N$, into the state evolution model, taking a new sample from the resulting probability density function to obtain $\mathbf{x}_t^{(i)}$, and computing the associated weight $w_t^{(i)}$ from $w_{t-1}^{(i)}$ based on the likelihood and dynamics models and an auxiliary importance function. A penalty function can be used to regulate particle coincidence.
6. Update the sample component labels $c_t^{(i)}$. This step is necessary if particle merging and splitting events are to be captured. It involves the application of a procedure to recluster the state samples from the M mixture components to yield M' new components. This procedure can be implemented in any convenient way, and it allows one to incorporate prior knowledge about merging and splitting events. In the simplest case one could apply a k-means clustering algorithm.
7. Update the mixture component set and the corresponding weights $\omega_{m,t}$. To determine whether particles have appeared or disappeared at any time t, one could apply some detection scheme as in step 4 to obtain a particle probability map, and compare this map to the current particle distribution, as follows from the posterior $p(\mathbf{x}_t|\mathbf{z}_{1:t})$. For each appearing particle, a new mixture component is added with predefined initial weight ω_b. Components with weights below some threshold ω_d are assumed to correspond to disappearing particles and are removed from the mixture. The weights $\omega_{m,t}$ are computed from the $\omega_{m,t-1}$ and the weights $w_t^{(i)}$ of the state samples.
8. Repeat steps 5–7 until all frames of the image sequence have been processed. The resulting posterior state distributions $p_m(\mathbf{x}_t|\mathbf{z}_{1:t})$ enable estimation of the states of the corresponding particles at any t.

Sample results with a specific implementation of this algorithm [57] applied to the tracking of microtubule plus-ends in the cytoplasm and androgen receptor proteins in the cell nucleus are shown in Fig. 14.6. This example illustrates the ability of the algorithm to deal with photobleaching, to capture newly appearing particles, to detect particle disappearance, and to handle closely passing particles. Alternative but related Bayesian filtering algorithms have also been applied to the tracking of cytoplasmic and nuclear HIV-1 complexes [152], viruses [129], neurofilaments in axons [153], and even cell nuclei [154].

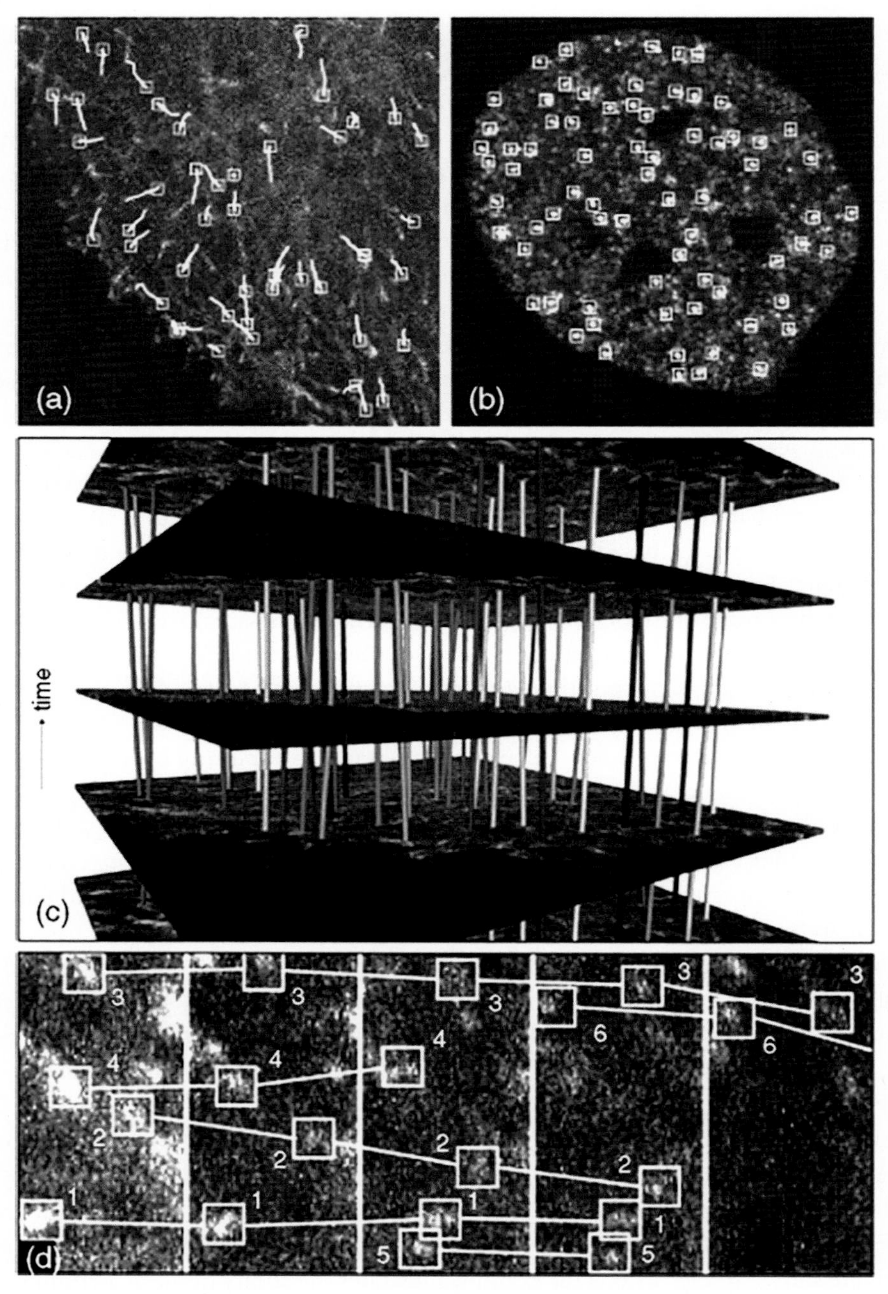

Fig. 14.6 Sample results from applying the SMC-based particle tracking algorithm. (a) and (b): Estimated current locations *(white squares)* and trajectories *(white curves)* up to the current time, for the most prominent particles in an arbitrary frame from the 2D+t image sequences shown in Figs. 14.1a and b, respectively. The trajectories in (b) are due to Brownian motion and are therefore strongly confined. (c): Artistic rendering of the trajectories linking multiple frames of dataset (a). (d): Illustration of the capability of the algorithm to deal with photobleaching, to capture newly appearing particles, to detect particle disappearance, and to handle closely passing particles.

In this chapter we have sketched the current state of the art in time-lapse microscopy imaging, in particular cell and particle tracking, from the perspective of automated image processing and subsequent motion analysis. While it is clear that biological research is increasingly relying on computerized methods, it also follows that such methods are still far from perfect, and there is much room for improvement. Although there is growing consensus on the strengths and weaknesses of specific approaches, there is currently no single best solution to the problem in general [144,145]. The main conclusion emerging from the literature is that, with currently available image processing and analysis methodologies, any given tracking application requires its own dedicated algorithm to achieve the best results.

14.7 Summary of Important Points

1. Living cells are sensitive to photodamage and require economizing their light exposure.
2. Success rates of automated image analysis generally increase with increasing SNR.
3. Time-lapse microscopy requires trading off SNR against spatial and temporal resolution.
4. Biological processes occur in 3D + t and should preferably be studied as such.
5. Studies in 2D + t should be accompanied by 3D + t experiments to confirm their validity.
6. Sampling theory applies not only to sampling in space but also to sampling in time.
7. Nonlinear filtering methods allow reducing noise while preserving strong gradients (edges).
8. Deconvolution of time-lapse microscopy imaging data is not always necessary.
9. Object motion is often a superposition of global and local displacements.
10. Global motion can be corrected by image registration.
11. Object tracking requires dedicated detection, segmentation, and association methods.
12. Optic flow methods can be used to compute collective cell motion and particle flows.
13. Image intensity thresholding alone is generally not adequate for segmentation.
14. Template matching works best for objects that do not change shape significantly as they move.
15. Watershed-based segmentation requires careful pre- and postprocessing strategies.
16. Deformable models allow easy use of both image information and prior knowledge.
17. By design, deformable models are very suitable for capturing morphodynamics.
18. Implicitly defined models are generally more flexible than explicitly defined models.

19. In fluorescence microscopy, objects are observed only indirectly, via fluorescent probes.
20. Fluorescent (nano)particles act as point light sources and appear as PSF-shaped spots.
21. Particle localization accuracy and resolvability depend strongly on photon count.
22. For single particles, a localization accuracy of about 10 nm is achievable in practice.
23. Two particles are resolvable with reasonable accuracy for separation distances greater than 50 nm.
24. Gaussian fitting is most suitable for detecting particles with diameter greater than one wavelength.
25. Template matching is most suitable for detecting particles with a diameter of several wavelengths.
26. For most particle detection methods the SNR must be greater than 5 to achieve satisfactory results.
27. Reliable temporal association usually requires the use of more criteria than just spatial distance.
28. Tracking of many closely spaced particles requires some form of joint association.
29. Probabilistic tracking methods reflect the uncertainty in the image data.
30. Models of object dynamics are helpful in tracking but should be used with care.
31. Displacement moments provide detailed information about diffusion characteristics.
32. The shape of the MSD-time curve of an object is indicative of its mode of motion.
33. In isotropic media the 2D diffusion coefficient is equal to the 3D diffusion coefficient.
34. Distinction must be made between microscopic, mesoscopic, and macroscopic diffusion.
35. Short trajectories show large statistical fluctuations in diffusivity measurements.
36. Subdiffusive patterns at short time scales may be due to noise in particle localization.
37. Computing the MSD over entire trajectories may obscure transitions in diffusivity.
38. Different velocity measures result from considering different distance measures.
39. Centroid-based velocity measurements can be very deceptive in cell tracking.
40. Velocity estimation errors depend on temporal resolution and localization uncertainty.

References

[1] E. Meijering, et al., Tracking in cell and developmental biology, Semin. Cell Dev. Biol. 20 (8) (2009) 894–902.
[2] C. Vonesch, et al., The colored revolution of bioimaging, IEEE Signal Process. Mag. 23 (3) (2006) 20–31.
[3] S.J. Sahl, S.W. Hell, S. Jakobs, Fluorescence nanoscopy in cell biology, Nat. Rev. Mol. Cell Biol. 18 (11) (2017) 685–701.
[4] F. Strobl, A. Schmitz, E.H.K. Stelzer, Improving your four-dimensional image: traveling through a decade of light-sheet-based fluorescence microscopy research, Nat. Protoc. 12 (6) (2017) 1103–1109.

[5] J.B. Pawley (Ed.), Handbook of Biological Confocal Microscopy, third ed., Springer, New York, 2006.
[6] E. Meijering, et al., Imagining the future of bioimage analysis, Nat. Biotechnol. 34 (12) (2016) 1250–1255.
[7] V. Ulman, et al., An objective comparison of cell-tracking algorithms, Nat. Methods 14 (2) (2017) 1141–1152.
[8] N. Chenouard, et al., Objective comparison of particle tracking methods, Nat. Methods 11 (3) (2014) 281–289.
[9] D. Gerlich, J. Ellenberg, 4D imaging to assay complex dynamics in live specimens, Nat. Cell Biol. 5 (2003) S14–S19.
[10] D.J. Stephens, V.J. Allan, Light microscopy techniques for live cell imaging, Science 300 (5616) (2003) 82–86.
[11] D.F. Katz, et al., Real-time analysis of sperm motion using automatic video image digitization, Comput. Methods Prog. Biomed. 21 (3) (1985) 173–182.
[12] S.O. Mack, D.P. Wolf, J.S. Tash, Quantitation of specific parameters of motility in large numbers of human sperm by digital image processing, Biol. Reprod. 38 (2) (1988) 270–281.
[13] F. Leymarie, M.D. Levine, Tracking deformable objects in the plane using an active contour model, IEEE Trans. Pattern Anal. Mach. Intell. 15 (6) (1993) 617–634.
[14] V. Awasthi, et al., Cell tracking using a distributed algorithm for 3-D image segmentation, Bioimaging 2 (2) (1994) 98–112.
[15] Y. Sato, et al., Automatic extraction and measurement of leukocyte motion in microvessels using spatiotemporal image analysis, IEEE Trans. Biomed. Eng. 44 (4) (1997) 225–236.
[16] C. De Hauwer, et al., In vitro motility evaluation of aggregated cancer cells by means of automatic image processing, Cytometry 36 (1) (1999) 1–10.
[17] S.T. Acton, K. Wethmar, K. Ley, Automatic tracking of rolling leukocytes in vivo, Microvasc. Res. 63 (1) (2002) 139–148.
[18] Z.N. Demou, L.V. McIntire, Fully automated three-dimensional tracking of cancer cells in collagen gels: determination of motility phenotypes at the cellular level, Cancer Res. 62 (18) (2002) 5301–5307.
[19] D. Dormann, et al., Simultaneous quantification of cell motility and protein-membrane-association using active contours, Cell Motil. Cytoskeleton 52 (4) (2002) 221–230.
[20] C. Zimmer, et al., Segmentation and tracking of migrating cells in video microscopy with parametric active contours: a tool for cell-based drug testing, IEEE Trans. Med. Imaging 21 (10) (2002) 1212–1221.
[21] D.J.E.B. Krooshoop, et al., An automated multi well cell track system to study leukocyte migration, J. Immunol. Methods 280 (1–2) (2003) 89–102.
[22] O. Debeir, et al., A model-based approach for automated in vitro cell tracking and chemotaxis analyses, Cytometry A 60 (1) (2004) 29–40.
[23] D.P. Mukherjee, N. Ray, S.T. Acton, Level set analysis for leukocyte detection and tracking, IEEE Trans. Med. Imaging 13 (4) (2004) 562–572.
[24] O. Debeir, et al., Tracking of migrating cells under phase-contrast video microscopy with combined mean-shift processes, IEEE Trans. Med. Imaging 24 (6) (2005) 697–711.
[25] A. Dufour, et al., Segmenting and tracking fluorescent cells in dynamic 3-D microscopy with coupled active surfaces, IEEE Trans. Image Process. 14 (9) (2005) 1396–1410.
[26] X. Chen, X. Zhou, S.T.C. Wong, Automated segmentation, classification, and tracking of cancer cell nuclei in time-lapse microscopy, IEEE Trans. Biomed. Eng. 53 (4) (2006) 762–766.
[27] N.N. Kachouie, et al., Probabilistic model-based cell tracking, Int. J. Biomed. Imaging 2006 (12186) (2006) 1–10.
[28] H. Shen, et al., Automatic tracking of biological cells and compartments using particle filters and active contours, Chemom. Intell. Lab. Syst. 82 (1–2) (2006) 276–282.
[29] C.A. Wilson, J.A. Theriot, A correlation-based approach to calculate rotation and translation of moving cells, IEEE Trans. Image Process. 15 (7) (2006) 1939–1951.
[30] X. Yang, H. Li, X. Zhou, Nuclei segmentation using marker-controlled watershed, tracking using mean-shift, and Kalman filter in time-lapse microscopy, IEEE Trans. Circuits Syst. Regul. Pap. 53 (11) (2006) 2405–2414.

[31] O. Dzyubachyk, et al., Advanced level-set-based cell tracking in time-lapse fluorescence microscopy, IEEE Trans. Med. Imaging 29 (3) (2010) 852–867.
[32] M. Maška, et al., Segmentation and shape tracking of whole fluorescent cells based on the Chan-Vese model, IEEE Trans. Med. Imaging 32 (6) (2013) 995–1006.
[33] S. Stylianidou, et al., SuperSegger: robust image segmentation, analysis and lineage tracking of bacterial cells, Mol. Microbiol. 102 (4) (2016) 690–700.
[34] M. Wang, et al., Multicell migration tracking within angiogenic networks by deep learning-based segmentation and augmented Bayesian filtering, J. Med. Imaging 5 (2) (2018), 024005.
[35] J.B. Lugagne, H. Lin, M.J. Dunlop, DeLTA: automated cell segmentation, tracking, and lineage reconstruction using deep learning, PLoS Comput. Biol. 16 (4) (2020), e1007673.
[36] C.M. Anderson, et al., Tracking of cell surface receptors by fluoresence digital imaging microscopy using a charge-coupled device camera, J. Cell Sci. 101 (2) (1992) 415–425.
[37] S.S. Work, D.M. Warshaw, Computer-assisted tracking of actin filament motility, Anal. Biochem. 202 (2) (1992) 275–285.
[38] R.N. Ghosh, W.W. Webb, Automated detection and tracking of individual and clustered cell surface low density lipoprotein receptor molecules, Biophys. J. 66 (5) (1994) 1301–1318.
[39] S.B. Marston, et al., A simple method for automatic tracking of actin filaments in the motility assay, J. Muscle Res. Cell Motil. 17 (4) (1996) 497–506.
[40] H. Bornfleth, et al., Quantitative motion analysis of subchromosomal foci in living cells using four-dimensional microscopy, Biophys. J. 77 (5) (1999) 2871–2886.
[41] W. Tvaruskó, et al., Time-resolved analysis and visualization of dynamic processes in living cells, Proc. Natl. Acad. Sci. U. S. A. 96 (14) (1999) 7950–7955.
[42] J. Apgar, et al., Multiple-particle tracking measurements of heterogeneities in solutions of actin filaments and actin bundles, Biophys. J. 79 (2) (2000) 1095–1106.
[43] M. Goulian, S.M. Simon, Tracking single proteins within cells, Biophys. J. 79 (4) (2000) 2188–2198.
[44] D. Gerlich, et al., Four-dimensional imaging and quantitative reconstruction to analyse complex spatiotemporal processes in live cells, Nat. Cell Biol. 3 (9) (2001) 852–855.
[45] T. Kues, R. Peters, U. Kubitscheck, Visualization and tracking of single protein molecules in the cell nucleus, Biophys. J. 80 (6) (2001) 2954–2967.
[46] D. Thomann, et al., Automatic fluorescent tag detection in 3D with super-resolution: application to the analysis of chromosome movement, J. Microsc. 208 (1) (2002) 49–64.
[47] P. Vallotton, et al., Recovery, visualization, and analysis of actin and tubulin polymer flow in live cells: a fluorescent speckle microscopy study, Biophys. J. 85 (2) (2003) 1289–1306.
[48] C.P. Bacher, et al., 4-D single particle tracking of synthetic and proteinaceous microspheres reveals preferential movement of nuclear particles along chromatin-poor tracks, BMC Cell Biol. 5 (45) (2004) 1–14.
[49] D. Li, et al., Three-dimensional tracking of single secretory granules in live PC12 cells, Biophys. J. 87 (3) (2004) 1991–2001.
[50] S. Bonneau, M. Dahan, L.D. Cohen, Single quantum dot tracking based on perceptual grouping using minimal paths in a spatiotemporal volume, IEEE Trans. Image Process. 14 (9) (2005) 1384–1395.
[51] J.F. Dorn, et al., Yeast kinetochore microtubule dynamics analyzed by high-resolution three-dimensional microscopy, Biophys. J. 89 (4) (2005) 2835–2854.
[52] E. Eden, et al., An automated method for analysis of flow characteristics of circulating particles from in vivo video microscopy, IEEE Trans. Med. Imaging 24 (8) (2005) 1011–1024.
[53] D. Sage, et al., Automatic tracking of individual fluorescence particles: application to the study of chromosome dynamics, IEEE Trans. Image Process. 14 (9) (2005) 1372–1383.
[54] I.F. Sbalzarini, P. Koumoutsakos, Feature point tracking and trajectory analysis for video imaging in cell biology, J. Struct. Biol. 151 (2) (2005) 182–195.
[55] A. Genovesio, et al., Multiple particle tracking in 3-D + t microscopy: method and application to the tracking of endocytosed quantum dots, IEEE Trans. Image Process. 15 (5) (2006) 1062–1070.
[56] K. Jaqaman, et al., Robust single-particle tracking in live-cell time-lapse sequences, Nat. Methods 5 (8) (2008) 695–702.

[57] I. Smal, et al., Particle filtering for multiple object tracking in dynamic fluorescence microscopy images: application to microtubule growth analysis, IEEE Trans. Med. Imaging 27 (6) (2008) 789–804.
[58] A. Jaiswal, et al., Tracking virus particles in fluorescence microscopy images using multi-scale detection and multi-frame association, IEEE Trans. Image Process. 24 (11) (2015) 4122–4136.
[59] C. Shi, et al., Micro-object motion tracking based on the probability hypothesis density particle tracker, J. Math. Biol. 72 (5) (2016) 1225–1254.
[60] Y. Zhong, et al., Developing noise-resistant three-dimensional single particle tracking using deep neural networks, Anal. Chem. 90 (18) (2018) 10748–10757.
[61] C. Ortiz-de-Solórzano, et al., Toward a morphodynamic model of the cell: signal processing for cell modeling, IEEE Signal Process. Mag. 32 (1) (2015) 20–29.
[62] C. Zimmer, et al., On the digital trail of mobile cells, IEEE Signal Process. Mag. 23 (3) (2006) 54–62.
[63] L.J. van Vliet, D. Sudar, I.T. Young, Digital fluorescence imaging using cooled CCD array cameras, in: J.E. Celis (Ed.), Cell Biology: A Laboratory Handbook, vol. 3, Academic Press, 1998, pp. 109–120.
[64] C. Wählby, et al., Combining intensity, edge and shape information for 2D and 3D segmentation of cell nuclei in tissue sections, J. Microsc. 215 (1) (2004) 67–76.
[65] D. Gerlich, J. Mattes, R. Eils, Quantitative motion analysis and visualization of cellular structures, Methods 29 (1) (2003) 3–13.
[66] D. Uttenweiler, et al., Spatiotemporal anisotropic diffusion filtering to improve signal-to-noise ratios and object restoration in fluorescence microscopic image sequences, J. Biomed. Opt. 8 (1) (2003) 40–47.
[67] P. Perona, J. Malik, Scale-space and edge detection using anisotropic diffusion, IEEE Trans. Pattern Anal. Mach. Intell. 12 (7) (1990) 629–639.
[68] W. Meiniel, J.C. Olivo-Marin, E.D. Angelini, Denoising of microscopy images: a review of the state-of-the-art, and a new sparsity-based method, IEEE Trans. Image Process. 27 (8) (2018) 3842–3856.
[69] M.B. Cannell, A. McMorland, C. Soeller, Image enhancement by deconvolution, in: J.B. Pawley (Ed.), Handbook of Biological Confocal Microscopy, Springer, 2006, pp. 488–500 (Chapter 25).
[70] W. Wallace, L.H. Schaefer, J.R. Swedlow, A workingperson's guide to deconvolution in light microscopy, BioTechniques 31 (5) (2001) 1076–1097.
[71] P. Sarder, A. Nehorai, Deconvolution methods for 3-D fluorescence microscopy images, IEEE Signal Process. Mag. 23 (3) (2006) 32–45.
[72] T.J. Holmes, D. Biggs, A. Abu-Tarif, Blind deconvolution, in: J.B. Pawley (Ed.), Handbook of Biological Confocal Microscopy, Springer, 2006, pp. 468–487 (Chapter 24).
[73] F. Aguet, D. Van De Ville, M. Unser, A maximum-likelihood formalism for sub-resolution axial localization of fluorescent nanoparticles, Opt. Express 13 (26) (2005) 10503–10522.
[74] S. Ram, E.S. Ward, R.J. Ober, Beyond Rayleigh's criterion: a resolution measure with application to single-molecule microscopy, Proc. Natl. Acad. Sci. U. S. A. 103 (12) (2006) 4457–4462.
[75] A. von Diezmann, Y. Shechtman, M.E. Moerner, Three-dimensional localization of single molecules for super-resolution imaging and single-particle tracking, Chem. Rev. 117 (11) (2017) 7244–7275.
[76] A.P. Goobic, J. Tang, S.T. Acton, Image stabilization and registration for tracking cells in the microvasculature, IEEE Trans. Biomed. Eng. 52 (2) (2005) 287–299.
[77] B. Rieger, et al., Alignment of the cell nucleus from labeled proteins only for 4D in vivo imaging, Microsc. Res. Tech. 64 (2) (2004) 142–150.
[78] I.H. Kim, Y.C.M. Chen, D.L. Spector, R. Eils, K. Rohr, Nonrigid registration of 2-D and 3-D dynamic cell nuclei images for improved classification of subcellular particle motion, IEEE Trans. Image Process. 20 (4) (2011) 1011–1022.
[79] J.V. Hajnal, D.L.G. Hill, D.J. Hawkes, Medical Image Registration, CRC Press, 2001.
[80] A. Sotiras, C. Davatzikos, N. Paragios, Deformable medical image registration: a survey, IEEE Trans. Med. Imaging 32 (7) (2013) 1153–1190.
[81] F.P.M. Oliveira, J.M.R.S. Tavares, Medical image registration: a review, Comput. Methods Biomech. Biomed. Eng. 17 (2) (2014) 73–93.
[82] C.Ó.S. Sorzano, P. Thévenaz, M. Unser, Elastic registration of biological images using vector-spline regularization, IEEE Trans. Biomed. Eng. 52 (4) (2005) 652–663.

[83] G. Rabut, J. Ellenberg, Automatic real-time three-dimensional cell tracking by fluorescence microscopy, J. Microsc. 216 (2) (2004) 131–137.
[84] T. Ragan, et al., 3D particle tracking on a two-photon microscope, J. Fluoresc. 16 (3) (2006) 325–336.
[85] S. Hou, K. Welsher, A protocol for real-time 3D single particle tracking, J. Vis. Exp. 131 (2018), e56711.
[86] F. Germain, et al., Characterization of cell deformation and migration using a parametric estimation of image motion, IEEE Trans. Biomed. Eng. 46 (5) (1999) 584–600.
[87] K. Miura, Tracking movement in cell biology, in: J. Rietdorf (Ed.), Advances in Biochemical Engineering/Biotechnology: Microscopy Techniques, Springer, Berlin, 2005.
[88] J. Delpiano, et al., Performance of optical flow techniques for motion analysis of fluorescent point signals in confocal microscopy, Mach. Vis. Appl. 23 (4) (2012) 675–689.
[89] D.R. Soll, The use of computers in understanding how animal cells crawl, Int. Rev. Cytol. 163 (1995) 43–104.
[90] D.J. Webb, A.F. Horwitz, New dimensions in cell migration, Nat. Cell Biol. 5 (8) (2003) 690–692.
[91] D. Dormann, C.J. Weijer, Imaging of cell migration, EMBO J. 25 (15) (2006) 3480–3493.
[92] S.T. Young, et al., Real-time tracing of spermatozoa, IEEE Eng. Med. Biol. Mag. 15 (6) (1996) 117–120.
[93] R.M. Donovan, et al., A quantitative method for the analysis of cell shape and locomotion, Histochemistry 84 (4–6) (1986) 525–529.
[94] D. Young, et al., Towards automatic cell identification in DIC microscopy, J. Microsc. 192 (2) (1998) 186–193.
[95] C. Chen, et al., A flexible and robust approach for segmenting cell nuclei from 2D microscopy images using supervised learning and template matching, Cytometry A 83 (5) (2013) 495–507.
[96] V. Grau, et al., Improved watershed transform for medical image segmentation using prior information, IEEE Trans. Med. Imaging 23 (4) (2004) 447–458.
[97] G. Lin, et al., A hybrid 3D watershed algorithm incorporating gradient cues and object models for automatic segmentation of nuclei in confocal image stacks, Cytometry A 56 (1) (2003) 23–36.
[98] C.F. Koyuncu, et al., Iterative h-minima-based marker-controlled watershed for cell nucleus segmentation, Cytometry A 89 (4) (2016) 338–349.
[99] N. Ray, S.T. Acton, Motion gradient vector flow: an external force for tracking rolling leukocytes with shape and size constrained active contours, IEEE Trans. Med. Imaging 23 (12) (2004) 1466–1478.
[100] M. Kass, A. Witkin, D. Terzopoulos, Snakes: active contour models, Int. J. Comput. Vis. 1 (4) (1988) 321–331.
[101] J.A. Sethian, Level Set Methods and Fast Marching Methods: Evolving Interfaces in Computational Geometry, Fluid Mechanics, Computer Vision, and Materials Science, Cambridge University Press, 1999.
[102] D.A. Van Valen, et al., Deep learning automates the quantitative analysis of individual cells in live-cell imaging experiments, PLoS Comput. Biol. 12 (11) (2016), e1005177.
[103] Y. Al-Kofahi, et al., A deep learning-based algorithm for 2-D cell segmentation in microscopy images, BMC Bioinf. 19 (1) (2018) 365.
[104] T. Falk, et al., U-net: deep learning for cell counting, detection, and morphometry, Nat. Methods 16 (1) (2019) 67–70.
[105] N. Ray, S.T. Acton, Data acceptance for automated leukocyte tracking through segmentation of spatiotemporal images, IEEE Trans. Biomed. Eng. 52 (10) (2005) 1702–1712.
[106] T. He, et al., Cell tracking using deep neural networks with multi-task learning, Image Vis. Comput. 60 (2017) 142–153.
[107] J. Wang, et al., Deep reinforcement learning for data association in cell tracking, Front. Bioeng. Biotechnol. 8 (2020) 298.
[108] J. Lippincott-Schwartz, G.H. Patterson, Development and use of fluorescent protein markers in living cells, Science 300 (5616) (2003) 87–91.
[109] B.N.G. Giepmans, et al., The fluorescent toolbox for assessing protein location and function, Science 312 (5771) (2006) 217–224.

[110] S.W. Hell, M. Dyba, S. Jakobs, Concepts for nanoscale resolution in fluorescence microscopy, Curr. Opin. Neurobiol. 14 (5) (2004) 599–609.
[111] L. Cognet, C. Leduc, B. Lounis, Advances in live-cell single-particle tracking and dynamic super-resolution imaging, Curr. Opin. Chem. Biol. 20 (2014) 78–85.
[112] S. Cox, Super-resolution imaging in live cells, Dev. Biol. 401 (1) (2015) 175–181.
[113] J. Gelles, B.J. Schnapp, M.P. Sheetz, Tracking kinesin-driven movements with nanometre-scale precision, Nature 331 (6155) (1988) 450–453.
[114] D. Thomann, et al., Automatic fluorescent tag localization II: improvement in super-resolution by relative tracking, J. Microsc. 211 (3) (2003) 230–248.
[115] B. Zhang, J. Zerubia, J.C. Olivo-Marin, Gaussian approximations of fluorescence microscope point-spread function models, Appl. Opt. 46 (10) (2007) 1819–1829.
[116] Y. Li, et al., Real-time 3D single-molecule localization using experimental point spread functions, Nat. Methods 15 (5) (2018) 367–369.
[117] J.C. Olivo-Marin, Extraction of spots in biological images using multiscale products, Pattern Recogn. 35 (9) (2002) 1989–1996.
[118] S. Helgadottir, A. Argun, G. Volpe, Digital video microscopy enhanced by deep learning, Optica 6 (4) (2019) 506–513.
[119] M.K. Cheezum, W.F. Walker, W.H. Guilford, Quantitative comparison of algorithms for tracking single fluorescent particles, Biophys. J. 81 (4) (2001) 2378–2388.
[120] B.C. Carter, G.T. Shubeita, S.P. Gross, Tracking single particles: a user-friendly quantitative evaluation, Phys. Biol. 2 (1) (2005) 60–72.
[121] I. Smal, et al., Quantitative comparison of spot detection methods in fluorescence microscopy, IEEE Trans. Med. Imaging 29 (2) (2010) 282–301.
[122] P. Ruusuvuori, et al., Evaluation of methods for detection of fluorescence labeled subcellular objects in microscope images, BMC Bioinf. 11 (2010) 248.
[123] K. Stěpka, et al., Performance and sensitivity evaluation of 3D spot detection methods in confocal microscopy, Cytometry A 87 (8) (2015) 759–772.
[124] M.A. Mabaso, D.J. Withey, B. Twala, Spot detection methods in fluorescence microscopy imaging: a review, Image Anal. Stereol. 37 (3) (2018) 173–190.
[125] K.T. Applegate, et al., plusTipTracker: quantitative image analysis software for the measurement of microtubule dynamics, J. Struct. Biol. 176 (2) (2011) 168–184.
[126] J.Y. Tinevez, et al., TrackMate: an open and extensible platform for single-particle tracking, Methods 115 (2017) 80–90.
[127] L. Feng, et al., Multiple dense particle tracking in fluorescence microscopy images based on multi-dimensional assignment, J. Struct. Biol. 173 (2) (2011) 219–228.
[128] I. Smal, E. Meijering, Quantitative comparison of multiframe data association techniques for particle tracking in time-lapse fluorescence microscopy, Med. Image Anal. 24 (1) (2015) 163–189.
[129] W.J. Godinez, K. Rohr, Tracking multiple particles in fluorescence time-lapse microscopy images via probabilistic data association, IEEE Trans. Med. Imaging 34 (2) (2015) 415–432.
[130] Y. Yao, et al., Deep learning method for data association in particle tracking, Bioinformatics 36 (19) (2020) 4935–4941.
[131] K. Chiba, et al., Simple and direct assembly of kymographs from movies using KYMOMAKER, Traffic 15 (1) (2014) 1–11.
[132] A.R. Chaphalkar, et al., Automated multi-peak tracking kymography (AMTraK): a tool to quantify sub-cellular dynamics with sub-pixel accuracy, PLoS One 11 (12) (2016), e0167620.
[133] E. Meijering, O. Dzyubachyk, I. Smal, Methods for cell and particle tracking, Methods Enzymol. 504 (9) (2012) 183–200.
[134] J.B. Beltman, A.F.M. Marée, R.J. de Boer, Analysing immune cell migration, Nat. Rev. Immunol. 9 (11) (2009) 789–798.
[135] D.S. Martin, M.B. Forstner, J.A. Käs, Apparent subdiffusion inherent to single particle tracking, Biophys. J. 83 (4) (2002) 2109–2117.
[136] H. Qian, M.P. Sheetz, E.L. Elson, Single particle tracking: analysis of diffusion and flow in two-dimensional systems, Biophys. J. 60 (4) (1991) 910–921.

[137] M.J. Saxton, K. Jacobson, Single-particle tracking: applications to membrane dynamics, Annu. Rev. Biophys. Biomol. Struct. 26 (1997) 373–399.
[138] J. Suh, M. Dawson, J. Hanes, Real-time multiple-particle tracking: application to drug and gene delivery, Adv. Drug Deliv. Rev. 57 (1) (2005) 63–78.
[139] X. Michalet, A.J. Berglund, Optimal diffusion coefficient estimation in single-particle tracking, Phys. Rev. E Stat. Nonlin. Soft Matter Phys. 85 (6–1) (2012), 061916.
[140] P. Struntz, M. Weiss, The hitchhiker's guide to quantitative diffusion measurements, Phys. Chem. Chem. Phys. 20 (45) (2018) 28910–28919.
[141] H. Shen, et al., Single particle tracking: from theory to biophysical applications, Chem. Rev. 117 (11) (2017) 7331–7376.
[142] Y. Tseng, et al., Micro-organization and visco-elasticity of the interphase nucleus revealed by particle nanotracking, J. Cell Sci. 117 (10) (2004) 2159–2167.
[143] N. Gal, D. Lechtman-Goldstein, D. Weihs, Particle tracking in living cells: a review of the mean square displacement method and beyond, Rheol. Acta 52 (5) (2013) 425–443.
[144] R. Ferrari, A.J. Manfroi, W.R. Young, Strongly and weakly self-similar diffusion, Phys. D 154 (1) (2001) 111–137.
[145] Y. Feng, J. Goree, B. Liu, Errors in particle tracking velocimetry with high-speed cameras, Rev. Sci. Instrum. 82 (5) (2011), 053707.
[146] A. Gharipour, A.W.C. Liew, Segmentation of cell nuclei in fluorescence microscopy images: an integrated framework using level set segmentation and touching-cell splitting, Pattern Recogn. 58 (2016) 1–11.
[147] A. Grushnikov, et al., 3D level set method for blastomere segmentation of preimplantation embryos in fluorescence microscopy images, Mach. Vis. Appl. 29 (1) (2017) 125–134.
[148] T. McInerney, D. Terzopoulos, Deformable models in medical image analysis: a survey, Med. Image Anal. 1 (2) (1996) 91–108.
[149] B. Ristic, S. Arulampalam, N. Gordon, Beyond the Kalman Filter: Particle Filters for Tracking Applications, Artech House, 2004.
[150] M. Isard, A. Blake, CONDENSATION—conditional density propagation for visual tracking, Int. J. Comput. Vis. 29 (1) (1998) 5–28.
[151] M.S. Arulampalam, et al., A tutorial on particle filters for online nonlinear/non-Gaussian Bayesian tracking, IEEE Trans. Signal Process. 50 (2) (2002) 174–188.
[152] N. Arhel, et al., Quantitative four-dimensional tracking of cytoplasmic and nuclear HIV-1 complexes, Nat. Methods 3 (10) (2006) 817–824.
[153] L. Yuan, et al., Object tracking with particle filtering in fluorescence microscopy images: application to the motion of neurofilaments in axons, IEEE Trans. Med. Imaging 31 (1) (2012) 117–130.
[154] O. Hirose, et al., SPF-CellTracker: tracking multiple cells with strongly-correlated moves using a spatial particle filter, IEEE/ACM Trans. Comput. Biol. Bioinform. 15 (6) (2018) 1822–1831.

CHAPTER FIFTEEN

Deep Learning

David Mayerich, Ruijiao Sun, and Jiaming Guo

15.1 Introduction

The development of artificial neural networks was inspired by brain function. Recent advances in high-performance computing enable the training of highly complex neural networks for image segmentation and analysis. Deep neural networks are an emerging technique demonstrating unprecedented performance in a number of microscopy tasks. In particular, convolutional neural networks are suitable for learning underlying spatial features in images and for using this information for segmentation, classification, and resolution improvement.

The following section provides a condensed timeline of neural network development, focusing on architectures commonly used for microscope image analysis. This includes an emphasis on convolutional neural networks (CNNs) and features commonly used to solve segmentation and classification problems with CNNs. Later sections address the basic concepts and practical applications of deep learning along with implementation frameworks.

15.1.1 Basic Components of Neural Networks

In 1943, McCulloch and Pitts [1] developed simulated interconnected neurons to produce highly complex patterns from basic binary elements (Fig. 15.1a). This model is the ancestor of the artificial neuron and a critical contribution to artificial neural networks. In 1949, Hebb [2] presented the Hebbian learning rule, which proposed adjustable weighting applied to each input to account for training over time. Based on these findings, Rosenblatt developed the first perceptron in 1958 [3] (Fig. 15.1b), which encoded learning through the modification of input weights. In 1960, Kelley [4] derived the basics of continuous backpropagation and in 1962 Dreyfus [5] proposed a simpler version based on the chain rule. Neural networks are discussed in Section 9.11.

15.1.2 A Timeline of Convolutional Neural Network Development

From the perspective of image processing, one of the most useful neural network architectures was developed in 1980 by Fukushima—the first convolutional neural network (CNN) [6]. This network, dubbed the neocognitron, recognized visual patterns. In 1985,

Microscope Image Processing
https://doi.org/10.1016/B978-0-12-821049-9.00015-0

Copyright © 2023 Elsevier Inc.
All rights reserved.

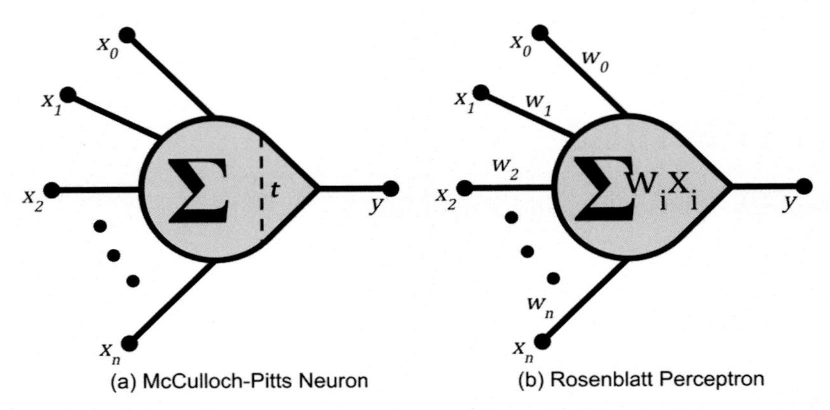

Fig. 15.1 Initial concepts for artificial neurons. (a) The McCulloch-Pitts neuron allowed the summation of binary inputs, followed by application of a threshold to evaluate the binary output. Appropriate thresholds allow the McCulloch-Pitts model to duplicate a wide range of binary operations. (b) The Rosenblatt perceptron combined the McCulloch-Pitts model with weights, as proposed by Hebb, and this allowed training the neuron's response to inputs.

Ackley et al. [7] invented the Boltzmann machine (BM), which was the first neural network capable of learning internal representations in complicated problems. One year later, Smolensky invented the restricted Boltzmann machine (RBM) [8], initially known as the harmonium, which restricted the number of intralayer connections between units to enable more efficient training.

In the decades that followed, several modifications made CNNs more viable from a theoretical perspective. In 1986, Jordan [9] introduced the concept of a recurrent neural network (RNN). This architecture contains at least one recurrent connection, or loop. These networks provide advantages for continuous and time-domain data, such as handwriting and speech recognition. Shortly afterward, Rumelhart et al. [10] developed, and Ballard [11] refined, the concept of autoencoders. Autoencoders produce images that are the same size as the input. This architecture enables unsupervised training for data compression algorithms, and it plays a key role in feature extraction and pixel-level classification. In 1990, LeCun et al. [12] proposed LeNet based on the neocognitron architecture. This was the first CNN successfully applied to the recognition of handwritten zip code digits. The vanishing gradient problem was a computational challenge that made deep neural networks impractical to train. In 1997, Hochreiter and Schmidhuber [13] introduced long short-term memory (LSTM), which allowed RNNs to partially solve this problem. The development of deep belief networks (DBNs) by Hinton et al. [14] in 2006 ushered in the modern era of deep learning. DBNs can be viewed as a combination of unsupervised networks with a layer-wise pretraining algorithm. Based on this pretraining scheme, Salakhutdinov and Hinton [15] introduced deep Boltzmann machines (BMs) in 2009.

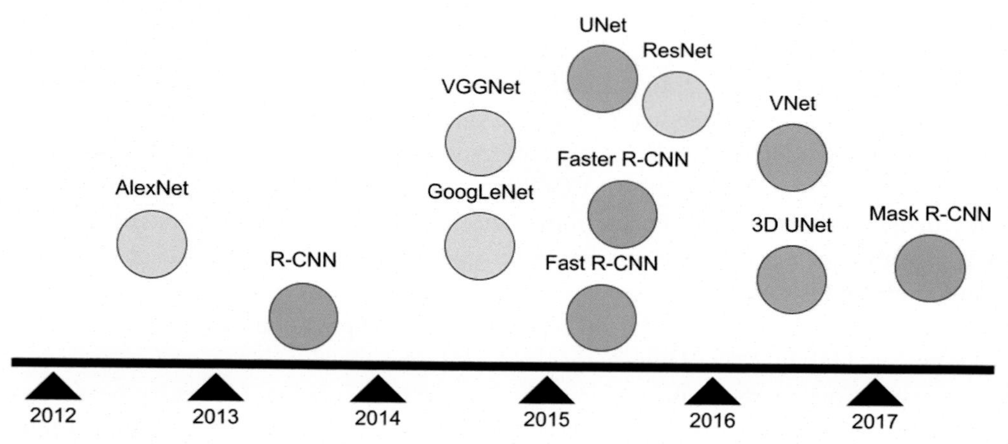

Fig. 15.2 The development timeline of those neural networks that are particularly useful for 2D and 3D processing and microscopy.

The era of CNNs began to flourish with the development of AlexNet by Krizhevsky et al. [16] in 2012, which won the annual ImageNet competition. Fig. 15.2 presents a timeline showing the rapid advance of CNNs. AlexNet was considered to be the most influential model in computer vision, and it spurred the development of various recent innovations in network models. In 2013, Girshick et al. [17] introduced R-CNN, which is a relatively simple and scalable algorithm for object detection using region proposals. Two milestones of deep learning introduced in 2014 were VGGNet proposed by Simonyan and Zisserman [18] and GoogleNet proposed by Szegedy et al. [19]. VGGNet demonstrated that network depth could significantly improve learning effectiveness, while GoogleNet proved that optimized nonsequential structures could perform better than sequential designs. In 2015, He et al. [20] proposed ResNet, which used residual blocks to alleviate another computational bottleneck, the degradation problem. R-CNN variants including Fast R-CNN [21], Faster R-CNN [22], and Mask R-CNN [23] were developed and have merged into the mainstream for object detection in recent years.

15.1.3 A Timeline of Deep Learning in Microscopy

Manual analysis is quite common in microscopy, but practical limitations, such as the required time and the error rate, limit its throughput. This has direct implications for the quality of the analysis, because it limits the number of images that can be practically studied. The appropriate use of machine learning has significantly improved the effectiveness and robustness of microscope image analysis over the last century [24,25]. These methods rely on sophisticated feature introspection and data representation, limiting their applications to relatively simple problems. Deep learning has given new life to microscope image analysis with an early implementation developed by Ciresan et al. [26] in 2012. In the early 2010s, deep neural networks were developed for various microscope image analysis applications, such as segmentation [26], detection [27], classification [28], and labeling [29]. Later, some classical networks that perform well in computer vision, such as AlexNet, VGGNet, and ResNet, were also applied in microscope image analysis [30,31]. Furthermore, the R-CNNs series first proposed in 2013 have demonstrated advantages in microscope image detection [32] and segmentation [33]. Transfer learning was introduced by Yosinski et al. [34] in 2014 to improve learning efficiency and reduce overfitting, and it has become an efficient pretraining tool for microscope image analysis [35]. In 2015, Ronneberger et al. [36] introduced U-Net as a novel architecture to achieve high performance in biomedical image segmentation. U-Net has been continuously developed to work with microscope images [37–39], and it has expanded into many related applications [40].

In recent decades, advances in microscopy have helped throughput keep pace with increasing workloads. Microscopy has become increasingly automated, particularly in clinical environments, often producing massive high-quality images that would be difficult to analyze using traditional methods. Deep learning has outperformed traditional image analysis in such scenarios by deriving underlying knowledge directly from raw data [41].

15.2 Deep Learning Concepts

In this section, we introduce some key concepts in deep learning and then review related applications in microscope image analysis.

15.2.1 Training

An artificial neuron, or a perceptron [3] (Fig. 15.1b), takes a weighted input, adds a regulated bias, and applies an activation function to produce an output. A neural network typically consists of three types of layers: the input layer, a hidden layer, and the output layer. Each layer is made up of a set of interconnected neurons, each with an activation function. During the training process, inputs from a known training set are repeatedly presented to the input layer. The weights are adjusted until the network's error rate on the training set is minimized and the known objects in the training set are properly labeled by the network.

15.2.2 Activation Functions

An activation function is a mathematical transformation used between layers to scale the output before passing it on to the next layer. While linear activation functions are sometimes used, nonlinear operations enable highly complex relationships between features and are therefore widely employed in real-world applications. Nonlinear transfer functions for microscopy tasks include the sigmoid, hyperbolic tangent (tanh), and rectified linear unit (ReLU). The sigmoid function offers a naturally smooth response with a normalized output interval, which is good for classification. However, the sigmoid function is not zero centered, resulting in an ineffective gradient descent due to potential "zig-zag" optimization. The tanh function offers a symmetric alternative to the sigmoid with a steeper gradient. However, it also saturates for large values, giving rise to a *vanishing gradient* that dramatically slows training in deep networks. The ReLU function rectifies the saturation problem by offering a constant derivative for all input values. At present, ReLU is the mainstream operation for microscopy due to its efficiency and rapid convergence. One controversial drawback of ReLU is known as "the dying ReLU problem" [42], in which neurons stop learning due to zero gradient. This is likely to happen when the learning rate is too high or there is a large negative bias term, and it may be mitigated by reducing the learning rate. ReLU variants [43] such as leaky ReLU and parametric ReLU eliminate the dying ReLU problem by using a small negative slope. But their learning performance is inconsistent and not necessarily better than ReLU.

There are several other operations that may be of interest to readers, such as exponential linear units (ELU) [44], softmax, and Swish [45]. In practice, one should select the activation function that is appropriate for the task (e.g., detection, segmentation, etc.), the type of network (CNN, RNN), and the type (intermediate or output) of the layer. In most applications, ReLU is a good place to start.

15.2.3 Cost Functions

A loss function, or cost function, is a nonnegative measure of the difference between the network output and ground truth [46]. The loss function distills the learning performance into a single value that decreases as the prediction accuracy increases, and it guides the training process. Depending on the learning task, loss functions commonly include mean squared error (MSE), mean absolute error (MAE), and cross-entropy. MSE, or L2 loss, measures the average of squared differences between predicted and actual observations. Due to squaring, MSE is mathematically simpler to optimize; however, it is sensitive to outliers since greater errors are penalized more heavily. Unlike MSE, MAE measures the average of absolute differences between predicted and actual observations and thus is less affected by outliers. However, MAE is less computationally efficient since its derivative is not continuous.

Cross-entropy, or log loss, is another loss function commonly used for classification. It computes a score that measures the average differences between predicted and actual probability distributions for predicting class 1. Its value increases as the predicted probability diverges from the actual label. Cross-entropy loss often leads to a better probability estimation at the cost of accuracy. There are other loss functions that interested readers could consider, such as hinge loss and Kullback-Leibler (KL) divergence. In practice, one should select a loss function that is appropriate for the tasks to be performed. MSE provides a good first choice.

15.2.4 Convolutional Neural Networks

A CNN is a common type of neural network that specializes in processing data sampled on a uniform grid, such as an image. It is widely used in microscopy for feature mapping [47]. A typical CNN is composed of three types of layers: the convolutional layer, the pooling layer, and the fully connected layer (Fig. 15.3). The convolutional layer applies a convolution filter to the input image and outputs a feature map that predicts the class to which each feature belongs. While different filters extract different features, the stride parameter controls the step-wise movement of the filter over the input image and establishes the size of the resulting feature map.

Following the convolutional layer, a pooling layer continuously reduces the number of parameters in the network by combining the pixels in selected windows. Complex

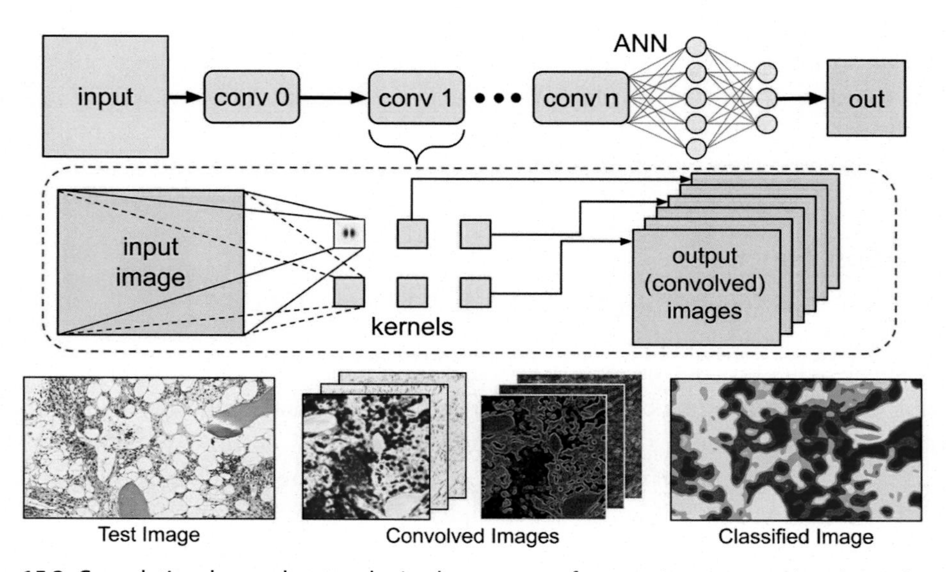

Fig. 15.3 Convolutional neural networks in the context of semantic segmentation. Input images are iteratively convolved with a series of kernels. Each kernel produces an output image that is typically smaller than the input image. Sequential convolutional layers produce larger numbers of smaller output images. The final convolved outputs are then passed to a traditional artificial neural network (ANN), which generates the classified output. In this example, an output is produced for each pixel of the input image to produce a semantically segmented image with a class label at each pixel position.

real-world image data may require a large number of different filters, and such downsampling operations shorten the training time and control overfitting. The most frequent type of pooling is max pooling, which takes the maximum pixel value in each selected window. Max pooling reduces the variance of the feature map and controls the output size to keep the training process computationally tractable. Average pooling, which calculates the average pixel value in each window, is useful only when smoothing an image is preferable. In general, one needs to decide the proper hyperparameters, including the filter size, the filter count, stride, and padding, when using a CNN.

The last major layer of a CNN is the fully connected layer, which is a traditional ANN with each neuron connected to every neuron in the next layer. The fully connected layer takes the results of the convolution/pooling operations and manipulates them for classification or regression.

15.3 Practical Applications

In microscope image analysis, deep learning is used as a feature extractor or a classifier in myriads of biomedical and histological applications. These include such traditional tasks as object detection, segmentation, and classification. In the past decade, deep learning has been widely applied to different histological entities, including human [48,49] and mouse brain [50–52], rat brain [53], zebrafish brain [54], drosophila brain [55], human liver [56], mouse liver [31], rat liver [57], human lung [58], mouse lung [59], human kidney [60], mouse kidney [61], human colon [62], human bone [63], mouse bone [64], human breast [35], human cervix [65], human prostate [66], human skin [67], and many more. Various CNN models have been deployed for these tasks. In this section, we review trending deep learning techniques with a focus on architecture innovations. Interested readers may refer to the aforementioned references for specific applications.

15.3.1 Classification

Classification (see Chapter 9) is one of the most fundamental tasks in microscope image analysis. Common classification methods compute characteristic feature values to describe objects (Chapter 8) and use them to assign labels to the objects. This process is relatively simple for humans but quite difficult to automate. For example, accurate protein classification in microscopy is difficult to achieve using conventional algorithms due to contrast disparities and the presence of punctate features. Deep learning offers a solution by objectively optimizing filters to identify features from training sets and translating the learned information into classification rules with minimum human intervention. Many CNN models have been successfully developed and applied to microscope images for cell and nuclear classification.

Deep learning made its mark in histology in the early 2010s. Buyssens et al. [68] proposed a multiscale convolutional neural network (MCNN) for classifying cytological pleural cancer cells using Feulgen staining. The MCNN model, similar to the multistage architecture introduced by Hubel and Weisel [69] in 1962, consists of multiple CNNs with different sizes. Each CNN is composed of convolutional and pooling layers as well as a classical fully connected layer for classification. A linear combination of all outputs achieves better classification rates than state-of-the-art methods. At around the same time, Cruz-Roa et al. [70] introduced a simple deep learning architecture to classify basal-cell carcinoma by automatically learning features from large skin histopathology images. Empirical evaluation of image representations has demonstrated the effectiveness of deep learning over traditional canonical representations.

Supervised classification relies heavily on precise labels, and comprehensive annotations often require a high level of expertise. This indicates a need for deep learning methods based on either coarse-grained labels or completely label-free methods for feature extraction and representation in microscopy. Xu et al. [71] proposed a multiple instance learning (MIL) framework to automatically extract fine-grained information from coarse-grained labels. The weakly supervised method provides better feature representations in colon cancer histopathology over fully supervised methods with fine labels.

Architectures popular in traditional computer vision applications were also deployed for microscope image classification. Gao et al. [72] developed an eight-layer CNN model based on the classical LeNet-5 model to classify human epithelial-2 cell staining patterns into six categories. This method has demonstrated that CNNs trained with a larger and nonaugmented dataset and then fine-tuned with a smaller and related dataset provide better performance than those trained directly with a smaller dataset. This strategy circumvents the problem of inadequate training data, and it has been verified in other studies [73]. In recent years, many other architectures have been applied to microscope image classification [56,74]. Xiao et al. [75] performed a comparative study on classification performance between several representative CNN methods and two ensemble methods on a moderately large cell dataset. They concluded that fine-tuned CNN models always outperform other methods. In particular, ResNet obtained the highest precision of prediction with the least training time.

15.3.2 Detection

The detection of microstructures such as cells and nuclei in digitized tissue is vital in microscopy. In particular, cell detection serves as a prerequisite for various quantitative tasks such as counting blood cells [76] and identifying cellular morphology [77]. For example, the mitotic count within a histology section is an important biomarker indicative of breast cancer. In clinical trials, this rate is determined manually by a pathologist or technologist, often subject to fatigue-induced errors. Ciresan et al. [27] developed an automated method to assess the prognostic value of absolute mitotic counts using a CNN, which handles square patches of pixels directly from raw RGB data. To cope with insufficient mitosis examples for training, they used data augmentation with arbitrary image mirroring and rotations. One of the first CNN implementations was applied to breast cancer histopathology images. It outperformed other contestants in the ICPR 2012 mitosis detection contest with an F-score of 0.718 [78]. Similar networks have been successfully applied in other tumor models for cell detection [79–81], and interested readers can explore these works.

One popular type of CNN is the Regional CNN (RCNN) model (Fig. 15.4). RCNN-based methods provide greater accuracy for object detection [17], but they are slow because they generate multiple region proposals. For faster performance, derivatives of RCNN (including Fast R-CNN [21], Faster R-CNN [22], and Mask R-CNN [23]) have been applied to microstructure detection in microscope images. For example, Hung et al. [82] applied Faster R-CNN to identify bounding boxes around objects and

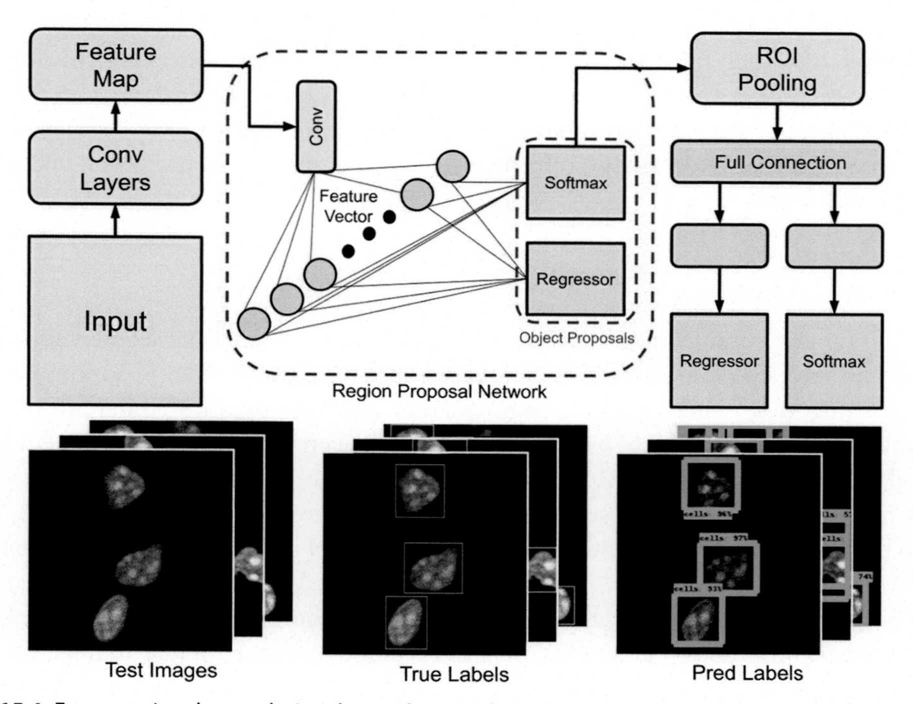

Fig. 15.4 Faster regional convolutional neural network (R-CNN) in the context of cell detection. Top: The faster R-CNN employs a region proposal network (RPN) to predict object boundaries and object scores simultaneously at each detection. ROI pooling is used to reshape all proposed regions into a fixed size and then feed them into a set of fully connected layers for further refinement. Bottom: In this example output labels are produced to locate cell boundaries.

classify them into different categories and then fine-tune the result using AlexNet. The proposed hybrid method achieved a more accurate result than humans in ambiguous cases. The R-CNN model series has been applied to many other microscopy detection applications and has been accordingly modified to provide optimal performance.

Most CNN-based methods use large, accurately labeled data sets for training, and this requires either laborious annotations or publicly available ground-truth data. Albarqouni et al. [83] developed a multiscale CNN AggNet that generates reliable ground-truth labels from nonexpert annotations by integrating an additional crowdsourcing layer into the model. This method first trained multiple CNN models with gold-standard annotations created by professionals and applied them for mitosis detection. All selected candidates were passed to the crowds for annotation and collected labels were fed to the trained networks with an aggregation layer for model refinement and ground-truth generation. This method was the first attempt to generate ground-truth training data by nonexpert users in a histological context.

In recent studies designed to cope with complicated and practical problems, some CNN-based deep learning methods have been developed for object detection in 3D volumetric data. For example, Weigert et al. [84] have proposed a modified 3D variant of ResNet to detect and segment cell nuclei in low signal-to-noise ratio 3D microscope images using a star-convex polyhedral representation. This approach, dubbed STARDIST-3D, uses U-Net and U-Net+in 3D cell detection and segmentation. It has been demonstrated to outperform a classical watershed-based method and deep learning methods.

15.3.3 Segmentation

Methods for performing the essential microscope image analysis task of segmentation can be divided into two categories. Semantic segmentation classifies each pixel into one of several categories, while instance segmentation also distinguishes different objects (Chapter 7). Ciresan et al. [26] have applied a unified deep CNN model to segment biological neuron membranes in electron microscope images. This early attempt not only demonstrated good performance in semantic segmentation of neuron membranes, but it also provided promising perspectives on applying CNN models to other segmentation tasks as well.

The U-Net series of approaches (Fig. 15.5), based on the fully convolutional network (FCN) model proposed by Long et al. [85], has become the most popular architecture for semantic segmentation. Ronneberger et al. [36] have developed an elegant U-Net architecture composed of a contracting path to capture context and a symmetric expanding path for precise localization. This model has yielded precise segmentation results on several different datasets with very reasonable training times. Cicek et al. [37] have

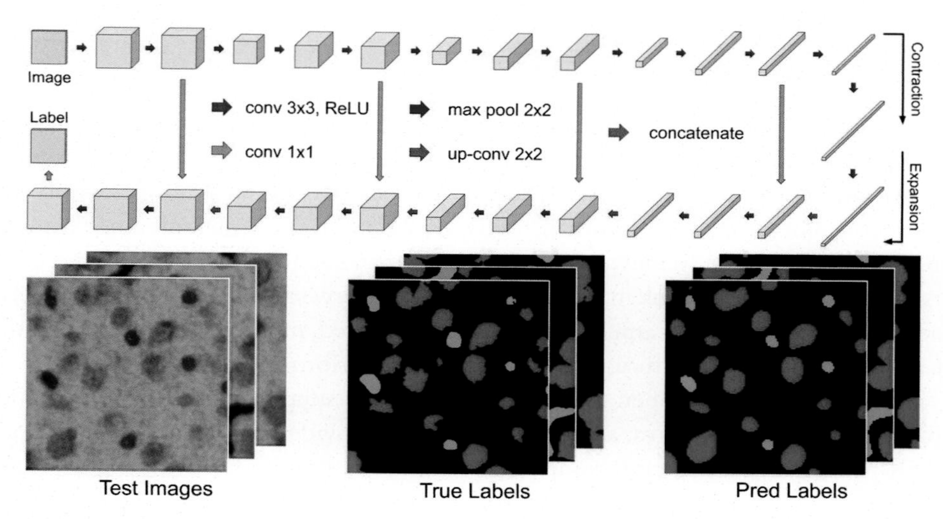

Fig. 15.5 U-Net in the context of semantic segmentation. Top: The U-Net architecture consists of a contracting path (top side) and an expanding path (bottom side) that visually form a U shape. The contracting path, which is built of standard convolutional networks, extracts high-level feature maps. The expanding path recovers the size of the segmentation map and combines it with the corresponding cropped feature map that is copied from the contracting path. Concatenation of feature maps helps regain localized information. In this example, an output mask is produced to semantically separate an image into vessel, cell, and background. Bottom: Test images produce output images with predicted labels that mostly match the true labels of the pixels.

introduced a 3D U-Net model by replacing all 2D operations with their 3D counterparts. Provided with some labeled 2D slices, the semiautomated setup achieved a high intersection of union (IoU) value in tubule segmentation, and the fully automated setup realized volumetric segmentation performance equivalent to 2D implementations. Further development on U-Net models like UNet++ [38], Micro-Net [86], and MultiResUNet [39] has been proposed to provide excellent performance in microscope image semantic segmentation.

Mask-RCNN has been shown to perform well in microscope image segmentation, particularly in instance segmentation. Johnson [87] used a Mask-RCNN model with a feature pyramid network backbone to detect and segment nuclei in microscopy images. In particular, the model using a ResNet-50-FPN backbone had low computational load while the model using a ResNet-101-FPN backbone obtained improved results. These two networks have proved that Mask-RCNN provides highly effective automated segmentation in a wide range of cell nuclei images. This work has also demonstrated that Mask-RCNN could be applied to improve segmentation performance for similar tasks with relatively little modification and customization.

15.4 Software Frameworks

Various deep learning architectures and algorithms are used in a myriad of microscopy applications, and they have proved to be considerably integral. This makes many deep learning software frameworks available to facilitate network construction and deployment. Since each framework is based on different cores for different purposes, choosing an appropriate framework is of great importance for data scientists and engineers to tackle a specific problem. In this section we review several trending open source frameworks, addressing such aspects as languages supported, multiple GPU capacity, support for a common architecture, and computational performance.

TensorFlow [88] is an open source framework that supports various programming languages such Python, C, Java, and Go. TensorFlow provides easy-to-use modular tools for high-performance numerical computation with robust multiple GPU support. It also provides useful built-in tools such as TensorBoard and TensorFlow, serving to facilitate data visualization and network deployment. It offers support for a wide range of neural networks such as CNNs, RNNs, RBMs, and DBNs. TensorFlow is considered to be the most reliable framework with top-notch documentation and community support maintained by Google.

PyTorch [89] is an open source framework that provides a straightforward interface suitable for rapid prototyping. PyTorch employs CUDA along with C/C++ libraries for tensor computation with strong support for GPUs. It also provides support for common neural networks and is considered to be the strongest competitor to TensorFlow.

Caffe [90] is an open source framework with a Python interface, and it supports GPU-based acceleration with CUDA. It also supports shared-memory multiprocessing in a cluster of systems with OpenMP. A newer release, named Caffe2, supports various network architectures such as CNNs, RBMs, and DBNs. However, support for RNNs is currently limited, and building complex layers must be done from lower levels. Despite this, Caffe2 offers a large number of pretrained models, allowing simple and fast demonstrations, particularly in mobile and cloud computing environments.

Keras [91] is an open source framework that can be integrated seamlessly into the TensorFlow workflow. Keras provides a simple interface for rapid deployment with multiple GPU support. It supports multiple deep learning backends, such as TensorFlow and Theano, along with compatible APIs. It also offers extensive documentation and community support with a large user base.

15.5 Training Deep Learning Networks

Deep learning studies have made tremendous progress in recent decades with advances in novel architectures, increased computation capability, and big data. But one of the persisting challenges in developing deep learning solutions to microscope imaging problems is posed by insufficient training dataset size. In most deep learning applications, a large training set is crucial for designing a robust network model. But large and diverse microscopy image datasets are difficult and expensive to acquire due to the complexity of image collection, variations in experimental conditions, and the effort required for annotation. During training, the network learns to recognize the training samples. But overfitting occurs when the network memorizes the details of the training data but cannot then generalize to the real-world population of objects that are encountered in widespread application. In this case the error rate in practice can be far worse than that predicted during training. This section addresses ways to overcome this problem.

15.5.1 Data Augmentation

In an effort to avoid overfitting and improve the generalization of trained models, methods such as dropout [92], batch normalization [93], and transfer learning [33] have been developed. Another useful method to reduce overfitting is data augmentation [94], which deals with training-set size directly and has been widely applied in microscopy. This section addresses data augmentation since it handles the overfitting problem at its root cause. Interested readers can explore other regularization methods to learn more about model optimization [95].

Data augmentation (Fig. 15.6) includes techniques that increase the size and generalization of training datasets [96]. Image-level augmentation methods discussed in this section include geometrical transformations and color transformations. Since choosing a proper augmentation technique is often problem specific, we review the scope of microscopy application for several common transformations.

Geometrical augmentation includes image-level transformations such as flipping, rotation, and cropping. Flipping and rotation are the most commonly used geometrical transformations in microscope image analysis to increase training data volume. Since most microscope images are orientationally invariant, structural features are preserved with little distortion. For example, neuronal structures such as dendrites and axons are rotation- and translation-invariant. Flipping and rotation have been shown to facilitate neuron tracing [97]. In addition, rotation also simulates incorrect horizontal calibration.

Another basic operation is cropping, which accounts for different magnifications. Cropping is useful for generating uniform images with normalized dimensions, and it has been widely used in microscopy for different applications, such as mice liver microscopy images [31]. Noise and blur additions are filter-based operations used to enhance the tolerance to quality variation of input images. Noise addition is realized by injecting the original image with a matrix of random values drawn from a predefined distribution, and blurring is achieved by convolving the original image with a predefined filter kernel. Since inadequate resolution and low signal-to-noise ratio

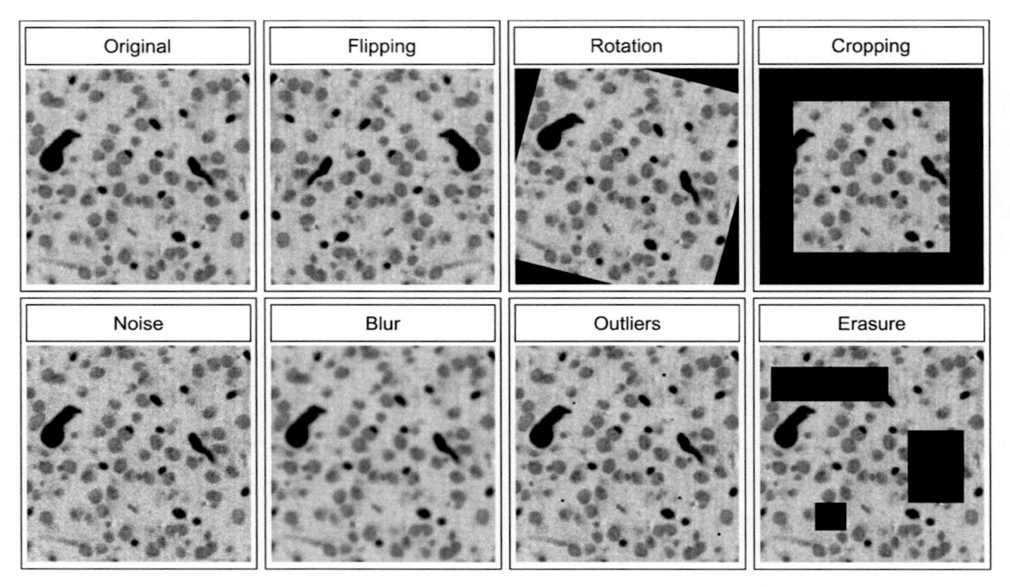

Fig. 15.6 Data augmentation provides additional examples for the training set. A larger training set allows the CNN to generalize by ignoring features that are components of common systemic imaging artifacts.

are common limitations in microscopy images, noise and blur additions synthesize new datasets under different imaging conditions to generalize the training set. For example, noise injection in microscopy images has been demonstrated to eliminate the bias toward brighter images that affects focus evaluation [98]. Other methods such as outlier addition and random erasing are controversial methods, because they may alter the image labels and therefore must be manually observed. Note that data is often augmented on the fly during training.

Since microscopy images differ from more than just positional and translational variation, color space transformations are also widely used in microscope image analysis. These include color shifts and contrast changes. However, photometric operations can cause fatal malfunction of the trained model in applications where color is a distinctive feature. Taylor et al. [94] performed a comparative study on the effectiveness of geometric and color transformations in public datasets.

No amount of data augmentation can compensate for object types that are simply missing from the training set. A system intended to differentiate abnormal from normal cells cannot be adequately trained on a dataset that does not contain representatives of all types of normal and abnormal cells. At present, there does not appear to be any generalized policy for training-set augmentation in microscopy, and one must select the proper methods based upon the specific application.

15.5.2 Transfer Learning

Another popular method to increase the generalization of trained models is called "transfer learning." This approach applies the weights learned from a related task (the source domain) as a starting point for a new task (the target domain) to improve both training efficiency and overall performance. Hence, the network model is not required to be trained from scratch in the target domain. The learned weights can be applied to extract features for object detection, segmentation, classification, and superresolution.

Based on a recent survey summarized by Tan et al. [99], deep transfer learning is classified into three categories: instance-based, network-based, and adversarial-based transfer learning. This section reviews transfer learning methods arising from this work.

Instance-based transfer learning involves using partial instances from the source domain. These instances are selected using a specific weight adjustment strategy to supplement the training set in the target domain. It allows tuning a high-quality network model with a few new datasets after having pretrained it with a large number of old datasets from the source domain. In microscope image analysis, Li et al. [100] have proposed enhanced TrAdaBoost, which can identify arbitrary numbers of classes and optionally modify instance weights to improve the efficiency of the identification of interregional sandstones.

The second category, network-based transfer learning, reuses parts of a pretrained network model, including its structure and connection parameters, in a new network model applied in the target domain. Studies have shown [101–103] that the first few layers of a deep neural network trained on a large dataset capture the transferable outline information. Some classical architectures, such as AlexNet [16], VGGNet [18], and ResNet [20], can pretrain a network model on a large dataset. For example, Nguyen et al. [74] have proposed a transfer learning network structure based on the feature integration of Inception-v3 [104], Resnet-152 [20], and Inception-ResNet-v2 [18]. They designed a two-layer fully connected structure for generic feature extraction in microscopy images. The proposed feature concatenation network showed superior performance compared to the training results of each single transfer learning network and five traditional classification algorithms.

The last category is adversarial-based transfer learning, which was first proposed by Ganin et al. [105]. It refers to learning a transferable representation that is predictive to the example labels but offers no information about the source domain and target domain. Majurski et al. [106] applied three methods (generative adversarial network (GAN)-based transfer learning, U-Net-based transfer learning, and U-Net trained only on annotated images) on absorbance microscopy images of human iRPE cells for comparative segmentation validation. Providing limited or no data augmentation, GAN-based transfer learning and U-Net-based transfer learning obtained more accurate segmentation results. However, there is a trade-off between segmentation accuracy and computational cost. Namely, U-Net-based transfer learning provided increased segmentation accuracy at the cost of more computational time.

15.6 Application of Deep Learning for Cell Nuclei Detection

Fig. 15.7 shows an example of using deep learning to detect cell nuclei in large brain images [107]. In this example, the authors used Faster RCNN (Section 15.3.2) [22] as the object detection model to detect cell nuclei in whole brain 2D sections. This network extracts the deep features using a set of convolutional layers, and it applies a region proposal network to the feature maps to obtain the location of proposed objects (Fig. 15.7). Dragan et al. [107] used DNA stain 4′,6-diamidino-2-phenylindole (DAPI) and pan-histones biomarkers to stain the cell nuclei and used them as input channels to the network (Fig. 15.7a).

This example demonstrates the ability of deep neural networks to cope with input data variabilities, such as the density variability seen in Fig. 15.7. Deep neural networks are able to extract abstract representations, and this gives them the potential to outperform conventional nuclear segmentation algorithms in difficult regions such as densely packed ensembles in hippocampus.

As discussed earlier, deep neural networks require large training datasets to achieve good performance. To reduce the labeling effort, the authors used transfer learning (Section 15.5.2) by initializing the network with pretrained weights from the ImageNet database [108]. This trained the network on more than 200,000 results from a conventional nuclear segmentation algorithm. Then the network was retrained on ~8,500 manually generated datapoints. This illustrates that with enough training samples, deep neural networks can learn sophisticated abstract representations from the input data and operate with a high degree of accuracy (see Fig. 15.7f). The authors have provided both code [109] and data [110] for testing by interested readers.

15.7 Challenges

This chapter presents some deep learning implementations, particularly deep CNNs, in microscope image analysis. Examples are provided based on both traditional tasks, including classification, detection, and segmentation, and on emerging tasks such as superresolution and digital staining. While deep learning has become the mainstream technique in various microscopy applications, there are remaining challenges and unsolved problems. One of the foremost problems in supervised deep learning is insufficient training dataset size, as discussed previously. Therefore unsupervised or weakly supervised learning [111] will most likely become an important trend for future development. Recent innovations in unsupervised deep learning, such as variational autoencoders (VAEs) [112] and GANs, [113] have been demonstrated to be effective

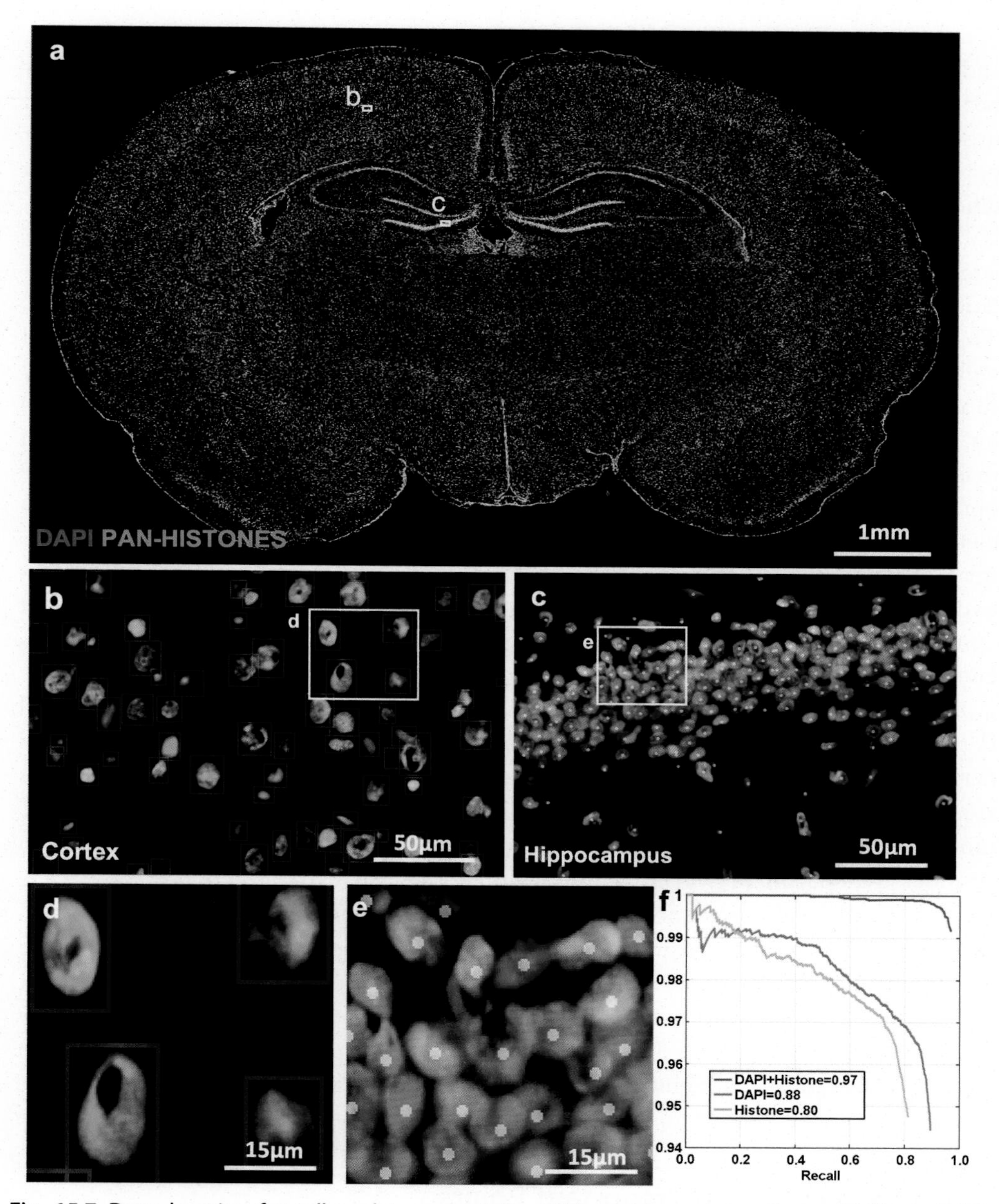

Fig. 15.7 Deep learning for cell nuclei segmentation. This montage shows the variability in DAPI (pan-nuclear DNA stain) and histone (pan-nuclear protein) expression labeling across a whole rat brain tissue slice. (a) The whole brain tissue slice. Close-up images show DAPI and pan-histone labeling variations in (b) a sparse and (c) a densely packed region of the brain. (d, e) The generated location (bounding box) results of the proposed model for reliable detection of the cell nuclei using transfer learning in conjunction with a Faster-RCNN network. (f) Receiver operating characteristic (ROC) curves comparing cell nuclei detection using single markers and using both markers together show that using both markers produces a significant improvement in performance. *(Dragan Maric, after [107])*

in different microscopy tasks [114–116]. However, current implementations are limited, and their advantages over traditional methods in microscope image analysis have not yet been well explored. Nevertheless, unlabeled data is much easier to acquire, and unsupervised deep learning is still a major focus in microscope image analysis. Another potential work-around is to develop networks that can extract information more efficiently [117].

Another common issue in deep learning is the computational complexity required for large datasets. In general, there is a trade-off between network performance and computational cost. Because of this, current tissue implementations are limited to small volumes. To the best of our knowledge, the largest tissue volume so far processed for cerebral cell segmentation is 1 mm^3 [118], a very small fraction of the entire mouse brain. The ability to analyze the entire brain volume would be of great importance for cerebral disease studies, such as those aimed at cerebrovascular quantification. Several attempts have been made to alleviate the computational load for high-performance learning. These include optimization algorithms [119], advanced computing techniques to improve training efficiency [120], and hardware acceleration using GPUs [121]. The current method for image-level quantification is to partition the raw images into a set of overlapping (or non-overlapping) patches and then selectively fuse the deep learning results [122]. Distributed computing [123] is also appropriate for this purpose, and it may well become an important future trend.

As mentioned earlier, some classical networks such as AlexNet, VGGNet, and ResNet have been used successfully in microscope image classification. But these models may not be optimal for other tasks, such as segmentation, because common CNN models are designed for small images. When applied to high-resolution microscopy images, model mismatch and lengthy training become problems. Therefore it is necessary to consider task-specific networks when implementing deep learning in microscope image analysis. To this end, several novel or modified CNN models have been proposed for cell and nucleus detection and segmentation [62,124]. These designs may well be the future trend to handle different tasks more effectively. A recent work has outlined a framework for automated selection of network models for structural identification in microscopy images using a cluster [125]. This opens the door to the automated generation of deep learning networks that are customized for particular scientific datasets.

15.8 Summary of Important Points

1. A typical neural network is composed of three types of layers: an input layer, one or more hidden layers, and an output layer. Each layer is made up of interconnected nodes (artificial neurons), each with an activation function.

2. An activation function is a mathematical operation used to rescale the output before passing it to the next layer. Nonlinear activation functions such as Sigmoid, tanh, and ReLU are widely used in practice.
3. A loss function is a nonnegative measure of the difference between ground truth and the network output. Common loss functions include MSE, MAE, and cross-entropy. The goal of training is to minimize the loss function.
4. A typical CNN is composed of three types of layers: a convolutional layer, a pooling layer, and a fully connected layer. The convolutional layer applies a convolution kernel to the input image to create a feature map that predicts the class to which each object belongs. The pooling layer reduces the number of parameters in the network by taking the maximum or average pixel value from selected windows.
5. CNNs such as VGGNet, GoogLeNet, and ResNet commonly used in computer vision applications can be applied to microscope image classification. These CNNs are usually trained with a large dataset and then fine-tuned with a smaller dataset to provide faster training and more accurate performance.
6. The regional CNNs, including Fast R-CNN, Faster R-CNN, and Mask R-CNN, have been widely applied for microscope image detection.
7. In addition to traditional CNNs, the U-Net implementation has been widely used for microscope image segmentation.
8. Tensorflow, Pytorch, Caffe, and Keras are trending software frameworks for deep learning.
9. A large dataset of labeled objects is required to train a deep learning network, and this can be difficult and expensive to obtain.
10. Inadequate (small) training sets can cause the network to memorize the training set and not be able to generalize and identify the full population of objects.
11. Data augmentation techniques are often used to expand the amount of available training data. They work by applying image transformations such as flipping, rotation, and cropping to generate additional training data.
12. Transfer learning is used to improve learning performance by using the weights trained for one task (the source domain) on a second related task (the target domain).
13. The training set must be representative of all the objects that the network may encounter in practice.

References

[1] W.S. McCulloch, W. Pitts, A logical calculus of the ideas immanent in nervous activity, Bull. Math. Biophys. 5 (4) (1943) 115–133, https://doi.org/10.1007/BF02478259.

[2] D.O. Hebb, The Organization of Behavior; A Neuropsychological Theory, Wiley, Oxford, England, 1949, p. xix (335).

[3] F. Rosenblatt, The perceptron: a probabilistic model for information storage and organization in the brain, Psychol. Rev. 65 (6) (1958) 386–408, https://doi.org/10.1037/h0042519.

[4] H.J. Kelley, Gradient theory of optimal flight paths, ARS J. 30 (10) (1960) 947–954, https://doi.org/10.2514/8.5282.
[5] S. Dreyfus, The numerical solution of variational problems, J. Math. Anal. Appl. 5 (1) (1962) 30–45, https://doi.org/10.1016/0022-247X(62)90004-5.
[6] K. Fukushima, Neocognitron: a hierarchical neural network capable of visual pattern recognition, Neural Netw. 1 (2) (1988) 119–130, https://doi.org/10.1016/0893-6080(88)90014-7.
[7] D.H. Ackley, G.E. Hinton, T.J. Sejnowski, A learning algorithm for boltzmann machines, Cogn. Sci. 9 (1) (1985) 147–169, https://doi.org/10.1016/S0364-0213(85)80012-4.
[8] P. Smolensky, Information Processing in Dynamical Systems: Foundations of Harmony Theory, Colorado Univ At Boulder Dept of Computer Science, Feb. 1986. [Online]. Available: https://apps.dtic.mil/sti/citations/ADA620727. (Accessed 20 August 2020).
[9] M.I. Jordan, J.W. Donahoe, V.P. Dorsel (Eds.), Chapter 25—Serial order: a parallel distributed processing approach, in: Advances in Psychology, vol. 121, North-Holland, 1997, pp. 471–495.
[10] D.E. Rumelhart, J.L. McClelland, Learning internal representations by error propagation, in: Parallel Distributed Processing: Explorations in the Microstructure of Cognition: Foundations, MITP, 1987, pp. 318–362.
[11] D.H. Ballard, "Modular learning in neural networks," in Proceedings of the Sixth National Conference on Artificial Intelligence—Volume 1, Seattle, Washington, Jul. 1987, pp. 279–284, (Accessed 20 August 2020).
[12] Y. LeCun, et al., D.S. Touretzky (Eds.), Handwritten digit recognition with a back-propagation network, in: Advances in Neural Information Processing Systems 2, Morgan-Kaufmann, 1990, pp. 396–404.
[13] S. Hochreiter, J. Schmidhuber, Long short-term memory, Neural Comput. 9 (8) (1997) 1735–1780, https://doi.org/10.1162/neco.1997.9.8.1735.
[14] G.E. Hinton, S. Osindero, Y.-W. Teh, A fast learning algorithm for deep belief nets, Neural Comput. 18 (7) (2006) 1527–1554, https://doi.org/10.1162/neco.2006.18.7.1527.
[15] R. Salakhutdinov, G. Hinton, An efficient learning procedure for deep Boltzmann machines, Neural Comput. 24 (8) (2012) 1967–2006, https://doi.org/10.1162/NECO_a_00311.
[16] A. Krizhevsky, I. Sutskever, G.E. Hinton, ImageNet classification with deep convolutional neural networks, in: F. Pereira, C.J.C. Burges, L. Bottou, K.Q. Weinberger (Eds.), Advances in Neural Information Processing Systems 25, Curran Associates, Inc, 2012, pp. 1097–1105.
[17] R. Girshick, J. Donahue, T. Darrell, J. Malik, Rich feature hierarchies for accurate object detection and semantic segmentation, in: 2014 IEEE Conference on Computer Vision and Pattern Recognition, Columbus, OH, USA, Jun. 2014, pp. 580–587, https://doi.org/10.1109/CVPR.2014.81.
[18] K. Simonyan, A. Zisserman, Very Deep Convolutional Networks for Large-Scale Image Recognition, Apr. 2015, ArXiv14091556 Cs. [Online]. Available: http://arxiv.org/abs/1409.1556. (Accessed 20 August 2020).
[19] C. Szegedy, et al., Going deeper with convolutions, in: 2015 IEEE Conference on Computer Vision and Pattern Recognition (CVPR), Boston, MA, USA, Jun. 2015, pp. 1–9, https://doi.org/10.1109/CVPR.2015.7298594.
[20] K. He, X. Zhang, S. Ren, J. Sun, Deep Residual Learning for Image Recognition, Dec. 2015, ArXiv151203385 Cs. [Online]. Available: http://arxiv.org/abs/1512.03385. (Accessed 21 August 2020).
[21] R. Girshick, Fast R-CNN, in: 2015 IEEE International Conference on Computer Vision (ICCV), Dec. 2015, pp. 1440–1448, https://doi.org/10.1109/ICCV.2015.169.
[22] S. Ren, K. He, R. Girshick, J. Sun, Faster R-CNN: towards real-time object detection with region proposal networks, in: C. Cortes, N.D. Lawrence, D.D. Lee, M. Sugiyama, R. Garnett (Eds.), Advances in Neural Information Processing Systems 28, Curran Associates, Inc, 2015, pp. 91–99.
[23] K. He, G. Gkioxari, P. Dollár, R. Girshick, Mask R-CNN, IEEE Trans. Pattern Anal. Mach. Intell. 42 (2) (2020) 386–397, https://doi.org/10.1109/TPAMI.2018.2844175.
[24] P.M. Nederlof, S. van der Flier, N.P. Verwoerd, J. Vrolijk, A.K. Raap, H.J. Tanke, Quantification of fluorescence in situ hybridization signals by image cytometry, Cytometry 13 (8) (1992) 846–852, https://doi.org/10.1002/cyto.990130807.

[25] M.V. Boland, M.K. Markey, R.F. Murphy, Classification of protein localization patterns obtained via fluorescence light microscopy, in: Proceedings of the 19th Annual International Conference of the IEEE Engineering in Medicine and Biology Society. "Magnificent Milestones and Emerging Opportunities in Medical Engineering" (Cat. No.97CH36136), vol. 2, Oct. 1997, pp. 594–597, https://doi.org/10.1109/IEMBS.1997.757680.
[26] D. Ciresan, A. Giusti, L.M. Gambardella, J. Schmidhuber, Deep neural networks segment neuronal membranes in electron microscopy images, in: F. Pereira, C.J.C. Burges, L. Bottou, K.Q. Weinberger (Eds.), Advances in Neural Information Processing Systems 25, Curran Associates, Inc, 2012, pp. 2843–2851.
[27] D.C. Cireşan, A. Giusti, L.M. Gambardella, J. Schmidhuber, Mitosis detection in breast cancer histology images with deep neural networks, in: K. Mori, I. Sakuma, Y. Sato, C. Barillot, N. Navab (Eds.), Medical Image Computing and Computer-Assisted Intervention—MICCAI 2013, vol. 8150, Springer Berlin Heidelberg, Berlin, Heidelberg, 2013, pp. 411–418.
[28] P. Foggia, G. Percannella, P. Soda, M. Vento, Benchmarking HEp-2 cells classification methods, IEEE Trans. Med. Imaging 32 (10) (2013) 1878–1889, https://doi.org/10.1109/TMI.2013.2268163.
[29] G.B. Huang, V. Jain, Deep and Wide Multiscale Recursive Networks for Robust Image Labeling, Dec. 2013, ArXiv13100354 Cs. [Online]. Available: http://arxiv.org/abs/1310.0354. (Accessed 23 August 2020).
[30] Y. Ishikawa, K. Washiya, K. Aoki, H. Nagahashi, Brain tumor classification of microscopy images using deep residual learning, in: SPIE BioPhotonics Australasia, vol. 10013, Dec. 2016, p. 100132Y, https://doi.org/10.1117/12.2242711.
[31] Y. Wang, et al., Classification of mice hepatic granuloma microscopic images based on a deep convolutional neural network, Appl. Soft Comput. 74 (2019) 40–50, https://doi.org/10.1016/j.asoc.2018.10.006.
[32] Y. Xiao, G. Yang, A fast method for particle picking in cryo-electron micrographs based on fast R-CNN, AIP Conf. Proc. 1836 (2017) 020080. https://doi.org/10.1063/1.4982020.
[33] K. Weiss, T.M. Khoshgoftaar, D. Wang, A survey of transfer learning, J. Big Data 3 (1) (2016) 9, https://doi.org/10.1186/s40537-016-0043-6.
[34] J. Yosinski, J. Clune, Y. Bengio, H. Lipson, How transferable are features in deep neural networks? in: Z. Ghahramani, M. Welling, C. Cortes, N.D. Lawrence, K.Q. Weinberger (Eds.), Advances in Neural Information Processing Systems 27, Curran Associates, Inc, 2014, pp. 3320–3328.
[35] S. Khan, N. Islam, Z. Jan, I. Ud Din, J.J.P.C. Rodrigues, A novel deep learning based framework for the detection and classification of breast cancer using transfer learning, Pattern Recogn. Lett. 125 (2019) 1–6, https://doi.org/10.1016/j.patrec.2019.03.022.
[36] O. Ronneberger, P. Fischer, T. Brox, U-net: convolutional networks for biomedical image segmentation, in: Medical Image Computing and Computer-Assisted Intervention—MICCAI 2015, Cham, 2015, pp. 234–241, https://doi.org/10.1007/978-3-319-24574-4_28.
[37] Ö. Çiçek, A. Abdulkadir, S.S. Lienkamp, T. Brox, O. Ronneberger, 3D U-net: learning dense volumetric segmentation from sparse annotation, in: Medical Image Computing and Computer-Assisted Intervention—MICCAI 2016, Cham, 2016, pp. 424–432, https://doi.org/10.1007/978-3-319-46723-8_49.
[38] Z. Zhou, M.M.R. Siddiquee, N. Tajbakhsh, J. Liang, UNet++: Redesigning Skip Connections to Exploit Multiscale Features in Image Segmentation, Jan. 2020, ArXiv191205074 Cs Eess. [Online]. Available: http://arxiv.org/abs/1912.05074. (Accessed 19 August 2020).
[39] N. Ibtehaz, M.S. Rahman, MultiResUNet : rethinking the U-net architecture for multimodal biomedical image segmentation, Neural Netw. 121 (2020) 74–87, https://doi.org/10.1016/j.neunet.2019.08.025.
[40] T. Falk, et al., U-net: deep learning for cell counting, detection, and morphometry, Nat. Methods 16 (1) (2019) 67–70, https://doi.org/10.1038/s41592-018-0261-2.
[41] F. Xing, Y. Xie, H. Su, F. Liu, L. Yang, Deep learning in microscopy image analysis: a survey, IEEE Trans. Neural Netw. Learn. Syst. 29 (10) (2018) 4550–4568, https://doi.org/10.1109/TNNLS.2017.2766168.

[42] L. Lu, Y. Shin, Y. Su, G.E. Karniadakis, Dying ReLU and Initialization: Theory and Numerical Examples, Nov. 2019, ArXiv190306733 Cs Math Stat. [Online]. Available: http://arxiv.org/abs/1903.06733. (Accessed 19 August 2020).

[43] B. Xu, N. Wang, T. Chen, M. Li, Empirical Evaluation of Rectified Activations in Convolutional Network, Nov. 2015, ArXiv150500853 Cs Stat. [Online]. Available: http://arxiv.org/abs/1505.00853. (Accessed 19 August 2020).

[44] D.-A. Clevert, T. Unterthiner, S. Hochreiter, Fast and Accurate Deep Network Learning by Exponential Linear Units (ELUs), Feb. 2016, ArXiv151107289 Cs. [Online]. Available: http://arxiv.org/abs/1511.07289. (Accessed 19 August 2020).

[45] S. Eger, P. Youssef, I. Gurevych, Is it Time to Swish? Comparing Deep Learning Activation Functions Across NLP Tasks, Jan. 2019, ArXiv190102671 Cs. [Online]. Available: http://arxiv.org/abs/1901.02671. (Accessed 19 August 2020).

[46] L. Rosasco, E.D. Vito, A. Caponnetto, M. Piana, A. Verri, Are loss functions all the same? Neural Comput. 16 (5) (2004) 1063–1076, https://doi.org/10.1162/089976604773135104.

[47] Y. LeCun, Y. Bengio, G. Hinton, Deep learning, Nature 521 (7553) (2015), https://doi.org/10.1038/nature14539, 7553.

[48] J. Gesperger, et al., Improved diagnostic imaging of brain tumors by multimodal microscopy and deep learning, Cancers 12 (7) (2020), https://doi.org/10.3390/cancers12071806, 7.

[49] P. Janjic, et al., Measurement-oriented deep-learning workflow for improved segmentation of myelin and axons in high-resolution images of human cerebral white matter, J. Neurosci. Methods 326 (2019), https://doi.org/10.1016/j.jneumeth.2019.108373, 108373.

[50] P. Teikari, M. Santos, C. Poon, K. Hynynen, Deep Learning Convolutional Networks for Multiphoton Microscopy Vasculature Segmentation, Jun. 2016, ArXiv160602382 Cs. [Online]. Available: http://arxiv.org/abs/1606.02382. (Accessed 19 August 2020).

[51] M.G. Haberl, et al., CDeep3M—plug-and-play cloud-based deep learning for image segmentation, Nat. Methods 15 (9) (2018), https://doi.org/10.1038/s41592-018-0106-z, 9.

[52] C.B. Kayasandik, W. Ru, D. Labate, A multistep deep learning framework for the automated detection and segmentation of astrocytes in fluorescent images of brain tissue, Sci. Rep. 10 (1) (2020), https://doi.org/10.1038/s41598-020-61953-9, 1.

[53] S. Wang, et al., Automated label-free detection of injured neuron with deep learning by two-photon microscopy, J. Biophotonics 13 (1) (2020), https://doi.org/10.1002/jbio.201960062, e201960062.

[54] E.C. Kugler, A. Rampun, T.J.A. Chico, P.A. Armitage, Segmentation of the zebrafish brain vasculature from light sheet fluorescence microscopy datasets, bioRxiv (2020), https://doi.org/10.1101/2020.07.21.213843. p. 2020.07.21.213843.

[55] L. Heinrich, J. Funke, C. Pape, J. Nunez-Iglesias, S. Saalfeld, Synaptic Cleft Segmentation in Non-Isotropic Volume Electron Microscopy of the Complete Drosophila Brain, May 2018, ArXiv180502718 Cs. [Online]. Available: http://arxiv.org/abs/1805.02718. (Accessed 27 August 2020).

[56] H. Lin, et al., Automated classification of hepatocellular carcinoma differentiation using multiphoton microscopy and deep learning, J. Biophotonics 12 (7) (2019), https://doi.org/10.1002/jbio.201800435, e201800435.

[57] Y. Yu, et al., Deep learning enables automated scoring of liver fibrosis stages, Sci. Rep. 8 (1) (2018), https://doi.org/10.1038/s41598-018-34300-2, 1.

[58] N. Coudray, et al., Classification and mutation prediction from non–small cell lung cancer histopathology images using deep learning, Nat. Med. 24 (10) (2018), https://doi.org/10.1038/s41591-018-0177-5, 10.

[59] F. Heinemann, G. Birk, T. Schoenberger, B. Stierstorfer, Deep neural network based histological scoring of lung fibrosis and inflammation in the mouse model system, PLoS One 13 (8) (2018), https://doi.org/10.1371/journal.pone.0202708, e0202708.

[60] M. Hermsen, et al., Deep learning–based histopathologic assessment of kidney tissue, J. Am. Soc. Nephrol. 30 (10) (2019) 1968–1979, https://doi.org/10.1681/ASN.2019020144.

[61] S. Kannan, et al., Segmentation of glomeruli within trichrome images using deep learning, Kidney Int. Rep. 4 (7) (2019) 955–962, https://doi.org/10.1016/j.ekir.2019.04.008.

[62] K. Sirinukunwattana, S.E.A. Raza, Y.-W. Tsang, D.R.J. Snead, I.A. Cree, N.M. Rajpoot, Locality sensitive deep learning for detection and classification of nuclei in routine colon cancer histology images, IEEE Trans. Med. Imaging 35 (5) (2016) 1196–1206, https://doi.org/10.1109/TMI.2016.2525803.
[63] A. Rehman, N. Abbas, T. Saba, S.I. ur Rahman, Z. Mehmood, H. Kolivand, Classification of acute lymphoblastic leukemia using deep learning, Microsc. Res. Tech. 81 (11) (2018) 1310–1317, https://doi.org/10.1002/jemt.23139.
[64] F. Buggenthin, et al., Prospective identification of hematopoietic lineage choice by deep learning, Nat. Methods 14 (4) (2017), https://doi.org/10.1038/nmeth.4182, 4.
[65] Y. Song, et al., A deep learning based framework for accurate segmentation of cervical cytoplasm and nuclei, in: 2014 36th Annual International Conference of the IEEE Engineering in Medicine and Biology Society, Aug. 2014, pp. 2903–2906, https://doi.org/10.1109/EMBC.2014.6944230.
[66] K. Nagpal, et al., Development and validation of a deep learning algorithm for improving Gleason scoring of prostate cancer, NPJ Digit. Med. 2 (1) (2019), https://doi.org/10.1038/s41746-019-0112-2, 1.
[67] P. Kaur, K.J. Dana, G.O. Cula, M.C. Mack, Hybrid deep learning for reflectance confocal microscopy skin images, in: 2016 23rd International Conference on Pattern Recognition (ICPR), Dec. 2016, pp. 1466–1471, https://doi.org/10.1109/ICPR.2016.7899844.
[68] P. Buyssens, A. Elmoataz, O. Lézoray, Multiscale convolutional neural networks for vision–based classification of cells, in: K.M. Lee, Y. Matsushita, J.M. Rehg, Z. Hu (Eds.), Computer Vision—ACCV 2012, vol. 7725, Springer Berlin Heidelberg, Berlin, Heidelberg, 2013, pp. 342–352.
[69] D.H. Hubel, T.N. Wiesel, Receptive fields, binocular interaction and functional architecture in the cat's visual cortex, J. Physiol. 160 (1) (1962) 106–154.2.
[70] A.A. Cruz-Roa, J.E. Arevalo Ovalle, A. Madabhushi, F.A. González Osorio, A deep learning architecture for image representation, visual interpretability and automated basal-cell carcinoma cancer detection, in: Medical Image Computing and Computer-Assisted Intervention—MICCAI 2013, Berlin, Heidelberg, 2013, pp. 403–410, https://doi.org/10.1007/978-3-642-40763-5_50.
[71] Y. Xu, T. Mo, Q. Feng, P. Zhong, M. Lai, E.I.-C. Chang, Deep learning of feature representation with multiple instance learning for medical image analysis, in: 2014 IEEE International Conference on Acoustics, Speech and Signal Processing (ICASSP), 2014, pp. 1626–1630, https://doi.org/10.1109/ICASSP.2014.6853873.
[72] Z. Gao, L. Wang, L. Zhou, J. Zhang, HEp-2 cell image classification with deep convolutional neural networks, IEEE J. Biomed. Health Inform. 21 (2) (2017) 416–428, https://doi.org/10.1109/JBHI.2016.2526603.
[73] N. Tajbakhsh, et al., Convolutional neural networks for medical image analysis: full training or fine tuning? IEEE Trans. Med. Imaging 35 (5) (2016) 1299–1312, https://doi.org/10.1109/TMI.2016.2535302.
[74] L.D. Nguyen, D. Lin, Z. Lin, J. Cao, Deep CNNs for microscopic image classification by exploiting transfer learning and feature concatenation, in: 2018 IEEE International Symposium on Circuits and Systems (ISCAS), 2018, pp. 1–5, https://doi.org/10.1109/ISCAS.2018.8351550.
[75] M. Xiao, X. Shen, W. Pan, Application of deep convolutional neural networks in classification of protein subcellular localization with microscopy images, Genet. Epidemiol. 43 (3) (2019) 330–341, https://doi.org/10.1002/gepi.22182.
[76] M.M. Alam, M.T. Islam, Machine learning approach of automatic identification and counting of blood cells, Healthc. Technol. Lett. 6 (4) (2019) 103–108, https://doi.org/10.1049/htl.2018.5098.
[77] A. Waisman, et al., Deep learning neural networks highly predict very early onset of pluripotent stem cell differentiation, Stem Cell Rep. 12 (4) (2019) 845–859, https://doi.org/10.1016/j.stemcr.2019.02.004\.
[78] L. Roux, et al., Mitosis detection in breast cancer histological images an ICPR 2012 contest, J. Pathol. Inform. 4 (2013) 8, https://doi.org/10.4103/2153-3539.112693.
[79] A. Cruz-Roa, et al., Accurate and reproducible invasive breast cancer detection in whole-slide images: a deep learning approach for quantifying tumor extent, Sci. Rep. 7 (2017) 46450, https://doi.org/10.1038/srep46450.

[80] T. Chen, C. Chefd'hotel, Deep learning based automatic immune cell detection for immunohistochemistry images, in: Machine Learning in Medical Imaging, Cham, 2014, pp. 17–24, https://doi.org/10.1007/978-3-319-10581-9_3.
[81] Y. Mao, Z. Yin, J.M. Schober, Iteratively training classifiers for circulating tumor cell detection, in: 2015 IEEE 12th International Symposium on Biomedical Imaging (ISBI), 2015, pp. 190–194, https://doi.org/10.1109/ISBI.2015.7163847.
[82] J. Hung, A. Carpenter, Proceedings of the IEEE Conference on Computer Vision and Pattern Recognition (CVPR) Workshops, 2017, pp. 56–61.
[83] S. Albarqouni, C. Baur, F. Achilles, V. Belagiannis, S. Demirci, N. Navab, AggNet: deep learning from crowds for mitosis detection in breast cancer histology images, IEEE Trans. Med. Imaging 35 (5) (2016) 1313–1321, https://doi.org/10.1109/TMI.2016.2528120.
[84] M. Weigert, U. Schmidt, R. Haase, K. Sugawara, G. Myers, Star-convex polyhedra for 3D object detection and segmentation in microscopy, in: 2020 IEEE Winter Conference on Applications of Computer Vision (WACV), Snowmass Village, CO, USA, Mar. 2020, pp. 3655–3662, https://doi.org/10.1109/WACV45572.2020.9093435.
[85] J. Long, E. Shelhamer, T. Darrell, Proceedings of the IEEE Conference on Computer Vision and Pattern Recognition (CVPR), 2015, pp. 3431–3440.
[86] S.E.A. Raza, et al., Micro-net: a unified model for segmentation of various objects in microscopy images, Med. Image Anal. 52 (2019) 160–173, https://doi.org/10.1016/j.media.2018.12.003.
[87] J.W. Johnson, Adapting Mask-RCNN for Automatic Nucleus Segmentation, ArXiv180500500 Cs, vol. 944, 2020, https://doi.org/10.1007/978-3-030-17798-0.
[88] M. Abadi, et al., TensorFlow: Large-Scale Machine Learning on Heterogeneous Distributed Systems, Mar. 2016, ArXiv160304467 Cs. [Online]. Available: http://arxiv.org/abs/1603.04467. (Accessed 5 August 2020).
[89] A. Paszke, et al., PyTorch: an imperative style, high-performance deep learning library, in: H. Wallach, H. Larochelle, A. Beygelzimer, F. d'Alché-Buc, E. Fox, R. Garnett (Eds.), Advances in Neural Information Processing Systems 32, Curran Associates, Inc, 2019, pp. 8026–8037.
[90] Y. Jia, et al., Caffe: convolutional architecture for fast feature embedding, in: Proceedings of the 22nd ACM International Conference on Multimedia, New York, NY, USA, Nov. 2014, pp. 675–678, https://doi.org/10.1145/2647868.2654889.
[91] N. Ketkar, Introduction to Keras, in: N. Ketkar (Ed.), Deep Learning with Python: A Hands-on Introduction, Apress, Berkeley, CA, 2017, pp. 97–111.
[92] N. Srivastava, G. Hinton, A. Krizhevsky, I. Sutskever, R. Salakhutdinov, Dropout: a simple way to prevent neural networks from overfitting, J. Mach. Learn. Res. 15 (1) (2014) 1929–1958.
[93] S. Ioffe, C. Szegedy, Batch Normalization: Accelerating Deep Network Training by Reducing Internal Covariate Shift, Mar. 2015, ArXiv150203167 Cs. [Online]. Available: http://arxiv.org/abs/1502.03167. (Accessed 19 August 2020).
[94] L. Taylor, G. Nitschke, Improving Deep Learning Using Generic Data Augmentation, Aug. 2017, ArXiv170806020 Cs Stat. [Online]. Available: http://arxiv.org/abs/1708.06020. (Accessed 19 August 2020).
[95] J. Kukačka, V. Golkov, D. Cremers, Regularization for Deep Learning: A Taxonomy, Oct. 2017, ArXiv171010686 Cs Stat. [Online]. Available: http://arxiv.org/abs/1710.10686. (Accessed 19 August 2020).
[96] C. Shorten, T.M. Khoshgoftaar, A survey on image data augmentation for deep learning, J. Big Data 6 (1) (2019) 60, https://doi.org/10.1186/s40537-019-0197-0.
[97] R. Li, T. Zeng, H. Peng, S. Ji, Deep learning segmentation of optical microscopy images improves 3-D neuron reconstruction, IEEE Trans. Med. Imaging 36 (7) (2017) 1533–1541, https://doi.org/10.1109/TMI.2017.2679713.
[98] S.J. Yang, et al., Assessing microscope image focus quality with deep learning, BMC Bioinf. 19 (1) (2018) 77, https://doi.org/10.1186/s12859-018-2087-4.
[99] C. Tan, F. Sun, T. Kong, W. Zhang, C. Yang, C. Liu, A survey on deep transfer learning, in: Artificial Neural Networks and Machine Learning—ICANN 2018, Cham, 2018, pp. 270–279, https://doi.org/10.1007/978-3-030-01424-7_27.

[100] N. Li, H. Hao, Q. Gu, D. Wang, X. Hu, A transfer learning method for automatic identification of sandstone microscopic images, Comput. Geosci. 103 (2017) 111–121, https://doi.org/10.1016/j.cageo.2017.03.007.
[101] J. Donahue, et al., DeCAF: A Deep Convolutional Activation Feature for Generic Visual Recognition, Oct. 2013, ArXiv13101531 Cs. [Online]. Available: http://arxiv.org/abs/1310.1531. (Accessed 21 August 2020).
[102] M.D. Zeiler, R. Fergus, Visualizing and Understanding Convolutional Networks, Nov. 2013, ArXiv13112901 Cs. [Online]. Available: http://arxiv.org/abs/1311.2901. (Accessed 21 August 2020).
[103] P. Sermanet, D. Eigen, X. Zhang, M. Mathieu, R. Fergus, Y. LeCun, OverFeat: Integrated Recognition, Localization and Detection Using Convolutional Networks, Feb. 2014, ArXiv13126229 Cs. [Online]. Available: http://arxiv.org/abs/1312.6229. (Accessed 21 August 2020).
[104] C. Szegedy, V. Vanhoucke, S. Ioffe, J. Shlens, Z. Wojna, Rethinking the Inception Architecture for Computer Vision, Dec. 2015, ArXiv151200567 Cs. [Online]. Available: http://arxiv.org/abs/1512.00567. (Accessed 21 August 2020).
[105] Y. Ganin, et al., Domain-adversarial training of neural networks, J. Mach. Learn. Res. 17 (1) (2016), 2096–2030.
[106] M. Majurski, et al., Cell image segmentation using generative adversarial networks, transfer learning, and augmentations, in: 2019 IEEE/CVF Conference on Computer Vision and Pattern Recognition Workshops (CVPRW), Long Beach, CA, USA, 2019, pp. 1114–1122, https://doi.org/10.1109/CVPRW.2019.00145.
[107] M. Dragan, et al., Whole-brain tissue mapping toolkit using large-scale highly multiplexed immunofluorescence imaging and deep neural networks, Nat. Commun. 12 (1) (2021) 1–12.
[108] J. Deng, et al., ImageNet: a large-scale hierarchical image database, in: Proc Cvpr Ieee, 2009, pp. 248–255.
[109] J. Jahanipour, et al., Whole-Brain Tissue Mapping Toolkit Using Large-Scale Highly Multiplexed Immunofluorescence Imaging and Deep Neural Networks (Code), Zenodo, 2021. https://zenodo.org/record/4415963#.YB2Ds6dKiUk.
[110] D. Maric, et al., Whole-Brain Tissue Mapping Toolkit Using Large-Scale Highly Multiplexed Immunofluorescence Imaging and Deep Neural Networks (Data), figshare, 2021, https://doi.org/10.6084/m9.figshare.13731585.v1.
[111] A.X. Lu, O.Z. Kraus, S. Cooper, A.M. Moses, Learning unsupervised feature representations for single cell microscopy images with paired cell inpainting, PLoS Comput. Biol. 15 (9) (2019), https://doi.org/10.1371/journal.pcbi.1007348, 9.
[112] D.P. Kingma, M. Welling, Auto-Encoding Variational Bayes, May 2014, ArXiv13126114 Cs Stat. [Online]. Available: http://arxiv.org/abs/1312.6114. (Accessed 24 August 2020).
[113] I. Goodfellow, et al., Generative adversarial nets, in: Z. Ghahramani, M. Welling, C. Cortes, N.D. Lawrence, K.Q. Weinberger (Eds.), Advances in Neural Information Processing Systems 27, Curran Associates, Inc, 2014, pp. 2672–2680.
[114] N. Miolane, F. Poitevin, Y.-T. Li, S. Holmes, Estimation of Orientation and Camera Parameters from Cryo-Electron Microscopy Images with Variational Autoencoders and Generative Adversarial Networks, 2020, pp. 970–971. [Online]. Available: https://openaccess.thecvf.com/content_CVPRW_2020/html/w57/Miolane_Estimation_of_Orientation_and_Camera_Parameters_From_Cryo-Electron_Microscopy_Images_CVPRW_2020_paper.html. (Accessed 24 August 2020).
[115] S. Lee, S. Han, P. Salama, K.W. Dunn, E.J. Delp, Three dimensional blind image deconvolution for fluorescence microscopy using generative adversarial networks, in: 2019 IEEE 16th International Symposium on Biomedical Imaging (ISBI 2019), Apr. 2019, pp. 538–542, https://doi.org/10.1109/ISBI.2019.8759250.
[116] H. Zhang, et al., High-throughput, high-resolution deep learning microscopy based on registration-free generative adversarial network, Biomed. Opt. Express 10 (3) (2019) 1044–1063, https://doi.org/10.1364/BOE.10.001044.
[117] P. Yuan, et al., Few Is Enough: Task-Augmented Active Meta-Learning for Brain Cell Classification, Jul. 2020, ArXiv200705009 Cs. [Online]. Available: http://arxiv.org/abs/2007.05009. (Accessed 19 August 2020).

[118] L. Saadatifard, A. Mobiny, P. Govyadinov, H. Nguyen, D. Mayerich, DVNet: A Memory-Efficient Three-Dimensional CNN for Large-Scale Neurovascular Reconstruction, Feb. 2020, ArXiv200201568 Cs Eess Stat. [Online]. Available: http://arxiv.org/abs/2002.01568. (Accessed 24 August 2020).

[119] Q.V. Le, J. Ngiam, A. Coates, A. Lahiri, B. Prochnow, A.Y. Ng, On optimization methods for deep learning, in: Presented at the ICML, Jan. 2011. [Online]. Available: https://openreview.net/forum?id=Sk4lD3W_bB. (Accessed 25 August 2020).

[120] J. Dean, et al., Large scale distributed deep networks, in: F. Pereira, C.J.C. Burges, L. Bottou, K.Q. Weinberger (Eds.), Advances in Neural Information Processing Systems 25, Curran Associates, Inc, 2012, pp. 1223–1231.

[121] S. Chetlur, et al., cuDNN: Efficient Primitives for Deep Learning, Dec. 2014, ArXiv14100759 Cs. [Online]. Available: http://arxiv.org/abs/1410.0759. (Accessed 25 August 2020).

[122] L. Hou, D. Samaras, T.M. Kurc, Y. Gao, J.E. Davis, J.H. Saltz, Patch-Based Convolutional Neural Network for Whole Slide Tissue Image Classification, Mar. 2016, ArXiv150407947 Cs. [Online]. Available: http://arxiv.org/abs/1504.07947. (Accessed 24 August 2020).

[123] M. Li, et al., Scaling Distributed Machine Learning with the Parameter Server, 2014, pp. 583–598. [Online]. Available: https://www.usenix.org/conference/osdi14/technical-sessions/presentation/li_mu. (Accessed 25 August 2020).

[124] Y. Xie, F. Xing, X. Kong, H. Su, L. Yang, Beyond classification: structured regression for robust cell detection using convolutional neural network, Med. Image Comput. Comput. Assist. Interv. 9351 (2015) 358–365, https://doi.org/10.1007/978-3-319-24574-4_43.

[125] R.M. Patton, et al., 167-PFlops deep learning for electron microscopy: from learning physics to atomic manipulation, in: SC18: International Conference for High Performance Computing, Networking, Storage and Analysis, Nov. 2018, pp. 638–648, https://doi.org/10.1109/SC.2018.00053.

CHAPTER SIXTEEN

Image Informatics

Kyle I.S. Harrington and Kevin W. Eliceiri

16.1 Introduction

Bioimage informatics is a rapidly evolving field focused on the handling and processing of image data of biological subjects with an emphasis on cellular and molecular data. While bioimaging data can be acquired with a wide range of instruments, from MRI and CT to microscopes, microscopy-based methods were the impetus behind the term "bioimage informatics" [1]. This is due in large part to the great challenges and opportunities in attempting to use computational methods to gain knowledge from complex and heterogeneous microscopy images and related metadata. Bioimage informatics covers the complete life cycle of a modern microscopy dataset, from image acquisition and curation to image analysis and visualization (see Fig. 16.1).

Image informatics methods may be tightly coupled with image acquisition to the extent of informing or even controlling the acquisition process, or may be completely decoupled, where all image processing and analysis is performed post hoc. Regardless of the interactivity of the imaging system, a modern image informatics pipeline can involve many steps and processes (Fig. 16.1).

Image acquisition involves converting the digital representation of imaging observations from imaging hardware into a data representation that can be used in the rest of the pipeline. Image storage addresses the digital representation of image data and its associated metadata, including data representation, compression, and storage hardware. Image processing, analysis, and visualization are often discussed in tandem because these processes focus on algorithms and data structures that operate on image data. Image-processing algorithms transform image data into other types of image data, while image analysis addresses the quantification of image data, often as nonimage data. Image curation is another critical image informatics process, wherein humans are involved in analysis and decision making about the rest of the image informatics pipeline. As a whole, image informatics involves a wide range of challenges that arise from not only the diverse imaging hardware, modalities, and image sizes, but the breadth of computational tasks in an image informatics pipeline.

Bioimage informatics pipelines often involve multiple software tools that are often separated into three broad categories: acquisition software, storage and curation suites, and image processing and analysis tools. Software tools to support image informatics

Microscope Image Processing
https://doi.org/10.1016/B978-0-12-821049-9.00002-2
Copyright © 2023 Elsevier Inc.
All rights reserved.

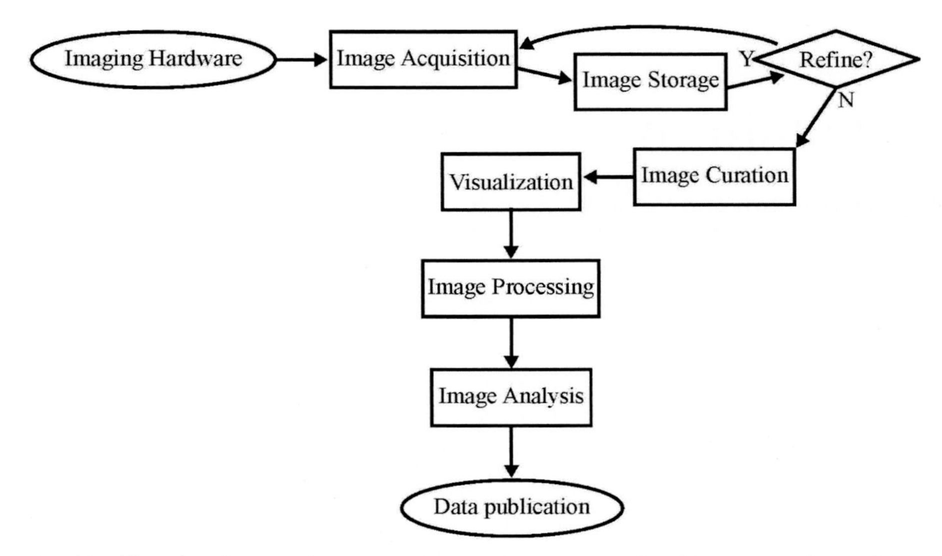

Fig. 16.1 The lifecycle of a typical imaging dataset. *(Image created with assistance from Ed Evans)*

are both closed- and open-source. Closed-source tools are proprietary to companies, can be expensive, and typically cannot be directly adapted by users. Open-source tools can be adapted by users and are generally available free of charge. Open-source tools can involve both commercial and academic entities. While both open- and closed-source tools have had major impacts on modern imaging, in this review we focus on open-source tools, given the opportunities for innovation through improved transparency and collaboration. We also focus on applications in biological microscopy, even though there are similar image informatics requirements and approaches in related imaging disciplines such as medical imaging and astronomy. Lastly, although there are myriad open-source image informatics tools, in this chapter we present a brief survey of open-source software tools that have been broadly adopted or have demonstrated unique capabilities.

Open-source software is increasingly utilized in modern science because it enables a complete review and adaptation of software tools, ensuring that future scientists can continue to innovate and build on these tools. Open-source tools are often free, which lowers the barrier for users and future researchers, and they are typically developed on public platforms, which adds the general public to the pool of potential developers. This also ensures that if the original developers of a tool cease to support the software, then others can pick up the maintenance and development. Ultimately, within the context of science, open-source aligns with the need for transparent research practices that enable reproducibility.

The term "open-source" has been applied to a variety of ideas, technologies, and platforms that are not open in the practical sense. A particularly problematic instance of this occurs when researchers add a footnote to a paper that says, "Software is available on request." This produces several accessibility concerns: the author's email may change, the author may act as a gatekeeper for research, etc. Truly open software is available in real time from publicly accessible sources, but the best open-source projects go well beyond this. They not only ensure the availability of source code, but they also provide executables that can be run on the target hardware platforms and operating systems, along with well-documented code and manuals, and a ticketing system for tracking bugs and feature requests. In fact, all of these features are readily available through existing version control system platforms, such as GitLab, GitHub, BitBucket, Sourceforge, and others. GitLab is a particularly notable example where both the platform itself and its source code are open-source. Closed-source proprietary solutions can have the advantage of more dedicated development resources and broader user support. But the bioimage informatics community has an incentive to use, develop, and support open-source tools. Broadly usable, free, open-source tools ensure that this community can grow.

16.2 Open-source Software Ecosystems

There are multiple overlapping software ecosystems in the open-source image informatics world. These are largely identified by the programming language in which they are implemented. Most major programming languages include some degree of support for image processing and visualization, but there are some key image informatics ecosystems that are largely responsible for driving and supporting the image informatics community.

16.2.1 Java Libraries and Tools

The Java image informatics ecosystem comprises a collection of libraries and tools, not all of which are fully compatible with each other, but the SciJava libraries discussed here seek to provide a unified toolkit. One of the reasons Java has been a popular choice in the image informatics world is that Java programs can be run on all major operating systems and are supported for essentially all hardware platforms. To use Java code, a user must install a Java Virtual Machine, which can execute a low-level code called Java bytecode. To create a Java program, a developer writes Java source code, which is then compiled into Java bytecode. This Java bytecode can then be run on any Java virtual machine that supports that version of bytecode. Java virtual machines have been developed for nearly all major hardware platforms and operating systems. Image informatics tools written in

Java can be run on all major operating systems and can continue to be used even as computing hardware standards are advanced.

The Java tool with the longest-standing history in image informatics is the original ImageJ program, which, in this chapter, is referred to as ImageJ1. It was originally developed in 1997 by Wayne Rasband at the US National Institutes of Health to address the need for a cross-platform tool. It was based on the success of a 1987 Pascal-based NIH Image effort by Rasband. ImageJ1's unique design enables extensibility and flexibility, and has attracted users for many years [2]. Several other Java-based tools have been developed since the introduction of ImageJ. For example, Icy (see Section 16.2.3) is a tool that leverages the ITK/VTK ecosystem [3].

Fiji (stands for "Fiji Is Just ImageJ") is a distribution of ImageJ1 that has significantly extended the ImageJ ecosystem since its debut in 2007 [4]. The ImgLib2 image data library [5] was introduced to overcome image size and dimensionality restrictions present in ImageJ1. ImageJ2 and SciJava [6] were developed to further alleviate limitations of ImageJ1, by providing improved support for modular code, automated user interface generation, and other features. ImageJ2 and ImageJ1 are both available in Fiji, allowing for modern features and support while still offering backward compatibility. Extensibility was enhanced by introducing a plugin update system allowing Fiji developers to readily share plugins and tools. Plugins are special-purpose software components (available in all versions of ImageJ) that can be integrated into a specific installation of Fiji, making it possible for the plugin to interoperate with the rest of the specific Fiji installation. Examples of popular plugins include TrackMate, a plugin for tracking spots (such as particles or cells) [7], and Weka Trainable Segmentation, an interactive machine learning tool for image segmentation [8]. Support for polyglot scripting is provided in Fiji, making it possible to write image informatics code in many programming languages, such as Groovy, Beanshell, Clojure, Python (the Jython dialect), and others. Some of the language extensions have grown into independent libraries [9].

An important element of the Java ecosystem is the Bio-Formats library [10], maintained by the Open Microscopy Environment (OME) team. This is a major effort that provides support for many popular image file formats. This open-source effort makes it possible to use bioimage informatics tools with most common bioimage file formats. The OME team is developing a new image format called OME-NGFF to support the growing complexity and size of bioimage datasets [11].

Many other bioimage informatics tools have been built with Java. An example is QuPath, [12] a new tool that supports quantitative pathology, focused primarily on whole slide images. MATLAB [13] is a widely used platform that makes extensive use of Java and is widely used by academic developers, particularly in the field of signal processing. It is, however, a commercial product and is not available as an open-source tool.

16.2.2 Python Tools

Python has a large programmer community, as well as tools for image informatics. At the core of the Python image informatics community is the NumPy library [14], which is used to represent image data as matrix-like arrays. Libraries, such as scikit-image [15], then implement image processing and analysis routines that use these image arrays. Popular tools have been built using these libraries. Cellprofiler [16,17] is a tool for creating image analysis pipelines in a very user-friendly format that was designed to target biologists with limited programming experience. Ilastik, [18] another powerful tool in the Python ecosystem, supports interactive machine learning. The three major machine learning frameworks that are commonly used for deep learning are Tensorflow, PyTorch, and jax. They also integrate into the NumPy and scikit toolkit by providing native support for NumPy arrays, and often mirroring APIs from scikit. The Python image informatics ecosystem also includes a powerful visualization tool, napari [19], that provides flexible visualization of Python image data.

16.2.3 C++ Tools

The leading C++-based open-source library for image analysis is the Insight Toolkit (ITK) [20]. This image analysis framework has a long-standing history and is often chosen because of the speed benefits of C++ versus other programming languages. A particularly accessible library for ITK, called Simple ITK, [21,22] has facilitated the use of ITK from other programming languages.

Vigra [23] is another C++ library with a long history of use in academic research. It has inspired other libraries, such as ImgLib2, by leveraging data structures that enable elegant algorithm development by extending a small number of adapters.

16.2.4 Tool Interoperation

As we mentioned in the introduction, an image informatics pipeline can often involve multiple tools. Since there is no single best tool or best open-source framework, interoperation between tools is particularly important. There are some key considerations about tool interoperation. The most common case of tool interoperation is performing a process in one tool, saving the result, and then opening the result in a different tool. An immediate concern with this type of interoperation is that saving a result can be a relatively slow process, especially in the case of large images. This generally becomes a concern when the time it takes to save data is longer than the time required to process the data in another tool. One way to work around this is to share the memory between tools.

There are multiple ways to support shared memory between tools. Simple ITK does this by providing other languages access to ITK data structures and algorithms as a library.

This is effective at enabling other programs to access ITK, but if those programs use a different data representation, the data itself must be copied from the ITK representation to the destination representation. On the other hand, the ImgLyb and PyImageJ libraries support interoperation between ImgLib2/ImageJ and Numpy/Scikit-image directly. They provide access to NumPy arrays as ImgLib2 data. Thus there is no need to copy or duplicate memory when using ImgLib2 and ImageJ code with Python data.

In general, interoperability between tools can be supported by Foreign Function Interfaces (FFIs), which provide a way to execute code in one programming language that was written in another language. Examples of this originate from Common Lisp and are found in Java as the Java Native Interface (JNI) and Java Native Access (JNA). Most major languages support some form of FFI, but the approaches can be quite specific and are outside the scope of this chapter.

Currently there is a trend toward polyglot solutions, where multiple programming languages are used to achieve a goal. Every programming language has its own strengths and weaknesses, and no language is superior in all aspects. However, it can be quite challenging to mix programming languages. Common approaches involve using programs or scripts written in each language that read and write their inputs from files. Then the programs/scripts are executed in a predefined order as a pipeline where the outputs of previous steps are used as inputs to subsequent steps. Interactive notebooks, such as Jupyter Notebook, have achieved the commendable goal of implementing polyglot solutions within single files. Notebook files are divided into cells, where each cell may be written in a different language. The cells of the notebook can be executed sequentially, or as needed by the programmer. These solutions are enabling developers to harness the best features from multiple languages simultaneously.

16.3 Image Acquisition

Micro-Manager [24] is a widely adopted open-source software platform for microscope control and image acquisition. It is free, and it has support for a wide range of hardware as well as strong extensibility/customizability/programmability/interoperability that is unmatched by commercial software. This has led to Micro-Manager being used for many different camera-based imaging modalities (e.g., wide-field, spinning disk confocal, and light-sheet [25] microscopy).

16.3.1 Image Processing and Analysis

Earlier chapters have covered the fundamentals of microscope image analysis, and the following chapter introduces usage examples. In this section we focus on open-source image informatics tools that offer state-of-the-art methods for common microscope image analysis tasks. The majority of these tools build upon or work as plugins to existing open-source software frameworks.

Cell segmentation is the task of identifying individual cells (see Chapter 7). The current top-performing algorithms for cell segmentation use supervised deep learning methods (Chapter 15). The StarDist approach uses an assumption of convex shape, where variable radii of a polar model are fit to image data by the trained model [26,27]. The model can be trained with either Python or ImageJ, depending on the user's preference. Another deep learning-based tool that addresses cell segmentation is CellPose [28]. It is a Python-based tool that performs single cell segmentation. CellPose introduces a new prior that involves preprocessing supervised training data by simulating diffusion to compute spatial gradients. The neural network is then trained on the spatial gradients, which allows for segmentations that can handle nonconvex shapes. There are an increasing number of deep learning-based approaches for cell segmentation, and a complete review is beyond the scope of this chapter.

Cell tracking is the task of observing the position of cells across time (Chapter 14). This can be accomplished by manually tracking cell locations, or it can leverage image segmentations. TrackMate is a powerful ImageJ tool that is popular because of its general usability [7]. Users can use detection algorithms to find candidate spots to track. Then, like many tracking programs, a tracking algorithm assigns tracks to cells across time points. A popular algorithm for doing this is the Linear Assignment Problem (LAP) tracker [29], which has been adapted in TrackMate to support cell division and more. However, with the increasing size of biological datasets, the number of cells that must be tracked has increased. Some algorithms and features of TrackMate work well for small numbers of cells, but can become prohibitive with large numbers of cells. The MaMuT and Mastodon tools [30] have been designed with these challenges in mind, and they share implementations with BigDataViewer [31].

While the previous examples are primarily from the ImageJ ecosystem, new tools are being developed in the Python ecosystem as well. The Arboretum plugin for napari [32] provides similar features to MaMuT, such as tracking algorithms, lineage tree plotting, and enhanced visualization.

Image registration and stitching are common tasks for large datasets, where regions of overlapping images are identified and aligned (see Chapter 4). TurboReg [33], a user-friendly plugin for ImageJ, is a popular open-source solution for quickly registering image stacks in ImageJ. It solves image registration problems under five different transformation constraints: (1) translation (find an offset between two images), (2) rigid body (translation with a rotation between the images), (3) scaled rotation (rigid body with a different scale between the two images), (4) affine (fully tunable affine transformation matrix), and (5) deformable (the transformation changes spatially, deforming at least one image). However, TurboReg is an ImageJ1 plugin, and has some limitations in the support of very large images. The BigStitcher tool for Fiji [34] is an image registration tool with additional features to support tiled, multiview light-sheet imaging. Both TurboReg and BigStitcher can solve for translation, rigid-body, and affine image registration models. BigStitcher provides a number of features that make it particularly appropriate for advanced image registration challenges that cannot be solved by aligning pairs of images.

Image processing algorithms can be very computationally intensive. Many algorithms require at least one iteration over all pixels in an image, others require multiple iterations, and some algorithms have even more costly scaling properties. As the size of the images increases, the resulting computational costs quickly grow, and the cost increases because all data in the image must be processed. This challenge has been addressed by utilizing new hardware technologies, specifically graphics processing units (GPUs). GPU technology has advanced primarily due to the demands of modern computer games, but it provides excellent computational properties for solving numerical problems. Advances in GPUs have directly spurred the growth of deep learning and other machine learning methods, and they have been widely used by the image informatics community. Tools such as CLIJ [35] and CLIJ2 [36] provide implementations of popular image processing algorithms for the GPU. This has led to significant reductions in runtime for image processing pipelines.

16.3.2 Machine Learning Platforms

Machine learning (ML) is now a fundamental part of modern microscope image analysis (see Chapter 15). For example, if traditional cell segmentation methods, such as the watershed algorithm (Chapters 6 and 7), do not work on a dataset, then the next attempt is likely to be the use of an ML tool. In additional to the user-friendly open-source cell segmentation tools previously discussed, there are ML platforms that provide broad support for image informatics.

ImJoy is a creative and flexible, cloud-based ML platform [37]. An example of ImJoy is shown in Fig. 16.2. ImJoy allows users to interface with the platform using a web browser, making ImJoy accessible on any device with sufficient browser support. Another cloud-based tool for ML is ZeroCostDL4Mic, which allows users to harness free cloud resources to process their images [38].

While cloud-based tools guarantee broad accessibility of tools, high-performance computers and clusters are often available to researchers performing image informatics, and this can be more cost-effective than cloud usage. The CSBDeep platform was initially developed for the CARE [39] and StarDist tools [26]. CSBDeep facilitates interoperation between tools based on Keras and ImageJ. DeepImageJ is another major effort to provide deep learning support in ImageJ, and it provides support for a large number of trained deep learning models [40].

KNIME is a standalone ML platform for general-purpose data science. The KNIME Image Processing plugin for KNIME provides ML support for image informatics [41]. It is worth emphasizing that in the development of the KNIME Image Processing plugin,

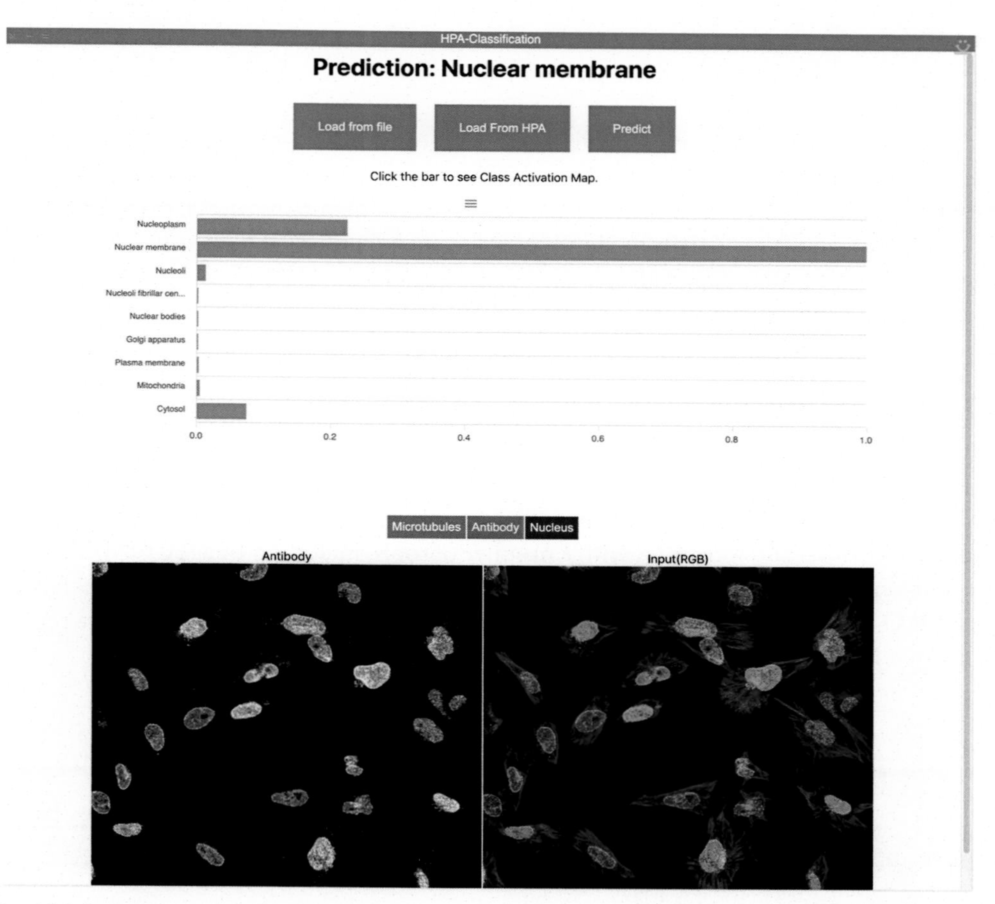

Fig. 16.2 ImJoy is a web-based platform for building interactive deep learning tools. This figure shows an ImJoy demo plugin for classifying protein localization patterns in microscopy images with deep neural networks. Users can select local images or load images from the Human Protein Atlas and predict protein localization labels with probability values. *(Image courtesy Wei Ouyang of the KTH Royal Institute of Technology)*

the KNIME team has made extensive contributions to the ImageJ and SciJava ecosystems by implementing algorithms and API features.

16.4 Image Storage and Curation

Image informatics applications are data intensive. Storage of images, as well as their curation, are particularly important for supporting data science and reproducibility. Images are encoded and compressed in a variety of file formats, some lossy (such as JPEG) and some lossless (such as PNG). Due to the large image volumes and the need to maintain the image data for extended periods of time, data storage facilities are a particularly critical concern.

Data repositories are being established to support the long-term storage of scientific image data. The Image Data Resource (IDR) is maintained by the Open Microscopy Environment (OME) team. The CellProfiler team maintains the Broad Bioimage Benchmark Collection (BBBC). Zenodo is an effort from CERN that stores any type of scientific data, including source code. It also provides a DOI for each data element, making the element citable.

16.4.1 Data Curation

The Open Microscopy Environment (OME) team created and supports the OMERO platform for managing microscope image data [42]. As discussed earlier, the OME team is known for maintaining the Bio-Formats library, which provides support for a large number of file formats and integrates with a number of tools, including ImageJ. OMERO has a client-server architecture that enables researchers to explore image data and metadata. OMERO is now used to manage the Image Data Resource public bioimage data repository (see Fig. 16.3).

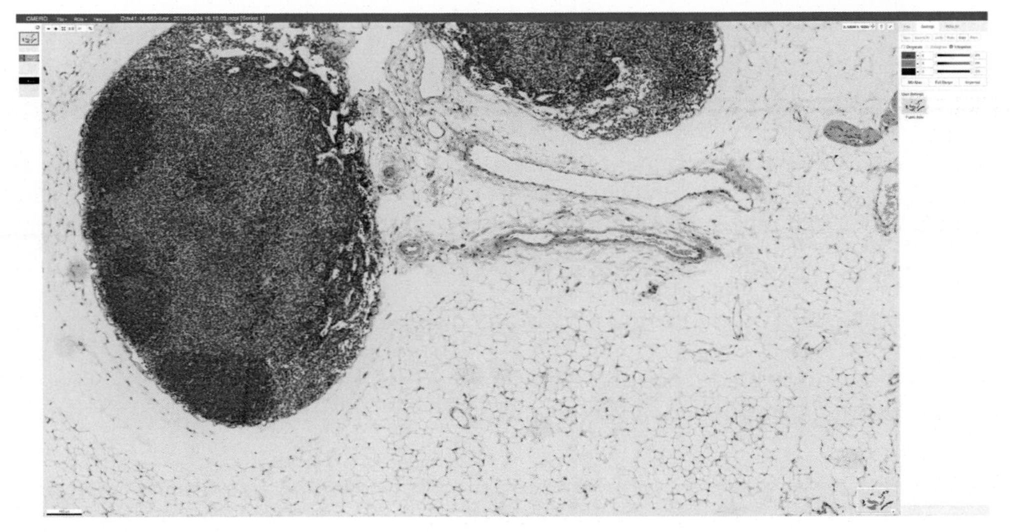

Fig. 16.3 The Image Data Resource is an OMERO-backed data repository. This figure shows a histopathology slide of a liver sample. The full dataset is publicly available and can be found at https://idr.openmicroscopy.org/webclient/img_detail/1919547/?dataset=336. *(Image acquired by the authors using the IDR data repository)*

16.4.2 Storage Backend

Storage of image data can become challenging due to the large volume of data taken up by experiments. The difficulties manifest themselves in many ways, including running out of space to store images and difficulties in sharing/transmitting data. Image data is primarily stored on three storage backends: local disk space, networked file systems, and the cloud. Local disks are often the cheapest backend for long-term storage. However, it is important to note that if an image informatics analyst chooses to use a local disk, then that user is also responsible for maintaining backups of the data. Proper backups should be maintained in triplicate in geographically distinct locations.

Network file systems (NFS) are often a desirable choice if they are available. NFS are often hosted by research organizations and can provide exceptionally large amounts of data storage, often with some support for maintaining backups. NFS will often be directly available from high-performance computing clusters, and they generally provide fast data transfer rates. However, some NFS are not publicly accessible, thus introducing obstacles to data sharing. Cloud-based storage is becoming increasingly popular, but pricing for cloud-based resources is based on data volume and usage, and it can become expensive to store data and transfer data.

16.5 Visualization

Open-source visualization tools are an important part of the open-source image informatics toolbox. Visualization tools come in several forms. The most common visualization approach for image informatics is a 2D rendering of image data. The classic open-source image informatics tool, ImageJ [2], provides this type of interface, with support for zooming, changing dimensions, overlaying annotations and colormaps, etc. In the Python community, napari is a popular, extensible tool that provides customizable interfaces. While the ImageJ interface provides many essential features for image informatics, it has limitations when visualizing large datasets. The BigDataViewer [31] tool provides a 2D interface for visualizing large N-dimensional datasets based on ImgLib2. It provides real-time visualization support for very large image sizes by leveraging image caching and multiresolution image pyramids. The Fiji distribution of ImageJ contains a 3D visualization plugin, 3D Viewer [43]. This plugin is still maintained but is not under active development. As a result, there has been a lack of modern 3D visualization for ImageJ. An ongoing effort to introduce modern 3D graphics and virtual reality into Fiji has yielded the SciView visualization tool [44,45] (Fig. 16.4).

There are other visualization tools available, including Allen Institute's 3D Cell Viewer [46], Vaa3D [47], Orbit [48], and napari [19]. Of these, napari has become the flagship user interface and visualization tool for the Python image informatics community.

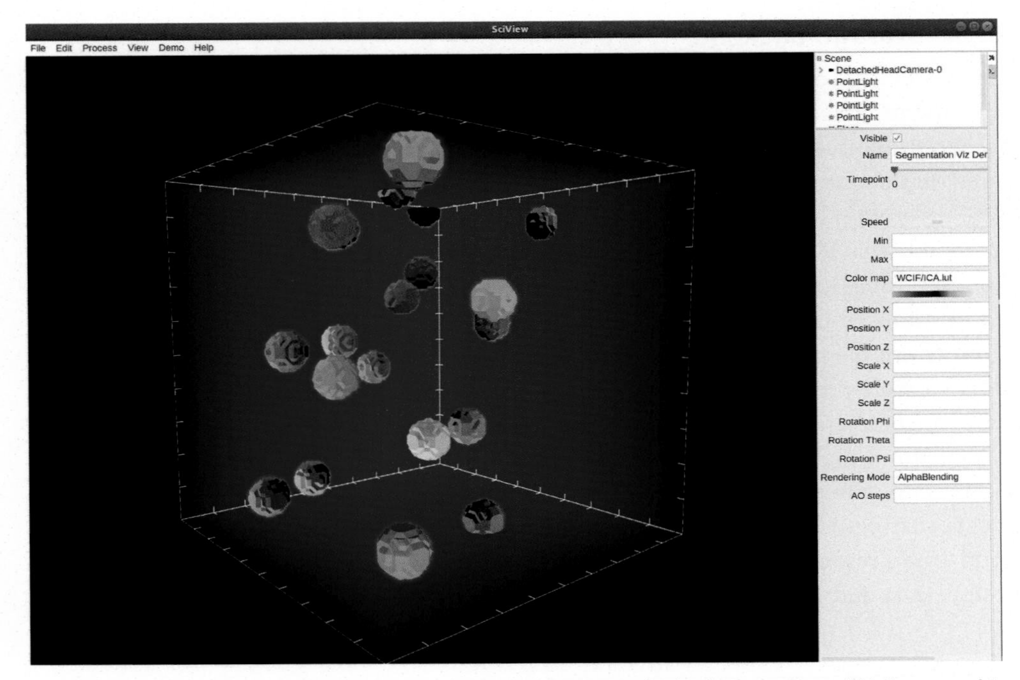

Fig. 16.4 SciView provides 3D visualization in Fiji with support for both OpenGL and Vulkan graphics libraries. This example shows a demo where synthetic 3D volumetric data of segmentation labels is rendered with surface meshes surrounding each segment.

16.6 Community

The open-source image informatics community is quite broad. A number of efforts to establish networks of image analysts have been met with enthusiasm and success. The Network of European Bioimage Analysts (NEUBIAS) is a major effort from the European Union to support meetings and provide education to image analysts across the EU. A number of other expertise-sharing networks have been formed as well, such as Global BioImaging, EuroBioImaging, and BioImaging North America. While these networks emphasize all microscopy techniques, including microscopy acquisition best practices, image informatics is an important topic.

The image analysis community has unified to establish a forum, forum.image.sc, for discussing common challenges. This forum is the primary mode of communication regarding the majority of open-source image informatics tools. A key aspect of the forum is that it enables researchers seeking answers to discover existing solutions. Researchers are encouraged to post detailed questions, along with example data, and then experienced image analysts generally respond by suggesting algorithms and tools that may be suited for the problem. The consequence is that the researcher asking the question is walked through the problem-solving process and is taught methods that will help in approaching new problems in the future.

A major challenge of open-source software is how to sustain and support further development. Unlike commercial tools, there is rarely a dedicated funding stream devoted to a given software platform. Funding agencies such as NSF and NIH put out regular grant calls for software development that includes bioimaging, but these grants emphasize innovation and not software maintenance. However, several groups, including the Chan Zuckerberg Initiative, have released specific software funding calls that can cover software development, curation, and maintenance. Such funding is vital to the success of the open-source imaging software community, and it needs to be acknowledged and supported. Just as importantly, the users need to support these tools by giving regular feedback and citations in their papers. Community lobbying of funding agencies regarding the importance of these tools is another critical need.

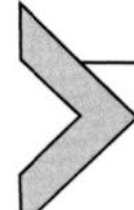

16.7 Conclusion

As a whole, open-source bioimage informatics spans multiple communities that are composed of users, developers, and advocates. The majority of technological advances in the field have been driven by developers creating new tools and methods with no direct, sustained support, or as a side activity to funded scientific studies. Nevertheless, the open-source bioimage informatics field has already had a huge impact on biological sciences by enabling FAIR [49] quantitative methods in the biological sciences. This impact is being increasingly recognized by funding agencies and foundations, which helps to ensure the stability of these vital software communities.

16.8 Summary of Important Points

1. Bioimage informatics covers the complete life cycle of a modern microscopy dataset and can involve many steps and processes.
2. Bioimage informatics pipelines involve multiple software tools that can be separated into three broad categories: acquisition software, storage and curation suites, and image processing and analysis tools.
3. Open-source tools are important, given the opportunities for innovation through improved transparency and collaboration.
4. Truly open software is available in real-time from publicly accessible sources.
5. Interoperation between tools is particularly important since there is no single best tool or best open-source framework.

6. The most common case of tool interoperation is performing a process in one tool, saving the result, and then opening it in a different tool.
7. Every programming language used to create a given tool has its own strengths and weaknesses, and no language is superior in all respects.
8. A major challenge of open-source software is how to sustain and support and further develop the tools.
9. Users of these tools can support them by giving regular feedback and citing them in their papers.

References

[1] H. Peng, Bioimage informatics: a new area of engineering biology, Bioinformatics 24 (17) (2008) 1827–1836.
[2] C.A. Schneider, W.S. Rasband, K.W. Eliceiri, NIH image to ImageJ: 25 years of image analysis, Nat. Methods 9 (7) (2012) 671–675.
[3] F. De Chaumont, et al., Icy: an open bioimage informatics platform for extended reproducible research, Nat. Methods 9 (7) (2012) 690–696.
[4] J. Schindelin, et al., Fiji: an open-source platform for biological-image analysis, Nat. Methods 9 (7) (2012) 676–682.
[5] T. Pietzsch, S. Preibisch, P. Tomančák, S. Saalfeld, ImgLib2—generic image processing in Java, Bioinformatics 28 (22) (2012) 3009–3011.
[6] C.T. Rueden, et al., ImageJ2: ImageJ for the next generation of scientific image data, BMC Bioinform. 18 (1) (2017) 1–26.
[7] J.Y. Tinevez, et al., TrackMate: an open and extensible platform for single-particle tracking, Methods 115 (2017) 80–90.
[8] I. Arganda-Carreras, et al., Trainable Weka Segmentation: a machine learning tool for microscopy pixel classification, Bioinformatics 33 (15) (2017) 2424–2426.
[9] K.I. Harrington, C.T. Rueden, K.W. Eliceiri, FunImageJ: a Lisp framework for scientific image processing, Bioinformatics 34 (5) (2018) 899–900.
[10] J. Moore, et al., OMERO and bio-formats 5: flexible access to large bioimaging datasets at scale, in: *Medical Imaging* 2015*: Image Processing*, 9413:941307, 2015.
[11] J. Moore, et al., OME-NGFF: scalable format strategies for interoperable bioimaging data, BioRxiv (2021).
[12] P. Bankhead, et al., QuPath: Open source software for digital pathology image analysis, Sci. Rep. 7 (1) (2017) 1–7.
[13] Matlab, MathWorks, 2021.
[14] C.R. Harris, et al., Array programming with NumPy, Nature 585 (7825) (2020) 357–362.
[15] S. Van der Walt, et al., scikit-image: image processing in Python, PeerJ 2 (2014), e453.
[16] A.E. Carpenter, et al., CellProfiler: image analysis software for identifying and quantifying cell phenotypes, Genome Biol. 7 (10) (2006) 1–11.
[17] C. McQuin, et al., CellProfiler 3.0: Next-generation image processing for biology, PLoS Biol. 16 (7) (2018) e2005970.
[18] S. Berg, et al., Ilastik: interactive machine learning for (bio) image analysis, Nat. Methods 16 (12) (2019) 1226–1232.
[19] napari, Chan Zuckerberg Initiative, 2021.
[20] W. Schroeder, L. Ng, J. Cates, The ITK software guide second edition updated for ITK version 2.4, FEBS Lett. 525 (2005) 53–58.
[21] B.C. Lowekamp, D.T. Chen, L. Ibáñez, D. Blezek, The design of SimpleITK, Front. Neuroinform. **7** (2013) 45.
[22] Z. Yaniv, B.C. Lowekamp, H.J. Johnson, R. Beare, SimpleITK image-analysis notebooks: a collaborative environment for education and reproducible research, J. Digit. Imaging 31 (3) (2018) 290–303.

[23] U. Köthe, The VIGRA image analysis library, Univ. Heidelb. Heidelb. Ger. (2013).
[24] A.D. Edelstein, M.A. Tsuchida, N. Amodaj, H. Pinkard, R.D. Vale, N. Stuurman, Advanced methods of microscope control using μManager software, J. Biol. Methods 1 (2) (2014).
[25] P.G. Pitrone, et al., OpenSPIM: an open-access light-sheet microscopy platform, Nat. Methods 10 (7) (2013) 598–599.
[26] U. Schmidt, M. Weigert, C. Broaddus, G. Myers, Cell detection with star-convex polygons, in: International Conference on Medical Image Computing and Computer-Assisted Intervention, 2018, pp. 265–273.
[27] M. Weigert, U. Schmidt, R. Haase, K. Sugawara, G. Myers, Star-convex polyhedra for 3d object detection and segmentation in microscopy, in: Proceedings of the IEEE/CVF Winter Conference on Applications of Computer Vision, 2020, pp. 3666–3673.
[28] C. Stringer, T. Wang, M. Michaelos, M. Pachitariu, Cellpose: a generalist algorithm for cellular segmentation, Nat. Methods 18 (1) (2021) 100–106.
[29] K. Jaqaman, et al., Robust single-particle tracking in live-cell time-lapse sequences, Nat. Methods 5 (8) (2008) 695–702.
[30] C. Wolff, et al., Multi-view light-sheet imaging and tracking with the MaMuT software reveals the cell lineage of a direct developing arthropod limb, Elife 7 (2018), e34410.
[31] T. Pietzsch, S. Saalfeld, S. Preibisch, P. Tomancak, BigDataViewer: visualization and processing for large image data sets, Nat. Methods 12 (6) (2015) 481–483.
[32] K. Ulicna, G. Vallardi, G. Charras, A.R. Lowe, Automated deep lineage tree analysis using a Bayesian single cell tracking approach, bioRxiv (2020).
[33] P. Thévenaz, TurboReg an ImageJ plugin automatic for the alignment of a source image or a stack to a target image, 2015.
[34] D. Hörl, et al., BigStitcher: reconstructing high-resolution image datasets of cleared and expanded samples, Nat. Methods 16 (9) (2019) 870–874.
[35] R. Haase, et al., CLIJ: GPU-accelerated image processing for everyone, Nat. Methods 17 (1) (2020) 5–6.
[36] R. Haase, CLIJ2, 2021.
[37] W. Ouyang, F. Mueller, M. Hjelmare, E. Lundberg, C. Zimmer, ImJoy: an open-source computational platform for the deep learning era, Nat. Methods 16 (12) (2019) 1199–1200.
[38] L. Von Chamier, et al., ZeroCostDL4Mic: an open platform to simplify access and use of deep-learning in microscopy, BioRxiv (2020).
[39] M. Weigert, et al., Content-aware image restoration: pushing the limits of fluorescence microscopy, Nat. Methods 15 (12) (2018) 1090–1097.
[40] E. Gómez-de-Mariscal, et al., DeepImageJ: A user-friendly environment to run deep learning models in ImageJ, bioRxiv (2021) 799270.
[41] C. Dietz, et al., Integration of the ImageJ ecosystem in KNIME analytics platform, Front. Comput. Sci. 2 (2020) 8.
[42] I.G. Goldberg, et al., The open microscopy environment (OME) data model and XML file: open tools for informatics and quantitative analysis in biological imaging, Genome Biol. 6 (5) (2005) 1–13.
[43] B. Schmid, J. Schindelin, A. Cardona, M. Longair, M. Heisenberg, A high-level 3D visualization API for Java and ImageJ, BMC Bioinform. 11 (1) (2010) 1–7.
[44] U. Günther, K.I. Harrington, Tales from the Trenches: Developing sciview, a new 3D viewer for the ImageJ community, *ArXiv Prepr. ArXiv200411897*, 2020.
[45] U. Günther, et al., Scenery: flexible virtual reality visualization on the Java VM, *ArXiv Prepr. ArXiv190606726*, 2019.
[46] 3D Cell Explorer, Allen Institute, 2021.
[47] H. Peng, A. Bria, Z. Zhou, G. Iannello, F. Long, Extensible visualization and analysis for multidimensional images using Vaa3D, Nat. Protoc. 9 (1) (2014) 193–208.
[48] M. Stritt, A.K. Stalder, E. Vezzali, Orbit image analysis: an open-source whole slide image analysis tool, PLoS Comput. Biol. 16 (2) (2020) e1007313.
[49] M.D. Wilkinson, et al., The FAIR guiding principles for scientific data management and stewardship, Sci. Data 3 (1) (2016) 1–9.

Glossary

This glossary of terms is provided to help the reader avoid confusion brought about by the usage of ordinary words given specialized meaning in this book. These definitions conform roughly to common usage in microscope image processing, but do not constitute a standard in the field.

Aberration in a lens, deviation from the ideal shape. In an image, the distortion that results (Chapter 2).
Algebraic operation an image processing operation involving the pixel-by-pixel sum, difference, product, or quotient of two images (Chapter 5).
Aliasing an artifact produced when the pixel spacing is too large in relation to the size of the detail in an image (Chapter 3).
Analytic continuation A computational method for extending the resolution of an imaging system beyond the diffraction limit (Section 12.4.1).
Arc (1) a continuous portion of a circle, (2) a connected set of pixels representing a portion of a curve (Chapter 7).
Background shading correction the process of eliminating nonuniformity of image background intensity by application of image processing (Chapter 5).
Background subtraction subtracting an image of the background to reduce shading (Chapter 5).
Binary image a digital image having only two gray levels (usually zero and one, black and white).
Blinking fluorophores fluorophores that show spontaneous fluctuations in emission, termed blinking/intermittency, where a discrete dark state can exist for milliseconds to hours (Section 13.2.2).
Blur a loss of image sharpness, introduced by defocus, low-pass filtering, camera motion, etc.
Border the first and last row and column of a digital image.
Boundary a closed path that surrounds an object in an image and separates it from the background (Chapter 7).
Boundary chain code a sequence of directions specifying the boundary of an object (Chapter 7).
Boundary pixel an interior pixel that is adjacent to at least one background pixel (contrast with "interior pixel," "exterior pixel").
Boundary tracking an image segmentation technique in which arcs are detected by searching sequentially from one arc pixel to the next (Chapter 7).
Brightness the value associated with a point in an image, representing the amount of light emanating or reflected from that point.
Cell tracking see object tracking.
Centroid the center of gravity of an object (Section 8.2.3.5).
Change detection an image processing technique in which the pixels of two registered images are compared (e.g., by subtraction) to detect differences in the objects therein.
Class one of a set of mutually exclusive, preestablished categories to which an object can be assigned (Chapter 9).
Classification the process of assigning an object to one of several preestablished classes, based on its measurements (Chapter 9).
Closed curve a curve whose beginning and ending points are at the same location (Chapter 7).
Closing a binary or grayscale morphological operation consisting of a dilation followed by an erosion (Chapter 6).
Cluster a set of points located close together in a space (e.g., in feature space) (Chapter 9).
Cluster analysis the detection, measurement, and description of clusters in a space (Chapter 9, Section 13.5.3).

Concave a characteristic of an object: at least one straight line segment between two interior points of the object is not entirely contained within the object (contrast with "convex") (Chapter 7).

Confocal microscopy an optical imaging technique that uses a spatial pinhole to block out-of-focus light in image formation, thereby yielding improved optical resolution and contrast (Section 10.3).

Confusion matrix a matrix showing, for objects belonging to every class, the probability that they will be assigned to each class (Chapter 9).

Connected pertaining to the pixels that make up an object or a curve: any two points within the object can be joined by an arc made up entirely of adjacent pixels that are also contained within the object (Chapter 7).

Contrast the amount of difference between the average brightness (or gray level) of an object and that of the surrounding background.

Contrast stretch a linear grayscale transformation (Chapter 5).

Convex a characteristic of an object: all straight line segments between two interior points of the object are entirely contained within the object (contrast with "concave") (Chapter 7).

Convolution a mathematical process for combining two functions to produce a third function. It models the operation of a shift-invariant linear system (Section 9.3).

Convolution kernel (1) the two-dimensional array of numbers used in convolution filtering of a digital image; (2) the function with which a signal or image is convolved (Section 9.3).

Convolutional neural network a neural network architecture suitable for learning underlying spatial features in images and using this information for segmentation, classification, and improved resolution (Chapter 15).

Curve (1) a continuous path through space; (2) a connected set of pixels representing a path (see "arc," "closed curve") (Chapter 7).

Curve fitting the process of estimating the best set of parameters for a mathematical function to approximate a curve.

Dark current intrinsic source of noise present in photodetectors that arises from sources such as thermal agitation of electrons (Chapter 3).

Data augmentation a collection of techniques that increase the size and generalization of training datasets in machine learning (Chapter 15).

Deblurring (1) an image processing operation designed to reduce blurring and sharpen the detail in an image; (2) removing or reducing the blur in an image, often one step of image restoration or reconstruction (Chapter 5).

Decision rule in pattern recognition, a rule or algorithm used to assign an object from an image to a particular class. The assignment is based on measurements of its features (Chapter 9).

Deconvolution the process of designing and implementing a convolution filter that compensates for a previously applied filter. It is typically used for deblurring an image (Chapter 5, Section 12.3).

Deep learning a subset of machine learning that includes neural networks with three or more layers that "learn" from large amounts of data (Chapter 15).

Diffraction limit the resolution of a perfect lens, limited only by the effects of diffraction (Chapters 2, 12, Section 13.2.1).

Digital image (1) an array of integers representing an image of a scene; (2) a sampled and quantized function of two or more dimensions, generated from and representing a continuous function of the same dimensionality; (3) an array generated by sampling a continuous function on a rectangular (or other) grid and quantizing its value at the sample points (Chapter 1).

Digital image processing the manipulation of pictorial information by computer.

Digitization the process of converting an image of a scene into digital form (Chapter 3).

Dilation a binary or grayscale morphological operation that uniformly increases the size of bright objects in relation to the background (Chapter 6).

Downsampling to make a digital image smaller by deleting pixels in the rows and columns. This process is also known as decimation.

Edge (1) a region of an image in which the gray level changes significantly over a short distance; (2) a set of pixels belonging to an arc and having the property that pixels on opposite sides of the arc have significantly different gray levels (Chapter 7).

Edge detection an image segmentation technique in which edge pixels are identified by examining neighborhoods (Chapter 7).

Edge enhancement any image processing technique in which edges are made to appear sharper by increasing the contrast between the gray levels of the pixels located on opposite sides of the edge (Chapter 5).

Edge image an image in which each pixel is labeled as either an edge pixel or a nonedge pixel (Chapter 7).

Edge linking a boundary finding technique that works by connecting edge pixels together (Chapter 7).

Edge operator a neighborhood operator that labels the edge pixels in an image (Chapter 7).

Edge pixel a pixel that lies on an edge (Chapter 7).

Emission filter a physical filter inserted in the emission path of a microscope to pass only light emitted from a specimen in a particular wavelength band (Section 10.3).

Enhance to increase contrast or subjective visibility of the contents of an image (Chapter 5).

Erosion a binary or grayscale morphological operation that uniformly reduces the size of bright objects in relation to the background (Chapter 6).

Excitation filter a physical filter inserted in the illumination path of a microscope to pass only light of wavelengths capable of exciting fluorescence (Chapter 10).

Expansion microscopy a superresolution technique based on using a polymer system to enlarge a tissue sample (Section 12.4.3).

Exterior pixel a pixel that falls outside all the objects in a binary image (contrast with "interior pixel") (Chapter 7).

False negative in two-class pattern recognition, a misclassification error in which a positive object is labeled as negative (Chapter 9).

False positive in two-class pattern recognition, a misclassification error in which a negative object is labeled as positive (Chapter 9).

Feature a characteristic of an object, something that can be measured, and that assists classification of the object (e.g., size, texture, shape) (Chapter 8).

Feature extraction a step in the pattern recognition process in which measurements of the objects are computed (Chapter 8).

Feature selection a step in the pattern recognition system development process in which measurements or observations are studied to identify those that can best be used to assign objects to classes (Chapter 9).

Feature space in pattern recognition, an n-dimensional vector space containing all possible measurement vectors (Chapter 9).

Fluorescence correlation spectroscopy (FCS) a highly sensitive method that provides high temporal and spatial resolution for single molecule imaging by monitoring the fluorescence fluctuations that arise from molecule diffusion within a small optically defined volume (on the order of femtoliters), following excitation by a focused laser beam (Section 10.7.9).

Fluorescence in situ hybridization (FISH) a molecular biology technique that utilizes DNA probes to allow the visualization of gene copy number and localization of specific DNA targets with fluorescence microscopy (Section 10.7.2).

Fluorescence lifetime imaging (FLIM) a technique that measures the average time a fluorescent molecule spends in the excited state before returning to ground state (Section 10.7.6).

Fluorescence microscopy a technique whereby fluorescent substances are examined in a microscope (Chapter 10).

Fluorescence recovery after photobleaching (FRAP) a technique that involves photobleaching a region of interest, thereby allowing the study of the consequent fluorescent recovery time due to the movement of nonbleached fluorescent molecules from the surrounding area (Section 10.7.7).

Fluorescence resonance energy transfer (FRET) a process involving the radiationless transfer of energy from a donor fluorophore to an appropriately positioned acceptor fluorophore (Section 10.7.5).

Fluorophore a molecule that absorbs light of a certain wavelength and emits light at a longer wavelength (Chapter 10).

Focal length the distance behind a lens at which an object located at infinity is in focus (Chapter 2).

Fourier transform a linear transform that decomposes an image into a set of sinusoidal frequency component functions (Chapter 5).

Frequency domain the Fourier transform relates a function of time or space to a function of frequency. The latter is said to exist in the frequency domain (Chapter 5).

Gaussian fitting determining the parameters of a Gaussian to best fit an observed spot image (Section 13.3.2).

Geometric correction an image processing technique in which a geometric transformation is used to remove geometric distortion (Chapter 4).

Geometric operation an image processing operation that modifies the spatial relationships of objects in an image (Chapter 4).

Gradient a vector that reflects the direction and magnitude of the slope of the gray level at a point in an image (Chapter 7).

Gray level (1) the value associated with a pixel in a digital image, representing the brightness of the original scene at the point represented by that pixel. (2) A quantized measurement of the local property of the image at a pixel location.

Gray scale the set of all possible gray levels in a digital image.

Gray scale transformation the function, employed in a point operation, that specifies the relationship between input and corresponding output gray level values (Chapter 5).

Hankle transform similar to the Fourier transform, a linear transformation that relates the (one-dimensional) profile of a circularly symmetric function of two dimensions to the (one-dimensional) profile of its two-dimensional (also circularly symmetric) Fourier transform (Section 10.4.3).

Hermite function a complex-valued function having an even real part and an odd imaginary part (Section 10.2.1).

High-pass filtering an image enhancement (usually convolution) operation in which the high spatial frequency components are emphasized relative to the low-frequency components (Chapter 5).

Histogram a graphical representation of the number of pixels in an image at each gray level.

Hole in a binary image, a connected region of background points that is surrounded by interior points (Chapters 6, 7).

Image any representation of a physical scene or of another image.

Image analysis extracting quantitative data from an image (Chapter 8).

Image calibration modifying an image so that its spatial or photometric values are related to standard units of measure (Chapter 8).

Image coding translating image data into another form from which it can be recovered, as for compression.

Image compression any process that eliminates redundancy from, or approximates, an image in order to represent it in a more compact form.

Image curation an image informatics process where humans are involved in the analysis and decision making about the rest of the image informatics pipeline (Chapter 16).

Image display making a digital image visible to the human eye (Chapter 3).

Image enhancement any process intended to improve the visual appearance of an image (Chapter 5).

Image filtering an operation, usually implemented by convolution, that modifies an image in a specific way (Chapter 5).

Image fusion combining two or more images into one (Chapter 5).

Image informatics the study of computational systems, especially those for storage and retrieval of images and imaging-derived information in research and clinical use (Chapter 16).

Image matching any process involving quantitative comparison of two images in order to determine their degree of similarity.

Image processing subjecting a digital image to a computation that alters the image in a predictable way.

Image processing operation a series of steps that transforms an input image into an output image.

Image reconstruction the process of constructing or recovering an image from data that occurs in non-image form.

Image registration a geometric operation intended to position one image of a scene with respect to another image of the same scene so that the objects in the two images coincide (Chapter 4).

Image restoration any process intended to return an image to its original condition by reversing the effects of prior degradations (Chapter 5).

Image segmentation (1) the process of detecting and delineating the objects of interest in an image. (2) The process of subdividing an image into disjoint regions. Normally these regions correspond to objects and the background upon which they lie (Chapter 7).

Image stitching combining two or more images into a larger mosaic image (Chapter 4).

Image warping see geometric operation (Chapter 4).

Immunofluorescence a molecular biology method that makes use of antibodies chemically labeled with fluorescent dyes to visualize molecules under a light microscope (Chapter 10).

Interior pixel in a binary image, a pixel that falls inside an object (contrast with "boundary pixel," "exterior pixel") (Chapter 7).

Interpolation the process of determining the value of a sampled function between its sample points. It is normally done by convolution with an interpolation function (Chapter 3).

Kernel (1) the two-dimensional array of numbers used in convolution filtering of a digital image; (2) the function with which a signal or image is convolved.

Linear discriminant analysis dimension reduction by a transformation that projects data onto a linear subspace that best discriminates among object classes (Section 9.10.1.2).

Local operation an image processing operation that assigns a gray level to each output pixel based on the gray levels of pixels located in a neighborhood of the corresponding input pixel. A neighborhood operation (contrast with "point operation").

Local property the interesting characteristic that varies with position in an image. It may be brightness, optical density, or color for microscope images.

Lossless image compression any image compression technique that permits exact reconstruction of the image.

Lossy image compression any image compression technique that inherently involves approximation and does not permit exact reconstruction of the image.

Low-pass filter a filter that attenuates the high-frequency detail in an image (Chapter 5).

Machine learning the study of computer algorithms that are capable of learning by the use of data and that improve automatically through experience (Chapter 15).

Magnification the relationship between the sizes of objects located on both sides of a lens (Chapter 2).

Maximum likelihood fitting establishing the parameters of a surface fit that maximizes the probability of a correct match (Chapter 13).

Measurement space in pattern recognition, an n-dimensional vector space containing all possible measurement vectors (Chapter 9).

Medial axis the centerline of an object in an image (Section 8.2.4.3).

Median filter a nonlinear spatial filter that replaces the gray value of the center pixel with the median gray value of the input group of pixels, in order to remove noise spikes and other single pixel anomalies (Chapter 5).

Minimum enclosing rectangle (MER) the bounding box of an object aligned such that it encloses that object with minimum area (Chapter 8).
Misclassification in pattern recognition, the assignment of an object to any class other than its true class (Chapter 9).
Morphological processing probing an image with a structuring element and either filtering or quantifying the image according to the manner in which the structuring element fits (or does not fit) within each object in the image (Chapter 6).
Motion analysis following the path of moving objects in a series of images (Chapter 14).
Multispectral image a set of images of the same scene, each formed by radiation from a different wavelength band (Chapter 10).
Neighborhood a set of adjacent pixels (Chapter 7).
Neighborhood operation an image processing operation that assigns a gray level to each output pixel based on the gray levels of pixels located in a neighborhood of the corresponding input pixel (see "local operation," contrast with "point operation").
Neural network an algorithm with the capacity to learn underlying relationships in a set of data through a process that mimics the operation of the human brain (Chapter 15).
Noise irrelevant components of an image that hinder recognition and interpretation of the data of interest.
Noise reduction any process that reduces the undesirable effects of noise in an image (Chapter 5).
Object in pattern recognition, a connected set of pixels in a binary image, usually corresponding to a physical object in the scene (Chapter 7).
Object label map an image of size equal to an original image, wherein the gray level of each pixel encodes the sequence number assigned to the object to which the corresponding pixel in the original image belongs (Chapter 7).
Object tracking following the path of moving objects in a series of images (Chapter 14).
Open-source source code that is made freely available for redistribution, modification, and use by others.
Opening a binary or grayscale morphological operation consisting of an erosion followed by a dilation (Chapter 6).
Optical image the result of projecting light emanating from a scene onto a surface, as with a lens (Chapter 2).
Optical sectioning a noninvasive method for obtaining a 3-D image of a specimen, without physical sectioning, by recording a series of images taken along the optical axis (Chapter 11).
Pattern a meaningful regularity that members of a class express in common, and which can be measured and used to assign objects to different classes (Chapter 9).
Pattern class one of a set of mutually exclusive, preestablished categories to which an object can be assigned (Chapter 9).
Pattern classification the process of assigning objects to pattern classes (Chapter 9).
Pattern recognition the detection, measurement, and classification of objects in an image by automatic or semiautomatic means (Chapter 9).
Perimeter the circumferential distance around the boundary of an object (Chapter 7).
Photoactivated localization microscopy (PALM) A single-molecule localization method that utilizes optical highlighter fluorescent proteins to stochastically switch on and off a subpopulation of molecules for sequential single-molecule readout (Sections 12.4.4, 13.1.1).
Photobleaching an inherent phenomenon in fluorophores that causes an effective reduction in, and ultimately a complete elimination of, fluorescence emission.
Photon shot noise noise in recorded images that results from the random nature of photon emission.
Picture element the smallest element of a digital image. The basic unit of which a digital image is composed.
Pixel contraction of "picture element."

Points accumulation for imaging in nanoscale topography (PAINT) a single-molecule localization method that achieves stochastic single-molecule fluorescence by molecular adsorption/absorption and photobleaching/desorption (Section 13.2.2).

Point operation an image processing operation that assigns a gray level to each output pixel based only upon the gray level of the corresponding input pixel. Contrast with "neighborhood operation" (Section 12.4.4).

Point spread function (psf) the image that a lens forms of a point source (Chapter 2).

Principal component analysis dimension reduction by a transformation that projects data onto a linear subspace that has been selected to produce maximum scatter (Chapter 9).

Probability density function a function specifying, for each possible value of a random variable, the probability of that value occurring (Chapter 9).

psf engineering designing an optical system to have a point spread function with specified characteristics (Section 13.4.3).

psf fitting determining the parameters of a function, typically a Gaussian, to best fit an observed point spread function (Chapter 13).

Quantitative image analysis any process that extracts quantitative data from a digital image (Chapter 8).

Quantization the process by which the local property of an image, at each pixel, is assigned one of a finite set of gray levels.

Ratio image an image obtained by dividing one image by another on a pixel-by-pixel basis.

Readout noise noise generated during the readout of an image from a photodetector.

Region a connected subset of an image (Chapter 7).

Region growing an image segmentation technique in which regions are formed by repeatedly combining subregions that are similar in gray level or texture (Chapter 7).

Registered (1) the condition of being in alignment; (2) when two or more images are in geometric alignment with each other, and the objects therein coincide (Chapter 4).

Registered images two or more images of the same scene that have been positioned with respect to one another so that each object in the scene occupies the same position (Chapter 4).

Resolution (1) in optics, the minimum separation distance between distinguishable objects; (2) in image processing, the degree to which closely spaced objects in an image can be distinguished from one another (Chapters 2, 12, 13).

Sampling the process of dividing an image into pixels (according to a sampling grid) and measuring the local property (e.g., brightness or color) at each pixel (Chapter 2).

Scene a particular arrangement of physical objects.

Shape analysis extracting measurements that reflect the shape (e.g., circularity) of an object (Chapter 8).

Sharp pertaining to the detail in an image, well defined and readily discernible.

Sharpening any image processing technique intended to enhance the detail in an image (Chapter 5).

Single-molecule localization microscopy (SMLM) a collection of techniques that can produce ultra-high-resolution images of biological specimens that have been labeled with a fluorescent dye (Sections 13.1.1, 13.2.2, 13.3).

Sinusoidal having the shape of the sine function.

Skeletonization a binary operation that reduces the objects in an image to a single pixel wide skeleton representation (Chapter 7).

Smoothing any image processing technique intended to reduce the amplitude of small detail in an image. This is often used for noise reduction (Chapter 5).

Spatial domain the Fourier transform relates a function of spatial variables to a function of frequency. The former is said to exist in the spatial domain (Chapter 5).

Spatial filter a convolution operation conducted in the spatial domain (Chapter 5).

Statistical pattern recognition an approach to pattern recognition that uses probability and statistical methods to assign objects to pattern classes (Chapter 9).

Steerable filter a directional filter whose orientation can be changed (Chapter 5).
Stimulated emission depletion microscopy (STED) a type of superresolution optical microscopy technique that creates superresolution images using the selective deactivation of fluorophores by minimizing the area of illumination at the focal point (Sections 12.4.2, 13.1.1, 13.5.3).
Stochastic optical reconstruction microscopy (STORM) a type of superresolution optical microscopy technique that is based on stochastic switching of single-molecule fluorescence signals off and on (Section 13.1.1).
Structuring element a matrix used in morphological processing that defines the neighborhood used in the processing of each pixel (Chapter 6).
Structured illumination microscopy (SIM) a fluorescence optical microscope imaging technique that increases resolution by exploiting interference patterns (Moiré patterns) created when two grids in the illumination pattern are overlaid at an angle (Section 12.4.5).
Super resolution microscopy a collection of techniques for imaging with resolution beyond the diffraction limit (Chapter 12).
Surface fitting the process of estimating the best set of parameters for a mathematical function to approximate a surface (Section 13.3.2).
System anything that accepts an input and produces an output in response.
Template matching comparing an unknown object to a standard object to determine the degree of similarity.
Temporal, radial-aperture-based intensity (TRABI) an estimation technique that uses the fluorescent on-states and nonfluorescent off-states of photoswitchable or photoactivatable fluorophores to achieve nanometer-scale axial precision (Section 13.4.4.2).
Texture an attribute representing the amplitude and spatial arrangement of the local variation of gray level in an image (Chapter 8).
Thinning a binary image processing technique that reduces objects to sets of thin (one-pixel-wide) curves (Chapter 6).
Threshold a specified gray level used for producing a binary image (Chapter 7).
Thresholding the process of producing a binary image from a grayscale image by assigning each output pixel the value 1 if the gray level of the corresponding input pixel is at or above the specified threshold gray level, and the value 0 if the input pixel is below that threshold (Chapter 7).
Time-lapse imaging recording a series of time-sequential images to determine the temporal behavior of objects (Chapter 14).
Trajectory the path followed by a moving object in a time-sequential set of images (Chapter 14).
Transfer function for a linear, shift-invariant system, the function of frequency that specifies the factor by which the amplitude of a sinusoidal input signal is multiplied in order to form the corresponding output signal (Chapter 5).
Transfer learning in machine learning, the improvement of learning in a new task through the transfer of knowledge from a related task that has already been learned (Chapter 15).
Upsampling making a digital image larger by adding pixels to the rows and columns, usually in an interleaved fashion.
Voxel contraction of "volume element," the extension of the 2-D pixel to 3-D (Chapter 11).
Watershed transform an image segmentation method that simulates the filling of a basin with water (Chapter 7).
Wiener filter a linear filter (convolution operation) designed to discriminate optimally against a known noise source (Chapter 5).

Further reading

[1] R.M. Haralick, Glossary and index to remotely sensed image pattern recognition concepts, Pattern Recogn. 5 (1973) 391–403.
[2] IEEE Std 610.4-1990, IEEE Standard Glossary of Image Processing and Pattern Recognition Terminology, IEEE, New York, 1990. 10017-2394.
[3] R.M. Haralick, L.G. Shapiro, Glossary of computer vision terms, Pattern Recogn. 24 (1991) 69–93.
[4] L. Darcy, L. Boston, Webster's New World Dictionary of Computer Terms, thire ed., Prentice-Hall, New York, 1988.
[5] B. Pfaffenberger, Que's Computer User's Dictionary, second ed., Que Corporation, Carmel, Indiana, 1991.
[6] C.J. Sippl, R.J. Sippl, Computer Dictionary, third ed., Howard W. Sams & Co., Indianapolis, IN, 1980.
[7] G.A. Baxes, Digital Image Processing, John Wiley & Sons, 1994.
[8] K.R. Castleman, Digital Image Processing, Prentice Hall, 1996.

Index

Note: Page numbers followed by *f* indicate figures and *t* indicate tables.

C

E

F

G

P

Q

R

S

T

Printed in the United States
by Baker & Taylor Publisher Services